Topics in Mining, Metallurgy and Materials Engineering

Series Editor

Carlos P. Bergmann, Federal University of Rio Grande do Sul, Porto Alegre, Rio Grande do Sul, Brazil

“Topics in Mining, Metallurgy and Materials Engineering” welcomes manuscripts in these three main focus areas: Extractive Metallurgy/Mineral Technology; Manufacturing Processes, and Materials Science and Technology. Manuscripts should present scientific solutions for technological problems. The three focus areas have a vertically lined multidisciplinarity, starting from mineral assets, their extraction and processing, their transformation into materials useful for the society, and their interaction with the environment.

More information about this series at http://www.springer.com/series/11054

Luis Felipe Verdeja González ·
Daniel Fernández González ·
José Ignacio Verdeja González

Operations and Basic Processes in Steelmaking

Springer

Luis Felipe Verdeja González
Escuela de Minas, Energía y Materiales
University of Oviedo
Oviedo, Asturias, Spain

Daniel Fernández González
Escuela de Minas, Energía y Materiales
University of Oviedo
Oviedo, Asturias, Spain

José Ignacio Verdeja González
Escuela de Minas, Energía y Materiales
University of Oviedo
Oviedo, Asturias, Spain

ISSN 2364-3293 ISSN 2364-3307 (electronic)
Topics in Mining, Metallurgy and Materials Engineering
ISBN 978-3-030-67999-6 ISBN 978-3-030-68000-8 (eBook)
https://doi.org/10.1007/978-3-030-68000-8

This Springer imprint is published by the registered company Springer Nature Switzerland AG
The registered company address is: Gewerbestrasse 11, 6330 Cham, Switzerland

This book is dedicated to the memory of José Antonio Pero-Sanz Elorz (1934–2012), who founded the research group in iron and steelmaking, metallurgy and materials at the University of Oviedo (Asturias, Spain) in the middle of the seventies of the last century. Several books have been published as a result of this long period of research and teaching, including Physical Metallurgy of Cast Irons (Springer, 2018), Structural Materials: Properties and Selection (Springer, 2019) and Operations and Basic Processes in Ironmaking (Springer, 2020).

Prologue

Iron together with copper are the only elements of the Periodic Table related with the four Human Ages: Stone, Copper, Bronze, and Iron. According to the history, approximately 1500 years BC, it is possible to locate the beginning of the Iron Age in the Hittites Kingdom, in the center of the Anatolia's peninsula, former Turkey.

Even when the art of the iron obtaining sinks its roots in the Human history, the science and technology of the steel as material used in structural applications has only had a marked development from the second half of the nineteenth century, when the steel production exceeded millions of tons in one year. Along the seven chapters of this book, we describe the operations and industrial processes associated to carbon steels, stainless steels, highly alloyed steels, and ferroalloys production.

This manuscript tries, above all, to be a text dedicated to the university studies for bachelor, master, and postgraduate students. At the same time, the book might be an useful instrument for the basic formation of those people interested in working in the ironmaking and steelmaking sector, as well as for the updating of the basic knowledge of the people that had knowledge of the iron and steel metallurgy.

The manuscript claims to show for the first time in a book of these characteristics, together with the most significative operating variables of the processes and operations of the steels and ferroalloys production, numerous solved exercises and problems (more than 80). The solved exercises and problems have as objective to conceptually support the theoretical fundamentals of the Basic Oxygen Furnace (BOF) process, the reactions and operations in the electric furnace, the secondary metallurgy, the practice in the solidification-casting molds, the operations in the tundish, the operations in the hot and cold plastic deformation of the metal and the study of all the physical and chemical changes produced in the different finishing operations and processes.

Habitually, the number of variables involved in the spontaneity and rate of the operations-processes are numerous ("big data" if we use the current terminology). We want to set up the basis for the development of models that, supported on the basic sciences, could provide-find answers, with the minimum number of variables, to the topics that are proposed and developed in each one of the chapters of the book. On the other hand, it is necessary to consider that the construction of models, based on the principles-laws of the science associated to the proposed exercises, leads to the "know-how" typical of the technological development. Definitely, we want to emphasize in the truth and the science when taking decisions, putting aside emotions and feelings.

It is possible that, for the professionals and iron and steelmaking specialists, the contents of the seven chapters of the book, for instance, in the process of pig iron refinement, the fusion-refinement of scrap, the production of ferroalloys or the molding, plastic deformation and finishing processes, might be limited. However, in the bibliography that is collected at the end of each chapter, we include monographs and specific reviews that allow the reader to strengthen knowledge about the concepts developed in the text, as well as to acquire a greater specialization that could propitiate the development and technological changes.

Finally, the authors want to thank Prof. José Antonio Pero-Sanz Elorz because he was responsible for creating the Research Group in Ironmaking, Steelmaking, Metallurgy and Materials of the University of Oviedo in the middle of the last century. We also express our gratitude to Profs. José Pedro Sancho Martínez and Ángel Alfonso Fernández of the University of Oviedo and Roberto González Ojeda of the Universidad Panamericana de México for their contributions both in the thermodynamic fundamentals and in the simulation and modeling of processes.

Oviedo, Asturias, Spain
March 2021

Luis Felipe Verdeja González
Daniel Fernández González
José Ignacio Verdeja González

Contents

Chapter 1
The Basic Oxygen Furnace to Obtain Steel

Abstract Pig iron is an intermediate product obtained in the blast furnace characterized by the high carbon content (3.5–4.5%) and the presence of other impurities (silicon, sulphur, phosphorus, manganese, etc.) that makes it a brittle material without applications. It is used as raw material in the Basic Oxygen Furnace (BOF) to produce steel and in the cupola furnace to produce cast irons. This chapter is dedicated to the production of steel in the Basic Oxygen Furnace. The technological development is presented as well as the main reactions involved in the treatment of pig iron in the converter. Additionally, the chapter is complemented with solved problems and exercises.

1.1 Steel in the History

Criteria that are usually considered for those raw materials used in the production of structural and functional metallic materials are three (Pero-Sanz et al. 2019):

- First: abundance and quality (limited presence of those elements that deteriorate the quality-properties of the finished product).
- Second: low operating costs of the extractive process used for the obtaining of the finished product. At the same time, it is necessary a good capacity of recycling of the different by-products generated in the extractive process.
- Third: good properties, prices, and environmental compatibility (long lifetime of the finished product, together with a suitable capacity of being recycled-reused).

From the 83 elements of the Periodic Table, once excluded from this number the elements belonging to the lanthanides, actinides, and radioactive sets, we should make a new process of selection of the elements that could accomplish the above-mentioned criteria. Making this operation, we might conclude that the final selection will fall on the eight most abundant elements (apart from the oxygen) of the earth crust, whose mean abundance is >0.5%. The first three elements of this list are: silicon, aluminum, and iron. The next five elements are: calcium, magnesium, sodium, potassium, and titanium. This first election, although the three criterions were considered,

L. F. Verdeja González et al., *Operations and Basic Processes in Steelmaking*,
Topics in Mining, Metallurgy and Materials Engineering,
https://doi.org/10.1007/978-3-030-68000-8_1

the most important factor would be the consequence of considering the first criterion: abundance and cost. On the other hand, it is well-known that elements such as the copper, zinc, lead, manganese, chromium, or nickel are important in the manufacture of structural and functional materials. Unfortunately, as they are scarce in nature, the pressure on the raw materials, production, and recycling processes are greater and greater every year (Pero-Sanz et al. 2019). This way, reports are being periodically elaborated in the USA and European Union about those elements-raw materials that are considered strategic for the technological-industrial development of these world areas.

From the eight most abundant elements, if we reconsider the second criterion as the most important (low operation costs), the final election will consider iron as the metal finally chosen (Pero-Sanz et al. 2019). If we apply the third criterion, we could also include aluminum and titanium as elements that are strong competitors and even displace iron from its predominant position, only if we would consider as criterions of more importance in the selection of materials those related with the corrosion, density, and electric conductivity, for instance.

In fact, steel was not historically considered as the most important structural material until the middle of the nineteenth century (Verdeja et al. 2020). Until that moment, even when the available raw material was abundant, it did not accomplish the second criterion indicated at the beginning of this section: the process of carbon refinement of the pig iron to obtain steel (crucible steel process, 1740, and the refining process performed in the puddling furnace, 1784, were long, expensive and difficultly controlled). The quality-properties of the obtained steel depended on the criterion of the foundrymen that, considering the color of the melt, had an idea of the steel that they were producing.

However, technological developments as the steam engine and its application to the transport (steam train and steamboat) were possible due to the steels manufactured in that period. This is the first period of the Industrial Revolution, from the beginning of the nineteenth century (Comellas 2000). During the first five decades of the nineteenth century, a noticeably increase in the number of foundries that transform the high carbon melts in steel took place. The evolution of the annual production of steel during the first seven decades of the nineteenth century was: <10,000 tons in 1800, 40,000 t in 1820, 100,000 t in 1840, 300,000 t in 1855, and 1,800,000 t in 1870.

The exaggerated expectations of growth of the railway concessions in Europe led in 1857 to an economic-financial crisis that would be neither the last nor the single of the modern history (Comellas 2000). The quality-cleaning of the produced steel was very deficient in that period. This way, there were frequent explosions in the steam boilers due to the catastrophic failure of materials. Anyway, steel demand was still growing in those years.

The second phase of the Industrial Revolution was narrowly related with the development of the Bessemer process (1856), which allowed to manufacture for the first time great quantities (tons) of homogeneous melts with low carbon (%C < 0.80%) in short times (tens of minutes). In the first moment, it was thought that the Bessemer steel had wide possibilities for the design of structures and industrial equipment.

Finally, it was checked that, even when resistance was acceptable, toughness was very deficient, and this steel was not efficient when it was necessary to achieve a narrow compromise between resistance and toughness.

Soon the ironmaking and steelmaking found a solution to the problem of the Bessemer steel (acid process of pig iron conversion that was not able to remove phosphorus and sulphur from the melt) using Thomas steel. Thomas process performed the conversion of the pig iron in a basic environment that was able to remove phosphorus from iron melts with great performance, although it was only possible to partially reduce sulphur content.

Even when the appearance of the Bessemer and Thomas processes gave an unknown celerity to the transformation of pig iron in a melt of chemical characteristics similar to those of the carbon steel, there were not analytical-instrumental techniques able to make an almost instantaneous tracking of the chemical quality of the obtained steel. It was reconsidered the convenience of increasing the times of conversion-transformation of the high carbon melts, and the capacity of the furnaces was also increased. Siemens steel was introduced in the last decades of the nineteenth century and the first half of the twentieth century. In the Siemens steelworks, there is time to not only track Mn, Si, P, and S contents but also dissolved gases, that were gaining importance in the quality of the final product (H, N, and O). Siemens steel quality was popular in that period and enjoyed a well-deserved prestige in the steel world because of being a quality guarantee.

At the beginning of the second half of the twentieth century (1956), a new process of pig iron conversion appeared with conversion rates even greater than those of the Bessemer and Thomas processes: the LD process (Linz-Donawitz). Regarding the Bessemer and Thomas converters, the air as oxidizing fluid is replaced by oxygen that is injected via a vertical lance at rates close to the speed of the sound. At the same time, the analytical technology considerably advanced and it was possible to track the quality of the steel in seconds (while these analytical techniques took hours in previous periods). Siemens steel was "fatal wound," it was only a matter of the time that the LD steel was the protagonist in the conversion of pig iron into carbon steels. This situation was observed in the last decades of the twentieth century.

On the other hand, during the last decades of the nineteenth century, the quantity of steel produced per year in the world exceeded one million tons. From that moment, it was possible to talk about a new product in the iron and steelmaking market: the recycled steel, colloquially known as scrap. As a consequence of this, the electric furnace also started its path in the ironmaking and steelmaking world (electric steelworks), aimed at reusing-remelting the scrap to transform it into steel again.

The electric and Siemens steelworks have competed during the first 60 years of the twentieth century in the production of both carbon steels and highly alloyed special steels. The times of elaboration in the electric steelworks have been always shorter than those of the Siemens steelworks, although in the last years, Siemens furnaces reached noticeably improvements in the recovery of heat and, with the introduction of lances to blow oxygen, it was possible to significantly reduce the times of conversion and melting of the scrap.

Other great stimulus for the efficiency and productivity of the ironmaking and steelmaking process started to develop in the decade of the sixties of the twentieth century: the continuous casting. However, continuous casting has not completely imposed until the moment when the properties-quality of the steel could match-exceed the properties of the product obtained by solidification in cast iron ingot molds. During the first decades of the twenty-first century, the objectives are, on the one hand, to reduce the thickness of the steel solidified by continuous casting and, on the other hand, to get close the geometry of the solidified product to the final characteristics of the finished product (near-net-shape profiles for building or rail for railway).

All the above-mentioned considerations would be the starting point of the science and technology of the steel that, from the middle of the nineteenth century, is not an art and has silently transformed, giving room to specific industrial revolutions (in the iron and steel field). It is possible, with view to a not far future, that the ironmaking and steelmaking process is involved in other "silent revolution" related with both the nature of the energy and the reductant used to manufacture steel. The question that is not in doubt is the utilization of steel as structural material in the future.

1.2 The Basic Oxygen Furnace: Introduction

The production of pig iron was around 2.5 million t in 1850, while the production of steel was around only 800,000 t that year. There was, for that reason, an obstacle to the production of steel due to the difficulties to refine the pig iron produced in the blast furnace.

In 1856, Henry Bessemer achieved the transformation, in only a few minutes, of thirty tons of pig iron into Bessemer steel that due to the high sulphur, phosphorus, and nitrogen contents could not be used in steels to be employed in drawing. The historic evolution of this ironmaking process and its impact in the modern science and technology can be read in the book of Bodsworth (1998).

Bessemer (acid), Thomas (from 1878, basic), and Martin-Siemens (reverberatory furnace) processes exercised their hegemony in the pig iron refinement since the fifties of the nineteenth century (beginning of the Industrial Revolution) until the second decade of the twentieth century.

The production of steel in the world has extraordinarily increased as a consequence of the new technologies of refinement, passed from the scale of thousands of tons in 1850 to the scale of millions (80 million t) in 1910. In this last year (1910) the productions of pig iron and steel were equivalents and, from that moment, the statistics of production of both materials has increased and separated.

The possibility of blowing oxygen instead of air (Bessemer and Thomas) causes that the kinetics of the refinement reactions become faster. From the three vertical blowing processes, Linz-Donawitz (LD) (1952), Rotor (1953), and Kaldo (1954), which were almost simultaneously developed, only the first one has consolidated as an economic and safe technology for the conversion of the pig iron.

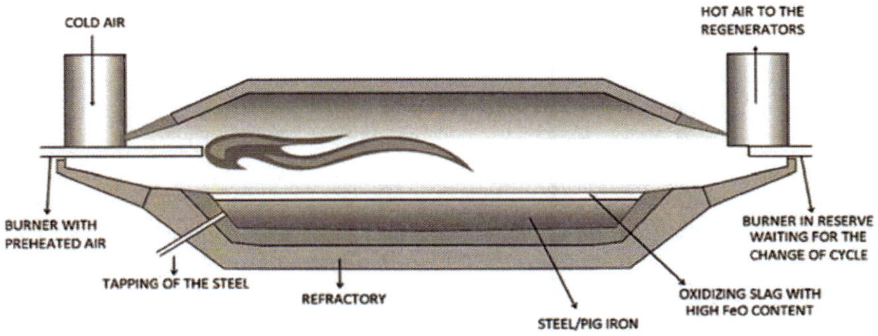

Fig. 1.1 Scheme of the open-hearth furnace (Martin-Siemens technology)

The first pneumatic bottom blowing converters Bessemer and Thomas competed, from their first developments, with the Martin-Siemens open-hearth furnace (Fig. 1.1). The oxygen supply to the bath in the Siemens furnace takes place through the metal-slag interphase and, thus it is required a great surface area of contact to achieve an acceptable production of steel.

The productivity of the Martin-Siemens furnaces was initially not high because the production of 250 t of steel required 900 min of refinement. On the contrary, in the bottom blowing converters, air is directly injected into the metal, and the residence time was reduced to only thirty minutes to treat 70 t of steel. However, the Martin-Siemens process was always very appreciated by the steelmakers due to the quality of the obtained steel. Martin-Siemens steel quality was a type of steel produced using the open-hearth process, independently of the type of steel: low carbon steel (ferritic), medium carbon (ferritic-pearlitic), or high carbon steel (pearlitic), being the rate of the refinement processes regulated and controlled with the analytic means existing in that period. Other advantage of the Martin-Siemens process was that the furnace could be fed with scrap as it had gas or fuel burner to melt it.

Bessemer and Thomas converters were autogenous systems as the energy required for the refinement was generated in the own process. All the energy to satisfy the heat losses and increase the temperature of the melt was provided by the reactions of oxidation of the C, Si, P, and Mn of the pig iron. The quantity of scrap was limited by the thermal balance that could be derived from the process.

In later developments, oxygen lances and fuel-oxygen burners were included in the open-hearth furnace to increase the thermal performance of the process and favor the kinetics of the reactions. Concretely, the Ajax process (1958) was a modification of the Martin-Siemens technology that used blowing of oxygen via lances and burners of air enriched with oxygen. It was possible to obtain 200 t of steel during the refinement process in only 350 min.

The improvements in the bottom blown converters were aimed at the development of materials for the tuyeres that allowed the injection of air enriched with oxygen. The problem of heating and wear of materials around the tuyeres had no immediate solution and the practice that was adopted consisted in cooling the tuyeres with oxidizing gases, H_2O and CO_2, using O_2–H_2O or O_2–CO_2 as injection mixtures. These developments resulted in the VLN (Very Low Nitrogen) process that started to operate in the south of Wales, but later it was displaced by the LD process.

In the Rotor and Kaldo processes, from the fifties of the twentieth century, the rotation rate of the converters (2 revolutions per minute in the Rotor and 35 rpm in the Kaldo) favored the contact metal-slag and the corresponding refinement rate of the pig iron. The rotation of the reactors allowed the liquid (metal-slag) to cold the refractory and to increase the thermal efficiency of the processes; in both converters, the carbon monoxide generated in the refinement reactions was burnt. However, both the outlay in refractories and the time required for the replacement of the refractory lining, as well as the frequent malfunctions and stoppages due to the failure in the rotation mechanism, implied a progressive reduction of the competitiveness of the process in front of the alternative LD.

The LD process was simultaneously developed in the Austrian cities of Linz and Donawitz in 1952. However, its initial conception was quickly modified to allow the injection of calcium oxide to operate with high phosphorus pig irons. In the BOS (Basic Oxygen Steelmaking) or BOF processes the oxygen is blown at a supersonic speed (Mach = 2) inside of the converter by means of a copper lance cooled with water. An emulsion metal-slag with more than 30% of the metal charged in the converter is produced because of the injection of oxygen in the metallic surface, which allows obtaining a very fast rate in the gasification or slagging of pig iron impurities. The main characteristics of the current BOF conversion processes are collected in Table 1.1.

In the sixties of the twentieth century, the process known as Bottom Oxygen Maxchutte (BOM) or Quick, quiet, quality, Basic Oxygen Process (Q-BOP) questioned the supremacy of the BOS-LD technology. The idea of Bessemer–Thomas to obtain liquid steel by bottom blowing of a mixture of gases comprising hydrocarbons, O_2, CO_2, Ar, and N_2 was restarted.

Table 1.1 Main characteristics of the current BOF conversion processes

Process	Company	Gases injected by the bottom	Nm^3/min
LBE	Arbed-Irsid	N_2, Ar (porous plug)	0.01–0.10
LD-CB	Nippon Steel	CO_2, N_2, Ar (porous plug)	0.02–0.06
LD-KGC	Kawasaki Steel	CO, N_2, Ar (porous plug) Tuyeres O_2	0.01–0.20 3.0–3.5
LD-OTB	Kobe Steel	CO, N_2, Ar (porous plug)	0.2–0.3
NK-CB	Nippon Kokan	CO_2, N_2, Ar (porous plug) Tuyeres O_2	0.04–0.10 3.0–3.3
BSC	BSC-UK	Tuyeres N_2, O_2	0.035–0.21
STB	Sumitomo	Tuyeres N_2, O_2, CO_2, Ar	0.10–0.15
LD-HC	CRM-Belgium	Tuyeres N_2, O_2, Ar	0.20–0.30
Q-BOP	US Steel	Tuyeres O_2, CaO and hydrocarbons	

Nowadays, all the conversion technologies have combined the injection of oxygen by lance, with sufficient rate to break the slag layer that covers the metal, and they supply oxidizing gases, decarburizing, nitriding, or reductant reagents, via tuyeres or porous plugs located at the bottom of the converter (Table 1.1).

Top-and-bottom blown converters have operational advantages not only with respect to the traditional LD converters but also with respect to the primitive technologies of bottom blowing. Regarding the converters with blowing using a vertical lance, top-and-bottom converters allow having a more homogeneous distribution of temperatures in the metallic bath and reduce the overflowing of slag and metal during the refinement (Fig. 1.2).

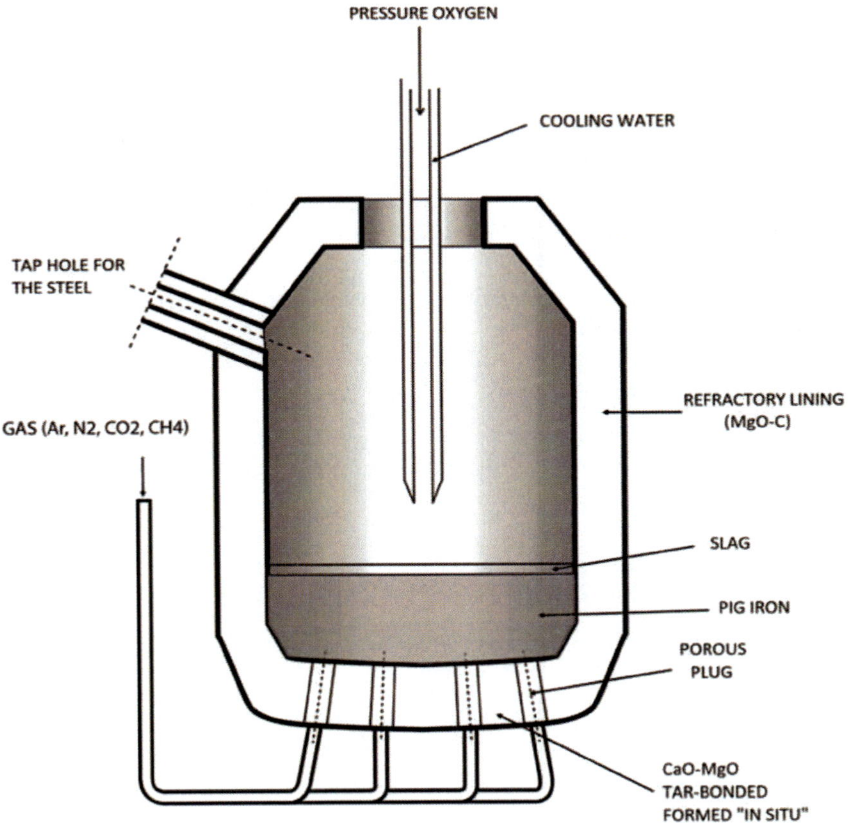

Fig. 1.2 Detail of a top-and-bottom blown converter

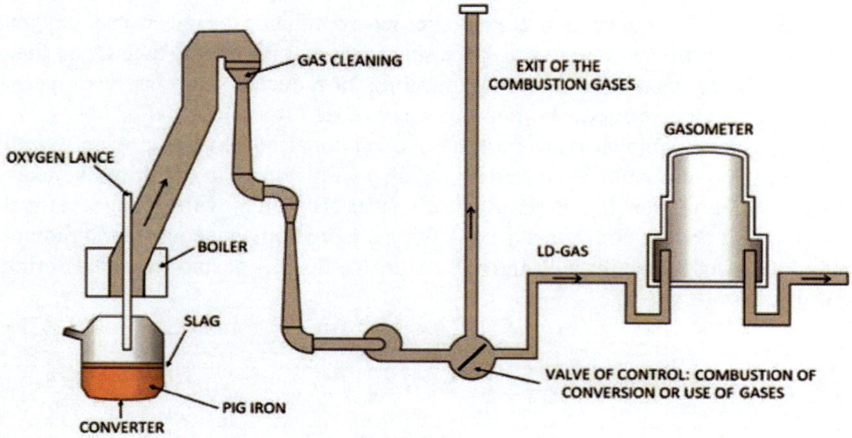

Fig. 1.3 Scheme of a BOF refinement installation with different alternatives to recover the gases of the process

On the other hand, top-and-bottom blown converters have advantages with respect to the BOF with only bottom blowing as, apart from allowing greater quantities of scrap in the charge, they achieve a significant reduction in the number of tuyeres (within 60 and 70%), which lead to frequent breakdowns during the refinement process.

One of the environmental problems of BOF plants is the emissions of suspended particulate matter to the atmosphere, Fig. 1.3. Apart from controlling the emissions of nitrogen oxides, NO_x, sulphur oxide, SO_2, and carbon oxides, CO/CO_2, the best available techniques should be used to pick up the suspended particulate matter:

- Wet cleaning-scrubber of gases.
- Dry scrubber of the suspended particulate matter by means of expansion-diminution of the gas speed.
- Gas cycloning—centrifugation.
- Retaining of particulate in bag filters.
- Electrostatic precipitation of either dry (ESP) or wet (WESP) gases.

It was possible to reduce in the last years, with the support of the best available techniques (BAT), the concentration of particulates in the Linz-Donawitz steelworks. It was possible to reduce from the 100 mg $\cdot$ Nm^{-3} of solid particles in the fumes of the chimney to a concentration of around 2.0 mg$\cdot Nm^{-3}$.

1.3 Reactions in the Converter

1.3.1 Thermodynamics of Solutions

A solution is a homogeneous mixture of two or more components. Throughout the text, solutions that will be subjected to further attention will be the following ones:

1. Solutions of different solutes (C, Mn, Ni, Cr, Ti, V, Mo, Nb, Co, Cu, Si, Al, P, S, O, H, and N) both in the molten iron and in the solid solutions of austenite and ferrite.
2. Slags formed by homogeneous mixtures of molten metallic oxides with the presence of variable quantities of C, S, P, N, and H.

Even when in most of metallic solutions there is a solvent (major component identified with the subscript one), in slags there is not a clear major component. Habitually, the behavior of slags is usually studied-simulated using three components of similar chemical characteristics (ternary diagrams). On another note, the major component of metallic solutions (both molten and solid) is habitually the iron.

The behavior of solutes in solutions can be expressed in two manners:

- Ideal behavior or Raoult's behavior, where the activity of the solute in the solution is identified with the concentration expressed as molar fraction:

$$a_i = x_i \tag{1.1}$$

- Real behavior, where the activity of the solute is identified with the product of the activity coefficient, γ_i, and the molar fraction, x_i:

$$a_i = \gamma_i \cdot x_i \tag{1.2}$$

The coefficient of activity of the solutes, γ_i, is function of both temperature, T, and concentration, x_i.The values that the coefficient of activity of solutes in solutions are two: positive deviations from the ideal situation, γ_i values greater than one; or negative deviations from the ideal situation, γ_i values smaller than one.

When the concentration of solutes in solution tends to zero, the thermodynamic behavior of these solutions is known with the name of diluted solutions. The behavior of solutes and solvent in diluted solutions has the following properties:

1. That the activity, a_i, capacity of reaction of solutes, in limited situations has an ideal behavior. The habitual situation is that solutes have positive or negative deviations with respect to the Raoult's.
2. The activity of solutes in diluted solutions is proportional to its concentration expressed as molar fraction:

$$a_i = \gamma_i^* \cdot x_i \tag{1.3}$$

where γ_i^* is the coefficient of activity at infinite dilution of the solute in the solution. The value of γ_i^*, as opposed to γ_i, is a variable that does not depend on the concentration of solute, it only depends on the solution temperature, T.

3. The activity of the solvents in diluted solutions, a_1, is identified with the ideal behavior:

$$a_1 = x_1 \tag{1.4}$$

4. The proportionality between the Raoult's activity and the concentration in diluted solutions leads to the definition of a new activity: Henry's activity, $a_{H,i}$,where it is directly identified the concentration expressed in percentage of solute, $\%w_i$, with this new expression of the activity:

$$a_{H,i} = \%w_i \tag{1.5}$$

The deviations from the ideality of the Henry's law caused by the increase of solute concentration, *i*, or by the presence of other components would be performed by means of the following equation:

$$a_{H,i} = f_i \cdot \%w_i \tag{1.6}$$

where f_i is the Henry's coefficient of activity.

With the aim of correcting the Raoult's activity of solutes in solutions due to the presence of other components, series are used:

$$\ln \gamma_i = \ln \gamma_i^* + \varepsilon_i^i \cdot x_i + \sum_2^n \varepsilon_i^j \cdot x_j \tag{1.7}$$

where ε_i^i and ε_i^j are, respectively, the interaction parameters of the solute, *i*, with itself (self-interaction) and the interaction parameters on the solute, *i*, due to the presence of other solutes, *j*, available in the solution. Generally, the interaction parameters are variables that depend on both the temperature and the concentration. However, in a range of concentrations not very large, the values of the interaction parameters are constant. The sign of the interaction parameters can be positive or negative depending on the increase or diminution of the solute, *i*, activity in the solution.

Similarly, the correction of the Henry's activity with respect to the ideal situation is performed by means of the coefficient of activity of the Henry's law, f_i, that can be quantified as follows:

$$\log f_i = e_i^i \cdot \%w_i + \sum_2^n e_i^j \cdot \%w_j \tag{1.8}$$

The physical meaning of the self-interaction parameter of Henry, e_i^i, and of the interaction parameter, e_i^j, would be the same than that previously indicated for the Raoult's solutions. The equation that relates the values of the Raoult's interaction parameters, ε_i^j, with those of Henry, e_i^j, is the following one:

$$\varepsilon_i^j = \frac{230.3}{A_1} \cdot A_j \cdot e_i^j + \left(\frac{A_1 - A_j}{A_1}\right) \tag{1.9}$$

where A_1 and A_j are, respectively, the atomic weights of the solvent and solutes, j and i of the solution.

Even when, the self-interaction and first order interaction parameters were used in most of the occasions, it would be possible to talk about the self-interaction and second order interaction that, according to the Henry's law, would comprise the following terms that should be added to the Eq. (1.8):

$$\frac{1}{2} \cdot r_i^i \cdot (\%w_i)^2 + \frac{1}{2} \cdot \sum_2^n r_i^j \cdot (\%w_j)^2 + \frac{1}{2} \cdot \sum_2^n r_i^{j,k} \cdot (\%w_j) \cdot (\%w_k) \tag{1.10}$$

where, r_i^i, r_i^j, and $r_i^{j,k}$ would be, respectively, the self-interaction and second order interaction parameters over the solute i of the rest of the components of the solution.

The standard free energy, $\Delta_r G^0_{a_i/a_{H,i}}$, associated to the change of state of reference of the elements dissolved in a certain solvent (molten iron, austenite, or ferrite) would be the same:

$$Me(s, l, g; a_i = 1) \leftrightarrow Me\big(dis; a_{H,i} = 1\big) \tag{1.11}$$

If we know that the relation between the Raoult's activity, a_i, and the Henry's activity, $a_{H,i}$, is equal to:

$$\frac{a_i}{a_{H,i}} = \frac{\gamma_i^* \cdot x_i}{\%w_i} \tag{1.12}$$

and if we know that, for diluted solutions, it is possible to consider the following approximation between the composition in weight percentage, $\%w_i$, and the molar fraction, x_i:

$$\frac{x_i}{\%w_i} = \frac{A_1}{100 \cdot A_i} \tag{1.13}$$

The value of the standard free energy, $\Delta_r G^0_{a_i/a_{H,i}}$, associated to the change of state of reference (from pure element-compound to the dissolved element-compound in a certain solvent until reaching a Henry's activity equal to one):

$$\Delta_r G^0_{a_i/a_{H,i}} = -R \cdot T \cdot \ln\left(\frac{a_{H,i}}{a_i}\right) = R \cdot T \cdot \ln\left(\frac{a_i}{a_{H,i}}\right)$$
$$= R \cdot T \cdot \ln\left(\frac{\gamma_i^* \cdot x_i}{\%w_i}\right) = R \cdot T \cdot \ln\left(\frac{\gamma_i^* \cdot A_1}{100 \cdot A_i}\right) \quad (1.14)$$

If we know the value of $\Delta_r G^0_{a_i/a_{H,i}}$, the value of $\Delta_r S^0_{a_i/a_{H,i}}$ would be equal to:

$$\Delta_r S^0_{a_i/a_{H,i}} = -\left(\frac{d\Delta_r G^0_{a_i/a_{H,i}}}{dT}\right)_P \quad (1.15)$$

and, finally, the enthalpy $\Delta_r H^0_{a_i/a_{H,i}}$ is calculated using the following equation:

$$\Delta_r H^0_{a_i/a_{H,i}} = \Delta_r G^0_{a_i/a_{H,i}} + T \cdot \Delta_r S^0_{a_i/a_{H,i}} \quad (1.16)$$

From now, it is possible to use two terminologies to express the dissolution of one element (for instance, carbon) in the liquid iron:

1. C(dis) or C(l), which indicates the dissolution of the carbon in the liquid iron and its activity, or capacity of reaction, is expressed by the Raoult's activity, a_C.
2. $\underline{C}$ or $C(\text{dis};\ a_{H,i} = 1)$, which also represents that the carbon is dissolved in the liquid iron, although its activity, in this case, will be expressed by the Henry's activity, $a_{H,C}$.

1.3.2 Thermodynamics and Kinetics

We present in Fig. 1.3 the two alternatives to use the gases in a BOF process:

- Total combustion of the CO in the boiler at the exit of the converter to produce water vapor.
- To collect the gas rich in CO in a gasometer (LD gas). Even though the composition of the LD gas varies at the same time that the refinement process progresses, a mean value is: 70% CO, 20% CO_2, 2% O_2, 1% H_2, and 7% N_2.

The main reaction of refinement in the BOF process is that corresponding to the carbon:

$$C(\text{dis; pig iron}) + FeO(\text{dis; slag}) \leftrightarrow CO(g) + Fe(\text{dis; pig iron}) \quad (1.17)$$

$$C(\text{dis; pig iron}) + \frac{1}{2}O_2(g) \leftrightarrow CO(g) \quad (1.18)$$

Although there are other secondary reactions of refinement:

– Silicon refinement:

$$Si\ (dis;\ pig\ \ iron) + 2O(dis;\ slag) \leftrightarrow SiO_2(dis;\ slag) \tag{1.19}$$

$$Si(dis;\ pig\,iron) + 2FeO(dis;\ slag) \leftrightarrow SiO_2(dis;\ slag) + 2Fe(dis;\ pig\,iron) \tag{1.20}$$

$$SiO_2(dis;\ slag) + 2CaO(dis;\ slag) \leftrightarrow Ca_2SiO_4(dis;\ slag) \tag{1.21}$$

– Manganese refinement:

$$Mn(dis;\ pig\ \ iron) + O(dis;\ slag) \leftrightarrow MnO(dis;\ slag) \tag{1.22}$$

$$Mn(dis;\ pig\ \ iron) + FeO(dis;\ slag) \leftrightarrow MnO(dis;\ slag) + Fe(dis;\ pig\ \ iron) \tag{1.23}$$

– Phosphorus refinement:

$$2P(dis;\ pig\ \ iron) + 5O(dis;\ slag) + 3CaO(dis;\ slag) \leftrightarrow P_2O_8Ca_3(dis;\ slag) \tag{1.24}$$

$$\begin{aligned} &2P(dis;\ pig\ \ iron) + 5FeO(dis;\ slag) + 3CaO(dis;\ slag) \\ &\quad \leftrightarrow P_2O_8Ca_3(dis;\ slag) + 5Fe(dis;\ pig\,iron) \end{aligned} \tag{1.25}$$

As it is possible to check, the mechanisms of refinement reactions can operate in the homogeneous metallic phase or through metal-slag reactions (heterogeneous phase).

We show in Fig. 1.4 the kinetics of the refinement processes of carbon, silicon, phosphorus, manganese, and sulphur. It is possible to define three steps:

– Step I (initial), which covers the first five minutes of the process and is characterized by:

  (a) Ignition of the bath (carbon combustion).
  (b) Oxidation of the silicon of the pig iron (Eq. 1.19).
  (c) Formation of the slag (CaO–FeO–SiO_2 ternary system, Fig. 1.5, edited from Levin et al. 1964) according to Eq. (1.21).

– Step II (oxidizing decarburization), which covers from the 5 min to the 15 min and is characterized by:

  (a) Carbon refinement (decarburization, around 0.25% C/min) according to Eqs. (1.17) and (1.18).
  (b) Manganese and phosphorus refinement.

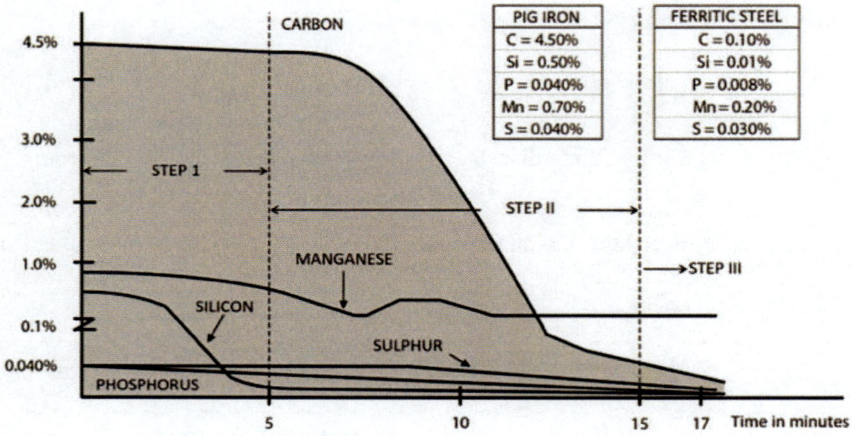

Fig. 1.4 Kinetics of refinement of a pig iron to obtain a ferritic steel

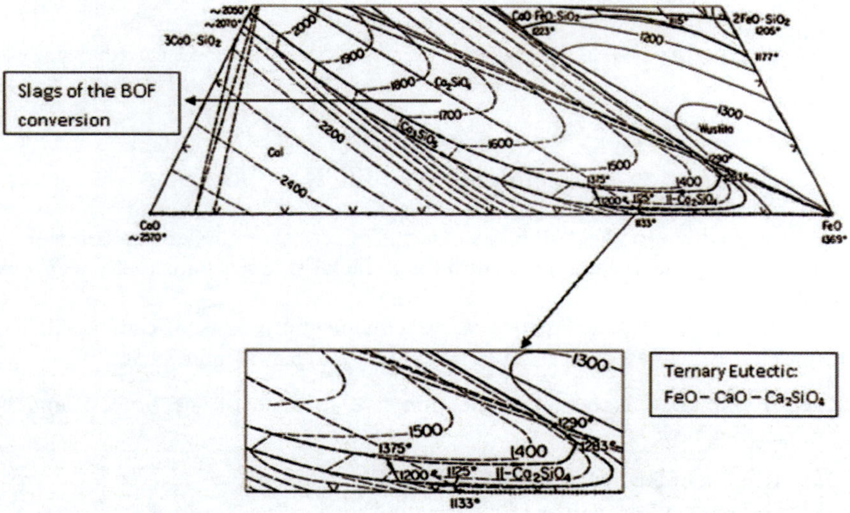

Fig. 1.5 Situation of the conversion slags in the CaO–FeO–SiO$_2$ ternary diagram

– Step III (decreasing decarburization and oxidation of the metal), which covers from the minute 15 to the end of the process, and is characterized by:

(a) Increasing of the FeO in the slag:

$$\text{Fe(dis; pig iron)} + \text{O(dis; pig iron)} \leftrightarrow \text{FeO(dis; slag)} \quad (1.26)$$

(b) Decreasing decarburization rates.
(c) Reaching maximum dephosphorizing rates: in the final moments of the blowing, the solubility of the CaO and FeO is maximum and, if the final temperature is not very high, it is possible to reach an optimum in the dephosphorizing of the steel.

Even though the main mechanism of carbon refinement during the Step II is via the metal-slag equilibrium (Eq. 1.17), other alternatives that exclusively affect the elimination of the carbon without involving (parallel reactions) the base metal (oxidation of the iron) are interesting in the period of decreasing decarburization (low carbon concentrations). If we analyze the system:

$$\mathrm{C(dis;\,pig\,iron)} + \mathrm{O(dis;\,pig\,iron)} \leftrightarrow \mathrm{CO(g)} \tag{1.27}$$

Reaction (1.27) can be displaced to the right (Sievert's law) when, from the minute 16 of blowing, it is introduced via tuyeres, or via the porous plug at the bottom of the converter, an inert gas (argon) that could make the pressure of the carbon monoxide in the gas-metal interface to reduce.

Equally, if carbon dioxide is injected in the liquid, it is possible to decarburize, without oxidizing the base metal, although in this situation the mechanism is:

$$\mathrm{C(dis;\,pig\,iron)} + \mathrm{CO_2(g)} \leftrightarrow \mathrm{2CO(g)} \tag{1.28}$$

Anyway, at the end of the refinement process in the BOF process (Step III), the concentration of carbon is reduced although the concentration of oxygen is increased. The dissolved carbon and oxygen concentrations in the liquid iron, as maximum, are those expressed by equilibrium (1.27). The standard free energy of this reaction is obtained using the data of the Appendices 1 and 3:

$$\mathrm{C(s)} + \frac{1}{2}\mathrm{O_2(g)} \leftrightarrow \mathrm{CO(g)} \tag{1.29}$$

$$\frac{1}{2}\mathrm{O_2(g)} \leftrightarrow \mathrm{O}\left(\mathrm{dis};\, a_{\mathrm{H,O}} = 1\right) \tag{1.30}$$

$$\mathrm{C(s)} \leftrightarrow \mathrm{C}\left(\mathrm{dis};\, a_{\mathrm{H,C}} = 1\right) \tag{1.31}$$

$$\begin{aligned}\Delta_r G^0_{\text{equilibrium 1.27}} &= \Delta_r G^0_{\text{equilibrium 1.29}} - \Delta_r G^0_{\text{equilibrium 1.30}} - \Delta_r G^0_{\text{equilibrium 1.31}} \\ &= -20900 - 39.76 \cdot \mathrm{T(K)}\left[\mathrm{J} \cdot (\mathrm{mol\,CO})^{-1}\right]\end{aligned} \tag{1.32}$$

$$\mathrm{K}_{\text{equilibrium 1.27}} = \frac{\mathrm{P_{CO}}}{a_{\mathrm{H,O}} \cdot a_{\mathrm{H,C}}} = \frac{\mathrm{P_{CO}}}{\mathrm{C_C} \cdot \mathrm{C_O}} \cdot \frac{1}{f_\mathrm{C} \cdot f_\mathrm{O}} \tag{1.33}$$

where $a_{H,i}$ are the Henry's activities, C_i is the quantity of the element dissolved in the pig iron in percentage and f_i is the Henry's coefficient of activity of the dissolved species.

The final solution of refinement is so diluted that the Henry's coefficient of activity of the oxygen, f_O, and of the carbon, f_C, are equal to one. At 1873 K, the equilibrium constant of the reaction (1.27) is:

$$\begin{aligned}\Delta_r G^0_{\text{equilibrium 1.27}} &= -R \cdot T \cdot \ln K_{\text{equilibrium 1.27}} \rightarrow K_{\text{equilibrium 1.27}} \\ &= \exp\left(\frac{-20900 - 39.76 \cdot 1873\left[J \cdot (\text{mol CO})^{-1}\right]}{-8.31 \frac{J}{\text{mol} \cdot K} \cdot 1873\ K}\right) \\ &\rightarrow K_{\text{equilibrium 1.27}} = 457 \end{aligned} \quad (1.34)$$

thus:

$$C_C \cdot C_O = \frac{P_{CO}}{457} \quad (1.35)$$

Or, if it is indicated the concentration of oxygen dissolved, C_O, in ppm (104 ppm $\simeq$ 1%):

$$C_C \cdot C_O(\text{ppm}) = 22 \cdot P_{CO} \quad (1.36)$$

If the temperature at the end of the blowing is 1873 K, $P_{CO} \simeq 1$ atm and the potential of oxygen is 850 ppm, the quantity of dissolved carbon is, according to the equilibrium (1.36), 0.026% (260 ppm C).

In the practice, for top-blowing converter and for a final condition of refinement similar to those previously mentioned, the relation of concentrations of carbon and oxygen dissolved in the metallic melt is:

$$C_C \cdot C_O(\text{ppm}) = 33 \cdot P_{CO} \quad (1.37)$$

thus $C_C = 0.039\,\%$ (with temperature of 1878 K, $P_{CO} \simeq 1$ atm and potential of 850 ppm), while for a top-and-bottom blown BOF converter (with the intervention of the decarburizing mechanism of Sievert) it is possible to obtain values of carbon dissolved in the liquid much closer to the equilibrium (1.36):

$$C_C \cdot C_O(\text{ppm}) = 24 \cdot P_{CO} \quad (1.38)$$

This way, the final carbon quantity is $C_C = 0.028\%$.

The assimilation (reversion) of the manganese of the slag into the steel, during the Step II of the refinement process, is justified by analyzing the equilibrium of the Eq. (1.23). The oxidation of the carbon by the FeO of the slag is so violent that the slag reduces the content of this oxide displacing the equilibrium of the reaction to the

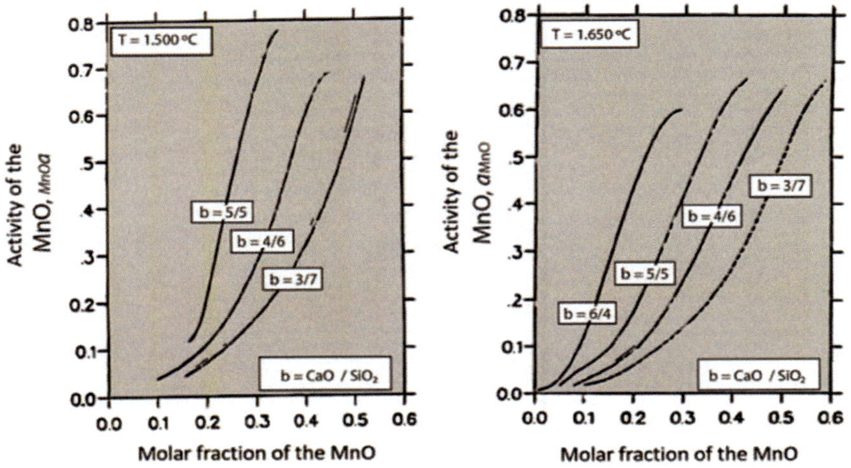

Fig. 1.6 Activities of the MnO (pure solid as state as reference) in the SiO_2–CaO–MnO system

left. On the other hand, at around the minute 10 of blowing, the slag is completely formed and, although the solubility of the calcium oxide does not reach its maximum value, the basicity index of the slag is acceptable. The activity of the manganese oxide in slags of the SiO_2–CaO–MnO system grows with the basicity index and decreases very slightly with the increase of the temperature (Fig. 1.6, edited from Springorum 1995).

Thus, equilibrium (1.23) will be displaced for a moment to the left when the basicity is increased but also with the activity of the manganese oxide of the BOF slag. Later, when the FeO content in the slag is recovered (low carbon gasification rates and high iron oxidation rates), the equilibrium will be again displaced to the right leading to the slagging of the manganese.

One of the future challenges of the BOF conversion is performing a continuous process with feeding of pig iron and slag-forming agents to obtain iron and slag as flows of exit. However, the idea of a continuous process is much more attractive if instead of pig iron, pre-reduced iron is used as raw material. DIOS process is based on this idea. It uses a continuous reactor to produce steel and operates without the basic installations of the integrated steel plants as the blast furnace of the BOF steelworks.

1.3.3 Application of the CaO–FeO–SiO_2 Ternary Diagram to the Study of the LD Slags

The most common method to represent a ternary alloy is the Roozebum method: over an equilateral triangle (Fig. 1.7), where each corner corresponds to 100% of

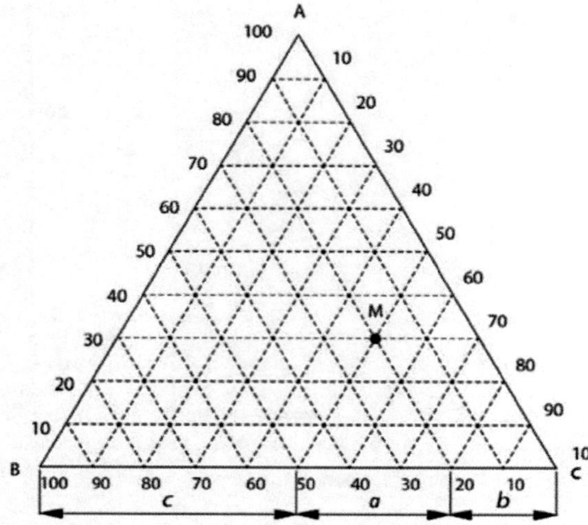

Fig. 1.7 Roozebum representation

each pure metal or compound, each side having a scale from 1 to 100% (AB, AC, and BC) (Pero-Sanz et al. 2017). An alloy M formed by a% of metal A, b% of metal B, and c% of metal C is represented by a point in Fig. 1.7. If an alloy has a constant proportion between B and C, this is:

$$\frac{b\%}{c\%} = \text{constant} \tag{1.39}$$

it will necessary be located over the AM line. Analogously, the geometric location of all alloys with percentage of A equal to a% are parallel to the side BC intersecting M. As temperature is considered an axis perpendicular to the plane ABC, different diagrams must be plotted to represent temperature variation.

Applying the phase rule for a three-component system, with a number of phases ϕ, will result in a maximum of four degrees of freedom (V), at constant pressure.

The Roozebum representation makes the generalization of the lever rule possible when, for a certain temperature, alloy H has two phases in equilibrium, with compositions I and J. Figure 1.8 shows, for the side AC, compositions *i*′, *h*′, and *j*′ of metal C. Supposing P_J and P_I are the respective weights of the phases J and I, the total weight of the alloy H (P_T) equals to $P_J + P_I$. On the other hand, the total weight of C can be expressed as:

$$P_T \cdot h' = P_J \cdot j' + P_I \cdot i' \tag{1.40}$$

and since:

$$P_T \cdot h' = P_J \cdot h' + P_I \cdot h' \tag{1.41}$$

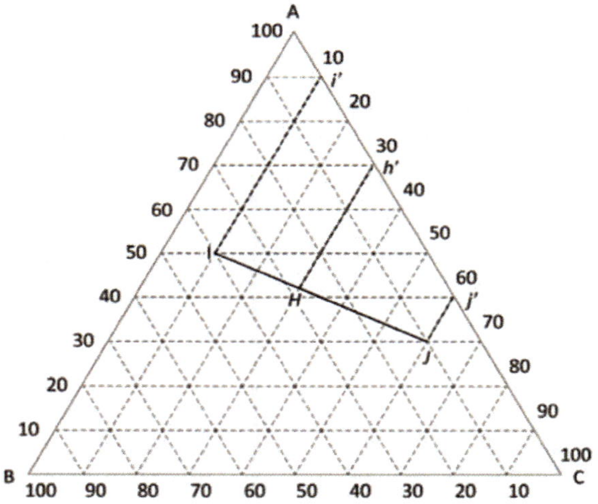

Fig. 1.8 Generalization of the lever rule for alloy H

then:

$$\frac{P_J}{P_I} = \frac{h' - i'}{j' - h'} = \frac{HI}{HJ} \tag{1.42}$$

When alloy H, at a certain temperature, is in equilibrium with three phases S, R, and L (one of the liquid) as shown in Fig. 1.9 in the triangle SRL (sometimes known as *tie-triangle*); equilibrium is of the invariant type (with no degrees of freedom) and, at that constant temperature, proportions in weight of the phases can be determined by repetitive application of the lever rule, being P_1, P_2, and P_3 the respective weights

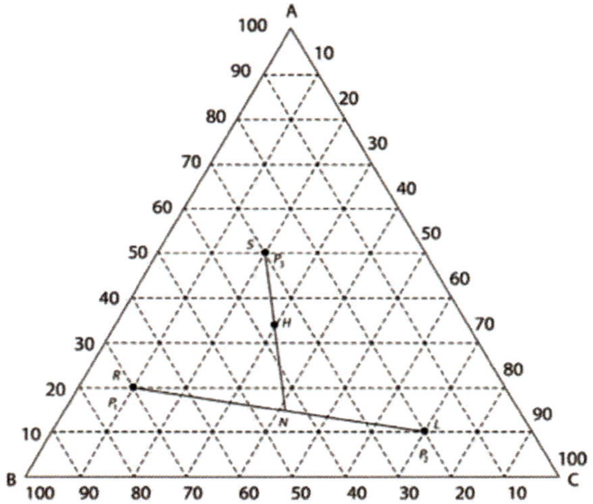

Fig. 1.9 Equilibrium between three phases S, R, and L

of the phases R, L, and S, and the total weight of the alloy H ($P_T = P_1 + P_2 + P_3$):

$$\frac{P_1}{P_1 + P_2} = \frac{NL}{LR} \tag{1.43}$$

$$\frac{P_2}{P_1 + P_2} = \frac{NR}{RL} \tag{1.44}$$

$$\frac{P_1 + P_2}{P_1 + P_2 + P_3} = \frac{SH}{SN} \tag{1.45}$$

$$\frac{P_1}{P_T} = \frac{NL}{LR} \cdot \frac{SH}{SN} \tag{1.46}$$

$$\frac{P_2}{P_T} = \frac{NR}{RL} \cdot \frac{SH}{SN} \tag{1.47}$$

$$P_3 = P_T - (P_1 + P_2) \tag{1.48}$$

The CaO–FeO–SiO_2 ternary diagram (Fig. 1.10) can be used to study the LD slags. In the Exercise 1.3, it is possible to see the application of the CaO–FeO–SiO_2 ternary diagram in the LD slags.

1.3.4 Solved Exercises

Exercise 1.1 Considering the Fe–O system, determine the maximum oxygen concentration in the liquid iron at 1600 °C.

Solution:

We consider the equilibrium of the oxygen dissolved in the molten iron in presence of iron oxide (II):

$$\mathrm{Fe(l)} + \mathrm{O}\left(\mathrm{dis};\, a_{\mathrm{H,O}} = 1\right) \leftrightarrow \mathrm{FeO(l)} \tag{1.49}$$

We use the data of the Appendices 1 and 3 to estimate the value of the standard free energy of the reaction (1.49):

$$\mathrm{Fe(l)} + \frac{1}{2}\mathrm{O_2(g)} \leftrightarrow \mathrm{FeO(l)} \tag{1.50}$$

$$\Delta_r G^0_{\text{reaction (1.50)}} = -224.0 + 0.050 \cdot \mathrm{T(K)}\left[\mathrm{kJ} \cdot (\mathrm{mol\,FeO})^{-1}\right] \tag{1.51}$$

$$\mathrm{O_2(g)} \leftrightarrow \mathrm{O}\left(\mathrm{dis};\, a_{\mathrm{H,O}} = 1\right) \tag{1.52}$$

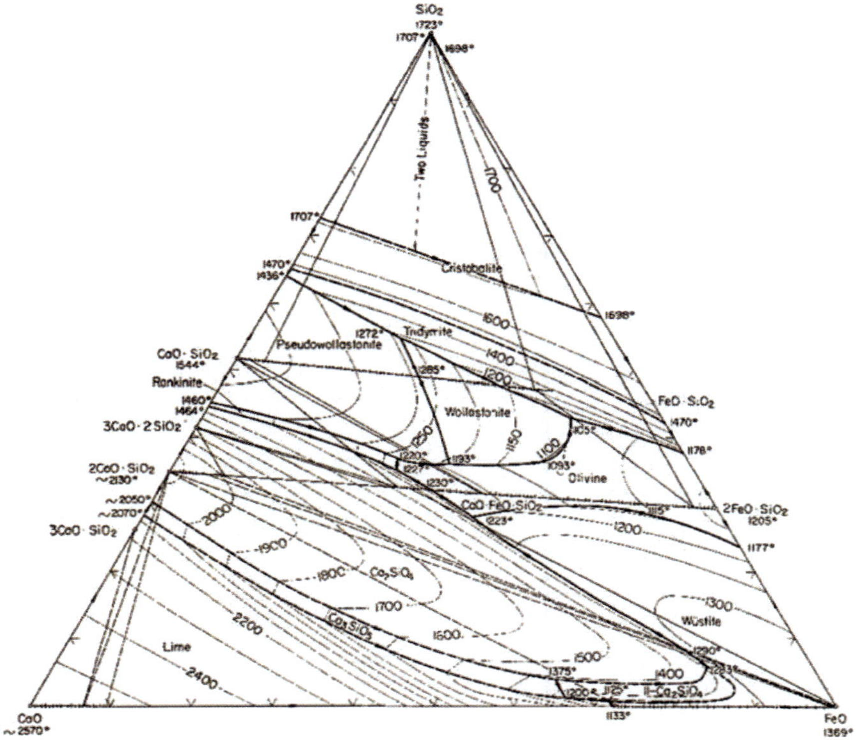

Fig. 1.10 CaO–FeO–SiO$_2$ ternary diagram (Levin et al. 1964)

$$\Delta_r G^0_{\text{reaction (1.52)}} = -117.0 - 3.14 \times 10^{-3} \cdot T(K)\left[kJ \cdot (mol\,FeO)^{-1}\right] \tag{1.53}$$

Consequently, the value of the standard free energy associated to the reaction (1.49):

$$\begin{aligned}\Delta_r G^0_{\text{reaction (1.49)}} &= \Delta_r G^0_{\text{reaction (1.50)}} - \Delta_r G^0_{\text{reaction (1.52)}} \\ &= -127.0 + 5.31 \times 10^{-2} \cdot T(K)\left[kJ \cdot (mol\,FeO)^{-1}\right]\end{aligned} \tag{1.54}$$

$$\Delta_r G^0_{\text{reaction (1.49)}}(1873\ K) = -27469\left[J \cdot (mol\,FeO)^{-1}\right] \tag{1.55}$$

However, in this exercise, we are going to use, in a parallel manner, the thermodynamic data of the reaction (1.49) taken from other sources: Deo and Boom (1993, p. 52) and Japan Society for the Promotion of Science (1988, p. 111):

– According to Deo and Boom (1993):

$$\Delta_r G^0_{\text{reaction (1.49)}}(1873\ \text{K}) = -24098\left[\text{J} \cdot (\text{mol FeO})^{-1}\right] \quad (1.56)$$

– According to the Japan Society for the Promotion of Science (1988):

$$\Delta_r G^0_{\text{reaction (1.49)}}(1873\ \text{K}) = -24368\left[\text{J} \cdot (\text{mol FeO})^{-1}\right] \quad (1.57)$$

The values of the equilibrium constant of the reaction (1.49) depending on the thermodynamic data are:

$$\Delta_r G^0 = -\text{R} \cdot \text{T} \cdot \ln \text{K}_{\text{eq}} \rightarrow \text{K}_{\text{eq}} = \exp\left(-\frac{\Delta_r G^0}{\text{R} \cdot \text{T}}\right) \quad (1.58)$$

– According to the Appendices 1 and 3:

$$\text{K}_{\text{eq}} = \exp\left(-\frac{-27469\left[\text{J} \cdot (\text{mol FeO})^{-1}\right]}{8.31 \cdot \left(\text{J} \cdot \text{mol}^{-1} \cdot \text{K}^{-1}\right) \cdot 1873\ \text{K}}\right) = 5.84 \quad (1.59)$$

– According to Deo and Boom (1993):

$$\text{K}_{\text{eq}} = \exp\left(-\frac{-24098\left[\text{J} \cdot (\text{mol FeO})^{-1}\right]}{8.31 \cdot \left(\text{J} \cdot \text{mol}^{-1} \cdot \text{K}^{-1}\right) \cdot 1873\ \text{K}}\right) = 4.70 \quad (1.60)$$

– According to the Japan Society for the Promotion of Science (1988):

$$\text{K}_{\text{eq}} = \exp\left(-\frac{-24368\left[\text{J} \cdot (\text{mol FeO})^{-1}\right]}{8.31 \cdot \left(\text{J} \cdot \text{mol}^{-1} \cdot \text{K}^{-1}\right) \cdot 1873\ \text{K}}\right) = 4.79 \quad (1.61)$$

Developing the terms involved in the equilibrium constant of the reaction (1.49), and assuming that: $a_{\text{Fe}} = a_{\text{FeO}} = 1$, the following nonlinear equation that allows calculating the percentage composition of oxygen dissolved in the molten iron is obtained:

$$\log(\%\,\text{O}) + e_{\text{O}}^{\text{O}} \cdot (\%\,\text{O}) + \log \text{K}_{\text{eq}}(\text{reaction (1.49)}) = 0 \quad (1.62)$$

Considering in all the cases that the self-interaction parameter of Henry of the oxygen in the molten iron is −0.185 (mean value of the data collected in the Appendix 4), it is possible to solve the nonlinear equation, for instance, calculating the oxygen concentration when the nonlinear function takes the value of zero. The obtained results are:

– According to the Appendices 1 and 3:

$$(\%O) = 0.185\%(1850\,\text{ppm}) \quad (1.63)$$

– According to Deo and Boom (1993):

$$(\%\,O) = 0.238\,\%(2380\,\text{ppm}) \quad (1.64)$$

– According to the Japan Society for the Promotion of Science (1988)

$$(\%\,O) = 0.231\,\%(2310\,\text{ppm}) \quad (1.65)$$

According to the Fe–O binary diagram (TAPP Database 1994, Massalski et al. 1990), the oxygen solubility in the iron in equilibrium with the FeO at 1600 °C is around 0.250% (2500 ppm).

On another note, the maximum solubility that the oxygen can reach in the molten iron at 1600 °C in equilibrium with a slag where the FeO Raoult's activity was smaller than one (iron (II) oxide dissolved in a slag) would be:

$$(\%\,O)_T = (\%\,O)_T^{=} \cdot a_{FeO} \quad (1.66)$$

where $(\%\,O)_T^{=}$ would represent the oxygen maximum solubility at any temperature for the equilibrium proposed in the reaction (1.49) when the FeO activity in the slag is equal to one.

Finally, it is possible to quantify, with this exercise, the deviations of the Henry's ideal behavior that a solute dissolved in the molten iron could have. For instance, if we considered that the dissolution of the oxygen in the molten iron could reach an ideal behavior, $f_O = 1$, the oxygen maximum solubility, considering the data of the Appendix 3, would be 0.171% (1710 ppm) instead of 0.185% (1850 ppm).

Commentaries: Deo and Boom (1993, p. 79) provide an equation to calculate the oxygen maximum solubility in molten iron as a function of the temperature and optical basicity of the LD slag:

$$\log(\%\,O)^{=} = -1.907 \cdot \Lambda_{slag} - \frac{6005}{T} + 3.570 \quad (1.67)$$

If the optical basicity was equal to one, slag formed by FeO, the solubility estimated for the oxygen at 1873 K would be equal to 0.02864% (or 286 ppm).

End of the exercise.

Exercise 1.2 Calculate the activity and the coefficient of activity of the CaO and the SiO_2 in a ternary slag of the CaO-FeO-SiO_2 system, Fig. 1.10, at the temperature of 1600 °C, when the melt is in equilibrium with two solid phases: 2CaO·SiO_2 (C_2S) and 3CaO·SiO_2 (C_3S).

Question starts:

Considering the data of the CaO–FeO–SiO_2 ternary diagram of Fig. 1.10, the system described by the enunciation is defined (degrees of freedom-variance of the system equal to zero), as there is only a ternary melt of defined composition that, at 1600 °C, is in equilibrium with the 3CaO·SiO_2 and the 2CaO·SiO_2. The final mass composition of the melt derived from the ternary diagram is: 49% CaO, 41% FeO, and 10% SiO_2. The corresponding composition of the slag expressed in molar fraction (the values of the atomic weights of the Ca, Fe, Si, and O are 40.08, 55.85, 28.09, and 16.00 g/mol) is, respectively:

$$x_{CaO} = \frac{49\,\text{g CaO}\cdot\frac{1\,\text{mol CaO}}{56.08\,\text{g CaO}}}{49\,\text{g CaO}\cdot\frac{1\,\text{mol CaO}}{56.08\,\text{g CaO}} + 41\,\text{g FeO}\cdot\frac{1\,\text{mol FeO}}{71.85\,\text{g FeO}} + 10\,\text{g SiO}_2\cdot\frac{1\,\text{mol SiO}_2}{60.09\,\text{g SiO}_2}} = 0.5424 \tag{1.68}$$

$$x_{FeO} = \frac{41\,\text{g FeO}\cdot\frac{1\,\text{mol FeO}}{71.85\,\text{g FeO}}}{49\,\text{g CaO}\cdot\frac{1\,\text{mol CaO}}{56.08\,\text{g CaO}} + 41\,\text{g FeO}\cdot\frac{1\,\text{mol FeO}}{71.85\,\text{g FeO}} + 10\,\text{g SiO}_2\cdot\frac{1\,\text{mol SiO}_2}{60.09\,\text{g SiO}_2}} = 0.3543 \tag{1.69}$$

$$x_{SiO_2} = \frac{10\,\text{g SiO}_2\cdot\frac{1\,\text{mol SiO}_2}{60.09\,\text{g SiO}_2}}{49\,\text{g CaO}\cdot\frac{1\,\text{mol CaO}}{56.08\,\text{g CaO}} + 41\,\text{g FeO}\cdot\frac{1\,\text{mol FeO}}{71.85\,\text{g FeO}} + 10\,\text{g SiO}_2\cdot\frac{1\,\text{mol SiO}_2}{60.09\,\text{g SiO}_2}} = 0.1033 \tag{1.70}$$

From the data of the Appendix 1, we calculate the standard free energies of the following reactions:

$$2CaO(s) + SiO_2(s) \leftrightarrow Ca_2SiO_4(s) \tag{1.71}$$

$$\Delta_r G^0_{\text{reaction (1.71)}} = -103.0 - 0.020\cdot T(K)\left[kJ\cdot(mol\,C_2S)^{-1}\right] \tag{1.72}$$

$$3CaO(s) + SiO_2(s) \leftrightarrow Ca_3SiO_5(s) \tag{1.73}$$

$$\Delta_r G^0_{\text{reaction (1.73)}} = -115.0 - 0.023\cdot T(K)\left[kJ\cdot(mol\,C_3S)^{-1}\right] \tag{1.74}$$

Consequently, the equilibrium between the three solid phases, C_2S–C_3S–CaO and the corresponding value of the standard free energy is the following one:

$$Ca_3SiO_5(s) \leftrightarrow Ca_2SiO_4(s) + CaO(s) \tag{1.75}$$

$$\Delta_r G^0_{\text{reaction (1.75)}} = \Delta_r G^0_{\text{reaction (1.71)}} - \Delta_r G^0_{\text{reaction (1.73)}} \tag{1.76}$$

$$\begin{aligned}\Delta_r G^0_{\text{reaction (1.75)}} &= 12.0 + 0.003 \cdot T(K)\left[kJ \cdot (mol\, C_2S)^{-1}\right] \\ &= 12000 + 3 \cdot T(K)\left[J \cdot (mol\, C_2S)^{-1}\right]\end{aligned} \tag{1.77}$$

The value of the standard free energy associated to the reaction (1.75) at 1873 K is 17619 J. Consequently, the value of the equilibrium constant, which is identified with the activity of the calcium oxide in the melt as the activities of the C_2S and C_3S (pure compounds) are equal to one, is:

$$\begin{aligned}\Delta_r G^0_{\text{reaction (1.75)}} &= -R \cdot T \cdot \ln K_{eq} \rightarrow K_{eq} \\ &= \exp\left(-\frac{\Delta_r G^0_{\text{reaction (1.75)}}}{R \cdot T}\right) \\ &= \exp\left(-\frac{17619\ J}{8.314\frac{J}{mol \cdot K} \cdot 1873}\right) = 0.3226\end{aligned} \tag{1.78}$$

$$K_{eq} = \frac{a_{CaO} \cdot a_{Ca_2SiO_4}}{a_{Ca_3SiO_5}} = a_{CaO} \rightarrow a_{CaO} = 0.3226 \tag{1.79}$$

On another note, the standard free energy associated to the reaction (1.71) at 1873 K is −140,460 J. The value of the corresponding equilibrium constant of the C_2S formation from the calcium and silicon oxides is:

$$K_{eq} = \exp\left(-\frac{-140460\ J}{8.314\frac{J}{mol \cdot K} \cdot 1873}\right) = 8263 \tag{1.80}$$

$$\begin{aligned}K_{eq} &= \frac{a_{Ca_2SiO_4}}{a^2_{CaO} \cdot a_{SiO_2}} = \frac{1}{a^2_{CaO} \cdot a_{SiO_2}} \rightarrow a_{SiO_2} \\ &= \frac{1}{K_{eq} \cdot a^2_{CaO}} = 1.16 \times 10^{-3}\end{aligned} \tag{1.81}$$

Finally, the activity coefficients of the calcium and silicon oxides at 1600 °C are calculated as follows:

$$\gamma_{CaO} = \frac{a_{CaO}}{x_{CaO}} = \frac{0.3226}{0.5424} = 0.5947 \tag{1.82}$$

$$\gamma_{SiO_2} = \frac{a_{SiO_2}}{x_{SiO_2}} = \frac{1.16 \times 10^{-3}}{0.1033} = 0.0112 \tag{1.83}$$

Commentaries: Data of the standard free energies of formation used for the C_2S and C_3S, Appendix 1, are not the most convenient. We have used values of the standard free energy at temperatures significantly higher (1873 K) than those recommended in Appendix 1: 1000 K for the C_3S and 673 K for the C_2S.

The physical meaning of the activity of one compound or element in a certain phase is identified with the capacity of reaction that might be attributed to it. If the compound/element has ideal behavior (Raoult's law), its capacity of reaction (activity) is identified with its concentration expressed as molar fraction. In this case, both calcium oxide and silicon oxide exhibit negative deviations with respect to the ideal behavior: their activities are smaller than the corresponding molar fractions. However, the most important deviation corresponds to the silicon oxide, most of it trapped by the calcium basic oxide to form the corresponding calcium silicates.

End of the exercise.

Exercise 1.3 Assuming applicable the ionic model of Temkin (Ballester et al. 2000, p. 280) to the following slag: 56% CaO, 27% FeO, 17% SiO_2, calculate the activities of both the calcium oxide and the ferrous oxide, as well as their respective coefficients of activity.

Question starts:

The activity of the basic oxides, CaO, or basic and oxidizing oxides, such as FeO, in a slag according to the Ionic Model of Temkin is calculated as follows:

$$a_{MeO} = N_{e;Me^{2+}} \cdot N_{e;O^{2-}} \tag{1.84}$$

where the Raoult's activity of the metallic oxide in the melt, a_{MeO}, is equal to the product of the equivalent fraction of the cation, Me^{2+}, in the mixture of cations and the equivalent fraction of the oxygen anion, O^{2-}, in the mixture of anions. Definitely, the Temkin model is reduced to executing a balance of charge (gram equivalent) in the ideal mixture of anions and cations that, presumably, form the slag according to the Ionic Model. As working hypothesis, before calculating the gram equivalents, it is necessary to identify the nature of the anions and cations that constitute the melt. In this case, and assuming a total degree of ionization of the reactions, there are two cations in the pseudoternary slag: calcium, Ca^{2+}, and iron, Fe^{2+}; and two anions: oxygen, O^{2-}, and silicate, $SiO_4{}^{4-}$, consequence of the following reactions:

$$CaO(f) \rightarrow Ca^{2+}(f) + O^{2-}(f) \tag{1.85}$$

$$FeO(f) \rightarrow Fe^{2+}(f) + O^{2-}(f) \tag{1.86}$$

$$SiO_2(f) + O^{2-}(f) \rightarrow SiO_4^{4-}(f) \tag{1.87}$$

The composition of the slag, expressed in molar fraction is:

$$x_{CaO} = \frac{56\,\text{g CaO} \cdot \frac{1\,\text{mol CaO}}{56.08\,\text{g CaO}}}{56\,\text{g CaO} \cdot \frac{1\,\text{mol CaO}}{56.08\,\text{g CaO}} + 27\,\text{g FeO} \cdot \frac{1\,\text{mol FeO}}{71.85\,\text{g FeO}} + 17\,\text{g SiO}_2 \cdot \frac{1\,\text{mol SiO}_2}{60.09\,\text{g SiO}_2}} = 0.6025 \quad (1.88)$$

$$x_{FeO} = \frac{27\,\text{g FeO} \cdot \frac{1\,\text{mol FeO}}{71.85\,\text{g FeO}}}{56\,\text{g CaO} \cdot \frac{1\,\text{mol CaO}}{56.08\,\text{g CaO}} + 27\,\text{g FeO} \cdot \frac{1\,\text{mol FeO}}{71.85\,\text{g FeO}} + 17\,\text{g SiO}_2 \cdot \frac{1\,\text{mol SiO}_2}{60.09\,\text{g SiO}_2}} = 0.2267 \quad (1.89)$$

$$x_{SiO_2} = \frac{17\,\text{g SiO}_2 \cdot \frac{1\,\text{mol SiO}_2}{60.09\,\text{g SiO}_2}}{56\,\text{g CaO} \cdot \frac{1\,\text{mol CaO}}{56.08\,\text{g CaO}} + 27\,\text{g FeO} \cdot \frac{1\,\text{mol FeO}}{71.85\,\text{g FeO}} + 17\,\text{g SiO}_2 \cdot \frac{1\,\text{mol SiO}_2}{60.09\,\text{g SiO}_2}} = 0.1707 \quad (1.90)$$

Using one mol of slag as reference for the balance of charges (anions and cations), the corresponding balance of cations and anions would be:

- Balance of cations:

  Calcium cations:

$$2 \cdot x_{CaO} = 2 \times 0.6025 = 1.2050\,\text{g} - \text{equivalent of Ca}^{2+}(\text{f}) \quad (1.91)$$

  Iron cations:

$$2 \cdot x_{FeO} = 2 \times 0.2267 = 0.4534\,\text{g} - \text{equivalent of Fe}^{2+}(\text{f}) \quad (1.92)$$

  Total g-equivalents of cations:

$$1.2050\,\text{g} - \text{equivalent of Ca}^{2+}(\text{f}) + 0.4534\,\text{g} - \text{equivalent of Fe}^{2+}(\text{f}) = 1.6584 \quad (1.93)$$

- Balance of anions:

  Silicate anions:

$$4 \cdot x_{SiO_2} = 4 \cdot 0.1707 = 0.6828\,\text{g} - \text{equivalent of SiO}_4^{4-}(\text{f}) \quad (1.94)$$

  Oxygen anions:

$$2 \cdot \left[(x_{CaO} + x_{FeO}) - 2 \cdot x_{SiO_2}\right] = 2 \cdot [(0.6025 + 0.2267) - 2 \cdot 0.1707] = 0.9756\,\text{g} - \text{equivalent of O}^{2-}(\text{f}) \quad (1.95)$$

Total g-equivalents of anions:

$$0.6828\,\mathrm{g} - \mathrm{equivalent\ of\ SiO_4^{4-}\,(f)} + 0.9756\,\mathrm{g} - \mathrm{equivalent\ of\ O^{2-}\,(f)} = 1.6584 \tag{1.96}$$

Finally, using the equation of Temkin for the CaO and FeO.

$$a_{\mathrm{CaO}} = \mathrm{N}_{e,\mathrm{Ca}^{2+}} \cdot \mathrm{N}_{e,\mathrm{O}^{2-}} = \left(\frac{1.2050}{1.6584}\right) \cdot \left(\frac{0.9756}{1.6584}\right) = 0.4274 \tag{1.97}$$

$$a_{\mathrm{FeO}} = \mathrm{N}_{e,\mathrm{Fe}^{2+}} \cdot \mathrm{N}_{e,\mathrm{O}^{2-}} = \left(\frac{0.4534}{1.6584}\right) \cdot \left(\frac{0.9756}{1.6584}\right) = 0.1608 \tag{1.98}$$

The values of their corresponding coefficients of activity would be:

$$\gamma_{\mathrm{CaO}} = \frac{a_{\mathrm{CaO}}}{x_{\mathrm{CaO}}} = \frac{0.4274}{0.6025} = 0.7094 \tag{1.99}$$

$$\gamma_{\mathrm{FeO}} = \frac{a_{\mathrm{FeO}}}{x_{\mathrm{FeO}}} = \frac{0.1608}{0.2267} = 0.7093 \tag{1.100}$$

Commentaries: As it is possible to check along this exercise, the temperature for which the calculation of the CaO and FeO activities could be valid was not mentioned. At least initially, the temperature should ensure the total melting of the slag (liquidus temperature) because below this temperature the composition will not be that mentioned in the enunciation of the exercise and, thus, the balance and the corresponding calculation of the Temkin's activities would not be the same.

According to Coudurier et al. (1985, pp. 242–243), the activity of the FeO and the coefficient of activity of the CaO at 1600 °C for the studied slag are:

$$a_{\mathrm{FeO}} = 0.55 \tag{1.101}$$

$$\gamma_{\mathrm{CaO}} = 0.63 \tag{1.102}$$

Even though the values estimated by Temkin for the capacity of reaction of the CaO in the slag are acceptable, those of the oxidizing capacity of the slag, associated to the content-activity of the FeO (0.55), are far from those obtained using the Temkin model (0.1608).

End of the exercise.

Exercise 1.4 The LD slag can be partially recycled in the blender of the sintering plant. Considering the pseudoternary slag of the Exercise 1.3: 56% CaO, 27% FeO, 17% SiO_2, calculate:

(a) The thermodynamically stable phases at room temperature.
(b) The temperature of starting of liquid phases formation (solidus temperature) and quantity of liquid and solid at this temperature.
(c) Proportion of solid and liquid phases at 1200 °C.

In all the cases, assume that the heating process can be analyzed using the CaO–FeO–SiO_2ternary diagram.

Question starts:

Even though in the ternary diagram it is possible to see several compatible triangles coherent with the global composition of the slag (56% CaO, 27% FeO, 17% SiO_2), we take those with the largest area: CaO–FeO–2CaO·SiO_2 (Ca_2SiO_4) or C_2S (see Fig. 1.11).

Applying the Roozebom representation, the proportion of these phases (free lime, ferrous oxide, and dicalcium silicate) at room temperature (in equilibrium) is 22.8%CaO, 27.7% FeO, and 49.5% Ca_2SiO_4 (Fig. 1.12).

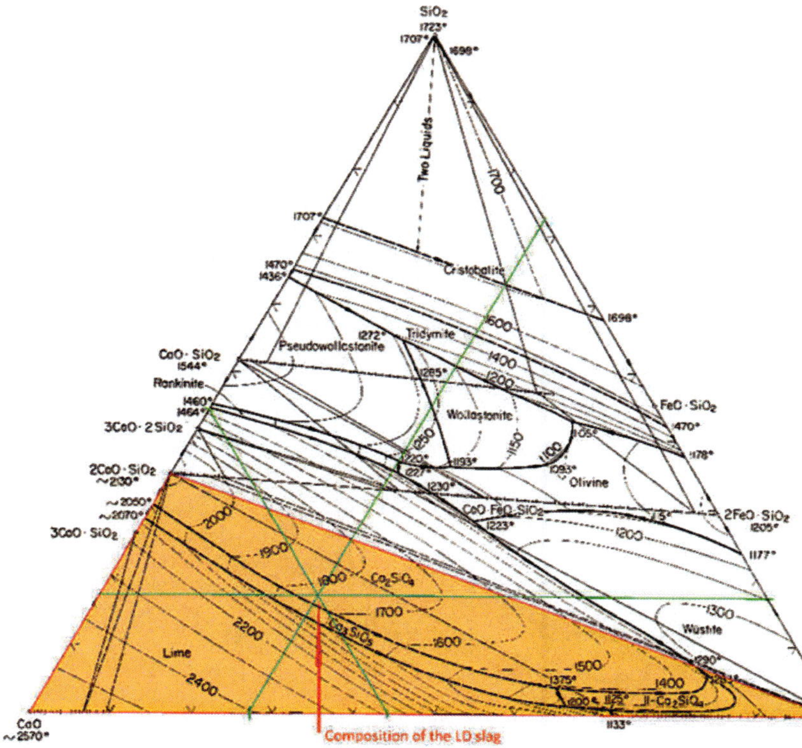

Fig. 1.11 Composition and compatibility triangle of the LD slag

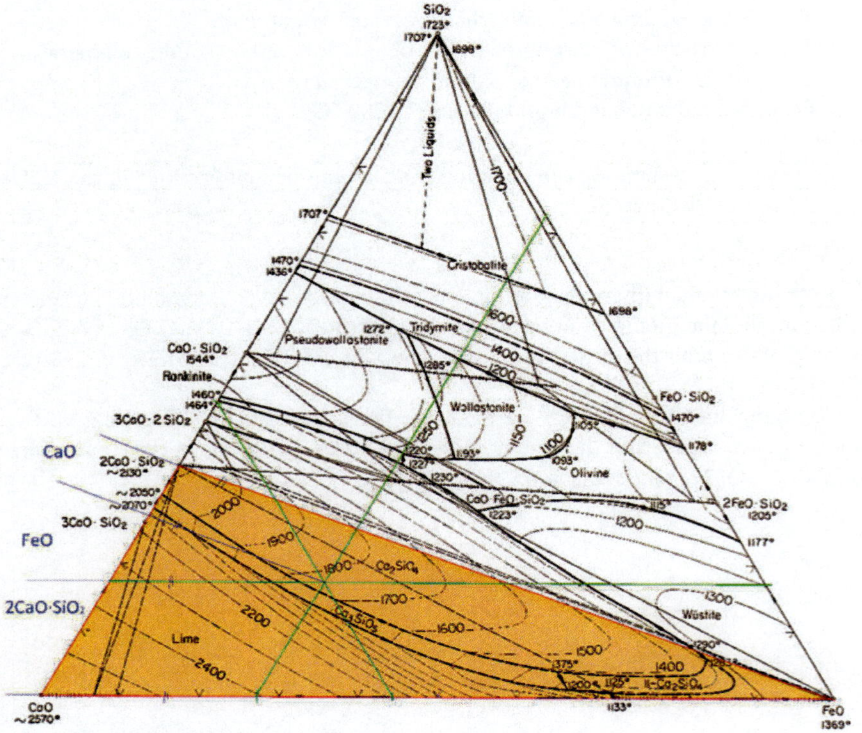

Fig. 1.12 Determination of the proportion of the three phases in the slag

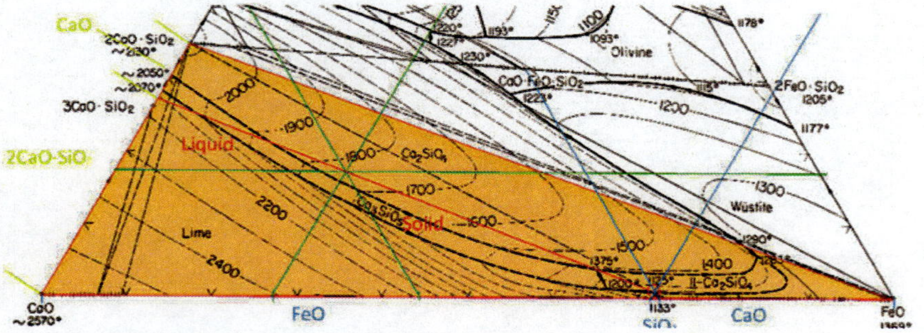

Fig. 1.13 Situation in the LD slag at the eutectic temperature

The appearance of liquid phases during the heating (solidus temperature) takes place at 1125 °C as a consequence of the following eutectic reaction (Fig. 1.13):

$$CaO(s) + FeO(s) + Ca_2SiO_4(s) \leftrightarrow \text{liquid} \quad (1.103)$$

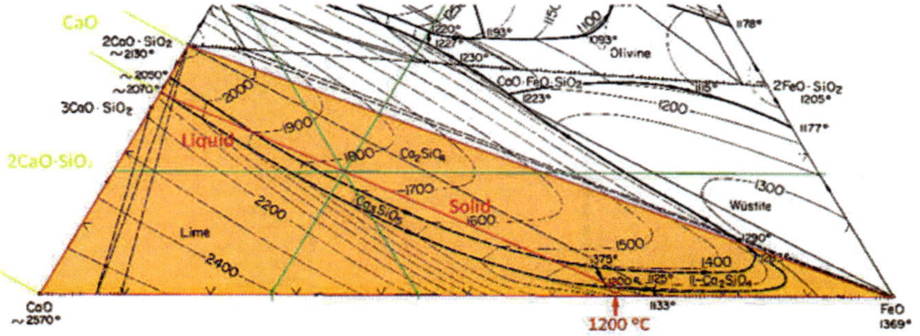

Fig. 1.14 Situation in the LD slag at the temperature of 1200 °C

The composition of the eutectic liquid (Fig. 1.13, light blue lines) is: 27.6% CaO, 71.2% FeO, and 1.2% SiO_2. At this temperature and considering the lever rule, 38.4% of liquid with the above-mentioned composition has formed. The 61.6% of solid in equilibrium with the liquid comprises 77.4% Ca_2SiO_4 and 22.6% CaO.

At 1200 °C, that might be the reference temperature for the sintering front of the plant, the quantity of liquid phase increases, there is 40.8% (Fig. 1.14). The solid represents 59.2%, and it comprises 81.2% of Ca_2SiO_4 and 18.8% CaO.

Commentaries: As it was observed in previous exercises, the pseudoternary model to study the thermal behavior of a slag has not the accurateness that is required for a good control of the operations. However, from the educational point of view, it can be useful to qualitatively explain some of the properties of LD slags; for instance, its chemical instability due to the presence of free lime. Thus, the pseudoternary model emphasizes in the fact that for the sintering front temperature, three liquid phases and other products of high refractoriness (as CaO and Ca_2SiO_4) are formed.

Finally, it is possible to state that the first steps of the process of the LD formation (Fig. 1.4, Step 1) can be studied with the CaO–FeO–SiO_2 ternary diagram (Fig. 1.10), which should be complemented with the data of the CaO–Fe_2O_3–SiO_2 ternary diagram, for those peritectic reactions of formation of new solid phases resulted from the incongruent melting of chemical compounds that can be formed during the heating of the ferric burden to be sintered.

End of the exercise.

1.4 Mass and Energy Balances

We consider the APL or Windows-APL-J, MATLAB, or HSC software in the simulation of the mass and energy balances of the BOF process.

As happened in the case of the example detailed in Fig. 1.1, the simulation also considers the obtaining of a ferric product from the pig iron (Table 1.2). However, we

Table 1.2 Mass and energy balances of a BOF process that operates with gas utilization: Ore-E: 90% Fe_2O_3, 10% SiO_2; Gas-S: 100% CO; Slag-S: 30% FeO, 46% Ca_2SiO_4, 24% CaO

Phase	T(K)	Weight (kg)	O_2 (%)	Fe (%)	C (%)	Si (%)	Fe_2O_3 (%)	CaO (%)
Oxygen- E	273	71	100					
Air-E	298	0	23.3					
Fluxes-E	298	81						100
Ore-E	298	17					90	
Pig iron-E	1620	881		94.5	4.5	1		
Scrap-E	298	192		99	0.646	0.354		
FeSi-E	298	2		21	4	75		
Gas-S	600	94						
Steel-S	1873	1000		99.9	0.05	0.05		
Slag-S	1873	150						24

present in this example the operation in the case of a refinement gas that is not burnt at the exit of the converter, on the contrary, it is collected in a gasometer (Fig. 1.3). Even though this gas has a high CO content, the LD gas does not comprise 100% CO, as it is shown in Table 1.2.

The variables defined to make the system consistent and dependent are:

- The energy supplied from inside of the system, zero in this case (for instance, this variable would have a defined value in the case of an electric furnace).
- Loss percentage with respect to the heat supplied by both the reaction and the gases of the process, 20.02% (Table 1.3).
- Temperature of reaction, 1873 K.
- Percentage of CO over the CO + CO_2 in the gases, 100% (all the gas is CO).

Table 1.3 Heat flows of the entrance and exit of the BOF process that operates with gas utilization (complement of Table 1.2)

Description of the flow	Heat (kJ)	
Heat of reaction	– 890.376	Supplied: 1036.025 kJ
Heat supplied by the gases	– 145.649	
Losses	207.262 (20.00%)	Consumed: 1036.025 kJ
Heat absorbed by the O_2 of the lance	130.223 (12.57%)	
Heat absorbed by the air	0 (0%)	
Heat absorbed by the slag-forming elements	180.506 (17.42%)	
Heat absorbed by the ore	25.535 (2.47%)	
Heat absorbed by the pig iron	212.924 (20.55%)	
Heat absorbed by the scrap	274.713 (26.52%)	
Heat supplied by the ferrosilicon	4.862 (0.47%)	

- Percentage of oxygen supplied by the lance with respect to the total oxygen (gas + ore), 81%.
- Basicity index of the slag in percentage (relation between the percentage of calcium oxide with respect to the silicon oxide):

$$46\text{g } Ca_2SiO_4(2\,CaO \cdot SiO_2) \cdot \frac{1\text{ mol } Ca_2SiO_4}{172\text{ g } Ca_2SiO_4} \cdot \frac{2\text{ mol CaO}}{1\text{ mol } Ca_2SiO_4} \cdot \frac{56\text{ g CaO}}{1\text{ mol CaO}} = 30\text{ g CaO} \tag{1.104}$$

$$46\text{g } Ca_2SiO_4(2CaO \cdot SiO_2) \cdot \frac{1\text{ mol } Ca_2SiO_4}{172\text{ g } Ca_2SiO_4} \cdot \frac{1\text{ mol } SiO_2}{1\text{ mol } Ca_2SiO_4} \cdot \frac{60\text{ g } SiO_2}{1\text{ mol } SiO_2} = 16\text{ g } SiO_2 \tag{1.105}$$

$$I_B(I) = \frac{\%\,CaO}{\%\,SiO_2} \cdot 100 = \frac{24 + 30}{16} = 3.375 \tag{1.106}$$

- Excess of combustion air (in this case, the weight of air is zero as all the refinement gas is collected).

Finally, we collect the results of the thermal balance for each one of the flows in Table 1.3.

Exercise 1.5 In a LD converter, to produce 1000 kg of an iron melt with 0.05% C and 0.05% Si, 80 kg of oxygen, and 86 kg of slag-forming agents-fluxes are required. The quantity of blown air is estimated in 179 kg and the produced gas in 264 kg. Calculate the requirements of pig iron, A, scrap, Ch, and slag, E, to close the balance.

Data: Composition of the pig iron, A, 4.0% C, and 1.00% Si; composition of the scrap, Ch, 0.6% C, and 0.30% Si; composition of the slag, E, 15% SiO_2. The percentage of CO_2 in the gas is 45 wt. %. The atomic weights of carbon, silicon, and oxygen are, respectively: 12.01, 28.09, and 16.00 g/mol.

Solution starts:

If we consider as entrance flows to the system oxygen, slag-forming agents, blown air, pig iron (A), and scrap (Ch), and as exit flows the metallic melt, gas, and slag (E), the balance can be expressed as follows (Fig. 1.15):

$$80 + 179 + 86 + A + Ch = 1000 + 264 + E \tag{1.107}$$

On the other hand, the partial balance to the carbon is:

$$A \cdot 4 + Ch \cdot 0.6 = 1000 \cdot 0.05 + 264 \cdot 0.45 \cdot \frac{12.01}{44.01} \tag{1.108}$$

Finally, the third equation is consequence of the partial balance to the silicon:

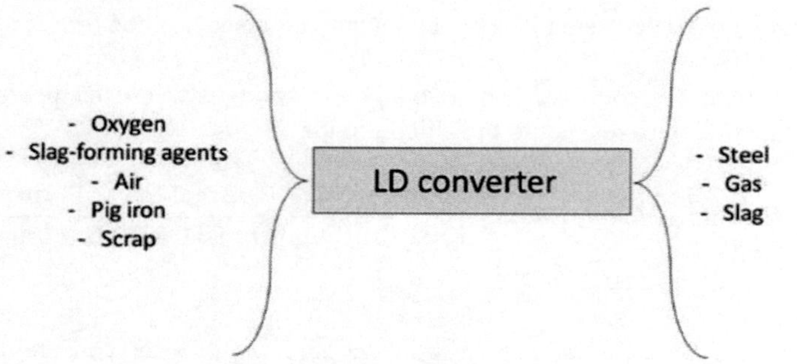

Fig. 1.15 Currents of entrance and exit to the system

$$A \cdot 1 + Ch \cdot 0.3 = 1000 \cdot 0.05 + E \cdot 15 \cdot \frac{28.09}{60.09} \tag{1.109}$$

We obtain from these three last equations:

$$A + Ch - E = 919 \tag{1.110}$$

$$4 \cdot A + 0.6 \cdot Ch = 3292 \tag{1.111}$$

$$A + 0.3 \cdot Ch - 7 \cdot E = 50 \tag{1.112}$$

We multiply Eq. (1.110) by seven and we subtract from it Eq. (1.111), and we obtain:

$$\begin{aligned} &7 \cdot (A + Ch - E - 919) - (4 \cdot A + 0.6 \cdot Ch - 3292) \\ &= 0 \rightarrow 6 \cdot A + 6.7 \cdot Ch = 6383 \end{aligned} \tag{1.113}$$

We subtract from Eq. (1.113) the Eq. (1.111) previously multiplied by the relation 6/4 with the purpose of eliminating the variable A. Thus, we obtain the value of the quantity of scrap, Ch, that was used in the process:

$$\begin{aligned} &(6 \cdot A + 6.7 \cdot Ch - 6383) - \frac{6}{4} \cdot (4 \cdot A + 0.6 \cdot Ch - 3292) = 0 \rightarrow 5.8 \cdot Ch \\ &= 1445 \rightarrow Ch = \frac{1445}{5.8} = 249\,\text{kg of scrap} \end{aligned} \tag{1.114}$$

The quantity of pig iron, A, can be calculated using Eq. (1.111):

$$A = \frac{3292 - 0.6 \cdot 249}{4} = 786\,\text{kg of pig iron} \tag{1.115}$$

Finally, it is possible to obtain the quantity of slag, E, produced in the conversion process using Eq. (1.107):

$$80 + 179 + 86 + 786 + 249 = 1000 + 264 + \mathrm{E} \rightarrow \mathrm{E} = 116\,\mathrm{kg\ of\ slag} \quad (1.116)$$

Commentaries: Even though it is important to reach absolute results for each one of the flows involved in the mass balance, in the case of the proposed exercise, all the used data are referred to the ton of melt produced in the process. However, the values in the case of the gases (for instance, oxygen consumption) are usually expressed in cubic meters measured under standard conditions of pressure and temperature (1.0 atm and 273 K). In this case, the oxygen consumption is 56 $Nm^3\ O_2 \cdot t^{-1}$ of melt or 3.30 $Nm^3\ O_2 \cdot t^{-1} \cdot min^{-1}$ for 17 min of blowing (see Fig. 1.4 and Table 1.1).

As the refinement is a process of exothermic nature, it is not common to pay special attention to the energy balance because the cooling element that equilibrates the excess of energy produced is the scrap. In the integrated steel plants, the requirements of scrap have increased with the improvement of the efficiency of the heat transfer in the converter. It was even noticeably improved the duration of the refractory lining with designs where high thermal conductivity materials are used, for instance magnesia-carbon refractories (Verdeja et al. 2014, pp. 283–286). However, integrated steel plants buy scrap in the market because the own process does not produce enough scrap to feed the LD converters.

Due to the above mentioned, there is not an excessive requirement on the increase of the temperature of the cold flows that enter in the system (oxygen and scrap) because this would require increasing the quantity of scrap to cold. In the future, integrated steel plants might consider the production of their own synthetic scrap-DRI using the reductant gases generated in the blast furnace, coke batteries, or the own LD process, especially if the tons of CO_2 produced per ton of steel must be reduced.

End of the exercise.

1.5 Kinetics of the Reactions in the Converter

Even when the physical–chemical metal-slag reactions (oxidizing basic slag) for the BOF process are similar to those indicated in Sect. 2.3, the kinetics of reactions in the electric furnace and those of the BOF process are not comparable.

If we specifically analyze the decarburizing of the steel:

$$\mathrm{C(dis;\ iron) + FeO(dis;\ slag) \leftrightarrow CO(g) + Fe(dis;\ iron)} \quad (1.117)$$

the controlling mechanism of the process (Fig. 1.16) can fall in one of the following steps:

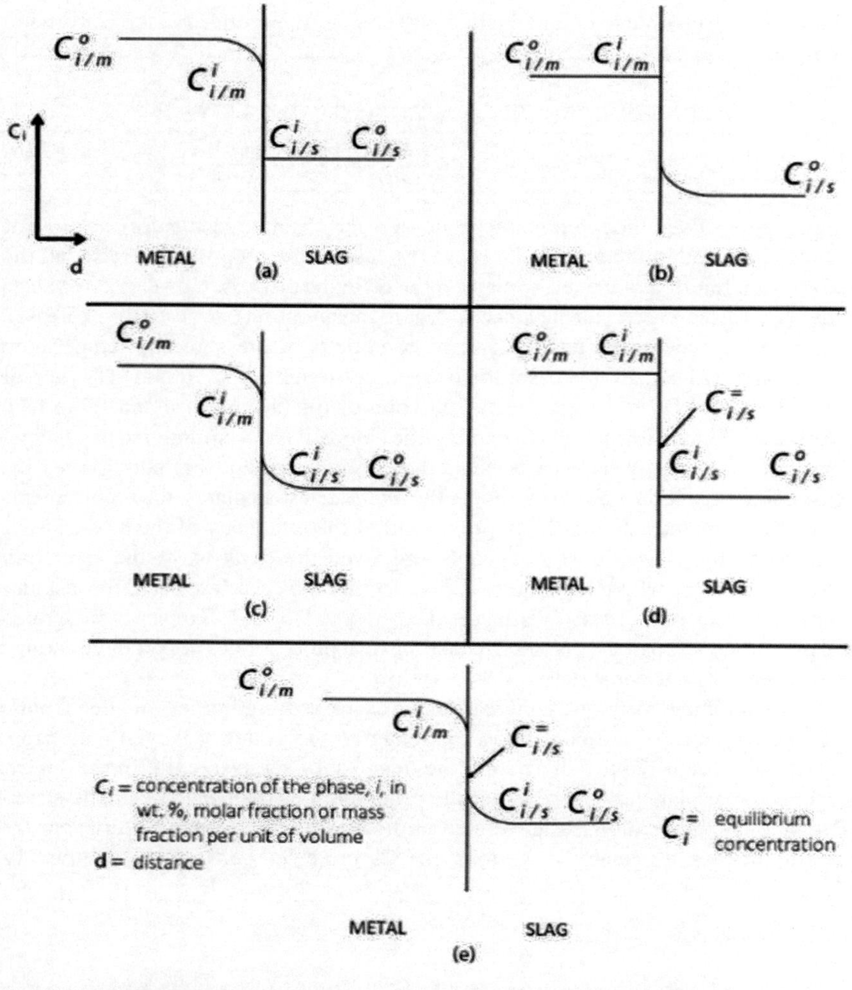

Fig. 1.16 Metal-slag reactions. Different concentration-penetration profiles as a function of the controlling step of the process. It is considered in all the cases that the mass transfer is performed from the metal to the slag (slag-formation of impurities): **a** kinetical-diffusional control in the metallic phase; **b** kinetical-diffusional control in the slag; **c** mixed control of the diffusional transfer; **d** kinetical-chemical control, and, **e** mixed kinetical control, by interface reaction

1. Diffusion of the FeO (in fact of the cation Fe^{2+} and of the anion O^{2-}) in the slag.
2. Chemical reaction in the metal-slag interphase.
3. Diffusion of the carbon in the metallic phase.
4. Chemical adsorption of the carbon on the metal-slag interphase, which is directly proportional to the fraction of surface area available and the concentration of carbon in the metal.

From all these steps, it is possible to discard the last three because of the following reasons:

- The chemical reaction is not sufficiently fast to consider it as controlling step at the temperature of the refinement reactions: $T \geq 1500\ °C$.
- As the diffusion of the carbon in the metallic phase is usually 10 times greater than the diffusion of the ionic species in the slag, the rate of carbon transfer in the liquid is not the controlling step of the process.
- The rate of adsorption-decarburization of the carbon (chemisorption), v, in the metal-gas interphase is:

$$v = k_q \cdot (1 - \theta) \cdot \left(C_i - C_i^{=}\right) \tag{1.118}$$

where k_q is the kinetics constant of the chemisorption process, θ is the fraction of surface area of the metal-gas interphase occupied by active elements (surfactant), C_i is the concentration of carbon in the interface, and $C_i^{=}$ is the equilibrium concentration of the reaction (1.117).

That is to say, the absorption can be the controlling step of the process when the carbon concentration in the metal is smaller than 0.10% and $(1 - \theta)$ is small due to, for instance, the presence of sulphur (surfactant element).

The flow of mass transfer due to the diffusion, J_C, first law of Fick, is identified with the product of the diffusion coefficient, D_i, and the gradient of concentration (variation of the concentration, ΔC_i, in a certain unity of length, Δl). The rate of mass transfer due to diffusion through a surface area, A, would be:

$$\frac{dC_i}{dt} = D_i \cdot \left(\frac{dC_i}{dl}\right) \cdot A \tag{1.119}$$

where the gradient of concentration of Eq. (1.119):

$$\left(\frac{dC_i}{dl}\right) = \frac{C_i^0 - C_i^{=}}{\Delta l} \tag{1.120}$$

is identified with the variation of the concentration that experiences the species that diffuses with respect to that of the equilibrium, $C_i^{=}$, along a certain length, Δl.

If the concentration of the species that diffuses, i, is not expressed in mass per unit of volume, V, but as molar fraction, x_i, then, it is possible to obtain the following equation:

$$\frac{dx_i}{dt} = D_i \cdot \left(\frac{dx_i}{dl}\right) \cdot \left(\frac{A}{V}\right) \tag{1.121}$$

$$\frac{d(\Delta x_i)}{\Delta t} = D_i \cdot \left(\frac{\Delta x_i}{\Delta l}\right) \cdot \left(\frac{A}{V}\right) \tag{1.122}$$

$$\frac{d(\Delta x_i)}{\Delta t} = \left(\frac{D_i}{\Delta l}\right) \cdot \Delta x_i \cdot \left(\frac{A}{V}\right) = k_d \cdot \Delta x_i \cdot \left(\frac{A}{V}\right) \tag{1.123}$$

where k_d is the mass transfer coefficient (dimensionally expressed in unities of length divided by time).

In a more specific manner, the rate of the decarburizing process of the pig iron melt with FeO dissolved in the slag is controlled by the diffusion of the ionic species Fe^{2+} and O^{2-} in the slag. Eq. (1.123) is transformed in:

$$\frac{dx_{FeO}}{dt} = k_d \cdot \left(x^0_{FeO} - x^=_{FeO}\right) \cdot \left(\frac{A}{V}\right) \tag{1.124}$$

The value of k_d depends on the characteristics of the contact between reacting phases (metal-slag-gas), x^0_{FeO} is the molar fraction of the oxide in the slag, and $x^=_{FeO}$ is the molar fraction corresponding to the equilibrium of the reaction (1.117).

To express the decarburizing rate as a function of the carbon loss in the melt instead of the disappearance of the FeO in the slag, it is necessary to modify Eq. (1.124) considering the following working hypothesis:

- According to reaction (1.117), one mol of FeO reacts with one mol of carbon.
- The average molecular weight of the slag (weighted average of the oxides that form 100 g of slag, calculation base habitually employed) is $65.00\ g \cdot mol^{-1}$. On another note, $12.01\ g \cdot mol^{-1}$ is the atomic weight of the carbon.
- The quantity of LD slag generated to produce 1000 kg of decarburized melt is of 125 kg.

The relation between the carbon percentage in the metallic melt, % C, and the iron molar fraction in the slag, x_{FeO}, would be:

$$1 = \left(\frac{x_{FeO}}{\text{mol slag}}\right) \times \left(\frac{\text{mol C}}{\text{mol FeO}}\right) \times \left(\frac{12.01\ \text{g C}}{\text{mol C}}\right) \times \left(\frac{\text{mol slag}}{65\ \text{g slag}}\right) \times \left(\frac{125\ \text{g slag}}{1000\ \text{g melt}}\right) \times \left(\frac{100\ \text{g melt}}{\%\ \text{C}}\right) \tag{1.125}$$

$$x_{FeO} = \frac{1000 \times 65 \cdot \%C}{100 \times 125 \times 12.01} = \frac{10 \times 65}{125 \times 12.01} \cdot \%C \tag{1.126}$$

Consequently, the kinetics for the decarburizing of the pig iron melt in the LD would be:

$$\frac{d(\%\ C)}{dt} = k_d \cdot \left(\%\ C^0 - \%\ C^=\right) \times \left(\frac{A}{V}\right) \times \left(\frac{65}{12.01 \cdot 100}\right) \times \left(\frac{1000}{125}\right) \tag{1.127}$$

For the estimation of the k_d, coefficient of diffusional mass transfer, empiric correlations obtained by means of heat transfer processes between beds of spherical particles and a fluid (liquid or gas) are habitually employed. Due to the analogies that exist between the equations used to study the heat and mass transfer, the mathematical equation is:

$$\frac{k_d \cdot L^*}{D_i} = 2.0 + 0.60 \cdot Re^{1/2} \cdot Sc^{1/3} \tag{1.128}$$

where the three dimensionless numbers involved are:

- Nusselt's number, Nu:

$$Nu = \frac{k_d \cdot L^*}{D_i} \tag{1.129}$$

- Reynolds's number, Re:

$$Re = \frac{L \cdot \rho \cdot v}{\eta} \tag{1.130}$$

- Schmidt's number, Sc:

$$Sc = \frac{\eta}{\rho \cdot D_i} \tag{1.131}$$

The characteristic linear dimension of the Reynolds's number, L, makes reference to the average size of the metallic drops of 1.50 cm (1.50×10^{-2} m) emulsified during the decarburizing process in the fluid (foamy slag) for a relative displacement rate of the molten phases in contact (metal-slag), v, of 3.5 $m \cdot s^{-1}$. From the physical–chemical characteristics of the BOF slags point of view, they are more basic, but with smaller oxidizing capacity, than those used in the electric steelworks. It is assumed that the density of the slag, ρ, is 3000 $kg \cdot m^{-3}$ and the viscosity, η, 0.100 $N \cdot s \cdot m^{-2}$ ($Pa \cdot s$) at 1650 °C.

The area-volume ratio (A/V), identified with the inverse of the height (0.8 m) that the reactive phases in the system reach, is of: 1.25 m^{-1}. As the concentration is expressed in percentage, the gradient of concentration, driving force of the reaction, is 4.5.

BOF slags are more basic, but less oxidizing, than those of the electric steelworks. If the density of the slag is 3000 $kg \cdot m^{-3}$, the viscosity is 0.100 $N \cdot s \cdot m^{-2}$ (Pa·s) at 1650 °C. Finally, a value of 1.00×10^{-8} $m^2 \cdot s^{-1}$ is assigned to the coefficient of diffusion of the oxygen in the slag (see Kawai and Shiraishi 1988, p. 196). Definitely, the value of the Nusselt's number is equal to:

$$\frac{k_d \cdot L^*}{D_i} = 2.0 + 0.60 \cdot Re^{1/2} \cdot Sc^{1/3}$$
$$= 2.0 + 0.60 \cdot \left(\frac{L \cdot \rho \cdot v}{\eta}\right)^{1/2} \cdot \left(\frac{\eta}{\rho \cdot D_i}\right)^{1/3} \tag{1.132}$$

$$\frac{k_d \cdot L^*}{D_i} = 2.0 + 0.60 \times \left(\frac{1.5 \times 10^{-2}\text{m} \cdot 3000\,\text{kg} \cdot \text{m}^{-3} \times 3.5\,\text{m} \cdot \text{s}^{-1}}{0.100\,\text{N} \cdot \text{s} \cdot \text{m}^{-2}}\right)^{1/2}$$
$$\times \left(\frac{0.100\,\text{N} \cdot \text{s} \cdot \text{m}^{-2}}{3000\,\text{kg} \cdot \text{m}^{-3} \cdot 1.00 \times 10^{-8}\,\text{m}^2 \cdot \text{s}^{-1}}\right)^{1/3} \tag{1.133}$$

$$\text{Nu} = \frac{k_d \cdot L^*}{D_i} = 357.70 \tag{1.134}$$

If we identify the value of L*, of the Nusselt's number with the thickness of the boundary layer, δ, that is developed within the metallic molten phases and the slag in the LD, then it is possible to estimate the value of δ using the following mathematical equation (Deo and Boom 1993, p. 114):

$$\delta = 1.5 \cdot D_i^{1/3} \cdot \left(\frac{\eta}{\rho}\right)^{1/6} \cdot \left(\frac{L}{v}\right)^{1/2} \tag{1.135}$$

In this case, the value of the characteristic linear dimension, L, is identified with the diameter of the converter (6.80 m):

$$\delta = 1.5 \times \left(1.00 \times 10^{-8}\text{m}^2 \cdot \text{s}^{-1}\right)^{1/3} \times \left(\frac{0.100\,\text{N} \cdot \text{s} \cdot \text{m}^{-2}}{3000\,\text{kg} \cdot \text{m}^{-3}}\right)^{1/6} \times \left(\frac{6.80\text{ m}}{3.50\,\text{m} \cdot \text{s}^{-1}}\right)^{1/2}$$
$$= 8.08 \times 10^{-4}\text{m } (808\,\mu m) \tag{1.136}$$

Obtained the estimation of the value of the boundary layer and its identification with the characteristic linear dimension of Nusselt, ($L^* = \delta$), the value of $k_d\left(\text{m} \cdot \text{s}^{-1}\right)$ using Eq. (1.134) is equal to:

$$k_d = \frac{357.70 \times D_i}{L^*} = \frac{357.70 \times 1.00 \times 10^{-8}\,\text{m}^2 \cdot \text{s}^{-1}}{8.08 \times 10^{-4}\text{ m}}$$
$$= 4.43 \times 10^{-3}\text{m} \cdot \text{s}^{-1}\left(0.443\,\text{cm} \cdot \text{s}^{-1}\right) \tag{1.137}$$

Finally, to calculate the decarburizing rate, it is necessary to specify two variables:

- The value of the driving force of the reaction: $\left(\%C^0 - \%C^=\right)$, which is equal to 4.50.
- The value of the area-volume ratio, (A/V), identified with the inverse of the height of the metal during the conversion process (0.80 m), that is to say:

$$\left(\frac{\mathrm{A}}{\mathrm{V}}\right)=\frac{1}{h_{\text{metal}}}=\frac{1}{0.80}=1.25\ \mathrm{m}^{-1} \tag{1.138}$$

To justify the estimation of the (A/V) ratio in the LD, we assume a cylindric symmetry, where the relation between the area of the upper surface of the cylinder and the volume of the cylinder is identified with the height that the metal reaches in the converter:

$$\left(\frac{\mathrm{A}}{\mathrm{V}}\right)=\frac{\pi\cdot r^2}{\pi\cdot r^2\cdot \mathrm{L}}=\frac{1}{h_{\text{metal}}} \tag{1.139}$$

However, in a converter of 220 t, the static height of the metal in the LD is 2.20 m (see Deo and Boom 1993, p. 34). This situation would be reached when the metal is charged into the LD by means of a ladle used to transfer the pig iron, for instance.

With the injection of oxygen, 40% of the metal can be emulsified with the slag. As a consequence of this situation, the effective height of the metal decreases and the matter interchange area between the metal and the slag grows. Thus, the (A/V) ratio in the LD can increase during a part of the blowing pattern-program, reaches a maximum and finishes with a decrease in the final stages. That is to say, we would be talking about an interval of values for the area/volume ratio in a LD between 1.67 and 0.80 m^{-1}, which would be compatible with an effective height of metal, h_{metal}, comprised within 60 and 125 cm. In the example that is being developed, we have considered an effective height of the metal of 0.8 m and a (A/V) ratio of 1.25 m^{-1}. Consequently, the decarburizing rate of the pig iron in the LD of the Step II, Fig. 1.4, Eq. (1.127), is equal to:

$$\begin{aligned}\frac{\mathrm{d}(\%\mathrm{C})}{\mathrm{d}t}&=k_d\cdot\left(\%\mathrm{C}^0-\%\mathrm{C}^{=}\right)\times\left(\frac{\mathrm{A}}{\mathrm{V}}\right)\times\left(\frac{65\times 10}{12.01\times 125}\right)\\&=4.43\times 10^{-3}\mathrm{m}\cdot\mathrm{s}^{-1}\times 4.50\times 1.25\mathrm{m}^{-1}\cdot\left(\frac{65\times 10}{12.01\times 125}\right)\times\frac{3600\,\mathrm{s}}{1\,\mathrm{h}}\\&=38.81\%\mathrm{C}\ \cdot\ \mathrm{h}^{-1}\end{aligned} \tag{1.140}$$

Decarburizing in BOF, Eq. (1.127), takes place at a rate of 38.81% $\mathrm{C}\cdot\mathrm{h}^{-1}$. The previous estimation has been made in the assumption that all the steel mass was emulsified in the slag. In the practice, it is considered that only around 40% of the total quantity of refined pig iron is found reacting in an effective way with the foamy slag. This way, the carbon refinement rate in the BOF would be of 15.52% $\mathrm{C}\cdot\mathrm{h}^{-1}$ or 0.26% $\mathrm{C}\cdot\mathrm{min}^{-1}$. The carbon gasification rate during the Step II (Fig. 1.4, Eq. 1.117) adjusts to the values estimated by the model proposed for high carbon concentrations ($0.1\leq\%\mathrm{C}\leq 4.5$) in the metallic melt.

Finally, we want to remember that the concept of boundary layer, δ, is identified with the thickness of fluid where the concentration of one element or compound does not reach a constant value characteristic of a phase chemically homogeneous. It is possible to demonstrate from the dimensional analysis of the Eq. (1.135) used to calculate the thickness of the boundary layer, δ, in the decarburizing of the LD that the exponents that affect the variables D_i, (η/ρ) and (L/v) are: 1/3, 1/6, and 1/2, respectively, and that all of them have kept a dimensional coherence with the unit of length of the boundary layer that was tried to be estimated.

Up to now, we have studied the Step II, Fig. 1.4, kinetics of the refinement-oxidation reactions in the converter and their possibilities to be modeled-simulated by means of the diffusional mass transfer mechanisms between metallic particles dispersed-emulsified in a nonmetallic molten medium (slag). In the Step III, Fig. 1.4, the slag is already formed from the physical–chemical point of view and the contacts with the metal can be linked with other reaction models as might be the kinetic model of permanent regime.

The kinetic equation of the permanent regime can be obtained from any element susceptible of being removed from a slag as a function of the time of contact, t, of the metallic melt with the slag. If the concentration of the element to be slagged expressed in percentage at the beginning of the process in the metal to be treated, $(\%w_i)^0$, the quantity of element to be eliminated-slagged in a quantity of metal to be treated, M_{metal}, would be:

$$\frac{(\%w_i)^0}{100} \cdot M_{metal} \tag{1.141}$$

In the kinetic model of permanent regime, the quantity of the element that we want to remove-slag is distributed, during the time that the process lasts, between the metal and the slag. A partial mass balance to the element that we want to slag would be:

$$\frac{(\%w_i)^0}{100} \cdot M_{metal} = \frac{(\%w_i)}{100} \cdot M_{metal} + \frac{(\%w_i)}{100} \cdot \eta_{s/m} \cdot \dot{V}_{slag} \cdot t \tag{1.142}$$

where t is the time of reaction-contact, $\dot{V}_{slag}$ is the rate of generation-production of slag (mass of slag per unit of time) and $\eta_{s/m}$ is the distribution coefficient of the element to be removed-refined-oxidized between the slag and metallic melt:

$$\eta_{s/m} = \frac{(\%w_i)^{=}_{metal}}{(\%w_i)^{=}_{slag}} \tag{1.143}$$

If we divide both terms of the Eq. (1.142) by the quantity of metal that was treated, M_{metal}, we finally obtain the equation that defines the evolution of the concentration, $(\%w_i)$, with the time in the kinetics model of the permanent regime:

$$(\%w_i)^0 = (\%w_i) + (\%w_i) \cdot \frac{\eta_{s/m} \cdot \dot{V}_{slag}}{M_{slag}} \cdot t \tag{1.144}$$

$$(\%w_i) = \frac{(\%w_i)^0}{1 + \frac{\eta_{s/m} \cdot \dot{V}_{slag}}{M_{slag}} \cdot t} \quad (1.145)$$

Finally, during the Step I, Fig. 1.4, (first 5 min of the process) the formation of the slag and the refinement-slagging of the silicon takes place. During the Step I there is still not emulsification of the metallic melt in the slag and, consequently, the calculation of the diffusional mass transfer is performed through a correlation that simulates the mass transfer from a surface of metallic melt that contacts with a fluid (gas, in this case). Thus, the relation between the Nusselt, Reynolds, and Schmidt numbers is the following one:

$$Nu = 0.332 \cdot Re^{1/2} \cdot Sc^{1/3} \quad (1.146)$$

Exercise 1.6 During the first five minutes of treating the pig iron in the converter (Figs. 1.2 and 1.4), the silicon refinement and the corresponding formation of slag take place. Suggest a kinetic model to predict the concentration of silicon in the metallic melt after five minutes of blowing in a LD converter that operates under the following conditions:

(a) The reference temperature of the system for the refinement period is1500 °C.
(b) The chemical composition of the initial metallic melt is the following: 4.50% C, 1.00% Mn, 1.00% Si, 0.06% P, and 0.02% S. The atomic weights of the iron, carbon, manganese, silicon, phosphorus, and sulphur are 55.85, 12.01, 54.94, 28.09, 30.97, and 32.06 g/mol, respectively.
(c) The relative movement rate between the gas and the metal is 4.50 $m \cdot s^{-1}$. The characteristic linear dimension of the LD of 220 t is 6.8 m, see Deo and Boom (1993, p. 34). Finally, the area/volume ratio of the conversion reactor in these first 5 min is 1.00 m^{-1}.
(d) The density and viscosity of the metallic melt at 1500 °C are 6854 $kg \cdot m^{-3}$ and 5.915×10^{-3} $Pa \cdot s$, respectively (Martin et al. 2009, p. 301). Finally, the diffusion coefficient of the silicon in the pig iron is 3.795×10^{-9} $m^2 \cdot s^{-1}$ at 1500 °C (Kawai and Shiraishi 1988, p. 188).

Question starts:

Silicon oxidation rate is controlled by the diffusion of the metal in the pig iron toward the liquid–gas interphase, where the maximum concentration of silicon takes place. The chemical reaction between silicon and oxygen atoms in the interphase is very fast and SiO_2 is formed. Later, reactions of calcium and iron silicates, which constitute the LD slag, take place.

To express the driving force of the reaction in Eq. (1.123) in weight percentage of silicon as replacement of the molar fraction, we must take into account the following equation of conversion (to change the form of expressing the silicon concentration in the pig iron, from the molar fraction to the weight percentage as a function of the atomic weight of the silicon (28.09 g/mol) and the mean molecular weight of the pig iron, $\overline{PM}_{\text{pig iron}}$):

$$x_{Si} = \frac{\overline{PM}_{\text{pig iron}} \cdot \%Si}{100 \times 28.09} = \left(\frac{\overline{PM}_{\text{pig iron}}}{2809}\right) \cdot \%Si \tag{1.147}$$

Consequently, the utilization of Eq. (1.123) in the process of silicon oxidation in the LD, expressing the silicon concentration in the pig iron in weight percentage, would give as result:

$$\frac{dx_{Si}}{dt} = k_d \cdot \left(x_{Si}^0 - x_{Si}^{=}\right) \cdot \left(\frac{A}{V}\right) \tag{1.148}$$

$$\left(\%Si^0 - \%Si\right) = k_d \cdot \left(\%Si^0 - \%Si^{=}\right) \cdot \left(\frac{A}{V}\right) \cdot \left(\frac{\overline{PM}_{\text{pig iron}}}{2809}\right) \cdot t \tag{1.149}$$

where $\% \, Si^0$ would be the silicon concentration in the pig iron at the beginning of the refinement process ($t = 0$), while % Si would represent the variation of silicon concentration in the pig iron while the refinement process progresses. On another note, it is necessary to consider that the value of the driving force of the reaction, difference of the silicon concentration, is 1.0%:

$$\left(\%Si^0 - \%Si^{=}\right) \simeq \%Si^0 = 1.0\% \tag{1.150}$$

because the quantity of silicon dissolved in the pig iron, compatible with a basic slag where most of the SiO_2 is combined with basic oxides (there is a small quantity of SiO_2 free to react), tends to zero.

The estimation of the average molecular weight of the pig iron, $\overline{PM}_{\text{pig iron}}$, using as base of calculus 100 g of pig iron, would be equal to:

$$\overline{PM}_{\text{pig iron}} = \frac{(4.50 \times 12.01 + 1 \times 54.94 + 1 \times 28.09 + 0.06 \times 30.97 + 0.02 \times 32.06 + 96.89 \times 55.85)}{100}$$
$$= 53.57 \text{ g} \cdot \text{mol}^{-1} \tag{1.151}$$

As in the steps developed in Sect. 1.4 to model the decarburizing-refinement of the carbon in the Step II of the LD process, assuming the Film Theory and stagnant film, the steps that is necessary to follow to calculate k_d, in Eq. (1.149) are the following ones:

(a) Determination of the thickness of the boundary layer of concentrations, Eq. (1.135), for the diffusional transfer of silicon in the pig iron:

$$\delta_{Si} = 1.5 \times \left(3.795 \times 10^{-9}\,\mathrm{m}^2 \cdot \mathrm{s}^{-1}\right)^{1/3} \times \left(\frac{5.915 \times 10^{-3}\mathrm{Pa} \cdot \mathrm{s}}{6854\,\mathrm{kg} \cdot \mathrm{m}^{-3}}\right)^{1/6} \times \left(\frac{6.80\,\mathrm{m}}{4.50\,\mathrm{m} \cdot \mathrm{s}^{-1}}\right)^{1/2}$$
$$= 2.8064 \times 10^{-4}\,\mathrm{m} \qquad (1.152)$$

(b) Once known the value of the boundary layer, through which, the silicon diffuses, we calculate the Nusselt's number as follows, Eq. (1.146):

$$\mathrm{Nu} = 0.332 \cdot (\mathrm{Re})^{0.5} \cdot (\mathrm{Sc})^{1/3} \qquad (1.153)$$

where:

$$\mathrm{Re} = \frac{\mathrm{L} \cdot \rho \cdot v}{\eta} = \frac{6.8\,\mathrm{m} \times 6854\,\mathrm{kg} \cdot \mathrm{m}^{-3} \times 4.5 \cdot \mathrm{m} \cdot \mathrm{s}^{-1}}{5.915 \times 10^{-3}\mathrm{Pa} \cdot \mathrm{s}} = 35457717.67 \qquad (1.154)$$

$$\mathrm{Sc} = \frac{\eta}{\rho \cdot \mathrm{D}_i} = \frac{5.915 \times 10^{-3}\mathrm{Pa} \cdot \mathrm{s}}{6854\,\mathrm{kg} \cdot \mathrm{m}^{-3} \times 3.795 \times 10^{-9}\,\mathrm{m}^2 \cdot \mathrm{s}^{-1}} = 227.46 \qquad (1.155)$$

and:

$$\mathrm{Nu} = 0.332 \times (35457717.67)^{0.5} \times (227.46)^{1/3} = 12067.89 \qquad (1.156)$$

If we know the value of Nusselt, we solve for the silicon coefficient of diffusional mass transfer, k_d, by identifying the linear characteristic dimension with the thickness of the boundary layer previously calculated ($\mathrm{L}^* = \delta_{Si}$):

$$k_{Si} = \mathrm{Nu} \times \left(\frac{\mathrm{D}_{Si}}{\delta_{Si}}\right) = 12067.89 \times \left(\frac{3.795 \times 10^{-9}\,\mathrm{m}^2 \cdot \mathrm{s}^{-1}}{2.8064 \times 10^{-4}\,\mathrm{m}}\right)$$
$$= 0.1632\,\mathrm{m} \cdot \mathrm{s}^{-1} \qquad (1.157)$$

(iii) Finally, if we know the value of k_d in Eq. (1.149) 5 min (300 s) after the starting of the blowing with oxygen, the value of the silicon concentration in the melt would be:

$$\left(\%\mathrm{Si}^0 - \%\mathrm{Si}\right) = 0.1632\,\mathrm{m} \cdot \mathrm{s}^{-1} \times 1 \times 1\,\mathrm{m}^{-1} \times \left(\frac{53.57\,\mathrm{g} \cdot \mathrm{mol}^{-1}}{2809}\right) \times 300\mathrm{s}$$
$$= 0.9337\% \rightarrow (1\% - \%\mathrm{Si})$$
$$= 0.9337\% \rightarrow \%\mathrm{Si} = 0.066\% \qquad (1.158)$$

That is to say, 93% of the silicon initially in the pig iron is at the end of the process in the slag.

Commentaries: The area-volume ratio of the process is estimated as a function of the height of the reactive metallic phase. In the blast furnace or in the electric furnace, the height of the hearth, equivalent to the height of the reactive phases of the pig iron-metal and slag, would be, in this case, replaced by the height of the metal in the converter that would be of around 2.0 m in resting state. When hitting the metal with the oxygen at a speed of around Mach 2, dimensionless number of Mach, Ma, that represents the relation between the speed of the gas and the speed of the sound, the value of the area-volume ratio might be reduced down to a half.

Finally, the value given to the displacement speed of the gaseous phase with the melt. This value depends on both the height at which the lance is located with respect to the metal and the design of the tuyeres at the exit of the oxygen in the extreme of it. Along the blowing process (blowing patterns or protocols), the position of the lance with respect to the height of the metal is variable: at the beginning, the distance between the lance and the metal is small and, at the same time that the blowing time progresses, this distance gradually increases (hard blowing); while in the final moments of the conversion process, this distance between the lance and the metal progressively grows (soft blowing, this favors the oxidation of the components of the slag).

Note: Molecular weight and mean molecular weight:

Conceptually, the molecular weight (PM_i) of a phase, *i*, and its mean molecular weight ($\overline{PM}_i$) are two different variables that can be used in the characterization of the chemical properties of a certain phase (system physically and chemically homogeneous).

The molecular weight (PM_i) of a phase with several components is equal to:

$$PM_i = \sum_{j=1}^{j=n} x_j \cdot PM_j \tag{1.159}$$

That is to say, the molecular weight (PM_i) is identified with the value of the summation of the products of the molar fractions and the molecular weight (or atomic weight if we are talking about molecules with only one atom) of the different components, *j*, that exist in the phase.

On the other hand, we can also chemically identify a phase with the mean molecular weight ($\overline{PM}_i$):

$$\overline{PM}_i = \sum_{j=1}^{j=n} \left(\frac{\% j}{100} \right) \times PM_j \tag{1.160}$$

In this case, the mean molecular weight ($\overline{PM}_i$) of phase is identified with the percentual mean of each one of the molecular-atomic weights of each one of the components that form the phase.

The values of the molecular weight (PM_i) of a phase, *i*, or its mean molecular weight ($\overline{PM}_i$) are not the same, although their values are very similar.

When we work with the composition of a certain component of a phase expressed in molar fraction and we want to express-change this concentration in percentage, we usually employ the following equation:

$$x_j = \frac{\frac{m_j}{PM_j}}{\frac{100}{PM_{phase}}} = \left(\frac{m_j}{100}\right) \times \left(\frac{PM_j}{\overline{PM}_{phase}}\right) = \left(\frac{\%j}{100}\right) \times \left(\frac{PM_j}{\overline{PM}_{phase}}\right) \quad (1.161)$$

where it is possible to use the mean molecular weight ($\overline{PM}_i$) of the phase as the most representative value for the molecular-atomic weights of the components that we have in 100 g.

On the other hand, when we estimate the properties: the specific volume, the density, or the specific heat of a certain phase from the partial molar properties of each one of the components of the phase, the values that must be used to calculate the specific volume, the density, or the specific heat in $m^3 \cdot mol^{-1}$, $kg \cdot m^{-3}$, or $J \cdot kg^{-1} \cdot K^{-1}$, would be the molecular weight (PM_i).

End of the exercise.

Exercise 1.7 Assuming that the kinetic model of the permanent regime can be applied to the dephosphorizing of a melt in the converter, calculate the evolution of the phosphorus concentration, from the minute 11 to the 17 (Fig. 1.4), under the following operating conditions:

(a) From the beginning of the refinement process to the minute 11, Fig. 1.4, there is not phosphorus removal, in an appreciably form, inside of the metallic melt.
(b) Phosphorus concentration of the metallic melt at the beginning of the slagging operation (minute 11) is 0.060%P.
(c) The quantity of treated metallic melt is of 200 t and the quantity of slag generated in the process is of 25 t (obtained in the production of 125 kg of LD slag per t of steel). However, the quantity of slag effectively involved in the metal-slag reaction is of 18 t.
(d) For the phosphorus partition coefficient between the slag and the metal we consider three possible values: 85.00, 96.00, and 115.00 (see later in the Exercises 1.8, 1.9 and 1.10).

Solution starts:

In previous sections we explained the kinetic model of permanent regime, Eq. (1.145). The adaptation of the kinetic model, to estimate the temporary evolution of the phosphorus content in the metallic melt during the last steps of the blowing in the conversion process in the LD, can be expressed as follows:

$$(\%\,P) = \frac{(\%\,P)^0}{1 + \eta_{s/m} \cdot Y} \quad (1.162)$$

where $(\% \mathrm{P})^0$ would represent the phosphorus concentration at the beginning of the process (minute 11), (% P) would be the phosphorus concentration in the melt after the minute eleven of conversion as a function of $\eta_{s/m}$, phosphorus partition coefficient between the slag and the metal, and Y is a dimensionless function that groups the following process variables: $\dot{V}_{slag}$, which represents the rate of slag-formation in the process; M_{metal}, which is the quantity of molten metal that was processed; t, is the duration of the treatment. That is to say:

$$Y = \frac{\dot{V}_{slag} \cdot t}{M_{metal}} \tag{1.163}$$

According to the data of the enunciation, the value of the variables involved in the model of transient regime are:

$$\dot{V}_{slag} = \frac{18\,t}{17\,min} = 1.05822\,t \cdot min^{-1} \tag{1.164}$$

$$M_{metal} = 200\,t \tag{1.165}$$

$$\eta_{s/m} = 85.00;\ \eta_{s/m} = 96.00;\ \eta_{s/m} = 115.00 \tag{1.166}$$

when the phosphorus concentration at the beginning of the treatment is:

$$(\% \mathrm{P})^0 = 0.060\% \mathrm{P} \tag{1.167}$$

With these data and the Eq. (1.163), Table 1.4 ($\eta_{s/m} = 85.00$), Table 1.5 ($\eta_{s/m} = 96.00$), and Table 1.6 ($\eta_{s/m} = 115.00$) are obtained.

As it is possible to appreciate, in the minute 17 of blowing, the phosphorus concentration in the melt, decreases down to ≃0.015%, when the partition coefficient takes the value of 115.00 that, on another note, is the highest level of phosphorus admitted nowadays in a steel. However, it is possible to prolongate the dephosphorizing treatment for several minutes with the aim of satisfying the phosphorus requirements.

Table 1.4 Phosphorus concentration when $\eta_{s/m} = 85.00$

Refinement time, t (min)	Reaction time, t (min)	Phosphorus concentration (%)
11	0	0.0600
12	1	0.0414
13	2	0.0316
14	3	0.0255
15	4	0.0214
16	5	0.0185
17	6	0.0162

Table 1.5 Phosphorus concentration when $\eta_{s/m} = 96.00$

Refinement time, t (min)	Reaction time, t (min)	Phosphorus concentration (%)
11	0	0.0600
12	1	0.0398
13	2	0.0298
14	3	0.0238
15	4	0.0198
16	5	0.0170
17	6	0.0148

Table 1.6 Phosphorus concentration when $\eta_{s/m} = 115.00$

Refinement time, t (min)	Reaction time, t (min)	Phosphorus concentration (%)
11	0	0.0600
12	1	0.0373
13	2	0.0271
14	3	0.0212
15	4	0.0175
16	5	0.0148
17	6	0.0129

In this situation, the oxygen flow injected via lance is reduced if compared with the flow during the first eleven minutes (hard blowing, which has as objective to oxidize the iron of the melt to form the slag). The height of the lance from the minute eleven, with respect to the level of the reacting melts, is normally greater if it is compared with its position during the first eleven minutes (soft blowing, which has as objective oxidizing the iron of the slag). Parallelly, from the minute eleven, the injection of inert gases via tuyeres or porous plugs, distributed at the bottom of the converter, Fig. 1.2, is increased. With this action, combined with the position of the lance and the injection of gases via porous plugs, the objective is that the slag involved in the process could be reaching the real value (in the exercise, that the 18 t could reach the 25 real tons).

Commentaries: If it would be possible to increase the effectivity of the slag involved in the process (to pass from 18 to 22 t) during the dephosphorizing process, the rate of slag-formation would be increased, $\dot{V}_{slag}$:

$$\dot{V}_{slag} = \frac{22\,t}{17\,min} = 1.2941\,t \cdot min^{-1} \tag{1.168}$$

Table 1.7 Phosphorus concentration when $\eta_{s/m} = 115.00$

Refinement time, t (min)	Reaction time, t (min)	Phosphorus concentration (%)
11	0	0.0600
12	1	0.0344
13	2	0.0241
14	3	0.0186
15	4	0.0151
16	5	0.0127
17	6	0.0110

and for a value of the partition coefficient, $\eta_{s/m} = 115.00$, we obtain the results collected in Table 1.7.

End of the exercise.

Exercise 1.8 During the thermal refinement (blowing) in the converter, Fig. 1.4, the slagging of the phosphorus takes place. Assuming that the slag has the following chemical composition: 54% CaO, 25% FeO, 19.5% SiO_2, and 1.5% P_2O_5, calculate the ability of trapping phosphorus that the slag has, determined by means of the gas-slag equilibrium, at 1680 °C, using the following equation, see Sect. 1.5:

$$\log C_{PO_4^{3-},g/s} = -18.184 + 35.84 \cdot \Lambda_{slag} - 23.35 \cdot \Lambda_{slag}^2 + A \quad (1.169)$$

where the value of the constant A is function of both the temperature and the concentration, expressed in weight percentage, of the FeO, MnO, and P_2O_5 in the slag, and is equal to:

$$A = \frac{22930}{T} - 0.0625 \cdot C_{Feo} - 0.04256 \cdot C_{Mno} + 0.359 \cdot C_{P_2O_5} \quad (1.170)$$

The molecular weight of the components of the slag, CaO, FeO, SiO_2, and P_2O_5, are, respectively: 56.08, 71.85, 60.09, and 141.94 g·mol^{-1}, respectively.

Question starts:

We consider as working hypothesis that, even when the temperature changes during the conversion-refinement of the pig iron (from 1350 to 1750 °C that can be reached at the end of the blowing), the reference temperature is 1680 °C.

As in previous exercises, we calculate the optical basicity of the slag formed by two basic oxides and two acid oxides. The first step is to calculate the number of moles in 100 g of slag:

$$\text{Moles} = \frac{19.5}{60.09} + \frac{1.50}{141.94} + \frac{54.0}{56.08} + \frac{25.0}{71.85}$$

$$= 1.6459 \text{ moles in 100 grams of slag} \tag{1.171}$$

The composition, expressed as molar fraction, for each one of the oxides of the slag, which is obtained by dividing the moles of each oxide by the total number of moles in 100 g of slag (1.6459 mol), is:

$$x_{SiO_2} = \frac{19.5/60.09}{1.6459} = 0.1972 \tag{1.172}$$

$$x_{P_2O_5} = \frac{1.5/141.94}{1.6459} = 0.0064 \tag{1.173}$$

$$x_{CaO} = \frac{54/56.08}{1.6459} = 0.5850 \tag{1.174}$$

$$x_{FeO} = \frac{25/71.85}{1.6459} = 0.2114 \tag{1.175}$$

The next step is to obtain the value of the summation resulting from the product of the molar fraction of each oxide of the slag, x_{MeO}, and the number of oxygen atoms in each oxide, $n(O)$:

$$\sum x_{MeO} \cdot n(O) = x_{SiO_2} \times 2 + x_{P_2O_5} \times 5 + x_{CaO} \times 1 + x_{FeO} \times 1$$
$$= 0.1972 \times 2 + 0.0064 \times 5 + 0.5850 \times 1 + 0.2114 \times 1 = 1.2228 \tag{1.176}$$

We calculate the new variable related with each one of the oxides in those that constitute the slag, X_{MeO}:

$$X_{MeO} = \frac{x_{MeO} \cdot n(O)}{\sum x_{MeO} \cdot n(O)} \tag{1.177}$$

and, thus:

$$X_{SiO_2} = \frac{0.1972 \times 2}{1.2228} = 0.3225 \tag{1.178}$$

$$X_{P_2O_5} = \frac{0.0064 \times 5}{1.2228} = 0.0263 \tag{1.179}$$

$$X_{CaO} = \frac{0.5850 \times 1}{1.2228} = 0.4784 \tag{1.180}$$

$$X_{FeO} = \frac{0.2114 \times 1}{1.2228} = 0.1729 \tag{1.181}$$

Finally, we determine the parameter of the optical basicity of the slag, Λ_{slag}, which results from the multiplication of the optical basicity assigned to each oxide, Λ_{MeO} (see Appendix 4) by the variable previously calculated for each one of the oxides of the slag, X_{MeO} (when there is more than one value for the optical basicity of one oxide, we use the average value):

$$\begin{aligned}\Lambda_{slag} &= \Lambda_{SiO_2} \cdot X_{SiO_2} + \Lambda_{P_2O_5} \cdot X_{P_2O_5} + \Lambda_{CaO} \cdot X_{CaO} + \Lambda_{FeO} \cdot X_{FeO} \\ &= 0.480 \times 0.3225 + 0.400 \times 0.0263 + 1.000 \\ &\times\ 0.4784 + 1.000 \times 0.1729 = 0.8166\end{aligned} \quad (1.182)$$

Finally, the value of the ability of the slag to pick up phosphorus in a slag with four components, at the temperature of 1680 °C (Eqs. 1.169 and 1.170):

$$A = \frac{22930}{1680 + 273} - 0.0625 \times 25 + 0.359 \times 1.50 = 10.7169 \quad (1.183)$$

$$\begin{aligned}\log\ C_{PO_4^{3-},g/s} &= -18.184 + 35.84 \times 0.8166 - 23.35 \times 0.8166^2 + 10.7169 \\ &= 6.2215 \rightarrow C_{PO_4^{3-},g/s} = 1.696 \times 10^6\end{aligned} \quad (1.184)$$

End of the exercise.

Exercise 1.9 Considering as starting information the ability of the slag to pick up phosphorus from the slag of the Exercise 1.8, $C_{PO_4^{3-},g/s} = 1.696 \times 10^6$, develop the mathematical equation that relates the partition coefficient of the phosphorus between the metal and the slag as a function of the Potential of Oxygen of the System during the process of conversion of the pig iron in an iron melt with low carbon content identified with the oxygen partial pressure at a temperature of 1680 °C.

The phosphorus pentoxide is a compound with acid characteristics and, in a basic slag like that previously studied in the Exercise 1.8, its coefficient of Raoult's activity is 5.0×10^{-3}.

The chemical characteristics of the metallic melt during the phosphorus slagging stage are: 2.00% C, 1.00% Mn, 0.01% Si, 0.06% P, and 0.040% S.

The molecular weights of the compounds that constitute the slag: CaO, FeO, SiO_2, P_2O_5, and the oxygen atom are 56.08, 71.85, 60.09, 141.94, and 16.00 $g \cdot mol^{-1}$, respectively.

Solution starts:

The ability of the phosphorus, calculated in the Exercise 1.8, is identified with phosphorus concentration in the slag (expressed as the value reached by the phosphorus pentoxide, P_2O_5), and the phosphorus and oxygen partial pressures in the gas, by means of the following equation:

$$C_{PO_4^{3-},\frac{g}{s}} = \frac{x_{P_2O_5} \cdot \gamma_{P_2O_5}}{\left(P_{P_2}\right)^{0.5} \cdot \left(P_{O_2}\right)^{1.25}} \quad (1.185)$$

Starting from the definition of the ability of a slag to pick up phosphorus, derived from the slag-gas equilibrium, it is possible to develop an equation to estimate the partition coefficient of the phosphorus between the slag and the iron metallic melt. In the definition of the coefficient of the ability of picking up phosphorus from a slag, the molar fraction of the P_2O_5 in the slag can be identified with its phosphorus percentage and the phosphorus partial pressure in the gas by the standard free energy associated to the change of state of reference of the phosphorus with the phosphorus percentage in the iron melt. This way, it is possible to transform the equation that is used to define the ability of an slag to pick up phosphorus in an equation that allows calculating the partition coefficient of the phosphorus between the slag and the iron melt for a certain temperature (1680 °C).

A. Relation between the P_2O_5 molar fraction in the slag, $x_{P_2O_5}$, and the phosphorus concentration in the slag expressed in percentage, $\%\,P_{slag}$.

By means of the following equality, where we change the method of expressing the phosphorus concentration in the slag, we find the functional relation between $x_{P_2O_5}$ and $\%\,P$:

$$x_{P_2O_5} \times \left(\frac{1\,\mathrm{mol\,P_2O_5}}{1\,\mathrm{mol\,slag}}\right) \times \left(\frac{\mathrm{mol\,slag}}{\mathrm{PM_{slag}}}\right) \times \left(\frac{2\,\mathrm{mol\,P}}{1\,\mathrm{mol\,P_2O_5}}\right) \times \left(\frac{30.97\,\mathrm{g\,P}}{1\,\mathrm{mol\,P}}\right) \times \left(\frac{100\,\mathrm{g\,slag}}{\%\,\mathrm{P_{slag}}}\right) = 1 \tag{1.186}$$

$$x_{P_2O_5} = \left(\frac{\%\,\mathrm{P_{slag}}}{100}\right) \times \left(\frac{\mathrm{PM_{slag}}}{2 \cdot 30.97}\right) \tag{1.187}$$

In Eq. (1.187), it is necessary to calculate the molecular weight of the slag, PM_{slag}, in the slag of four components: 54% CaO, 25% FeO, 19.5% SiO_2 and 1.5% P_2O_5. The molar fractions of the different oxides were calculated in the Exercise 1.8: $x_{SiO_2} = 0.1972$, $x_{P_2O_5} = 0.0064$, $x_{CaO} = 0.5850$, and $x_{FeO} = 0.2114$. Thus, the molecular weight of the slag is:

$$\begin{aligned} \mathrm{PM_{slag}} &= x_{SiO_2} \cdot \mathrm{PM_{SiO_2}} + x_{P_2O_5} \cdot \mathrm{PM_{P_2O_5}} + x_{CaO} \cdot \mathrm{PM_{CaO}} + x_{FeO} \cdot \mathrm{PM_{FeO}} \\ &= 0.1972 \times 60.09 + 0.0064 \times 141.94 + 0.5850 \times 56.08 + 0.2114 \times 71.85 \\ &= 60.75\mathrm{g} \cdot \mathrm{mol}^{-1} \end{aligned} \tag{1.188}$$

B. Calculation of the relation between the phosphorus partial pressure in the gas and the percentual concentration of phosphorus, $\%P_{metal}$, in the iron melt of the composition indicated in the enunciation.

The relation between the partial pressure of phosphorus in the gas and the percentual concentration (Henry's activity) of the phosphorus in the iron melt, data of the Appendix 3, is the following one:

$$\frac{1}{2}P_2(g) \leftrightarrow P\left(\text{dis};\ a_{H,P} = 1\right) \quad (1.189)$$

$$\Delta_r G^0_{\text{reaction (1.189)}}(T) = -140000 - 6.92 \cdot T(K)\left[J \cdot (\text{mol P})^{-1}\right] \quad (1.190)$$

At 1680 °C, the value of the equilibrium constant associated to the equilibrium of the reaction (1.189), $K_{\text{reaction (1.189)}}$, is:

$$\begin{aligned}\Delta_r G^0_{\text{reaction (1.189)}}(T) &= -R \cdot T \cdot \ln K_{\text{reaction (1.189)}} \rightarrow K_{\text{reaction (1.189)}} \\ &= \exp\left(-\frac{\Delta_r G^0_{\text{reaction (1.189)}}(T)}{R \cdot T}\right) \\ &= \exp\left[-\frac{-140000 - 6.92 \times 1953\left(J \cdot (\text{mol P})^{-1}\right)}{8.314\left(J \cdot (\text{mol} \cdot K)^{-1}\right) \times 1953\,K}\right] \\ &= 12765.3 \end{aligned} \quad (1.191)$$

Consequently, the relation between the phosphorus partial pressure in the gas and the phosphorus concentration in the metal is equal to:

$$P^{0.5}_{P_2} = \left(\frac{a_{H,P}}{K_{\text{reaction (1.289)}}}\right) = \left(\frac{\%\,P_{\text{metal}} \cdot f_P}{12765.3}\right) \quad (1.192)$$

where f_P is the coefficient of activity that corrects the deviations of the Henry's law. In the case of a melt of minority five solutes-compounds: C, Mn, Si, P, and S, the value of f_P, data of Appendix 4, can be estimated considering the following stages:

1. Calculation of the value of f_P at 1600 °C as a function of the interaction parameters of Appendix 4, that define the influence of the carbon, manganese, silicon, phosphorus, and sulphur on the capacity of reaction of the phosphorus in the iron melt (when there are more than one value we use the arithmetic mean). Table 1.8 collects the values of the interaction parameters and the percentages of the different elements in the metallic melt.

Thus, the value of f_P is calculated as follows:

$$\log f_P = e^C_P \cdot \%C + e^{Mn}_P \cdot \%Mn + e^{Si}_P \cdot \%Si + e^P_P \cdot \%P + e^S_P \cdot \%S \quad (1.193)$$

Table 1.8 Interaction parameters and percentages of the different elements in the metallic melt

$e^C_P = 0.128$	$e^{Mn}_P = -0.016$	$e^{Si}_P = 0.110$	$e^P_P = 0.058$	$e^S_P = 0.031$
%C = 2.00	%Mn = 1.00	%Si = 0.01	%P = 0.06	%S = 0.04

$$\log f_P = 0.128 \times 2.00 - 0.016 \times 1.00 + 0.110 \times 0.01 + 0.058 \times 0.06 + 0.031 \times 0.04 \quad (1.194)$$

$$\log f_P = 0.24582 \rightarrow f_P = 1.7612 \quad (1.195)$$

End of the exercise.

Exercise 1.10 Considering as starting data the equation that relates the phosphorus partition coefficient between the metal and the slag, as a function of the Potential of Oxygen of the System during the conversion process of the pig iron in an iron melt with low carbon content, Exercise 1.9, calculate the minimum quantity of phosphorus that could be obtained in the iron melt. The oxygen partial pressure (Potential of Oxygen of the System, POS) at 1680 °C is 2.5 times the oxygen partial pressure that exists for the equilibrium between the oxidized phases FeO and Fe_3O_4 at 1680 °C.

Solution starts:

The POS during the final stages of the thermal refinement process of the pig iron depends on the quantity of ferrous oxide that can be oxidized (into magnetite, Fe_3O_4):

$$3FeO(s) + \frac{1}{2}O_2(g) \leftrightarrow Fe_3O_4(s) \quad (1.196)$$

The value of the standard free energy associated to the FeO/Fe_3O_4 equilibrium, using the data of Appendix 1, is:

$$\Delta_r G^0_{\text{reaction (1.196)}}(T) = -296.0 + 0.112 \cdot T(K)\left[kJ \cdot (\text{mol}\,Fe_3O_4)^{-1}\right] \quad (1.197)$$

At 1680 °C (1953 K), the value of the standard free energy is:

$$\begin{aligned}\Delta_r G^0_{\text{reaction (1.196)}}(1953\,K) &= -296.0 + 0.112 \times 1953 \\ &= -77.264\left[kJ \cdot (\text{mol}\ Fe_3O_4)^{-1}\right]\end{aligned} \quad (1.198)$$

$$\begin{aligned}\Delta_r G^0_{\text{reaction (1.196)}}(T) &= -R \cdot T \cdot \ln K_{\text{reaction (1.196)}} \rightarrow K_{\text{reaction (1.196)}} \\ &= \exp\left(-\frac{\Delta_r G^0_{\text{reaction (1.196)}}(T)}{R \cdot T}\right) \\ &= \exp\left[-\frac{-77264\left(J \cdot (\text{mol}\,Fe_3O_4)^{-1}\right)}{8.314\left(J \cdot (\text{mol} \cdot K)^{-1}\right) \cdot 1953\,K}\right] \\ &= 116.56\end{aligned} \quad (1.199)$$

The value of the equilibrium constant, $K_{\text{reaction (1.196)}}$, is 116.56 and the oxygen partial pressure in equilibrium with the condensed phases FeO/Fe_3O_4 is equal to:

$$K_{reaction\,(1.196)} = \frac{a_{Fe_3O_4}}{a_{FeO}^3 \cdot P_{O_2}^{1/2}} = 116.56 \tag{1.200}$$

where:

$$a_{Fe_3O_4} = a_{FeO} = 1 \tag{1.201}$$

and:

$$P_{O_2}^{=} = \left(\frac{1}{116.56}\right)^2 = 7.3604 \times 10^{-5}\text{atm} \tag{1.202}$$

On the other hand, and as it is indicated in the enunciation of the exercise, the effective pressure for the conditions in the LD process is 2.5 times the equilibrium pressure, $P_{O_2}^{=}$:

$$P_{O_2}^{=} = 7.3604 \times 10^{-5}\text{atm} \times 2.5 = 1.8401 \times 10^{-4}\text{atm} \tag{1.203}$$

If we consider the equations obtained in the Exercises 1.8 and 1.9 for the relation-distribution of the phosphorus between the slag and the metal:

$$P_{P_2}^{0.5} = \left(\frac{a_{H,P}}{K_{reaction\,(1.196)}}\right) = \left(\frac{\%\,P_{metal} \cdot f_P}{12765.3}\right) \rightarrow \%\,P_{metal} = \frac{P_{P_2}^{0.5} \cdot 12765.3}{f_P} \tag{1.204}$$

$$x_{P_2O_5} = \left(\frac{\%P_{slag}}{100}\right) \times \left(\frac{PM_{slag}}{2 \times 30.97}\right) \rightarrow \%P_{slag} = \frac{100 \times x_{P_2O_5} \times 2 \times 30.97}{PM_{slag}} \tag{1.205}$$

$$C_{PO_4^{3-},g\backslash s} = \frac{x_{P_2O_5} \cdot \gamma_{P_2O_5}}{\left(P_{P_2}\right)^{0.5} \cdot \left(P_{O_2}\right)^{1.25}} \rightarrow P_{P_2} = \left[\frac{x_{P_2O_5} \cdot \gamma_{P_2O_5}}{\left(P_{O_2}\right)^{1.25} \cdot C_{PO_4^{3-},g/s}}\right]^2 \tag{1.206}$$

$$\frac{\%P_{slag}}{\%P_{metal}} = \frac{\frac{100 \times x_{P_2O_5} \times 2 \times 30.97}{PM_{slag}}}{\frac{P_{P_2}^{0.5} \times 12765.3}{f_P}} = \frac{\frac{100 \times x_{P_2O_5} \times 2 \times 30.97}{PM_{slag}}}{\frac{\left[\left(\frac{x_{P_2O_5} \times \gamma_{P_2O_5}}{\left(P_{O_2}\right)^{1.25} \times C_{PO_4^{3-},g/s}}\right)^2\right]^{0.5} \times 12765.3}{f_P}}$$

$$= \frac{\frac{100 \times x_{P_2O_5} \times 2 \times 30.97}{PM_{slag}}}{\frac{\frac{x_{P_2O_5} \times \gamma_{P_2O_5}}{\left(P_{O_2}\right)^{1.25} \times C_{PO_4^{3-},g/s}} \times 12765.3}{f_P}} = \frac{100 \times x_{P_2O_5} \times 2 \times 30.97 \times f_P}{PM_{slag} \cdot \frac{x_{P_2O_5} \times \gamma_{P_2O_5}}{\left(P_{O_2}\right)^{1.25} \times C_{PO_4^{3-},g/s}} \cdot 12765.3}$$

$$= \frac{100 \times (P_{O_2})^{1.25} \times C_{PO_4^{3-},g/s} \times 2 \times 30.97 \times f_P}{PM_{slag} \times \gamma_{P_2O_5} \times 12765.3} \tag{1.207}$$

$$\frac{\%P_{slag}}{\%P_{metal}} = \frac{100 \times \left(1.84 \times 10^{-4}\right)^{1.25} \times 1.696 \times 10^6 \times 2 \times 30.97 \times 1.7612}{60.75 \times 5 \times 10^{-3} \times 12765.3} = 102.25 \tag{1.208}$$

The minimum quantity of phosphorus in the metal that can be reached in the dephosphorizing process of the converter is corresponding to that obtained by the thermodynamic equilibrium at 1680 °C with a slag that has molar fraction of P_2O_5 of 0.0064 and a molecular weight of 60.75 $g \cdot mol^{-1}$. The phosphorus percentage reached in a slag compatible with a molar fraction of phosphorus pentoxide, P_2O_5, of 0.0064 would be:

$$\%P_{slag} = \frac{100 \times 0.0064 \times 2 \times 30.97}{60.75} = 0.6525\,\% \tag{1.209}$$

$$\frac{\%P_{slag}}{\%P_{metal}} = 102.25 \rightarrow \%\,P_{metal} = \frac{\%P_{slag}}{102.25} = \frac{0.6525}{102.25} = 6.3814 \times 10^{-3}\,\% \tag{1.210}$$

Commentaries: The ability of a slag to pick up phosphorus can be calculated by means of either the slag-gas or metal-slag equilibriums. From the experimental point of view, the ideal behavior of the gases facilitates the obtaining of more accurate data for the equilibrium reached between the gas and the slag, due to the better knowledge about the behavior of atoms and molecules in the ideal gas. The inconvenient of being necessary to calculate the ability of the slag to pick up phosphorus from a slag with data of the metal-slag equilibrium is working with two phases, where the behavior of the different solutes is far, in most of the occasions, from the ideal behavior.

Finally, the value of the phosphorus partition coefficient between the slag and the iron melt is particularly sensitive to the POS that is used. If we consider a value of the oxygen pressure, P_{O_2}, equal to 2.4 times that of the equilibrium between the condensed phases FeO/Fe_3O_4, $P^{=}_{O_2}$, the the value of the partition coefficient is:

$$P^{=}_{O_2} = 7.3604 \times 10^{-5}\text{atm} \times 2.4 = 1.7665 \times 10^{-4}\text{atm} \tag{1.211}$$

$$\begin{aligned}\frac{\%P_{slag}}{\%P_{metal}} &= \frac{100 \times (2.4 \times P_{O_2})^{1.25} \times C_{PO_4^{3-},g/s} \times 2 \times 30.97 \times f_P}{PM_{slag} \cdot \gamma_{P_2O_5} \cdot 12765.3} \\ &= \frac{100 \times (1.7665 \times 10^{-4})^{1.25} \times 1.696 \times 10^6 \times 2 \times 30.97 \times 1.7612}{60.75 \times 5 \times 10^{-3} \times 12765.3} \\ &= 97.17\end{aligned} \tag{1.212}$$

while if the value of P_{O_2} is 2.6 times that of the $P^{=}_{O_2}$, the value of the partition coefficient is:

$$P^{=}_{O_2} = 7.3604 \times 10^{-5}\text{atm} \times 2.6 = 1.9137 \times 10^{-4}\text{atm} \tag{1.213}$$

$$\begin{aligned}\frac{\%P_{slag}}{\%P_{metal}} &= \frac{100 \times (2.4 \times P_{O_2})^{1.25} \times C_{PO_4^{3-},g/s} \times 2 \times 30.97 \times f_P}{PM_{slag} \times \gamma_{P_2O_5} \times 12765.3} \\ &= \frac{100 \times (1.9137 \times 10^{-4})^{1.25} \times 1.696 \times 10^6 \times 2 \times 30.97 \times 1.7612}{60.75 \times 5 \times 10^{-3} \times 12765.3} \\ &= 107.40\end{aligned} \tag{1.214}$$

End of the exercise.

1.6 Physical–Chemical Characteristics of the LD Slags

If we consider that a slag is a homogeneous mixture of molten metallic oxides with small additions of P, S, C, N, F, and H, its chemical composition is one of the fundamental data as starting point to value the possibilities of recirculation and reutilization of it.

Habitually, the chemical composition of a slag is expressed as a function of the weight percentages of the major oxides in it. The combination of the basic oxides with the acid oxides propitiates the formation of chemical oxides that, together with the metallic oxides that have still not reacted, can propitiate the appearance of liquid phases at temperatures noticeable lower than that of the starting chemical compounds and metallic oxides.

A property derived from the chemical composition is the average molecular weight of the slag, $\overline{PM}_{slag}$, whose details are indicated in the Eq. (1.160).

Although from the practical-industrial point of view, the chemical characterization of slags is performed by the weight percentage of the elements that form the slag as oxides, from the thermodynamic point of view (calculation of the activities of the oxides) and also from the estimation-prediction of the properties of the slag (molar volume, $V_{m/slag}$, heat capacity, $c_{m/slag}$, or liquid–gas surface tension, γ^{lg}_{slag}) using the values of the partial molar properties of the oxides point of view, it is necessary to express the composition of the slag in the form of molar fraction, $x_{Me_xO_y}$.

To calculate-estimate the molar volume of the slag as a function of the partial molar volumes of its oxides, it is possible to use the following equation:

$$V_{m/slag} = \sum_{1}^{n} \left(x_{Me_xO_y}\right)_i \cdot \left(\overline{V}\left(Me_xO_y\right)_i\right) m^3 \cdot mol^{-1} \tag{1.215}$$

where $\overline{V}\left(Me_xO_y\right)_i$ is the partial molar volume of each one of the oxides in $m^3 \cdot mol^{-1}$.

Similarly, it is possible to make an estimation of both the heat capacity of the slag and the liquid–gas surface tension:

$$C_{m/slag} = \sum_{1}^{n} \left(x_{Me_xO_y}\right)_i \cdot \left(\overline{c_p}\left(Me_xO_y\right)_i\right) \ J \cdot mol^{-1} \cdot K^{-1} \tag{1.216}$$

$$\gamma_{slag}^{lg} = \sum_{1}^{n} \left(x_{Me_xO_y}\right)_i \cdot \left(\overline{\gamma}\left(Me_xO_y\right)_i^{lg}\right) mN \cdot m^{-1} \tag{1.217}$$

where $\overline{c_p}\left(Me_xO_y\right)_i$ is the partial molar heat capacity of each one of the oxides in the slag in $J \cdot mol^{-1} \cdot K^{-1}$ and $\overline{\gamma}\left(Me_xO_y\right)_i^{lg}$ is partial molar contribution of the liquid–gas surface energy of each one of the oxides in the slag.

For the estimation-calculation of the slag viscosity, η, expressions of Arrhenius type are usually employed:

$$\eta = A \cdot T \cdot \exp\left(\frac{B}{T}\right) \tag{1.218}$$

where A and B are two variables that depend on the composition of the oxidized components of the slag, particularly Na_2O or CaF_2, that even when they are in small percentages, they have a great influence on the viscosity of the slag.

From the activity's point of view, capacity of reaction of each one of the oxidized components of the slag, it is possible to find several options for its study and characterization:

– Thermodynamic model: one of the options is considering each one of the oxidized compounds and taking into account that their activities are proportional to the product of the molar fraction and the coefficient of activity:

$$a_{Me_xO_y} = x_{Me_xO_y} \cdot \gamma_{Me_xO_y} \tag{1.219}$$

– Molecular model: other option is considering that the oxides and compounds of the slag have ideal behavior and, their activities can be identified with their molar fraction:

$$a_{FeO} = x_{FeO}; \quad a_{Ca_2SiO_4} = x_{Ca_2SiO_4} \tag{1.220}$$

– Ionic model: it is assumed for the slag formed by a homogeneous mixture of anions and cations that the activity of each neutral molecule would be equal to the product of the equivalent concentration of the cation, $N_{e,\,Me^{n+}}$, in the mixture of cations, and the equivalent concentration of the anion, $N_{e,\,X^{n-}}$, in the mixture of anions:

$$a_{MeX} = N_{e,Me^{n+}} \cdot N_{e,X^{n-}} \quad (1.221)$$

Slags in metallurgical processes, apart from representing essential instruments for the canalization of the refinement process, constitute a fundamental phase for the entrapment and control of the impurities in the metallic melts. In the case of the iron metallurgy, the elements that are necessary to control are phosphorous, sulphur, nitrogen, and hydrogen. Consequently, it is interesting to know the ability of picking up these elements that the slags could have. For example, in the BOF steelworks (LD converter), the thermodynamic variables of operation lead to that the removal of phosphorus is the only efficient option using the proposed slags.

The objective of any dephosphorizing process is calculating the distribution that the phosphorus might reach between the slag and the metal under certain conditions of pressure (around the atmospheric pressure) and temperature. With this purpose, models are developed to predict the values that are obtained for the phosphorus partition coefficient between the slag and the metal supported either on thermodynamic, molecular, ionic, and/or empiric equations or on, certain occasions, mixture of estimation-calculation proposals.

From the ionic models that were developed to predict the partition coefficients, the model proposed by Flood, Forland, and Grjotheim (FFG) is considered. The FFG model considers that a basic slag comprises an ensemble of ionic species (anions and cations) of ideal behavior that reach the behavior of the ideal solutions: the formation of the slag is achieved for a mixture enthalpy zero and an ideal entropy of mixture:

$$a_{O^{2-}} = x_{O^{2-}};\ a_{Ca^{2+}} = x_{Ca^{2+}};\ a_{CaO} = N_{e,\,Ca^{2+}} \cdot N_{e,\,O^{2-}} \quad (1.222)$$

where $N_{e,\,Ca^{2+}}$ and $N_{e,\,O^{2-}}$ are, respectively, the equivalent fraction of calcium cations and oxygen anions in the ideal mixture of anions and cations that form the slag and $x_{Ca^{2+}}$ and $x_{O^{2-}}$ are the molar fractions of the calcium and oxygen ions, respectively.

The partition coefficient of the phosphorus between the metal and the slag is function of the greater or smaller participation of the so-called *elemental reactions of interchange* in the global process.

That is to say, according to the FFG model, the phosphorus slagging reaction in the ionic form is:

$$2P\ (dis;\ pig\ iron) + 5O(dis;\ pig\ iron) + 3O^{2-}(dis;\ slag) \leftrightarrow P_2O_8^{6-}(dis;\ slag) \quad (1.223)$$

If we consider the participation of the different cations, coming from the dissociation of the basic oxides, such as Ca^{2+}, Mg^{2+}, Mn^{2+}, Fe^{2+}, and Na^{+}, it is possible to write all possible simultaneous reactions of phosphorus slagging and its corresponding equilibrium constants:

$$2P(dis;\ pig\ iron) + 5O(dis;\ pig\ iron) + 3CaO(dis;\ slag) \leftrightarrow Ca_3P_2O_8(dis;\ slag) \rightarrow K_{Ca^{2+}} \quad (1.224)$$

$$2P(\text{dis; pig iron}) + 5O(\text{dis; pig iron}) + 3MgO(\text{dis; slag}) \leftrightarrow Mg_3P_2O_8(\text{dis; slag}) \rightarrow K_{Mg^{2+}} \quad (1.225)$$

$$2P(\text{dis; pig iron}) + 5O(\text{dis; pig iron}) + 3MnO(\text{dis; slag}) \leftrightarrow Mn_3P_2O_8(\text{dis; slag}) \rightarrow K_{Mn^{2+}} \quad (1.226)$$

$$2P(\text{dis; pig iron}) + 5O(\text{dis; pig iron}) + 3FeO(\text{dis; slag}) \leftrightarrow Fe_3P_2O_8(\text{dis; slag}) \rightarrow K_{Fe^{2+}} \quad (1.227)$$

$$2P(\text{dis; pig iron}) + 5O(\text{dis; pig iron}) + 3Na_2O(\text{dis; slag}) \leftrightarrow Na_6P_2O_8(\text{dis; slag}) \rightarrow K_{Na^{+}} \quad (1.228)$$

According to Flood, the calculation of the equilibrium constant of Eq. (1.223) is:

$$K_P = \frac{x_{P_2O_8^{6-}}}{x_{O^{2-}}^3 \cdot a_P^2 \cdot a_O^5} \quad (1.229)$$

which leads to:

$$\log K_P = N_{e,Ca^{2+}} \cdot \log K_{Ca^{2+}} + N_{e,Mg^{2+}} \cdot \log K_{Mg^{2+}} + \cdots \quad (1.230)$$

where $N_{e,Ca^{2+}}$ and $N_{e,Mg^{2+}}$ are the equivalent fractions of calcium and magnesium cations in the ideal mixture of cations.

Using as reference the model of FFG, the IRSID (Gaye 1982) has considered-proposed the following equation to calculate the value of K_P:

$$\log K_P = N_{e,\,Ca^{2+}} \cdot \log K_{Ca^{2+}} + N_{e,Mg^{2+}} \cdot \log K_{Mg^{2+}} + \cdots + g_P \quad (1.231)$$

where:

$$g_P = -8.18 + 1.41 \cdot S + \frac{13200}{T} + D \cdot x_{O^{2-}} + E \cdot x_{O^{2-}}^2 \quad (1.232)$$

where:

$$D = 40.55 - 6.95 \cdot S - \frac{64500}{T} \quad (1.233)$$

$$E = -13.10 + 5.91 \cdot S + \frac{22100}{T} \quad (1.234)$$

$$S = \frac{x_{SiO_4^{4-}}}{x_{SiO_4^{4-}} + x_{PO_4^{3-}}} \tag{1.235}$$

Finally, the values provided by the IRSID for the possible *elemental reactions of interchange* of the dephosphorizing process would be:

$$\log K_{Ca^{2+}} = \frac{82200}{T} - 34.2 \tag{1.236}$$

$$\log K_{Mn^{2+}} = \frac{64370}{T} - 31.4 \tag{1.237}$$

$$\log K_{Na^{+}} = \frac{100600}{T} - 31.7 \tag{1.238}$$

$$\log K_{Mg^{2+}} = \frac{75200}{T} - 34.2 \tag{1.239}$$

$$\log K_{Fe^{2+}} = \frac{59300}{T} - 33.9 \tag{1.240}$$

For the calculation of the ferrous oxide activity in the slag, a_{FeO}, the IRSID uses the classic thermodynamic model to correct the Raoult's activity of the FeO of the slag with the coefficient of activity. However, to calculate the molar fraction-activity of FeO, it is suggested that the equations-estimations proposed by the ionic models (products of the concentration of the oxygen anion and the concentration of the iron cation):

$$\log a_{FeO} = \log\left(x_{O^{2-}} \cdot x_{Fe^{2+}}\right) + \log \gamma_{FeO} \tag{1.241}$$

where:

$$\log \gamma_{FeO} = A + B \cdot x_{O^{2-}} + C \cdot x_{O^{2-}}^2 \tag{1.242}$$

where values of A, B, and C are calculated as follows:

$$A = 5.50 - 1.12 \cdot S - \frac{5600}{T} \tag{1.243}$$

$$B = -16.07 + 2.85 \cdot S + \frac{20800}{T} \tag{1.244}$$

$$C = 10.97 - 1.62 \cdot S - \frac{16100}{T} \tag{1.245}$$

Eq. (1.223) can displace to the phosphorus slagging if high oxygen concentrations are used in the melt. However, it is possible that this source of oxygen, which habitually the FeO provides to the slag, might not be exclusively used by the phosphorus, but also by the oxidation-slagging of other metals available in the melt, for instance carbon, silicon, and manganese. The dephosphorizing yield of the pig iron in the BOF steelworks, consequently, is affected by the existence of parallel mechanisms of oxidation-refinement of elements different from the phosphorus (C, Si, and Mn).

Other model of prediction supported on an empiric equation of three parameters: temperature, CaO percentage in the slag and total iron in the slag is the following one:

$$\log \frac{C_{P,\ slag}}{C_{P,\ pig\ iron}} = \frac{22350}{T} + 7.0 \cdot \log C_{CaO} + 2.5 \cdot \log C_{Fe} - 24 \tag{1.246}$$

Eq. (1.246) can be used to estimate the distribution of the phosphorus between the metal and the slag always and when the CaO content in the slag was >24% CaO. According to Eq. (1.246), when the phosphorus quantity in the slag is 1.30% P_2O_5 (0.57% P) at the temperature of 1973 K (1700 °C), phosphorus percentage in the metal reaches a value of 0.00297% (29.7 ppm), if the composition of the slag is: 50% CaO and 16.79% Fe total (18% FeO + 4% Fe_2O_3).

More recently, correlations between the ability of phosphorus removal of slags, $C_{PO_4^{3-},g/s}$, and its optical basicity, Λ_{slag} (Ballester et al. 2000, p. 272) were proposed:

$$\log C_{PO_4^{3-},g/s} = -18.184 + 35.84 \cdot \Lambda_{slag} - 23.35 \cdot \Lambda_{slag}^2 + A \tag{1.247}$$

where:

$$A = \frac{22930}{T} - 0.0625 \cdot C_{FeO} - 0.04256 \cdot C_{MnO} + 0.359 \cdot C_{P_2O_5} \tag{1.248}$$

where C_i is the percentage of the different oxides in the slag. Definitely, the objective is to predict the partition coefficient of the phosphorus between the metal and the slag.

The ability of one slag for the phosphorus removal, $C_{PO_4^{3-},g/s}$, can be defined on the following gas-slag equilibrium (we should think that the elements available in a gas always reach an ideal behavior, while in metallic melts or slags they usually show strong deviations from the ideal situation):

$$\frac{1}{2}P_2(g) + \frac{5}{4}O_2(g) + \frac{3}{2}O^{2-}(dis;\ slag) \leftrightarrow PO_4^{3-}(dis;\ slag) \tag{1.249}$$

and, thus:

$$C_{PO_4^{3-},g/s} = \frac{C_{slag}}{P_{P_2}^{0.5} \cdot P_{O_2}^{1.25}} \tag{1.250}$$

where the phosphorus concentration in the slag can be expressed as a function of considering the ionic or the molecular models in different manners. For instance, for the ionic models, it is possible to use as expression of the concentration, the molar fraction of the phosphate anion, $x_{PO_4^{3-}}$, or the equivalent fraction of the anion in the mixture of anions, $N_{e,PO_4^{3-}}$. The utilization of molecular models habilitates the expression of the concentrations of the slag in the form of the molar fraction of the P_2O_5, $x_{P_2O_5}$, or of the tricalcium phosphate, $x_{Ca_3P_2O_8}$. Besides, developing the terms-variables involved in the equilibrium constant of the reaction (1.249):

$$K_{reaction\ (1.249)} = \frac{C_{slag} \cdot \gamma_{slag}}{P_{P_2}^{0.5} \cdot P_{O_2}^{1.25} \cdot a_{O^{2-}}^{1.5}} \tag{1.251}$$

where γ_{slag} is the coefficient of activity of the phosphorus ions or compounds in the slag. Thus, the ability of phosphorus removal of a certain slag will be:

$$C_{PO_4^{3-},g/s} = \frac{K_{reaction\ (1.249)} \cdot a_{O^{2-}}^2}{\gamma_{slag}} \tag{1.252}$$

The optical basicity of one slag, Λ_{slag}, (that connects the oxides that form a slag with Pauling's electronegativity, Ballester et al. 2000, p. 272) can be calculated using the data of the optical basicity associated to the oxides that form the slag by means of the following equation:

$$\Lambda_{slag} = \Lambda_{M,1} \cdot X(M_1) + \Lambda_{M,2} \cdot X(M_2) + \cdots + \Lambda_{M,i} \cdot X(M_i) \tag{1.253}$$

where $\Lambda_{M,i}$ is the optical basicity of the oxides of the slag and $X(M_i)$ is the new variable defined by the following equation:

$$X(M_i) = \frac{x_i \cdot n_O}{\sum_1^i x_i \cdot n_O} \tag{1.254}$$

where x_i is the molar fraction of the oxide, M_i, in the slag, and n_O is the number of oxygens in the oxide molecule, M_i.

Kinetics of the dephosphorizing process can be controlled either by the diffusion in the slag or in the metallic phase. It is probable that neither the chemical reaction nor the interphase reaction are the slow stages or controlling stages of the process. The temperature in the BOF process, higher than 1500 °C, makes the chemical reactions in the metal-slag interphase very fast. On another note, phosphorus is not an element superficially active although P_2O_5 and FeO of slags are superficially actives (diminution of the metal-slag interface tension, γ_{slag}^{m-s}, due to the presence of FeO and P_2O_5 in the formulation of the slag). Thus, resistances to the formation of calcium phosphates in the metal-slag interface are not expected as a consequence of the low temperature or the limited concentration of reagents involved in the phosphorus oxidation (reaction 1.223).

1.6.1 *Solved Exercises and Problems*

Exercise 1.11 A LD slag has the following chemical composition: 50% CaO, 6.0% MgO, 18% FeO, 5% MnO, 15% SiO_2, 1.30% P_2O_5, 0.70% Al_2O_3, 4.0% Fe_2O_3. Calculate the molecular weight of the slag, its basicity, and its optical basicity.

Data: The values used for the atomic weights ($g \cdot mol^{-1}$) of the Ca, Mg, Fe, Mn, Si, P, Al, and O are, respectively: 40.08, 24.31, 55.85, 54.94, 28.09, 30.97, 26.98, and 16.00.

Question starts:

First, we calculate the molecular weight of the slag. We collect the molecular weights of each one of the oxides that constitute the slag in Table 1.9.

Now, we calculate the molar fraction of each oxide:

$$\begin{aligned} A = {} & 15.00\,g\,SiO_2 \cdot \frac{1\,mol\,SiO_2}{60.09\,g\,SiO_2} + 1.30\,g\,P_2O_5 \cdot \frac{1\,mol\,P_2O_5}{141.94\,g\,P_2O_5} \\ & + 0.70\,g\,Al_2O_3 \cdot \frac{1\,mol\,Al_2O_3}{101.84\,g\,Al_2O_3} + 4.00\,g\,Fe_2O_3 \cdot \frac{1\,mol\,Fe_2O_3}{159.70\,g\,Fe_2O_3} \\ & + 50.00\,g\,CaO \cdot \frac{1\,mol\,CaO}{56.08\,g\,CaO} + 6.00\,g\,MgO \cdot \frac{1\,mol\,MgO}{40.31\,g\,MgO} \\ & + 18.00\,g\,FeO \cdot \frac{1\,mol\,FeO}{71.85\,g\,FeO} + 5.00\,g\,MnO \cdot \frac{1\,mol\,MnO}{70.94\,g\,MnO} \\ = {} & 1.6521 \end{aligned} \tag{1.255}$$

$$x_{SiO_2} = \frac{15.00\,g\,SiO_2 \cdot \frac{1\,mol\,SiO_2}{60.09\,g\,SiO_2}}{A} = 0.1511 \tag{1.256}$$

$$x_{P_2O_5} = \frac{1.30\,g\,P_2O_5 \cdot \frac{1\,mol\,P_2O_5}{141.94\,g\,P_2O_5}}{A} = 0.0055 \tag{1.257}$$

Table 1.9 Molecular weight of each one of the oxides of the slag

SiO_2	P_2O_5	Al_2O_3	Fe_2O_3	CaO	MgO	FeO	MnO
60.09	141.94	101.84	159.70	56.08	40.31	71.85	70.94

$$x_{Al_2O_3} = \frac{0.70\,g\,Al_2O_3 \cdot \frac{1\,mol\,Al_2O_3}{101.84\,g\,Al_2O_3}}{A} = 0.0042 \quad (1.258)$$

$$x_{Fe_2O_3} = \frac{4.00\,g\,Fe_2O_3 \cdot \frac{1\,mol\,Fe_2O_3}{159.70\,g\,Fe_2O_3}}{A} = 0.0152 \quad (1.259)$$

$$x_{CaO} = \frac{50.00\,g\,CaO \cdot \frac{1\,mol\,CaO}{56.08\,g\,CaO}}{A} = 0.5397 \quad (1.260)$$

$$x_{MgO} = \frac{6.00\,g\,MgO \cdot \frac{1\,mol\,MgO}{40.31\,g\,MgO}}{A} = 0.0901 \quad (1.261)$$

$$x_{FeO} = \frac{18.00\,g\,FeO \cdot \frac{1\,mol\,FeO}{71.85\,g\,FeO}}{A} = 0.1516 \quad (1.262)$$

$$x_{MnO} = \frac{5.00\,g\,MnO \cdot \frac{1\,mol\,MnO}{70.94\,g\,MnO}}{A} = 0.0427 \quad (1.263)$$

The molecular weight of the slag is calculated as follows:

$$\begin{aligned} PM_{slag} &= PM_{SiO_2} \cdot x_{SiO_2} + PM_{P_2O_5} \cdot x_{P_2O_5} + PM_{Al_2O_3} \cdot x_{Al_2O_3} \\ &+ PM_{Fe_2O_3} \cdot x_{Fe_2O_3} + PM_{CaO} \cdot x_{CaO} + PM_{MgO} \cdot x_{MgO} \\ &+ PM_{FeO} \cdot x_{FeO} + PM_{MnO} \cdot x_{MnO} \\ &= 60.09 \times 0.1511 + 141.94 \times 0.0055 + 101.84 \times 0.0042 \\ &+ 159.70 \times 0.0152 + 56.08 \times 0.5397 + 40.31 \times 0.0901 \\ &+ 71.85 \times 0.1516 + 70.94 \times 0.0427 \\ &= 60.54\,g \cdot mol^{-1} \end{aligned} \quad (1.264)$$

The basicity of the slag according to the index, $I_B(I)$, most widely used in the industrial practice is:

$$I_B(I) = \frac{\%CaO}{\%SiO_2} = \frac{50}{15} = 3.33 \quad (1.265)$$

However, in the control of the dephosphorizing processes, the following relation might be used:

$$I_B(II) = \frac{\%CaO - 4 \cdot \%P_2O_5}{\%SiO_2} = \frac{44.8}{15} = 2.99 \quad (1.266)$$

Equally, in those slags with significant quantities of dolostone:

$$I_B(III) = \frac{\%CaO + \%MgO}{\%\,iO_2 + \%Al_2O_3} = \frac{56}{15.7} = 3.57 \quad (1.267)$$

To calculate the optical basicity of the slag, the following steps are followed:

1. Calculate the number of moles that there are in 100 g of slag:

$$\begin{aligned} \text{Number of moles} &= \frac{15.0}{60.09} + \frac{1.30}{141.94} + \frac{0.70}{101.84} + \frac{4.00}{159.70} \\ &+ \frac{50.0}{56.08} + \frac{6.00}{40.31} + \frac{18.00}{71.85} + \frac{5.00}{70.94} \\ &= 1.6521 \text{ moles in 100 g of slag} \end{aligned} \quad (1.268)$$

2. Composition, expressed as molar fraction, for each one of the oxides of the slag, as a result of dividing the moles that, for each one of the oxides that there are in 100 g of slag by the total number of moles, 1.6521 mol. Results calculated at the beginning of the exercise are now collected in Table 1.10.

3. We proceed to obtain the value of the summation resulted from the product of the molar fraction of each oxide of the slag, x_{MeO}, and the number of oxygen atoms of each oxide of the slag, $n(O)$.

$$\begin{aligned} \sum x_{MeO} \cdot n(O) &= 0.1511 \times 2 + 0.055 \times 5 + 0.0042 \times 3 + 0.0152 \times 3 \\ &+ 0.5397 \times 1 + 0.0901 \times 1 + 0.1516 \times 1 + 0.0427 \times 1 \\ &= 1.2119 \end{aligned} \quad (1.269)$$

4. We make the calculation of the new variable connected to each one of the oxides that constitute the slag, X_{MeO}:

$$X_{MeO} = \frac{x_{MeO} \cdot n(O)}{\sum x_{MeO} \cdot n(O)} \quad (1.270)$$

$$X_{SiO_2} = \frac{x_{SiO_2} \cdot n(O)}{\sum x_{MeO} \cdot n(O)} = \frac{0.1511 \times 2}{1.2119} = 0.2493 \quad (1.271)$$

$$X_{P_2O_5} = \frac{x_{P_2O_5} \cdot n(O)}{\sum x_{MeO} \cdot n(O)} = \frac{0.0055 \times 5}{1.2119} = 0.0229 \quad (1.272)$$

$$X_{Al_2O_3} = \frac{x_{Al_2O_3} \cdot n(O)}{\sum x_{MeO} \cdot n(O)} = \frac{0.0042 \times 3}{1.2119} = 0.0103 \quad (1.273)$$

Table 1.10 Composition of the slag expressed as molar fraction

x_{SiO_2}	$x_{P_2O_5}$	$x_{Al_2O_3}$	$x_{Fe_2O_3}$	x_{CaO}	x_{MgO}	x_{FeO}	x_{MnO}
0.1511	0.0055	0.0042	0.0152	0.5397	0.0901	0.1516	0.0427

$$X_{Fe_2O_3} = \frac{x_{Fe_2O_3} \cdot n(O)}{\sum x_{MeO} \cdot n(O)} = \frac{0.0152 \times 3}{1.2119} = 0.0375 \quad (1.274)$$

$$X_{CaO} = \frac{x_{CaO} \cdot n(O)}{\sum x_{MeO} \cdot n(O)} = \frac{0.5397 \times 1}{1.2119} = 0.4453 \quad (1.275)$$

$$X_{MgO} = \frac{x_{MgO} \cdot n(O)}{\sum x_{MeO} \cdot n(O)} = \frac{0.0901 \times 1}{1.2119} = 0.0743 \quad (1.276)$$

$$X_{FeO} = \frac{x_{FeO} \cdot n(O)}{\sum x_{MeO} \cdot n(O)} = \frac{0.1516 \times 1}{1.2119} = 0.1251 \quad (1.277)$$

$$X_{MnO} = \frac{x_{MnO} \cdot n(O)}{\sum x_{MeO} \cdot n(O)} = \frac{0.0427 \times 1}{1.2119} = 0.0352 \quad (1.278)$$

5. Finally, we calculate the parameter of the optical basicity of the slag, Λ_{slag}, resulted form the multiplication of the optical basicity attributed to each oxide, Λ_{MeO} (see the corresponding data in Appendix 4 corresponding to the physical and chemical characteristics of the slag) and the above-calculated variable for each one of the oxides of the slag, X_{MeO} (when there is more than one value assigned to the optical basicity of each oxide, we use the average value). Results are collected in Table 1.11.

This way, the optical basicity of the slag is:

$$\begin{aligned}\Lambda_{slag} &= \Lambda_{SiO_2} \cdot X_{SiO_2} + \Lambda_{P_2O_5} \cdot X_{P_2O_5} + \Lambda_{Al_2O_3} \cdot X_{Al_2O_3} + \Lambda_{Fe_2O_3} \cdot X_{Fe_2O_3} \\ &\quad + \Lambda_{CaO} \cdot X_{CaO} + \Lambda_{MgO} \cdot X_{MgO} + \Lambda_{FeO} \cdot X_{FeO} + \Lambda_{MnO} \cdot X_{MnO} \\ &= 0.480 \times 0.2493 + 0.400 \times 0.0229 + 0.608 \times 0.0103 \\ &\quad + 0.800 \times 0.0375 + 1.000 \times 0.4453 + 0.865 \times 0.0743 \\ &\quad + 1.000 \times 0.1251 + 0.980 \times 0.352 = 0.8342 \end{aligned} \quad (1.279)$$

Commentaries: The capacity of the calcium oxide, CaO, as compound able to supply oxygen anions, O^{2-}, to a slag can attract the attention of the reader, being this sustained by the value given to its optical basicity, Λ_{CaO}, that is same as that of the iron (II) oxide. However, while the acid–base characteristics of the CaO and FeO are equivalent, according to the ionic model, the oxidizing ability of one slag is determined by its FeO content but not by the CaO content.

This way, if we consider the data of the formation standard free energy of the following iron and calcium silicates (data of Appendix 1):

Table 1.11 Optical basicity of the oxides in the slag

Λ_{SiO_2}	$\Lambda_{P_2O_5}$	$\Lambda_{Al_2O_3}$	$\Lambda_{Fe_2O_3}$	Λ_{CaO}	Λ_{MgO}	Λ_{FeO}	Λ_{MnO}
0.480	0.400	0.608	0.800	1.000	0.865	1.000	0.975

$$CaO(s) + 2SiO_2(s) \leftrightarrow 2CaO \cdot SiO_2(2CS)(s) \quad (1.280)$$

$$FeO(s) + 2SiO_2(s) \leftrightarrow 2FeO \cdot SiO_2(2FeS)(s) \quad (1.281)$$

$$\Delta_r G^0_{\text{reaction (1.280)}} = -103.0 - 0.020 \cdot T \left[kJ \cdot (mol\,CaO)^{-1}\right] \text{ or } \left[kJ \cdot (2\,mol\,SiO_2)^{-1}\right] \quad (1.282)$$

$$\Delta_r G^0_{\text{reaction (1.281)}} = -55.0 + 0.028 \cdot T \left[kJ \cdot (mol\,FeO)^{-1}\right] \text{ or } \left[kJ \cdot (2\,mol\,SiO_2)^{-1}\right] \quad (1.283)$$

The most stable silicate, considering the same unity of matter in both reactions (2 mol of SiO_2) is the 2CS: the minimum free energy for any temperature is that of the $2CaO{\cdot}SiO_2$. Consequently, the acid–basic character of CaO in front of FeO should be greater and, for that reason, having CaO greater basicity than FeO. However, it is possible to see from the value given to the optical basicity of Ca and Fe oxides (Table 1.11) that the basicity of both oxides is the same, but in the industrial practice we have always considered that CaO is more basic than FeO.

End of the exercise.

Exercise 1.12 A LD slag has the following chemical composition: 50% CaO, 6.0% MgO, 18% FeO, 5% MnO, 15% SiO_2, 1.30% P_2O_5, 0.70% Al_2O_3, 4.0% Fe_2O_3. Obtain the pseudoternary composition (CaO–FeO–SiO_2) most adjusted to the data of the liquidus temperature provided by Mills and Keene (1987, p. 13). Besides, calculate for the chosen pseudoternary slag: average molecular weight, basicity, density, specific heat, and viscosity, at 1650 °C.

Data: The partial molar properties of the three components of the slag, see Ballester et al. (2000, pp. 275–276) are collected in Tables 1.12 and 1.13.

Finally, we use an Arrhenius's type equation for the calculation of the viscosity expressed in Pa·s:

$$\mu = A \cdot T \cdot exp\left(\frac{B}{T}\right) \quad (1.284)$$

where A and B are:

Table 1.12 Partial molar volumes at 1500 °C, $\overline{V}_i \times 10^6$ in $m^3 \cdot mol^{-1}$

CaO	FeO	SiO_2
20.70	15.80	$19.55 + 7.966 \cdot x_{SiO_2}$

Table 1.13 Partial molar specific heat in molten state, $\overline{C}_{P-i}$ in $J \cdot mol^{-1} \cdot K^{-1}$

CaO	FeO	SiO_2
80.80	76.6	87.0

$$ln\ A = -19.81 + 1.73 \cdot x_{CaO} + 5.82 \cdot x_{CaF_2} + 7.02 \cdot x_{Na_2O} - 33.76 \cdot x_{Al_2O_3} \tag{1.285}$$

$$B = 31140 - 23896 \cdot x_{CaO} - 46356 \cdot x_{CaF_2} - 39159 \cdot x_{Na_2O} + 68833 \cdot x_{Al_2O_3} \tag{1.286}$$

Finally, the values of the atomic weight (g · mol^{-1}) of the Ca, Mg, Fe, Mn, Si, P, Al, and O are, respectively: 40.08, 24.31, 55.85, 54.94, 28.09, 30.97, 26.98, and 16.00.

Question starts:

To estimate a pseudoternary composition from a slag comprised of eight oxides, we take the three main oxides and, around them, we associate those with similar acid–base characteristics. We present two alternatives for the construction of the pseudoternary:

- Alternative A:

$$CaO = CaO + MgO = 56\% \tag{1.287}$$

$$FeO = FeO + MnO = 23\% \tag{1.288}$$

$$SiO_2 = SiO_2 + P_2O_5 + Al_2O_3 + Fe_2O_3 = 21\% \tag{1.289}$$

- Alternative B:

$$CaO = CaO + MgO = 56\% \tag{1.290}$$

$$FeO = FeO + MnO + Fe_2O_3 = 27\% \tag{1.291}$$

$$SiO_2 = SiO_2 + P_2O_5 + Al_2O_3 = 17\% \tag{1.292}$$

If we represent the pseudoternary compositions in the CaO–FeO–SiO_2 ternary diagram, Fig. 1.17, we obtain a liquidus temperature of 1830 °C for the slag A and a liquidus temperature of 1760 °C for the slag B.

According to Mills and Keene (1987, p. 13), the slag with the closest composition to that proposed in the exercise is: 49% CaO, 6.0% MgO, 14.2% FeO, 8% MnO, 18% SiO_2, 2.0% Al_2O_3, 2.8% Fe_2O_3, which has a liquidus temperature of 1702 °C. Other of the proposed LD slags, with greater FeO and Fe_2O_3 content, due to the decreasing of the CaO and MgO content is, according to Mill and Keene (1987): 48% CaO, 2.0% MgO, 18.3% FeO, 8% MnO, 18% SiO_2, 2.0% Al_2O_3, 3.7% Fe_2O_3, which has a liquidus temperature of 1604 °C.

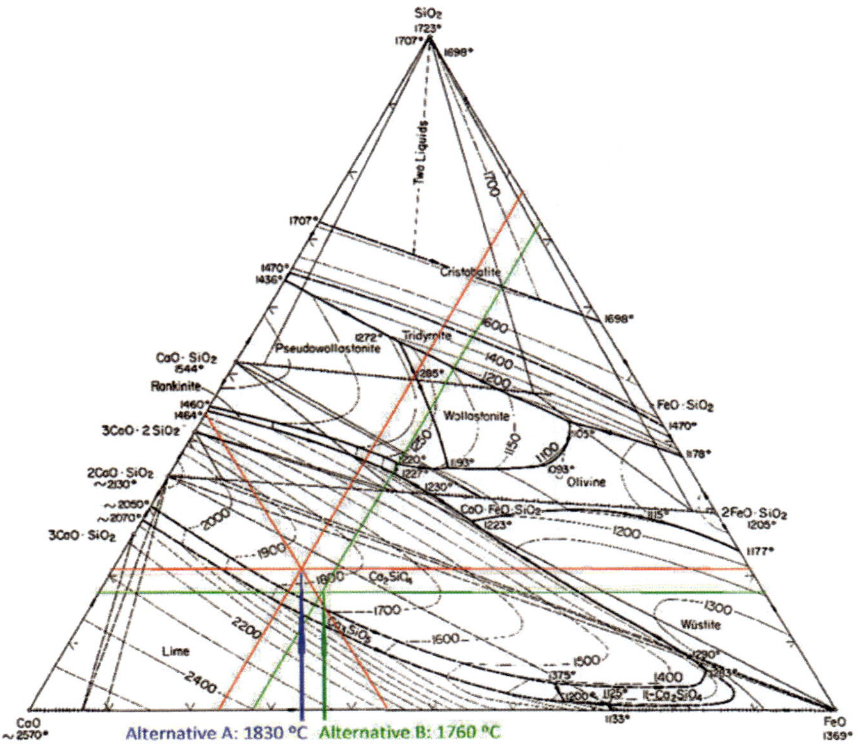

Fig. 1.17 Representation of the two slags in the CaO–FeO–SiO_2 ternary diagram

Consequently, the model proposed for the slag B (56% CaO, 27% FeO, 17% SiO_2) with an average molecular:

$$
\begin{aligned}
x_{SiO_2} &= \frac{17\,\mathrm{g\,SiO_2} \cdot \frac{1\,\mathrm{mol\,SiO_2}}{60.09\,\mathrm{g\,SiO_2}}}{17\,\mathrm{g\,SiO_2} \cdot \frac{1\,\mathrm{mol\,SiO_2}}{60.09\,\mathrm{g\,SiO_2}} + 56\,\mathrm{g\,CaO} \cdot \frac{1\,\mathrm{mol\,CaO}}{56.08\,\mathrm{g\,CaO}} + 27\,\mathrm{g\,FeO} \cdot \frac{1\,\mathrm{mol\,FeO}}{71.85\,\mathrm{g\,FeO}}} \\
&= 0.1707 \qquad (1.293)
\end{aligned}
$$

$$
\begin{aligned}
x_{CaO} &= \frac{56\,\mathrm{g\,CaO} \cdot \frac{1\,\mathrm{mol\,CaO}}{56.08\,\mathrm{g\,CaO}}}{17\,\mathrm{g\,SiO_2} \cdot \frac{1\,\mathrm{mol\,SiO_2}}{60.09\,\mathrm{g\,SiO_2}} + 56\,\mathrm{g\,CaO} \cdot \frac{1\,\mathrm{mol\,CaO}}{56.08\,\mathrm{g\,CaO}} + 27\,\mathrm{g\,FeO} \cdot \frac{1\,\mathrm{mol\,FeO}}{71.85\,\mathrm{g\,FeO}}} \\
&= 0.6025 \qquad (1.294)
\end{aligned}
$$

$$\begin{aligned} x_{FeO} &= \frac{27\,g\,FeO \cdot \frac{1\,mol\,FeO}{71.85\,g\,FeO}}{17\,g\,SiO_2 \cdot \frac{1\,mol\,SiO_2}{60.09\,g\,SiO_2} + 56\,g\,CaO \cdot \frac{1\,mol\,CaO}{56.08\,g\,CaO} + 27\,g\,FeO \cdot \frac{1\,mol\,FeO}{71.85\,g\,FeO}} \\ &= 0.2268 \end{aligned} \tag{1.295}$$

$$\begin{aligned} PM_{slag} &= PM_{SiO_2} \cdot x_{SiO_2} + PM_{CaO} \cdot x_{CaO} + PM_{FeO} \cdot x_{FeO} \\ &= 60.09 \times 0.1707 + 0.6025 \times 56.08 + 0.2268 \times 71.85 \\ &= 60.34 g \cdot mol^{-1} \end{aligned} \tag{1.296}$$

is the model that more approaches to the experimental data of Mills and Keene (1987) (even with a difference of 58 °C in the liquidus temperature).

For the calculation of the basicity, we proceed as in previous exercises. However, to use the values of the optical basicity, Appendix 4, we consider as criterion to use the average of the values indicated for each oxide. The basicity of the slag according to the index, $I_B(I)$:

$$I_B(I) = \frac{\%CaO}{\%SiO_2} = \frac{56}{17} = 3.29 \tag{1.297}$$

while the optical basicity is:

1. Number of moles in 100 g of slag:

$$\begin{aligned} Moles &= \frac{56}{56.08} + \frac{17}{60.09} + \frac{27}{71.85} \\ &= 1.6573 \text{ moles in } 100\,g \text{ of slag} \end{aligned} \tag{1.298}$$

2. Composition, expressed as molar fraction, of each one of the oxides of the slag is: $x_{CaO} = 0.6025$, $x_{SiO_2} = 0.1707$; $x_{FeO} = 0.2268$.
3. Summation of the product of the molar fraction of each oxide of the slag, x_{MeO}, and the number of oxygen atoms in each oxide, $n(O)$:

$$\sum x_{MeO} \cdot n(O) = 0.6025 \times 1 + 0.1707 \times 2 + 0.2267 \times 1 = 1.1706 \tag{1.299}$$

4. We calculate X_{MeO}:

$$X_{CaO} = \frac{0.6025 \times 1}{\sum x_{MeO} \cdot n(O)} = \frac{0.6025}{1.1706} = 0.5147 \tag{1.300}$$

$$X_{SiO_2} = \frac{0.1707 \times 2}{\sum x_{MeO} \cdot n(O)} = \frac{0.3414}{1.1706} = 0.2916 \tag{1.301}$$

$$X_{FeO} = \frac{0.2267 \times 1}{\sum x_{MeO} \cdot n(O)} = \frac{0.2267}{1.1706} = 0.1937 \tag{1.302}$$

5. Finally, we calculate the optical basicity of the slag, Λ_{slag}:

$$\begin{aligned}\Lambda_{slag} &= X_{CaO} \cdot \Lambda_{CaO} + X_{SiO_2} \cdot \Lambda_{SiO_2} + X_{FeO} \cdot \Lambda_{FeO} \\ &= 0.5147 \times 1 + 0.2916 \times 0.48 + 0.1937 \times 1 \\ &= 0.8484\end{aligned} \quad (1.303)$$

The density of the slag of 56% CaO, 27% FeO, and 17% SiO_2, whose composition in molar fraction would be $x_{CaO} = 0.6025$, $x_{FeO} = 0.2268$, $x_{SiO_2} = 0.1707$, from the partial molar volumes of the three components at 1500 °C, according to the data of Table 1.12, would be:

$$\begin{aligned}V_m^{slag} &= \left[20.70 \cdot x_{CaO} + 15.8 \cdot x_{FeO} + \left(19.55 + 7.966 \cdot x_{SiO_2}\right) \cdot x_{SiO_2}\right] \times 10^{-6} \\ &= [20.70 \times 0.6025 + 15.8 \times 0.2268 + (19.55 + 7.966 \times 0.1707) \times 0.1707] \times 10^{-6} \\ &= 19.6229 \times 10^{-6} m^3 \cdot mol^{-1}\end{aligned} \quad (1.304)$$

The value of the molar volume of the slag at 1500 °C is 19.6229×10^{-6} $m^3 \cdot mol^{-1}$. If we know that the variation of the molar volume of the slag, V_m^{slag}, with the temperature is equal to:

$$\left(\frac{dV_m^{slag}}{dT}\right) = 0.01 \quad (1.305)$$

the molar volume of the pseudoternary slag, alternative B, at 1650 °C is:

$$\begin{aligned}V_m^{slag}(1650\ °C) &= V_m^{slag}(1500\ °C) + 0.01 \times (1650\ °C - 1500\ °C) \\ &= 21.1229 \times 10^{-6} m^3 \cdot mol^{-1}\end{aligned} \quad (1.306)$$

Finally, the density of the slag at 1650 °C is:

$$\begin{aligned}\rho_{slag} &= \frac{1}{V_m^{slag}} \times \frac{PM_{slag}}{1} = \frac{mol}{21.1229 \cdot 10^{-6}\, m^3} \times \frac{60.34\, g}{mol} \times \frac{1\, kg}{1000\, g} \\ &= 2856.6\, kg \cdot m^{-3}\end{aligned} \quad (1.307)$$

The estimation of the specific heat at constant pressure, c_p^{slag}, at 1650 °C, would be (Table 1.13):

$$\begin{aligned}c_p^{slag} &= (80.8 \times 0.6025 + 76.6 \times 0.2268 + 87.0 \times 0.1707) J \cdot mol^{-1} \cdot K^{-1} \\ &= 80.8981\, J \cdot mol^{-1} \cdot K^{-1}\end{aligned} \quad (1.308)$$

While the density of the slags varies with the temperature, the estimations about the specific heat are valid in a wide interval of temperature. If we consider that the average molecular weight of the slag, PM_{slag}, is 60.34 g · mol^{-1}, the value of the specific heat of the slag at 1650 °C would be:

$$\begin{aligned} c_p^{slag}(1650\ ^\circ\mathrm{C}) &= c_p^{slag}(1500\ ^\circ\mathrm{C}) \cdot \frac{1}{PM_{slag}} \times \frac{1000\ \mathrm{g}}{1\ \mathrm{kg}} \\ &= 80.8981 \frac{\mathrm{J}}{\mathrm{mol} \cdot \mathrm{K}} \times \frac{1\ \mathrm{mol}}{60.34\ \mathrm{g}} \times \frac{1000\ \mathrm{g}}{1\ \mathrm{kg}} \\ &= 1340.7 \frac{\mathrm{J}}{\mathrm{kg} \cdot \mathrm{K}} \end{aligned} \quad (1.309)$$

Finally, for the viscosity, we use the Arrhenius equation indicated in the enunciation (Eqs. 1.284, 1.285, and 1.286):

$$\mu = \mathrm{A} \cdot \mathrm{T} \cdot \exp\left(\frac{\mathrm{B}}{\mathrm{T}}\right) \quad (1.310)$$

where A and B are:

$$\begin{aligned} \ln \mathrm{A} &= -19.81 + 1.73 \times x_{CaO} = -19.81 + 1.73 \times 0.6025 \\ &= -18.7676 \rightarrow \mathrm{A} = 7.0680 \times 10^{-9} \end{aligned} \quad (1.311)$$

$$\mathrm{B} = 31140 - 23896 \times x_{CaO} = 31140 - 23896 \times 0.6025 = 16741.6 \quad (1.312)$$

Consequently, the value of the viscosity of the pseudoternary slag for the alternative B, at 1650 °C, is:

$$\mu = 7.0680 \times 10^{-9} \times (1650 + 273) \times \exp\left(\frac{16741.6}{1650 + 273}\right) = 0.082\ \mathrm{Pa} \cdot \mathrm{s} \quad (1.313)$$

Commentaries: We indicate now the values of the physical–chemical properties of the slag of the enunciation at 1650 °C (50.0% CaO, 6.0% MgO, 18.0% FeO, 5.0% MnO, 15.0% SiO_2, 1.3% P_2O_5, 0.7% Al_2O_3, 4.0% Fe_2O_3). The basicity of the slag according to the index, $I_B(I)$:

$$I_B(I) = \frac{\%CaO}{\%SiO_2} = \frac{50}{15} = 3.33 \quad (1.314)$$

while the optical basicity is 0.8342 (Exercise 1.11).

The molar volume of the pseudoternary slag, alternative B, at 1500 °C is (molar fractions are collected in Table 1.10):

$$
\begin{aligned}
V_m^{slag}(1500\,°C) &= V_m^{SiO_2} \cdot x_{SiO_2} + V_m^{P_2O_5} \cdot x_{P_2O_5} + V_m^{Al_2O_3} \cdot x_{Al_2O_3} + V_m^{Fe_2O_3} \cdot x_{Fe_2O_3} \\
&+ V_m^{CaO} \cdot x_{CaO} + V_m^{MgO} \cdot x_{MgO} + V_m^{FeO} \cdot x_{FeO} + V_m^{MnO} \cdot x_{MnO} \\
&= \left(19.55 + 7.966 \cdot x_{SiO_2}\right) \cdot x_{SiO_2} + (65.7) \cdot x_{P_2O_5} \\
&+ \left(28.31 + 31.32 \cdot x_{Al_2O_3} - 31.45 \cdot x_{Al_2O_3}^2\right) \cdot x_{Al_2O_3} + (38.4) \cdot x_{Fe_2O_3} \\
&+ (20.7) \cdot x_{CaO} + (16.1) \cdot x_{MgO} + (15.8) \cdot x_{FeO} + (15.6) \cdot x_{MnO} \\
&= (19.55 + 7.966 \times 0.1511) \times 0.1511 + (65.7) \times 0.0055 \\
&+ \left(28.31 + 31.32 \times 0.0042 - 31.45 \times 0.0042^2\right) \times 0.0042 \\
&+ (38.4) \times 0.0152 + (20.7) \times 0.5397 + (16.1) \times 0.0901 \\
&+ (15.8) \times 0.1516 + (15.6) \times 0.0427 \\
&= 19.8831 \times 10^{-6} m^3 \cdot mol^{-1}
\end{aligned}
\tag{1.315}
$$

at 1650 °C is:

$$
\begin{aligned}
V_m^{slag}(1650\ °C) &= V_m^{slag}(1500\ °C) + 0.01 \times (1650\ °C - 1500\ °C) \\
&= \left[19.8831 + 0.01 \times (1650\ °C - 1500\ °C)\right] \times 10^{-6} m^3 \cdot mol^{-1} \\
&= 21.3831 \times 10^{-6} m^3 \cdot mol^{-1}
\end{aligned}
\tag{1.316}
$$

Finally, the density of the slag at 1650 °C is ($PM_{slag} = 60.54\,g \cdot mol^{-1}$, Exercise 1.11):

$$
\begin{aligned}
\rho_{slag} &= \frac{1}{V_m^{slag}} \cdot \frac{PM_{slag}}{1} = \frac{mol}{21.3831 \times 10^{-6}\ m^3} \times \frac{60.54\,g}{mol} \times \frac{1\,kg}{1000\,g} \\
&= 2831.21\,kg \cdot m^{-3}
\end{aligned}
\tag{1.317}
$$

The estimation of the specific heat at constant pressure, c_p^{slag}, at 1650 °C, would be:

$$
\begin{aligned}
c_p^{slag} &= c_p^{SiO_2} \cdot x_{SiO_2} + c_p^{P_2O_5} \cdot x_{P_2O_5} + c_p^{Al_2O_3} \cdot x_{Al_2O_3} + c_p^{Fe_2O_3} \cdot x_{Fe_2O_3} \\
&+ c_p^{CaO} \cdot x_{CaO} + c_p^{MgO} \cdot x_{MgO} + c_p^{FeO} \cdot x_{FeO} + c_p^{MnO} \cdot x_{MnO} \\
&= 87 \times 0.1511 + 242.7 \times 0.0055 + 146.4 \times 0.0042 + 191.2 \times 0.0152 \\
&+ 80.8 \times 0.5397 + 90.4 \times 0.0901 + 76.6 \times 0.1516 + 79.9 \times 0.0427 \\
&= 84.7705\,J \cdot mol^{-1} \cdot K^{-1}
\end{aligned}
\tag{1.318}
$$

While the density of the slags varies with the temperature, the estimations about the specific heat are valid in a wide interval of temperature. If we consider that the average molecular weight of the slag, PM_{slag} is 60.54 g · mol^{-1}, the value of the specific heat of the slag at 1650 °C would be:

$$c_p^{slag}(1650\ °C) = c_p^{slag}(1500\ °C) \cdot \frac{1}{PM_{slag}} \cdot \frac{1000\,g}{1\,kg} = 84.7705 \frac{J}{mol \cdot K} \cdot \frac{1\,mol}{60.54\,g} \cdot \frac{1000\,g}{1\,kg}$$
$$= 1400.24 \frac{J}{kg \cdot K} \tag{1.319}$$

Finally, for the viscosity, we use the Arrhenius equation indicated in the enunciation (Eqs. 1.284, 1.285, and 1.286):

$$\mu = A \cdot T \cdot \exp\left(\frac{B}{T}\right) \tag{1.320}$$

where A and B are:

$$\begin{aligned} \ln A &= -19.81 + 1.73 \cdot x_{CaO} - 33.76 \cdot x_{Al_2O_3} \\ &= -19.81 + 1.73 \times 0.5397 - 33.76 \times 0.0042 \\ &= -19.1813 \rightarrow A = 5.50 \times 10^{-9} \end{aligned} \tag{1.321}$$

$$\begin{aligned} B &= 31140 - 23896 \cdot x_{CaO} + 68833 \cdot x_{Al_2O_3} \\ &= 31140 - 23896 \times 0.5397 + 68833 \times 0.0042 = 18532.4 \end{aligned} \tag{1.322}$$

Consequently, the value of the viscosity of the pseudoternary slag for the alternative B, at 1650 °C, is:

$$\mu = 5.5 \times 10^{-9} \cdot (1650 + 273) \times \exp\left(\frac{18532.4}{1650 + 273}\right) = 0.1621\ Pa \cdot s \tag{1.323}$$

If we compare these results with those obtained for the simplification of the pseudoternary slag alternative B (56% CaO, 27% FeO, 17% SiO_2), almost all the values might be considered acceptable with the exception of two: the liquidus temperature (difference of 58 °C) and the optical basicity (a difference of ± 0.0142 means a noticeable change in the chemical composition).

End of the exercise.

Exercise 1.13 Using the data provided by Mills and Keene (1987, p. 13), which refer to the liquidus temperature of twelve LD slags (Tables 1.14 and 1.15), analyze if there is correlation between the optical basicity and the liquidus temperature.

For all the LD slags, from the A to the F, the SiO_2, Al_2O_3, MnO, and MgO contents are constant and equal to 23% SiO_2, 2.0% Al_2O_3, 8.0% MnO, and 6.0% MgO.

SiO_2, Al_2O_3, MnO and MgO contents are also constant for slags from G to L and equal to that of slags A–F, but MgO content is equal to 2%.

Question starts:

Table 1.14 Composition of the LD slag and líquidus temperature (slags A–F)

Type of slag	%Fe_2O_3	%FeO	%CaO	Temperature liquidus (°C)
A	5.3	26.7	29.0	1334
B	4.5	22.5	34.0	1369
C	3.7	18.3	39.0	1446
D	2.8	14.2	44.0	1514
E	2.0	10.0	49.0	1598
F	1.2	5.8	54.0	1715

Table 1.15 Composition of the LD slag and liquidus temperature (slags G–L)

Type of slag	%Fe_2O_3	%FeO	%CaO	Temperature liquidus (°C)
G	5.3	26.7	33.0	1366
H	4.5	22.5	38.0	1473
I	3.7	18.3	43.0	1561
J	2.8	14.2	48.0	1623
K	2.0	10.0	53.0	1660
L	1.2	5.8	58.0	1777

In the data provided by Mills and Keene (1987), there is not grouping of the slags in two groups as it is expressed in the enunciation of the exercise. The reason for the separation into two is that when we represent the optical basicity of the slag, Λ_{slag} (calculated as in previous exercises), in front of the liquidus temperature, T_L, it is possible to check that there is correlation between these variables (Fig. 1.18):

$$T_L = 199857 \cdot \Lambda_{slag}^2 - 289436 \cdot \Lambda_{slag} + 105562 \tag{1.324}$$

However, the estimations of the liquidus temperature that could be calculated from the optical basicity are greater than ±40 °C (Table 1.16) with respect to the data provided by Mills and Keene (1987).

On the contrary, if we group the twelve samples in the two groups indicated in the enunciation of the exercise (Tables 1.14 and 1.15), we see that correlations between the Λ_{slag} and the T_L with deviations well below the 50 °C. We show in Tables 1.17 and 1.18, the values used in the obtaining of the correlations.

The best correlation between the liquidus temperature and the optical basicity of the slags would be obtained with the values of the slags A–F, which is a polynomial equation of second order (the graphical representation appears in Fig. 1.19):

$$T_L = 8.78 \times 10^5 \cdot \Lambda_{slag}^2 - 1.352 \times 10^6 \cdot \Lambda_{slag} + 521772 \tag{1.325}$$

This equation allows maximum deviations of ±9 °C.

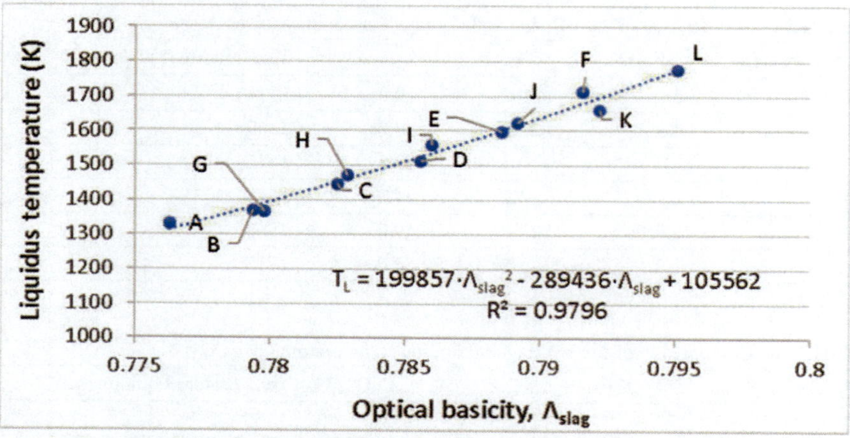

Fig. 1.18 Representation of the optical basicity, Λ_{slag}, and the liquidus temperature, T_L (slags A–L)

Table 1.16 Results of the estimation of the liquidus temperature using the Eq. (1.324)

	A	B	C	D	E	F
Liquidus temperature, T_L	1334	1369	1446	1514	1598	1715
Optical basicity, Λ_{slag}	0.7763	0.7794	0.7825	0.7856	0.7886	0.7916
Estimation of the liquidus temperature	1315	1382	1452	1526	1602	1681
	G	H	I	J	K	L
Liquidus temperature, T_L	1366	1473	1516	1623	1660	1777
Optical basicity, Λ_{slag}	0.7798	0.7829	0.7860	0.7892	0.7922	0.7951
Estimation of the liquidus temperature	1390	1461	1536	1617	1697	1778

Table 1.17 Values of the liquidus temperature, T_L, and the optical basicity, Λ_{slag}, for the slags A–F

	A	B	C	D	E	F
Liquidus temperature, T_L	1334	1369	1446	1514	1598	1715
Optical basicity, Λ_{slag}	0.7763	0.7794	0.7825	0.7856	0.7886	0.7916

Table 1.18 Values of the liquidus temperature, T_L, and the optical basicity, Λ_{slag}, for the slags G–L

	G	H	I	J	K	L
Liquidus temperature, T_L	1366	1473	1516	1623	1660	1777
Optical basicity, Λ_{slag}	0.7798	0.7829	0.7860	0.7892	0.7922	0.7951

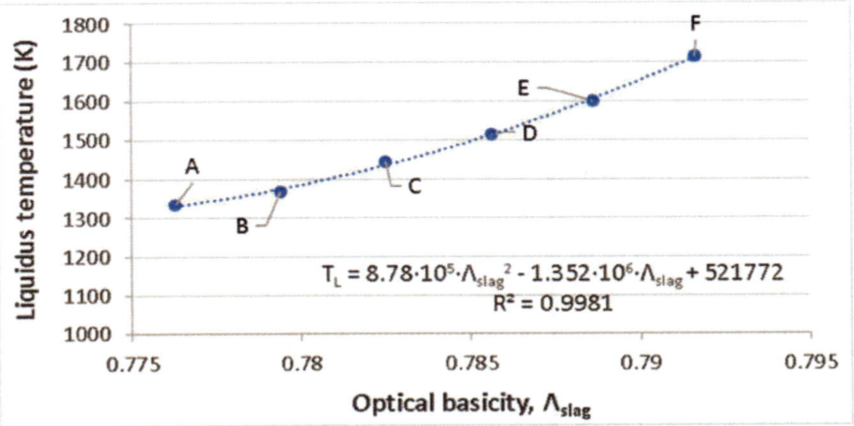

Fig. 1.19 Representation of the optical basicity, Λ_{slag}, and the liquidus temperature, T_L (slags A–F)

Table 1.19 Results of the estimation of the liquidus temperature using the Eq. (1.325)

	A	B	C	D	E	F
Liquidus temperature, T_L	1334	1369	1446	1514	1598	1715
Optical basicity, Λ_{slag}	0.7763	0.7794	0.7825	0.7856	0.7886	0.7916
Estimation of the liquidus temperature	1334	1377	1437	1514	1604	1710

We show in Table 1.19 the estimation of the liquidus temperature using the Eq. (1.325), and we compare it with the results of Mills and Keene (1987).

In the case of the samples of slag from G to L, the linear correlation is slightly more favorable than the polynomic correlation, as a deviation of ±31 °C is obtained (Fig. 1.20). The equation of the linear correlation is:

$$T_L = 2.49 \times 10^4 \cdot \Lambda_{slag} - 18035 \quad (1.326)$$

We collect in Table 1.20 the values of the liquidus temperature estimated using the Eq. (1.326) and the values of Mills and Keene (1987).

Commentaries: The presence of MgO in the slag, replacing CaO, leads to a reduction of the liquidus temperature of slags.

It might be interesting to reach a better knowledge and control of the BOF process by means of the liquidus temperature of the slags. Instead of using the experimental calculating of the liquidus temperature, T_L, which is a long and expensive test, it is possible, by means of multivariate correlations, to provide useful information for the control of the process. On another note, one of the variables that nowadays appears as an alternative to characterize the slags is the optical basicity.

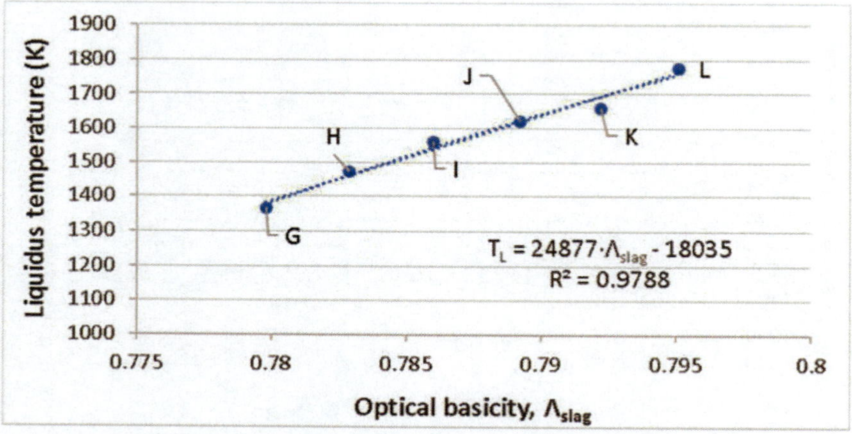

Fig. 1.20 Representation of the optical basicity, Λ_{slag}, and the liquidus temperature, T_L (slags G–L)

Table 1.20 Results of the estimation of the liquidus temperature using the Eq. (1.326)

	G	H	I	J	K	L
Liquidus temperature, T_L	1366	1473	1561	1623	1660	1777
Optical basicity, Λ_{slag}	0.7798	0.7829	0.7860	0.7892	0.7922	0.7951
Estimation of the liquidus temperature	1382	1459	1536	1616	1691	1763

Using the optical basicity of the slag and polynomic equations of second order, it is possible to estimate thermodynamic parameters related with ability of picking up phosphorus or sulphur from the slags, as well as to estimate the partition coefficient between the metal and the slag. Finally, it is also possible to take advantage of the contributions of the thermodynamic software, as FactSage, and the correlations derived from the optical basicity of the slag to estimate the liquidus temperature, T_L, of the BOF slags.

End of the exercise.

References

Ballester A, Verdeja LF, Sancho JP (2000) Metalurgia Extractiva, Fundamentos. Síntesis, Madrid, Spain

Bodsworth C (1998) Sir Henry Bessemer: father of steel industry. IOM Communications, USA

Comellas JL (2000) Historia Breve del Mundo, Contemporáneo. Rialp, Madrid, Spain

Coudurier L, Hopkins DW, Wilkomisky I (1985) Fundamentals of metallurgical processes, 2nd edn. Pergamon Press, Oxford, UK

Deo B, Boom R (1993) Fundamentals of steelmaking metallurgy. Prentice Hall International, Hertfordshire, UK

Gaye H (1982) Equilibria and kinetics in steelmaking. Journées d'etude LBE. First LBE Workshop, Avignon, 10–14 October 1982, p 4

Kawai Y, Shiraishi Y (1988) Handbook of physico-chemical properties at high temperatures. The Iron and Steel Institute of Japan-Committee of Japan Society for Promotion of Science, Tokyo, Japan.

Levin E, Robins CR, McMurdie HF (1964) Phase diagrams for ceramists. The American Ceramic Society, Columbus, Ohio, USA (and supplement of figures from the 2067 to the 4149 edited in 1969).

Martín R, Barbés MA, Barbés MF, Marinas E, Ayala J, Mochón J, Verdeja LF, García F (2009) Características sólido-fluido en el crisol del horno alto según el modelo de desgaste nodal (MDN). Revista De Metalurgia De Madrid 45(4):295–304

Massalski TB, Okamoto H, Subramanian PR, Kacprzac L (1990) Binary alloy phase diagrams. ASM International, Materials Park, Ohio, USA

Mills KC, Keene BJ (1987) Physical properties of BOS slags. Int Mater Rev 32(1–2):1–120

Pero-Sanz JA, Quintana MJ, Verdeja LF (2017) Solidification and solid-state transformations of metals and alloys, 1st edn. Elsevier, Boston, USA.

Pero JA, Fernández-González D, Verdeja LF (2019) Structural materials: properties and, selection. Springer International Publishing, Cham, Switzerland

Springorum D (1995) Slag Atlas. Düsseldorf, Germany, Verein Deutscher Eisenhüttenleute (VDEh)

TAPP Database (1994) Thermochemical and physical properties (Software). E. S. Microwave Inc., Cincinnati, OH, USA

The Japan Society for the promotion of science. The 19th Committee on Steelmaking (1988) Steelmaking data sourcebook. Gordon and Breach Science Publishers, New York, USA

Verdeja LF, Ballester A, Sancho JP, González R (2014) Refractory and ceramic materials. Ed. Síntesis, 1st edn., Madrid, Spain

Verdeja JI, Fernández-González D, Verdeja LF (2020) Operations and basic processes in ironmaking. Springer International Publishing, Cham, Switzerland

Chapter 2
The Electric Steelworks

Abstract Electric furnace has been gaining importance in the production of steel since the first decades of the twentieth century. It is possible to produce pig iron in an electric furnace, but this option is very far from being a competitor for the blast furnace or Corex process, even in those countries with cheap electric power. Other option is to produce steel, and particularly to produce highly alloyed or stainless steels, where it is necessary to add significant quantities of elements such as chromium, nickel, manganese, silicon, niobium, titanium, vanadium, molybdenum, tungsten, or boron. We also deal in this chapter with the production of ferroalloys, as they are an important material in the iron and steelmaking industry, and particularly, we describe the design of furnaces for their production. Theoretical part described in this manuscript is complemented with solved exercises.

2.1 Introduction

The demand of steel in the world in 1900 was of 27 million of tons and a decade later, to cover the demand, the production of steel tripled, which promoted the introduction of scrap in the market of the ironmaking raw materials.

The alternatives that were proposed to use scrap were two: the open-hearth furnace (Martin-Siemens steelworks) and the electric steelworks.

From 1914, electric furnace was prevailing in the iron and steelmaking industry as one of the most efficient alternatives in the production of steel. While in the BOF steelworks the temperature of the molten metal is kept constant by oxidation of the elements dissolved in the pig iron, in the electric furnace, electric power is used as instrument for the thermal control of the process.

Due to the operating conditions in the electric furnace, it is not necessary to maintain high oxidizing potentials during both the fusion process and the heating of the metal (the stage that generates an oxidizing slag). With a suitable control of the oxygen potential in the system, it is possible to create a slightly reductant environment that allows retaining high percentages of alloying elements easily oxidable in the molten iron. On the other hand, if the last stage of the refinement process (process

L. F. Verdeja González et al., *Operations and Basic Processes in Steelmaking*,
Topics in Mining, Metallurgy and Materials Engineering,
https://doi.org/10.1007/978-3-030-68000-8_2

with double slag) is performed with a reductant slag to remove the sulphur or to assimilate in the melt valuable alloying elements retained in the oxidizing slag, only small quantities of deoxidizing elements will be required to deoxidize the steel, which noticeably reduces the quantity of inclusions in the steel (clean steels).

The heat generated by the electric arc (direct or indirect) was more efficiently controlled whenever the technological progress was incorporated to the ironmaking sector. This allowed to even overcome the problems of oxidation of either the fuel or active metals alloyed with iron.

The heating of the metal takes place preferentially in the upper part of the bath were very high temperature can be locally reached. It is required, for that reason, a strict control of the process to avoid an excessive heating of the refractory lining.

From 1970, a progressive increase of both the power and capacity of production of the three-phase electric arc (direct) furnaces was taken into account, and specific consumptions of 550 kWh/t of liquid iron were reached (using only scrap as ferric charge of the furnace).

Electric steelworks progress toward the implementation of technologies aimed at both reducing the specific consumption of electrical energy and protecting the environment, as for instance:

- Preheating the scrap.
- Auxiliary burners in furnaces.
- Pulverized coal as combustible.
- Replacing alternating current by direct current (reduction of acoustics pollution). The number of graphite electrodes in the furnace was reduced from three to one.
- Burden formed by pig iron (30%), scrap (40%), and pre-reduced iron (30%) or by iron carbide, scrap, and pre-reduced iron.
- Reduction of the times between charge and tapping of the furnace as well as reduction of the specific consumptions of refractories and electrodes.

The evolution of the electric furnace process makes thinking that the electric furnace will not be a simple instrument to melt scrap, and it will be more similar to a chemical reactor of complete mixture. It will incorporate the most sophisticated systems to control the refining and tapping. The stirring of the metal will be performed by means of the injection of gas via lance or via porous plugs (tuyeres) located in the hearth of the furnace.

2.2 The Electric Furnace: Characteristics and Applications

The special features of the electric furnace point out it as a suitable instrument to manufacture a wide range of products for the iron and steelmaking industry:

- Mild steel, competing with the BOF steelworks.
- Special and stainless steels, highly alloyed (without the concurrence of the BOF process).

- Ferroalloys.
- Ceramic products as silicon carbide, calcium silicide, calcium carbide, or silicon nitride.

2.2.1 Production of Pig Iron

The manufacture of pig iron using an electric furnace is very far from being a competitor for the blast furnace or the Corex process, even in those countries with cheap electric power. For the same reason, the possibilities of using the electric furnace are optimal when:

- The capacity of production of its possible competitors is not high.
- The products have a high added value.
- The products have a participation in the market of thousands of tons instead of millions of tons.

We represent in Fig. 2.1 a scheme of the electric furnace used in the production of pig iron (it is possible to use a similar scheme for the ferroalloys production, Sect. 2.2.3).

The most important difference with respect to the blast furnace is that the heat necessary for the direct reduction of the iron ore comes from the electrical energy instead of from the ironmaking coke:

$$Fe_2O_3(s) + 3C(s) \leftrightarrow 2Fe(l) + 3CO(g) \quad (2.1)$$

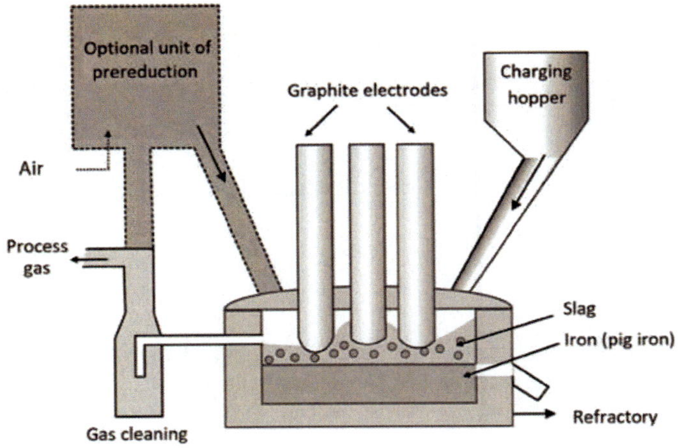

Fig. 2.1 Electric furnace used in the production of pig iron

The gas that leaves the zone of smelting reduction, 3 molecules of CO per any 2 atoms of iron, reaction (2.1), comprises mainly CO with a 5–10% N_2, which comes from the entrance of air during the process of charge (it is usual in electric furnaces to operate under a pressure lower than the atmospheric). Out of the direct reduction zone, the potential of oxygen of the system (POS) is incompatible with a great rate of CO without the presence of CO_2. On the other hand, and from the environmental point of view, it is not possible to emit gases with great CO concentrations to the atmosphere. Consequently, the solution would be the oxidation of the gases and taking advantage of the heat produced to heat the ferric charge (iron ore), the reductant charge (ironmaking coke), and the slagging agents (ironmaking lime). Definitely, the energy that might be recovered, assuming that the temperature of the gases that leave the zone of smelting reduction is 1200 °C and that the reference room temperature is 25 °C, is:

1. On the one hand, the combustion of CO is represented by the following reaction:

$$CO(g) + O_2(g) \rightarrow CO_2(g) \tag{2.2}$$

The standard enthalpy of reaction, $\Delta_r H^0(T)$, can be obtained with the data of Appendix 2 and represents the heat of combustion of the CO:

$$C(s) + \frac{1}{2}O_2(g) \rightarrow CO(g) \tag{2.3}$$

$$\Delta_r H^0(\text{reaction 2.3}) = -104.0 - 0.0075 \cdot T\left[kJ \cdot (mol\ CO)^{-1}\right] \tag{2.4}$$

$$C(s) + O_2(g) \rightarrow CO_2(g) \tag{2.5}$$

$$\Delta_r H^0(\text{reaction 2.5}) = -393.0 - 0.0017 \cdot T\left[kJ \cdot (mol\ CO_2)^{-1}\right] \tag{2.6}$$

consequently, the value of the standard enthalpy of CO combustion is:

$$\begin{aligned}\Delta_r H^0(\text{reaction 2.2}) &= \Delta_r H^0(\text{reaction 2.5}) - \Delta_r H^0(\text{reaction 2.3})\\ &= -289.0 + 0.0058 \cdot T\left[kJ \cdot (mol\ CO)^{-1}\right]\end{aligned} \tag{2.7}$$

The value of the standard enthalpy of the reaction (2.1) at 1200 °C, considering that the atomic weight of the iron is 55.85 g mol^{-1}, is:

$$\begin{aligned}\Delta_r H^0(\text{reaction 2.1}) &= -280.46\frac{\text{kJ}}{\text{mol CO}}\cdot\frac{3\text{ mol CO}}{2\text{ mol Fe}}\\ &= -420.69\frac{\text{kJ}}{\text{mol Fe}}\cdot\frac{1\text{ mol Fe}}{55.85\text{ g Fe}}\cdot\frac{10^6\text{ g Fe}}{1\text{ t Fe}}\\ &= -7.53\cdot 10^6\frac{\text{kJ}}{\text{t Fe}}\end{aligned} \tag{2.8}$$

2. On another note, the sensitive heat accumulated by the CO_2 generated in the combustion of the CO (reaction (2.2)) from the 1200 °C to the reference temperature of 25 °C is determined. Considering as average specific heat of the CO_2 (TAPP Database 1994) 46.353 J mol^{-1} K^{-1}:

$$\begin{aligned}\text{Sensitive heat} - CO_2 &= 46.353\frac{\text{J}}{\text{mol } CO_2\cdot\text{K}}\cdot(1473\text{K} - 298\text{K})\\ &\cdot\frac{1\text{ mol } CO_2}{1\text{ mol CO}}\cdot\frac{3\text{ mol CO}}{2\text{ mol Fe}}\cdot\frac{1\text{ kJ}}{10^3\text{ J}}\\ &= 82.185\frac{\text{kJ}}{\text{mol Fe}}\cdot\frac{1\text{ mol Fe}}{55.85\text{ g Fe}}\cdot\frac{10^6\text{ g Fe}}{1\text{ t Fe}} = 1.47\cdot 10^6\frac{\text{kJ}}{\text{t Fe}}\end{aligned} \tag{2.9}$$

Summarizing, the calorific heat susceptible of being recovered would be: 9.00×10^6 kJ (t Fe)$^{-1}$.

Assuming that the ore of the charge would comprise hematite, Fe_2O_3, and that the reduction with CO would only allow the reduction of the hematite into wustite (FeO), only 10% of the oxygen of the ferric charge could be transferred to the gas in this situation.

Coke consumption (the quality required for the coke is not the same than that required for use in the blast furnace) is around 375 kg t^{-1} of pig iron, while the consumption of electric power rises up to 2000 kWh. However, if the preheating, pre-reduction of raw materials, effective transfer of heat to the charge, and activation of the reduction-smelting kinetics processes are optimized, it is possible to obtain values of around 1000 kWh t^{-1} of pig iron.

To conclude, the electric furnace cannot compete with the blast furnace even when the installation costs are much smaller. It will become competitive when the added value of the resulting product will be high and its participation in the market (tons) will be small. An example are the ferroalloys as high added value products if compared with the pig iron, which have a market very smaller than that of the pig iron (hundreds of millions of tons of pig iron are yearly produced).

2.2.2 Steel Production

Even though BOF steelworks reach very satisfactory performances in the production of mild steels, it does not happen the same when we want to manufacture highly alloyed or stainless steels where it is necessary to add significant quantities of chromium, nickel, manganese, silicon, niobium, titanium, vanadium, molybdenum, tungsten, or boron.

The electric steelworks are the only ones that can produce high-quality steel and can compete in the production of mild steels with the integrated steel plants (with BOF process). We show in Fig. 2.2 a scheme of the operations in the electric furnace.

However, a nozzle or tap hole in the hearth of the furnace located at the height of the main tap hole is usually employed in the current operation of tapping of

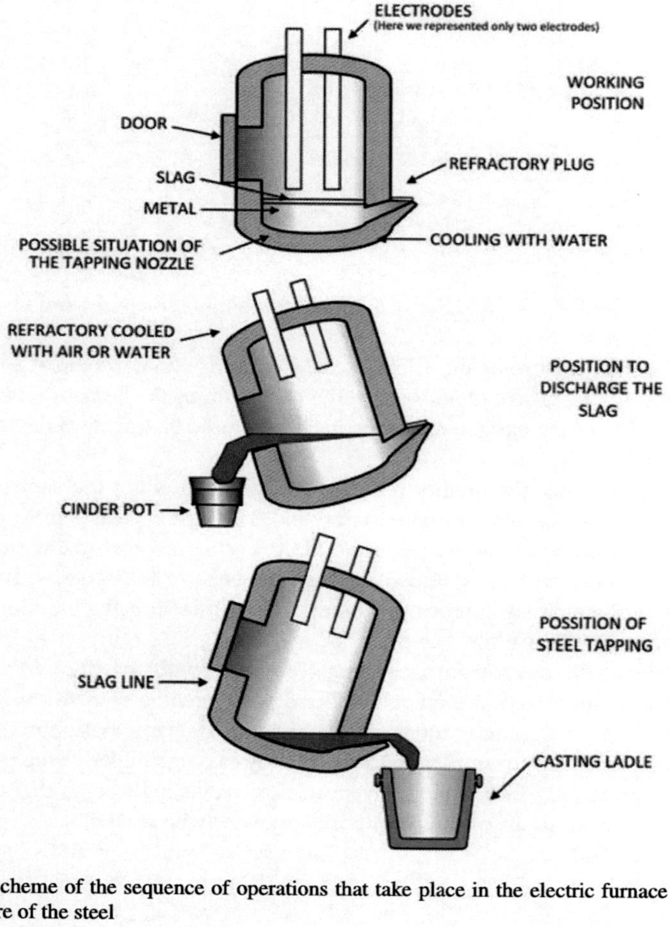

Fig. 2.2 Scheme of the sequence of operations that take place in the electric furnace during the manufacture of the steel

the furnace. This allows emptying the furnace in the working position (Fig. 2.2); equally, the connection of nozzles for the protection of the pouring stream, avoiding the reoxidation or re-nitriding of the metal is also possible.

Scrap or pre-reduced iron is charged in the upper zone of the furnace after the complete opening of the roof of the furnace. The consumption of electric power per ton of liquid iron decreases if we operate with preheated burdens or if variable quantities of pig iron or iron carbide are available in the burden.

Consumptions of electrical energy of 390 kWh t^{-1} of liquid iron can be obtained when the feeding is formed by different proportions of scrap, pig iron, and iron carbide. An approximate distribution of the costs of steel production in the electric arc furnace might be:

- Scrap: 60%.
- Electric power: 10%.
- Workforce: 5%.
- Oxygen consumption: 5%.
- Consumption of electrodes: 5%.
- Maintenance costs: 15%.

In the production of highly alloyed steels, parameters of interaction of solutes in the metal are used for the analysis of the metal-slag-gas equilibrium. Normally, Henry's activity of solutes in liquid iron cannot be joined with concentration expressed in percentage. At least, interactions of first order must be considered to obtain results close to the experimental data.

Interaction parameters of the Henry's law, e_i^j, are collected in Table 2.1 for the most common alloying elements in special steels.

According to the CaO–FeO–SiO_2 ternary diagram, the composition of a refining oxidizing slag of electric steelworks corresponds to the point B of Fig. 2.3 (edited from Levin et al. 1969). In the fusion oxidizing stage (refining), thermodynamic conditions are favorable for the slagging of both silicon and phosphorus. Initially, the charge of slag-forming agents comprises iron ore, mill scale, and lime (point A in the ternary diagram). As the refining process progresses, silicon slags and, the slag is formed (point B). During the process of slag-formation, the locus of all the intermediate products (intermediate slags of refining) is located along the line AB. This way, it is possible to simulate in the ternary diagram what would be the composition, liquidus temperature, quantity of solid to be dissolved, and equilibrium phases during the formation of the slag.

2.2.3 *Ferroalloys*

Ferroalloys are impure metallic products that, apart from the valuable active metal, contain considerable quantities of iron. They are raw materials of special importance

Table 2.1 Interaction parameters, e_i^j, in the liquid iron at 1600 °C according to (a) Sigworth and Elliot (1974) and (b) The Japan Society for the Promotion of Science (1988)

Solutes (i)	Elements (j)									
	C	Cr	Ni	Ti	V	Mo	Nb	O	N	H
C	0.14[a]	−0.024[a]	0.012[a]		0.077[a]	−0.008[a]	−0.06[a]	−0.34[a]	0.11[a]	0.67[a]
	0.243[b]	−0.023[b]	0.0097[b]		0.029[b]	−0.014[b]	−0.056[b]	−0.32[b]	0.11[b]	0.67[b]
Cr	−0.12[a]	0.0003[a]	0.0002[a]	0.059[a]		0.0018[a]		−0.14[a]	−019[a]	−0.33[a]
	−0114[b]	0.0003[b]	0.0002[b]		0.012[b]	0.0018[b]		−0.16[b]	−0.182[b]	−0.34[b]
Ni	0.042[a]	−0.0003[a]	0.001[a]					0.01[a]	0.028[a]	−0.25[a]
	0.032[b]	0.0003[b]	0.001[b]					0.01[b]	0.015[b]	−0.36[b]
Ti		0.055[a]		0.013[a]				−1.8[a]	−1.8[a]	−1.1[a]
				0.042[b]				−3.4[b]	−2.06[b]	−1.1[b]
V	−0.34[a]				0.015[a]			−0.97[a]	−0.35[a]	−0.59[a]
	−0.14[b]	0.0119[b]			0.031[b]			−0.46[b]	−0.46[b]	−0.59[b]
Mo	0.0097[a]	0.0008[a]						0.001[a]	−0.1[a]	−0.2[a]
	−0.14[b]	−0.13[b]				0.012[b]		0.008[b]	−0.1[b]	−0.13[b]
Nb							0[a]	−0.83[a]	−0.42[a]	−0.61[a]
	0.01[b]						0[b]	−0.72[b]	−0.48[b]	−0.7[b]
O	−0.45[a]	−0.013[a]	0.006[a]	−0.4[a]	−0.3[a]	0.0035[a]	−0.14[a]	−0.2[a]	0.057[a]	−3.1[a]
	−0.421[b]	−0.055[b]	0.006[b]	−1.12[b]	−0.14[b]	0.005[b]	−0.12[b]	−0.17[b]	−0.14[b]	0.73[b]
N	0.13[a]	-0.047[a]	0.01[a]	-0.53[a]	0.093[a]	-0.011[a]	-0.06[a]	0.05[a]	0[a]	
	0.13[b]	-0.046[b]	0.007[b]	-0.6[b]	0.111[b]	-0.011[b]	-0.068[b]	-0.12[b]	0[b]	
H	0.06[a]	0.0022[a]	0[a]	-0.019[a]	0.007[a]	0.0022[a]	0.0023[a]	-0.19[a]		0[a]
	0.06[b]	0.0024[b]	-0.002[b]	-0.019[b]	0.007[b]	0.0029[b]	0.0033[b]	0.05[b]		0[b]

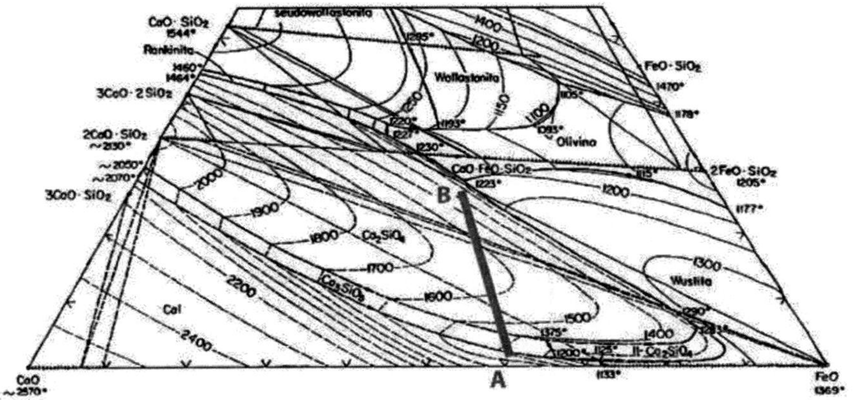

Fig. 2.3 Oxidizing slags of refining of electric furnace located in the CaO–FeO–SiO_2 ternary system. A-Initial composition of the nonmetallic burden, B-Composition of the basic slag at the end of the process

in the production of mild steels and special steels, and they are used either as deoxidizing or desulphurizing elements but also to add elements in the manufacture of special steels.

Carbon content in stainless steel is so small that, if the used ferroalloys were produced by carbothermic reaction, the ferroalloy will introduce carbon in the steel as undesirable impurity. This way, special ferroalloys, with low carbon content, must be prepared using methods different from the carbothermic reduction, for instance, the metallothermic reduction, or the molten salts electrolysis.

The obtaining of ferrosilicon can be performed in electric furnace using quartz as raw material:

$$SiO_2(s, l) + 2C(s) \leftrightarrow Si(dis; iron) + 2CO(g) \tag{2.10}$$

The temperature of operation is always greater than the melting point of the silica with the purpose of favoring the kinetics of the process. However, apart from the condensed phases (metallic silicon, carbon, and silicon oxide), silicon carbide, SiC, can be formed and, the volatile compound silicon monoxide (silicon (II) oxide), SiO, can be also found in the gas.

In a three components system, silicon-oxygen-carbon, the degrees of freedom, L, are:

$$L = 5 - F(\text{equilibrium phases}) \tag{2.11}$$

and L is zero when, at constant total pressure ($P_T = P_{CO} + P_{SiO}$), the following equilibriums are established:

- Between solid phases (SiO_2, SiC, and C) and the gas.
- Between condensed phases (SiO_2, SiC, and Si) and the gas.

The first of these equilibriums can be expressed by the following reaction:

$$3C(s) + SiO_2(s) \leftrightarrow SiC(s) + 2CO(g) \quad (2.12)$$

while the second equilibrium can be expressed as follows:

$$3SiO_2(l) + 2SiC(s) \leftrightarrow Si(dis; iron) + 4SiO(g) + 2CO(g) \quad (2.13)$$

It is interesting to produce ferrosilicon to displace reaction (2.13) to the right and, for that reason, the following variables must be considered:

- working at temperatures higher than 1850 °C.
- silicon carbide dissolves in the metal in presence of liquid iron:

$$SiC(s) \leftrightarrow Si(dis; iron) + C(dis; iron) \quad (2.14)$$

- the activity of the silicon noticeably decreases in presence of liquid iron (Raoult's coefficient of activity smaller than one).

However, despite the favorable effects of the temperature and liquid iron in the reaction (2.13), silicon monoxide, SiO, is always lost in the gas: greater losses when increasing the quantity of silicon in the ferroalloy.

Raw material for the obtaining of ferromanganese is pyrolusite, MnO_2. As the oxides of manganese with valence greater than two (MnO_2, Mn_2O_3, and Mn_3O_4) are unstable at low temperatures, they decompose according to the following reactions:

$$2MnO_2(s) \leftrightarrow Mn_2O_3(s) + \frac{1}{2}O_2(g) \quad (2.15)$$

$$3Mn_2O_3(s) \leftrightarrow 2Mn_3O_4(s) + \frac{1}{2}O_2(g) \quad (2.16)$$

$$Mn_3O_4(s) \leftrightarrow 3MnO(s) + \frac{1}{2}O_2(g) \quad (2.17)$$

or they are easily reduced into MnO with gases of great oxygen potential; MnO is the oxide that should be taken into account for the carbothermal obtaining of ferroalloys:

$$MnO(dis; slag) + C(s) \leftrightarrow Mn(dis; iron) + CO(g) \quad (2.18)$$

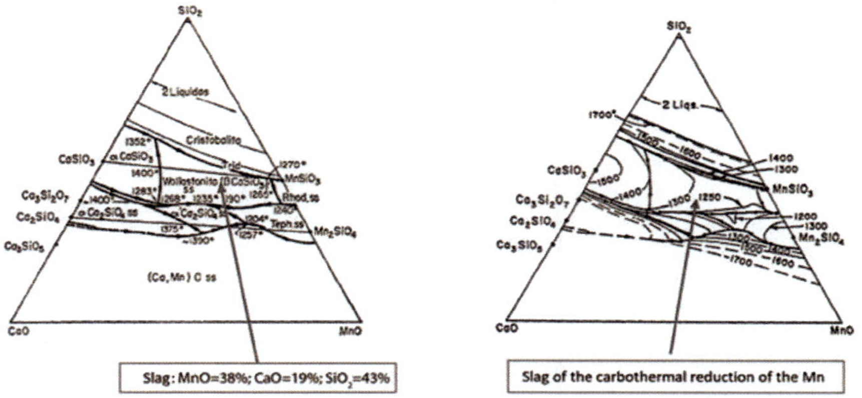

Fig. 2.4 Slag product of the carbothermal reduction of the manganese in the CaO–MnO–SiO_2 ternary system

Ferromanganese obtained by carbothermal reduction has high carbon content (8%C). The existing alternatives to obtain low carbon ferromanganese are:

(A) Treating high carbon ferromanganese with an oxidizing slag produced in the carbothermal reduction of pyrolusite, with the following pseudoternary composition (Fig. 2.4, adapted from Levin et al. 1969): 38% MnO, 19% CaO, and 43% SiO_2.

Thus, the carbon dissolved either in the ferroalloy or in the carbide of manganese (Mn_7C_3) dispersed in the metal react with the MnO of the slag and carbon is released in the form of CO:

$$\text{C(dis; iron)} + \text{MnO(dis; slag)} \leftrightarrow \text{Mn(dis; iron)} + \text{CO(g)} \tag{2.19}$$

$$\text{Mn}_7\text{C}_3\text{(s)} + 3\text{MnO(dis; slag)} \leftrightarrow 10\text{Mn(dis; iron)} + 3\text{CO(g)} \tag{2.20}$$

B. By metallothermic reduction of the manganese ore with Fe-35% Mn-40% Si.
C. Using injection of inert (Ar) or oxidizing (CO_2) gases in ferromanganese previously refined with oxidizing slags, as it is indicated by reactions (2.19) and (2.20). An oxidizing gas, as CO_2, decarburizes the metal according to the following reaction:

$$\text{C(dis; iron)} + \text{CO}_2\text{(g)} \leftrightarrow 2\text{CO(g)} \tag{2.21}$$

D. Using vacuum degasification techniques in ferromanganese refined by any of the previously mentioned mechanisms. The solubility of both carbon and manganese diminish the total pressure of the system:

$$\mathrm{C(dis;\,iron) + O(dis;\,iron) \leftrightarrow CO(g)} \tag{2.22}$$

Equilibrium (2.22) is displaced to the formation of carbon monoxide when the total pressure of the system is reduced.

One alternative to produce carbon-free manganese is the metallothermic reduction of MnO synthetically obtained from slags, powders (residues) and low manganese ores, or the electrolytic deposition by the decomposition of aqueous solutions of manganese sulphate.

The production of ferrochromium by carbothermic reduction of chromium oxide allows obtaining an alloy 70% chromium and 8% carbon. A significant part of the carbon can be liberated by oxidation of the metal with air at 2100 K:

$$\mathrm{Cr_2O_3(dis;\,slag) + 3C(dis;\,iron) \leftrightarrow 2Cr(dis;\,iron) + 3CO(g)} \tag{2.23}$$

Equally, it is possible to obtain ferrotitanium, ferrovanadium, ferromolybdenum, ferroniobium, and ferrotungsten by carbothermal reduction. Carbon content in titanium and vanadium ferroalloys is important as a consequence of the affinity of these metals by the carbon, while the affinity is smaller in the case of ferromolybdenum and ferrotungsten.

2.2.4 Ceramic Materials

The synthesis of metallic carbides of ceramic characteristics is performed via reduction with carbon of their corresponding refractory oxides:

$$\frac{2}{3}\mathrm{Al_2O_3(s) + 3C(s) \leftrightarrow} \frac{1}{3}\mathrm{Al_4C_3(s) + 2CO(g)} \tag{2.24}$$

$$\mathrm{CaO(s) + 3C(s) \leftrightarrow CaC_2(s) + CO(g)} \tag{2.25}$$

$$\mathrm{SiO_2(s) + 3C(s) \leftrightarrow SiC(s) + 2CO(g)} \tag{2.26}$$

Even more, a zone of high temperature around the electric arc furnace (>1800 °C) is necessary to produce silicon carbide, where silicon monoxide is formed:

$$\mathrm{Si(l) + SiO_2(s) \leftrightarrow 2SiO(g)} \tag{2.27}$$

Besides, there are other zones far from the previous one, at 1500 °C, toward where SiO moves to react with carbon (Sect. 2.2.3, production of ferrosilicon):

$$SiO(g) + 2C(s) \leftrightarrow SiC(c) + CO(g) \tag{2.28}$$

2.2.5 Solved Exercises

Exercise 2.1 An electric furnace slag, coming from the production of carbon steels ($0.02 \leq \%C \leq 0.80$), has the following chemical composition: 30.0% CaO, 5.0% MgO, 32.0% FeO; 5% MnO; 23.0% SiO_2; 1.50% P_2O_5; 0.5% Al_2O_3; 4.0% Fe_2O_3.Transform the chemical composition of this slag in a system of three components: CaO–FeO–SiO_2 to study it in the ternary diagram of Fig. 2.3. Moreover, calculate for the obtained pseudoternary slag: molecular weight of the slag, basicity, optical basicity, density, specific heat, and viscosity at 1650 °C. To calculate the above-mentioned properties, use the data collected in the exercises of the Chap. 1 (Exercise 1.12) for the BOF slag.

Solution Starts

To obtain the formulation of a pseudoternary slag from a slag formed by eight oxidized components, we take the three main oxides: CaO, FeO, and SiO_2, and around them we group those oxides with similar acid–basic characteristics. We considered the alternative of grouping B considered in the Exercise 1.12, which gave in that case a better approximation to the experimental data of the liquidus temperature. That is to say:

$$CaO = CaO + MgO = 35.0\% \tag{2.29}$$

$$FeO = FeO + MnO + Fe_2O_3 = 40.0\% \tag{2.30}$$

$$SiO_2 = SiO_2 + P_2O_5 + Al_2O_3 = 25.0\% \tag{2.31}$$

If we represent the obtained pseudoternary composition in the ternary of Fig. 2.5, we obtain a value of the liquidus temperature of around 1460 °C.

The molar fraction of the slag is calculated as follows:

$$x_{CaO} = \frac{35\text{ g CaO} \cdot \frac{1\text{ mol CaO}}{56.08\text{ g CaO}}}{35\text{ g CaO} \cdot \frac{1\text{ mol CaO}}{56.08\text{ g CaO}} + 40\text{ g FeO} \cdot \frac{1\text{ mol FeO}}{71.85\text{ g FeO}} + 25\text{ g SiO}_2 \cdot \frac{1\text{ mol SiO}_2}{60.09\text{ g SiO}_2}} = 0.3908 \tag{2.32}$$

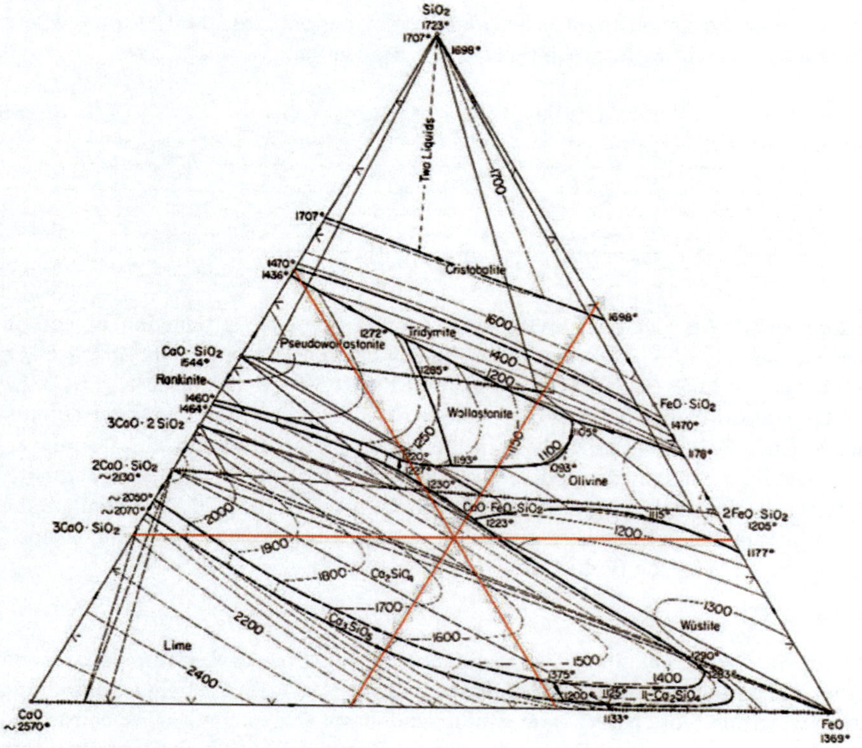

Fig. 2.5 Representation of the slag in the CaO–FeO–SiO$_2$ ternary diagram

$$x_{FeO} = \frac{40\ \text{g FeO} \cdot \frac{1\ \text{mol FeO}}{71.85\ \text{g FeO}}}{35\ \text{g CaO} \cdot \frac{1\ \text{mol CaO}}{56.08\ \text{g CaO}} + 40\ \text{g FeO} \cdot \frac{1\ \text{mol FeO}}{71.85\ \text{g FeO}} + 25\ \text{g SiO}_2 \cdot \frac{1\ \text{mol SiO}_2}{60.09\ \text{g SiO}_2}} = 0.3486 \tag{2.33}$$

$$x_{SiO_2} = \frac{25\ \text{g SiO}_2 \cdot \frac{1\ \text{mol SiO}_2}{60.09\ \text{g SiO}_2}}{35\ \text{g CaO} \cdot \frac{1\ \text{mol CaO}}{56.08\ \text{g CaO}} + 40\ \text{g FeO} \cdot \frac{1\ \text{mol FeO}}{71.85\ \text{g FeO}} + 25\ \text{g SiO}_2 \cdot \frac{1\ \text{mol SiO}_2}{60.09\ \text{g SiO}_2}} = 0.2605 \tag{2.34}$$

$$\begin{aligned} PM_{slag} &= x_{CaO} \cdot PM_{CaO} + x_{FeO} \cdot PM_{FeO} + x_{SiO_2} \cdot PM_{SiO_2} \\ &= 0.3908 \cdot 56.08 + 0.3486 \cdot 71.85 + 0.2605 \cdot 60.09 = 62.62\ \text{g} \cdot \text{mol}^{-1} \end{aligned} \tag{2.35}$$

Now, we calculate the basicity, first according to the $I_B(I)$ index:

$$I_B(I) = \frac{\%CaO}{\%SiO_2} = \frac{35}{25} = 1.40 \tag{2.36}$$

and then we calculate the optical basicity considering the data of the Appendix 4:

1. Number of moles in 100 g of slag:

$$\text{Moles} = \frac{35}{56.08} + \frac{40}{71.85} + \frac{25}{60.09} = 1.5969 \text{ moles in 100 grams of slag} \tag{2.37}$$

2. Composition expressed in molar fraction, of each one of the oxides of the slag. We have the molar fractions at the beginning of the exercise.
3. Summation of the product of the molar fraction of each oxide of the slag, x_{MeO}, and the number of oxygen atoms in each oxide, n(O):

$$\sum x_{MeO} \cdot n(O) = 0.3906 \cdot 1 + 0.3486 \cdot 1 + 0.2605 \cdot 2 = 1.2602 \tag{2.38}$$

4. We calculate X_{MeO}:

$$X_{CaO} = \frac{x_{CaO} \cdot n(O)}{\sum x_{MeO} \cdot n(O)} = \frac{0.3906 \cdot 1}{1.2602} = 0.3099 \tag{2.39}$$

$$X_{FeO} = \frac{x_{FeO} \cdot n(O)}{\sum x_{MeO} \cdot n(O)} = \frac{0.3486 \cdot 1}{1.2602} = 0.2766 \tag{2.40}$$

$$X_{SiO_2} = \frac{x_{SiO_2} \cdot n(O)}{\sum x_{MeO} \cdot n(O)} = \frac{0.2605 \cdot 2}{1.2602} = 0.4134 \tag{2.41}$$

5. Finally, we calculate the optical basicity of the slag, Λ_{slag} :

$$\begin{aligned} \Lambda_{slag} &= X_{CaO} \cdot \Lambda_{CaO} + X_{FeO} \cdot \Lambda_{FeO} + X_{SiO_2} \cdot \Lambda_{SiO_2} \\ &= 0.3099 \cdot 1 + 0.2766 \cdot 1 + 0.4134 \cdot 0.48 \\ &= 0.7849 \end{aligned} \tag{2.42}$$

We proceed as in the Exercise 1.12 to calculate the density of the slag of 35% CaO, 40% FeO, and 25% SiO_2 from the partial molar volumes of the three components at 1500 °C (Table 1.12):

$$\begin{aligned} V_m^{slag} &= \left[20.70 \cdot x_{CaO} + 15.8 \cdot x_{FeO} + \left(19.55 + 7.966 \cdot x_{SiO_2}\right) \cdot x_{SiO_2}\right] \cdot 10^{-6} \\ &= [20.70 \cdot 0.3906 + 15.8 \cdot 0.3486 + (19.55 + 7.966 \cdot 0.2605) \cdot 0.2605] \cdot 10^{-6} \\ &= 19.2266 \cdot 10^{-6} m^3 \cdot mol^{-1} \end{aligned} \tag{2.43}$$

The value of the molar volume of the slag at 1500 °C is $19.2266 \times 10^{-6}\ m^3\ mol^{-1}$. If we know that the variation of the molar volume of the slag, V_m^{slag}, with the temperature is equal to:

$$\left(\frac{dV_m^{slag}}{dT}\right) = 0.01 \tag{2.44}$$

the molar volume of the pseudoternary slag at 1650 °C is:

$$\begin{aligned} V_m^{slag}(1650\,°C) &= V_m^{slag}(1500\,°C) + 0.01 \cdot (1650°C - 1500°C) \\ &= 20.7266 \cdot 10^{-6} m^3 \cdot mol^{-1} \end{aligned} \tag{2.45}$$

Finally, the density of the slag at 1650 °C is:

$$\rho_{slag} = \frac{1}{V_m^{slag}} \cdot \frac{\overline{PM}_{slag}}{1} = \frac{mol}{20.7266 \cdot 10^{-6} m^3} \cdot \frac{62.62\ g}{mol} \cdot \frac{1\ kg}{1000\ g} = 3021.23 kg \cdot m^{-3} \tag{2.46}$$

The estimation of the specific heat at constant pressure, c_p^{slag}, at 1650 °C, would be:

$$\begin{aligned} c_p^{slag} &= (80.8 \cdot 0.3906 + 76.6 \cdot 0.3486 + 87.0 \cdot 0.2605) J \cdot mol^{-1} \cdot K^{-1} \\ &= 80.9267 J \cdot mol^{-1} \cdot K^{-1} \end{aligned} \tag{2.47}$$

While the density of the slags varies with the temperature, the estimations about the specific heat are valid in a wide interval of temperature. If we consider that the average molecular weight of the slag, PM_{slag}, is 62.62 g mol^{-1}, the value of the specific heat of the slag at 1650 °C would be:

$$\begin{aligned} c_p^{slag}\left(1650^\circ C\right) &= c_p^{slag}\left(1500^\circ C\right) \cdot \frac{1}{\overline{PM}_{slag}} \cdot \frac{1000\ g}{1\ kg} = 80.9267 \frac{J}{mol \cdot K} \cdot \frac{1\ mol}{62.62\ g} \cdot \frac{1000\ g}{1\ kg} \\ &= 1292.35 \frac{J}{kg \cdot K} \end{aligned} \tag{2.48}$$

Finally, for the viscosity, we use the Arrhenius equation indicated in the enunciation (Eq. (1.284)):

$$\mu = A \cdot T \cdot \exp\left(\frac{B}{T}\right) \tag{2.49}$$

where A and B are:

Table 2.2 Composition of the slag expressed as molar fraction

x_{SiO_2}	$x_{P_2O_5}$	$x_{Al_2O_3}$	$x_{Fe_2O_3}$	x_{CaO}	x_{MgO}	x_{FeO}	x_{MnO}
0.2416	0.0067	0.0031	0.0158	0.3377	0.0783	0.2812	0.0356

$$\ln A = -19.81 + 1.73 \cdot x_{CaO} = -19.81 + 1.73 \cdot 0.3906$$
$$= -19.1343 \rightarrow A = 4.8989 \cdot 10^{-9} \quad (2.50)$$

$$B = 31140 - 23896 \cdot x_{CaO} = 31140 - 23896 \cdot 0.3906 = 21806.2 \quad (2.51)$$

Consequently, the value of the viscosity of the pseudoternary slag for the alternative B, at 1650 °C, is:

$$\mu = 4.8989 \cdot 10^{-9} \cdot (1650 + 273) \cdot \exp\left(\frac{21806.2}{1650 + 273}\right) = 0.7922\ \text{Pa} \cdot \text{s} \quad (2.52)$$

Commentaries : As it was already pointed out in the Exercise 1.12, the conversion of one slag with eight components (30.0% CaO; 5.0% MgO; 32.0%FeO; 5% MnO; 23.0% SiO_2; 1.50% P_2O_5; 0.50% Al_2O_3; 4.0%Fe_2O_3) in a system of three components have as advantage the simplification of the calculations and also an academic component. In the estimation of some of the properties, predictions are adjusted to the real values but in other cases, the prediction is not sufficiently accurate. Even, as it would be later checked, the system of eight components does not provide sufficient accurateness, especially in the case of the value of the viscosity at 1650 °C. We give the results for the physical–chemical characteristics to evaluate the differences between the slag of eight components and the slag of three components:

The composition of the slag expressed as molar fraction is collected in Table 2.2.

The molecular weight of the slag is calculated as follows:

$$\begin{aligned} PM_{slag} &= x_{CaO} \cdot PM_{CaO} + x_{FeO} \cdot PM_{FeO} + x_{SiO_2} \cdot PM_{SiO_2} + x_{P_2O_5} \cdot PM_{P_2O_5} \\ &\quad + x_{Al_2O_3} \cdot PM_{Al_2O_3} + x_{Fe_2O_3} \cdot PM_{Fe_2O_3} + x_{MgO} \cdot PM_{MgO} + x_{MnO} \cdot PM_{MnO} \\ &= 0.3377 \cdot 56.08 + 0.2812 \cdot 71.85 + 0.2416 \cdot 60.09 + 0.0067 \cdot 141.94 \\ &\quad + 0.0031 \cdot 101.96 + 0.0158 \cdot 159.69 + 0.0783 \cdot 40.3 + 0.0356 \cdot 70.94 \\ &= 63.13\ \text{g} \cdot \text{mol}^{-1} \end{aligned} \quad (2.53)$$

Now, we calculate the basicity, first according to the $I_B(I)$ index:

$$I_B(I) = \frac{\%CaO}{\%SiO_2} = \frac{30}{23} = 1.30 \quad (2.54)$$

and then we calculate the optical basicity considering the data of the Appendix 4:

1. Number of moles in 100 g of slag:

$$\begin{aligned}\text{Moles} &= \frac{30}{56.08} + \frac{32}{71.85} + \frac{23}{60.09} + \frac{1.5}{141.94} + \frac{0.5}{101.96} + \frac{4.0}{159.69} \\ &\quad + \frac{5}{40.3} + \frac{5}{70.94} \\ &= 1.5981 \text{ moles in 100 grams of slag}\end{aligned} \tag{2.55}$$

2. Composition expressed in molar fraction, of each one of the oxides of the slag. We have the molar fractions in Table 2.2.
3. Summation of the product of the molar fraction of each oxide of the slag, x_{MeO}, and the number of oxygen atoms in each oxide, n(O):

$$\begin{aligned}\sum x_{MeO} \cdot n(O) &= 0.3377 \cdot 1 + 0.2812 \cdot 1 + 0.2416 \cdot 2 + 0.0067 \cdot 5 + 0.0031 \cdot 3 \\ &\quad + 0.0158 \cdot 3 + 0.0783 \cdot 1 + 0.0356 \cdot 1 = 1.3062\end{aligned} \tag{2.56}$$

4. We calculate X_{MeO}:

$$X_{CaO} = \frac{x_{CaO} \cdot n(O)}{\sum x_{MeO} \cdot n(O)} = \frac{0.3377 \cdot 1}{1.3062} = 0.2585 \tag{2.57}$$

$$X_{FeO} = \frac{x_{FeO} \cdot n(O)}{\sum x_{MeO} \cdot n(O)} = \frac{0.2812 \cdot 1}{1.3062} = 0.2153 \tag{2.58}$$

$$X_{SiO_2} = \frac{x_{SiO_2} \cdot n(O)}{\sum x_{MeO} \cdot n(O)} = \frac{0.2416 \cdot 2}{1.3062} = 0.3699 \tag{2.59}$$

$$X_{P_2O_5} = \frac{x_{P_2O_5} \cdot n(O)}{\sum x_{MeO} \cdot n(O)} = \frac{0.0067 \cdot 5}{1.3062} = 0.0256 \tag{2.60}$$

$$X_{Al_2O_3} = \frac{x_{Al_2O_3} \cdot n(O)}{\sum x_{MeO} \cdot n(O)} = \frac{0.0031 \cdot 3}{1.3062} = 0.0071 \tag{2.61}$$

$$X_{Fe_2O_3} = \frac{x_{Fe_2O_3} \cdot n(O)}{\sum x_{MeO} \cdot n(O)} = \frac{0.0158 \cdot 3}{1.3062} = 0.0363 \tag{2.62}$$

$$X_{MgO} = \frac{x_{MgO} \cdot n(O)}{\sum x_{MeO} \cdot n(O)} = \frac{0.0783 \cdot 1}{1.3062} = 0.0599 \tag{2.63}$$

$$X_{MnO} = \frac{x_{MnO} \cdot n(O)}{\sum x_{MeO} \cdot n(O)} = \frac{0.0356 \cdot 1}{1.3062} = 0.0273 \tag{2.64}$$

5. Finally, we calculate the optical basicity of the slag, Λ_{slag} :

$$\begin{aligned}\Lambda_{slag} &= X_{CaO}\cdot\Lambda_{CaO} + X_{FeO}\cdot\Lambda_{FeO} + X_{SiO_2}\cdot\Lambda_{SiO_2} + X_{P_2O_5}\cdot\Lambda_{P_2O_5} + X_{Al_2O_3}\cdot\Lambda_{Al_2O_3}\\ &\quad + X_{Fe_2O_3}\cdot\Lambda_{Fe_2O_3} + X_{MgO}\cdot\Lambda_{MgO} + X_{MnO}\cdot\Lambda_{MnO}\\ &= 0.2585\cdot 1 + 0.2153\cdot 1 + 0.3699\cdot 0.48 + 0.0256\cdot 0.40 + 0.0071\cdot 0.608\\ &\quad + 0.0363\cdot 0.80 + 0.0599\cdot 0.865 + 0.0273\cdot 0.98 = 0.7735 \end{aligned} \tag{2.65}$$

the molar volume of the pseudoternary slag at 1500 °C is:

$$\begin{aligned}V_m^{slag}\left(1500^\circ C\right) &= V_m^{SiO_2}\cdot x_{SiO_2} + V_m^{P_2O_5}\cdot x_{P_2O_5} + V_m^{Al_2O_3}\cdot x_{Al_2O_3} + V_m^{Fe_2O_3}\cdot x_{Fe_2O_3}\\ &\quad + V_m^{CaO}\cdot x_{CaO} + V_m^{MgO}\cdot x_{MgO} + V_m^{FeO}\cdot x_{FeO} + V_m^{MnO}\cdot x_{MnO}\\ &= \left(19.55 + 7.966\cdot x_{SiO_2}\right)\cdot x_{SiO_2} + (65.7)\cdot x_{P_2O_5}\\ &\quad + \left(28.31 + 31.32\cdot x_{Al_2O_3} - 31.45\cdot x_{Al_2O_3}^2\right)\cdot x_{Al_2O_3} + (38.4)\cdot x_{Fe_2O_3}\\ &\quad + (20.7)\cdot x_{CaO} + (16.1)\cdot x_{MgO} + (15.8)\cdot x_{FeO} + (15.6)\cdot x_{MnO}\\ &= (19.55 + 7.966\cdot 0.2416)\cdot 0.2416 + (65.7)\cdot 0.0067\\ &\quad + \left(28.31 + 31.32\cdot 0.0031 - 31.45\cdot 0.0031^2\right)\cdot 0.0031 + (38.4)\cdot 0.0158\\ &\quad + (20.7)\cdot 0.3377 + (16.1)\cdot 0.0783 + (15.8)\cdot 0.2812 + (15.6)\cdot 0.0356\\ &= 19.5726\cdot 10^{-6}\mathrm{m}^3\cdot\mathrm{mol}^{-1} \end{aligned} \tag{2.66}$$

at 1650 °C is:

$$\begin{aligned}V_m^{slag}(1650^\circ C) &= V_m^{slag}(1500^\circ C) + 0.01\cdot(1650^\circ C - 1500^\circ C)\\ &= \left[19.5726 + 0.01\cdot\left(1650^\circ C - 1500^\circ C\right)\right]\cdot 10^{-6}\mathrm{m}^3\cdot\mathrm{mol}^{-1}\\ &= 21.0726\cdot 10^{-6}\mathrm{m}^3\cdot\mathrm{mol}^{-1} \end{aligned} \tag{2.67}$$

Finally, the density of the slag at 1650 °C is:

$$\rho_{slag} = \frac{1}{V_m^{slag}}\cdot\frac{PM_{slag}}{1} = \frac{\mathrm{mol}}{21.0726\cdot 10^{-6}\mathrm{m}^3}\cdot\frac{63.13\ \mathrm{g}}{\mathrm{mol}}\cdot\frac{1\ \mathrm{kg}}{1000\ \mathrm{g}} = 2995.84\ \mathrm{kg}\cdot\mathrm{m}^{-3} \tag{2.68}$$

The estimation of the specific heat at constant pressure, c_p^{slag}, at 1650 °C, would be:

$$\begin{aligned}c_p^{slag} &= c_p^{SiO_2}\cdot x_{SiO_2} + c_p^{P_2O_5}\cdot x_{P_2O_5} + c_p^{Al_2O_3}\cdot x_{Al_2O_3} + c_p^{Fe_2O_3}\cdot x_{Fe_2O_3} + c_p^{CaO}\cdot x_{CaO}\\ &\quad + c_p^{MgO}\cdot x_{MgO} + c_p^{FeO}\cdot x_{FeO} + c_p^{MnO}\cdot x_{MnO}\\ &= 87\cdot 0.2416 + 242.7\cdot 0.0067 + 146.4\cdot 0.0031 + 191.2\cdot 0.0158\\ &\quad + 80.8\cdot 0.3377 + 90.4\cdot 0.0783 + 76.6\cdot 0.2812 + 79.9\cdot 0.0356\\ &= 84.8689\ \mathrm{J}\cdot\mathrm{mol}^{-1}\cdot\mathrm{K}^{-1} \end{aligned} \tag{2.69}$$

While the density of the slags varies with the temperature, estimations about specific heat are valid in a wide interval of temperature. If we consider that the average molecular weight of the slag, PM_{slag}, is 63.13 g mol^{-1}, the value of the specific heat of the slag at 1650 °C would be:

$$\begin{aligned} c_p^{slag}(1650^\circ C) &= c_p^{slag}(1500^\circ C) \cdot \frac{1}{PM_{slag}} \cdot \frac{1000\ g}{1\ kg} \\ &= 84.8689 \frac{J}{mol \cdot K} \cdot \frac{1\ mol}{63.13\ g} \cdot \frac{1000\ g}{1\ kg} = 1344.35 \frac{J}{kg \cdot K} \end{aligned} \tag{2.70}$$

Finally, for the viscosity, we use the Arrhenius equation indicated in the enunciation (Eq. (1.284)):

$$\mu = A \cdot T \cdot \exp\left(\frac{B}{T}\right) \tag{2.71}$$

where A and B are:

$$\begin{aligned} \ln A &= -19.81 + 1.73 \cdot x_{CaO} - 33.76 \cdot x_{Al_2O_3} \\ &= -19.81 + 1.73 \cdot 0.3377 - 33.76 \cdot 0.0031 \\ &= -19.3304 \rightarrow A = 4.0262 \cdot 10^{-9} \end{aligned} \tag{2.72}$$

$$\begin{aligned} B &= 31140 - 23896 \cdot x_{CaO} + 68833 \cdot x_{Al_2O_3} \\ &= 31140 - 23896 \cdot 0.3377 + 68833 \cdot 0.0031 \\ &= 23283.70 \end{aligned} \tag{2.73}$$

Consequently, the value of the viscosity of the pseudoternary slag, at 1650 °C, is:

$$\mu = 4.0262 \cdot 10^{-9} \cdot (1650 + 273) \cdot \exp\left(\frac{23283.70}{1650 + 273}\right) = 1.4038\ Pa \cdot s \tag{2.74}$$

As it is possible to check, apart from the differences in the optical basicity, the values of the viscosity vary from 0.8 Pa s in the simplified slag to 1.4038 Pa s in the real slag. However, we did not consider the presence in the eight components slag of two phases that have significant influence on the value of the viscosity: CaF_2 and Na_2O. Habitually, both LD (Linz-Donawitz) slag and EF (Electric Furnace) slag usually contain these two phases with the aim of reducing the liquidus temperature and activating the kinetics of the refinement reactions. The value of the viscosity has decisively influence in the kinetics of the mass transfer processes between the metal and the slag.

We finally indicate that the value of the viscosity obtained at 1650 °C for a slag with the following composition, considering the model of Riboud: 33.0% CaO, 2.0%

MgO, 32.0% FeO, 23.0% SiO_2, 1.50% P_2O_5, 0.50% Al_2O_3, 4.0% Fe_2O_3, 3.0% CaF_2, 1% Na_2O is 0.508 Pa s. Equally, the corresponding pseudoternary slag, where the 4% of FeO was replaced by the addition of calcium fluoride and sodium oxide, to finally obtain a system of five components: 35% CaO, 36% FeO, 25% SiO_2, 3.0% CaF_2, and 1.0% Na_2O, has a viscosity at 1650 °C of 0.409 Pa s. It is possible to see that there are several active components that have noticeably influence in the calculation of the viscosity.

Solution ends.

Exercise 2.2 Analyze the solidification process of a pseudoternary slag of electric furnace (that was characterized in the Exercise 2.1) of the following chemical composition: 35.0% CaO, 40% FeO, and 25.0% SiO_2, represented in the ternary diagram of Fig. 2.5.

Solution Starts

As we have seen in the Exercise 2.1, liquidus temperature is around 1460 °C (Fig. 2.5) and the first solid that precipitates as a result of the cooling of the melt is dicalcium silicate, Ca_2SiO_4 (primary field of crystallization of the electric furnace slag).

During the interval of temperatures comprised within 1460 and 1250 °C, only Ca_2SiO_4 precipitates (Fig. 2.6). Only when we cold the system below 1250 °C, the precipitation of FeO starts, together with Ca_2SiO_4 (which began to precipitate at 1460 °C, liquidus temperature of the slag). To obtain the two extremes of the "horizontal-tie" line, it is necessary to draw a line that joins the point corresponding to the global composition of the slag and the point corresponding to the Ca_2SiO_4 (located in the side CaO–SiO_2 of the triangle). When this line is prolongated, from the global point of the slag, until crossing the thick line that delimitates the primary fields of crystallization of Ca_2SiO_4 and FeO, we obtain the phases in equilibrium (a solid and a liquid), compatible with the global composition of the slag at 1250 °C.

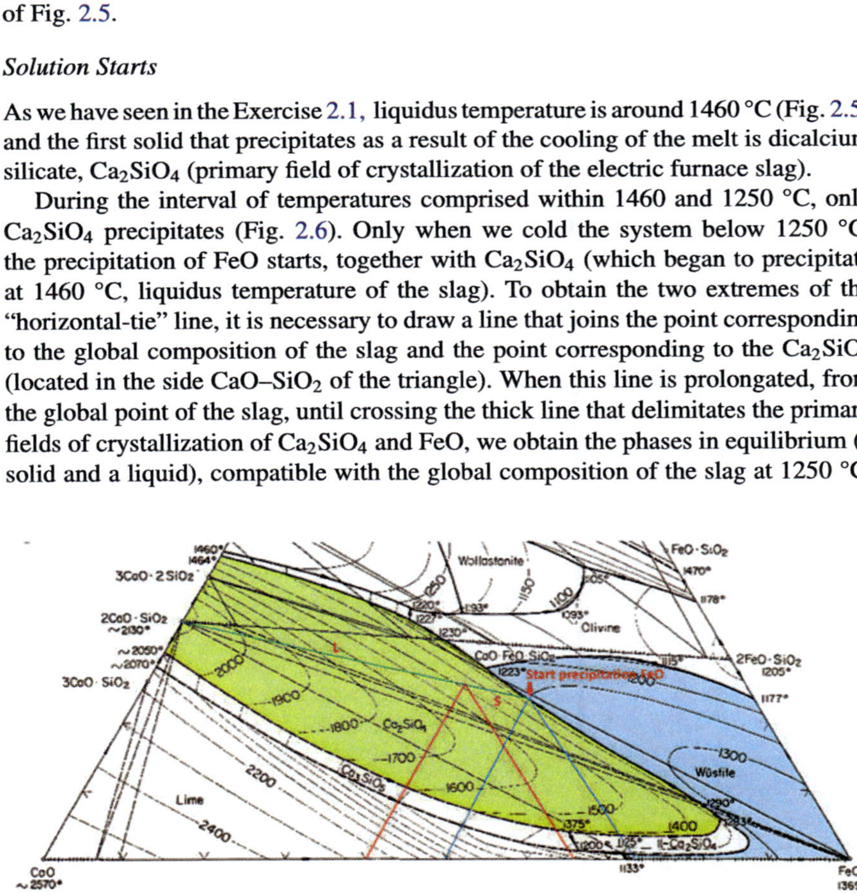

Fig. 2.6 Representation of the start of the FeO precipitation

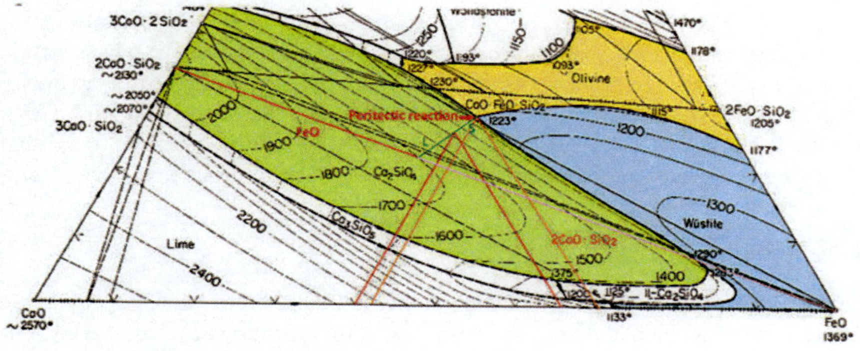

Fig. 2.7 Representation of the situation when the peritectic reaction takes place

The composition of the melt in equilibrium with the Ca_2SiO_4 at 1250 °C would be: 28.5% CaO, 48.5% FeO, 23.0% SiO_2 (Fig. 2.6). The quantity of solid that precipitated above 1250 °C would be 18%, while there are still 82% of liquid.

As the cooling progresses, an invariant point is reached at 1223 °C, which is consequence of the following first or second type of peritectic reaction. This peritectic reaction generates a new solid of variable stoichiometric characteristics, olivine:

$$FeO(s) + Ca_2SiO_4(s) + liquid \leftrightarrow Ca_xFe_ySiO_{(2+x+y)}(s) \tag{2.75}$$

where the composition of the peritectic liquid would be: 31.5% CaO, 41.5% FeO, and 27% SiO_2 (Fig. 2.7). Before the peritectic reaction, there is 61% of liquid in equilibrium with 39% of solid (formed by 62% Ca_2SiO_4 and 38% FeO) (Fig. 2.7).

When the solidification-cooling process of the melt progresses below 1223 °C, the new solid phase, olivine, which had appeared surrounding the grains of Ca_2SiO_4 and FeO in the peritectic reaction, dissolves again in the melt and disappears. The liquid phase in the cooling below 1223 °C would progress along the thick curve that delimitates the primary fields of crystallization of the olivine and wustite, FeO. Consequently, the liquid would impoverish in CaO and enrich in FeO: the solids that precipitate from the melt, below 1223 °C, present greater quantities of Ca_2SiO_4 than that of FeO.

Finally, at the solidus temperature of 1177 °C (Fig. 2.8), the solidification of the electric furnace slag concludes. The last liquid in solidifying would be free of CaO, and would produce a new solid phase, fayalite, Fe_2SiO_4. Before the solidification of the fayalite, at 1177 °C there is 28% of liquid in equilibrium with 72% of solid (this solid is formed by 76% Ca_2SiO_4 and 24% FeO).

Solidified slag, according to the thermodynamic equilibrium, has a mineralogical composition compatible with the elemental composition of 35% CaO, 40% FeO, and 25% SiO_2, represented in the compatibility triangle (Ca_2SiO_4–Fe_2SiO_4–FeO) (Fig. 2.9): 53% Ca_2SiO_4, 24% Fe_2SiO_4, and 23% FeO.

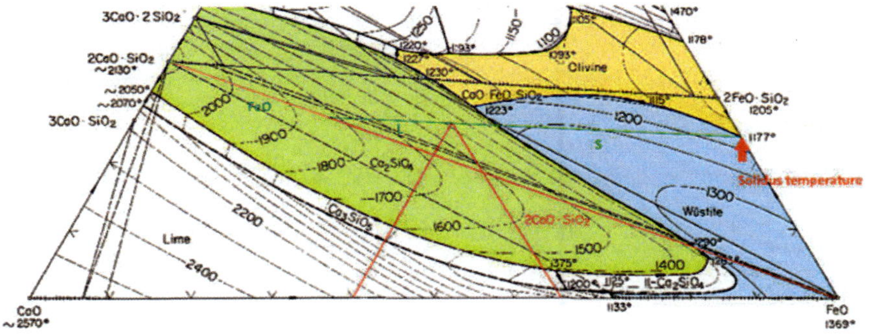

Fig. 2.8 Representation of the situation when the solidus temperature is reached

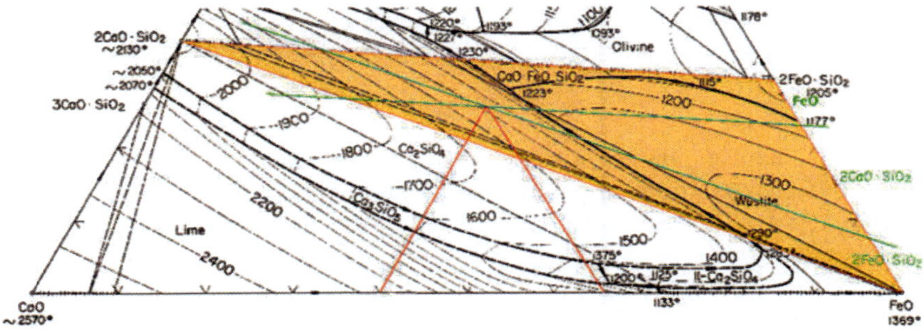

Fig. 2.9 Percentage of the different phases at room temperature deduced from the Ca_2SiO_4–Fe_2SiO_4–FeO compatibility triangle

Commentaries: From the structural point of view, the microscopic observation of a polished layer of slag would reflect the participation of some primary constituents of Ca_2SiO_4 and FeO inside of a matrix (28%) formed, almost all, by fayalite.

Habitually, the industrial cooling is fast with the aim of propitiating the transport-storage (cooling with water) and the structures of the compounds that are formed are usually few ordered-crystallized (low crystallinity in the X-ray diffraction).

Concentrated solar thermal energy could be an instrument-technology to stabilize the magnetite, Fe_3O_4, formed at around 1500 °C from wustite and fayalite. Equally, the treatment of these slags with concentrated solar thermal energy would stabilize the CaO in the form of calcium silicates and a vitreous phase of high SiO_2 content, where the nonvolatile impurities could be concentrated (Fernández-González et al. 2018a, b, c, 2019a, b, c).

Solution ends.

Exercise 2.3 Assuming applicable the ionic model of Temkin to the following pseudoternary slag of electric furnace: 35% CaO, 40% FeO, 25% SiO_2 (Exercise 2.1).

Calculate the activity of both calcium oxide and ferrous oxide, as well as their respective coefficients of activity. The molecular weights of the three components of the slag, CaO, FeO, and SiO_2 are 56.08, 71.85, and 60.09 g mol^{-1}, respectively.

Solution Starts

In 100 g of slag, the number of moles of CaO, FeO, and SiO_2 are:

$$n_{CaO} = \frac{35\ g\ CaO}{100\ g\ slag} \cdot \frac{1\ mol\ CaO}{56.08\ g\ CaO} = 0.6241\ moles\ CaO\ in\ 100\ grams\ of\ slag \quad (2.76)$$

$$n_{FeO} = \frac{40\ g\ FeO}{100\ g\ slag} \cdot \frac{1\ mol\ FeO}{71.85\ g\ FeO} = 0.5567\ moles\ FeO\ in\ 100\ grams\ of\ slag \quad (2.77)$$

$$n_{SiO_2} = \frac{40\ g\ SiO_2}{100\ g\ slag} \cdot \frac{1\ mol\ SiO_2}{60.09\ g\ SiO_2} = 0.4160\ moles\ SiO_2\ in\ 100\ grams\ of\ slag \quad (2.78)$$

The total number of moles in 100 g of slag is 1.5969 mol.

The composition of the slag, expressed in molar fraction, is: $x_{CaO} = 0.3908$; $x_{FeO} = 0.3486$; $x_{SiO_2} = 0.2605$, and the molecular weight of the slag, PM_{slag}, is 62.62 g mol^{-1}.

Proceeding as in the Exercise 1.3, as the ideal mixture of anions and cations that form the slag is the same: calcium and iron (II) cations and oxygen (O^{2-}) and silicate ($SiO_4{}^{4-}$) anions; assuming the same reactions of total ionization, of supply of oxygen anions and consumption of oxygen anions, the balance of the anions-cations equivalents, in 100 g of electric furnace slag, is the following one:

—Balance of cations:

Calcium cations:

$$2 \cdot n_{CaO} = 2 \cdot 0.6241 = 1.2482\ g - equivalent\ of\ Ca^{2+}(f) \quad (2.79)$$

Iron cations:

$$2 \cdot n_{FeO} = 2 \cdot 0.5567 = 1.1134\ g - equivalent\ of\ Fe^{2+}(f) \quad (2.80)$$

Total g-equivalents of cations:

$$1.2482\ g - equivalent\ of\ Ca^{2+}(f) + 1.1134\ g - equivalent\ of\ Fe^{2+}(f) = 2.3616 \quad (2.81)$$

—Balance of anions:

Silicate anions:

$$4 \cdot n_{SiO_2} = 4 \cdot 0.4160 = 1.6642 \text{ g} - \text{equivalent of } SiO_4^{4-}(f) \tag{2.82}$$

Oxygen anions:

$$\begin{aligned} 2 \cdot \left[(n_{CaO} + n_{FeO}) - 2 \cdot n_{SiO_2}\right] &= 2 \cdot [(0.6241 + 0.5567) - 2 \cdot 0.4160] \\ &= 0.6974 \text{ g} - \text{equivalent of } O^{2-}(f) \end{aligned} \tag{2.83}$$

Total g-equivalents of anions:

$$1.6642 \text{ g} - \text{equivalent of } SiO_4^{4-}(f) + 0.6974 \text{ g} - \text{equivalent of } O^{2-}(f) = 2.3616 \tag{2.84}$$

According to Temkin, the activity of the CaO and FeO is equal to the fraction of g-equivalent of cations, calcium or iron (II), in the ideal mixture of cations multiplied by the fraction of g-equivalent of anions oxygen in the ideal mixture of anions:

$$a_{CaO} = N_{e,Ca^{2+}} \cdot N_{e,O^{2-}} = \left(\frac{1.2482}{2.3616}\right) \cdot \left(\frac{0.6974}{2.3616}\right) = 0.1561 \tag{2.85}$$

$$a_{FeO} = N_{e,Fe^{2+}} \cdot N_{e,O^{2-}} = \left(\frac{1.1134}{2.3616}\right) \cdot \left(\frac{0.6974}{2.3616}\right) = 0.1392 \tag{2.86}$$

The values of the coefficients of activity would be:

$$\gamma_{CaO} = \frac{a_{CaO}}{x_{CaO}} = \frac{0.1561}{0.3908} = 0.3994 \tag{2.87}$$

$$\gamma_{FeO} = \frac{a_{FeO}}{x_{FeO}} = \frac{0.1392}{0.3486} = 0.3993 \tag{2.88}$$

Commentaries: According to Coudurier et al. (1985, pp. 242–243), the activity of FeO and CaO at 1600 °C for the electric furnace slag are:

$$a_{CaO} = 7.08 \cdot 10^{-2} \tag{2.89}$$

$$a_{FeO} = 0.72 \tag{2.90}$$

These values are very far from those previously calculated using the model of Temkin. The ionic model of Temkin is acceptable in the following two situations:

- High level of ionization of the slag without enhancing the affinity for the formation of chemical compounds.
- The utilization of the slag at high temperatures, where the ionization of the atomic groupings that are more stable at low temperatures can be favored.

End of the exercise.

Exercise 2.4 Calculate the activity of the CaO and FeO in a pseudoternary slag of electric furnace that has the following composition: 35% CaO, 40% FeO, and 25% SiO_2 at a temperature of 1650 °C. Use the molecular model of Schneck in the following situations:

Situation A: Assume that the melt is formed by the following compounds: Ca_2SiO_4 (2CS), $CaSiO_3$ (CS), FeO, and CaO.

Situation B: Assume that there are the following compounds in the melt: Ca_2SiO_4 (2CS), FeO, CaO, SiO_2.

Use the data of the Exercise 2.3 to make the corresponding mass balances.

Solution Starts

The working hypothesis of Schenck is considering the slag as an ideal homogeneous mixture of molecules, Raoult's law, where it is possible to find both elemental oxidized compounds and binary or ternary combinations of oxides to form chemical compounds. Consequently, if we calculate the corresponding molar fraction of each one of the molecular species of the melt, the activity of each one of the species is identified with their molar fraction.

Considering the working hypothesis A, the balance in moles for any quantity used as reference or base of calculus (for instance, 100 g of slag) would be:

$$n_{CaO} = n^*_{CaO} + n_{CS} + 2 \cdot n_{2CS} \tag{2.91}$$

$$n_{SiO_2} = n_{CS} + 2 \cdot n_{2CS} \tag{2.92}$$

$$n_{FeO} = n^*_{FeO} \tag{2.93}$$

$$n_T = n_{CaO}{}^* + n_{FeO}{}^* + n_{CS} + n_{2CS} \tag{2.94}$$

On the other hand, the relation between the number of moles of the molecular species 2CS and CS is calculated by means of the following thermodynamic equilibrium:

$$Ca_2SiO_4(s) \leftrightarrow CaO(s) + CaSiO_3(s) \tag{2.95}$$

$$K_{reaction(2.95)} = \frac{{n_{CaO}}^{*} \cdot n_{CS}}{n_{2CS} \cdot n_T} = \frac{x_{CaO} \cdot x_{CaSiO_3}}{x_{Ca_2SiO_4}} = \frac{\left(\frac{n_{CaO}}{n_T}\right) \cdot \left(\frac{n_{CaSiO_3}}{n_T}\right)}{\left(\frac{n_{Ca_2SiO_4}}{n_T}\right)} \quad (2.96)$$

The data of the standard free energy showed in Appendix 1 correspond to the values of the standard energy of formation of the 2CS and CS:

$$2CaO(s) + SiO_2(s) \leftrightarrow Ca_2SiO_4(s) \quad (2.97)$$

$$CaO(s) + SiO_2(s) \leftrightarrow CaSiO_3(s) \quad (2.98)$$

$$\Delta_r G^0_{reaction(2.97)}(T) = -103.0 - 0.020 \cdot T \left[kJ \cdot (mol\ 2CS)^{-1}\right] \quad (2.99)$$

$$\Delta_r G^0_{reaction(2.98)}(T) = -88.7 + 0.001 \cdot T\left[kJ \cdot (mol\ CS)^{-1}\right] \quad (2.100)$$

Concretely, the variation of the standard free energy, Appendix 1, and the equilibrium constant of the reaction (2.95) at 1650 °C are:

$$\begin{aligned}\Delta_r G^0_{reaction(2.95)}(T) &= -\Delta_r G^0_{reaction(2.97)}(T) + \Delta_r G^0_{reaction(2.98)}(T)\\ &= 14.3 - 0.021 \cdot T\left[kJ \cdot (mol\ CaO)^{-1}\right]\end{aligned} \quad (2.101)$$

$$\Delta_r G^0_{reaction(2.95)}(1650\,°C) = 54.683\left[kJ \cdot (mol\ CaO)^{-1}\right] \quad (2.102)$$

$$\begin{aligned}\Delta_r G^0_{reaction(2.95)}(1650\,°C) &= 54.683\left[kJ \cdot (mol\ CaO)^{-1}\right]\\ &= -R \cdot T \cdot \ln K_{reaction(2.95)} \rightarrow K_{reaction(2.95)} = 3.271 \cdot 10^{-2}\end{aligned} \quad (2.103)$$

We consider that in all the equations, n_{CaO}, n_{FeO}, and n_{SiO_2} are the number of moles of calcium oxide, iron oxide, and silicon oxide in 100 g of slag of the above-indicated chemical composition (35% CaO, 40% FeO, and 25% SiO_2):

$$n_{CaO} = \frac{35\ g\ CaO}{100\ g\ slag} \cdot \frac{1\ mol\ CaO}{56.08\ g\ CaO} = 0.6241 \quad (2.104)$$

$$n_{FeO} = \frac{40\ g\ FeO}{100\ g\ slag} \cdot \frac{1\ mol\ FeO}{71.85\ g\ FeO} = 0.5567 \quad (2.105)$$

$$n_{SiO_2} = \frac{25\ g\ SiO_2}{100\ g\ slag} \cdot \frac{1\ mol\ SiO_2}{60.09\ g\ SiO_2} = 0.4160 \quad (2.106)$$

where the variables that must be calculated using the molecular model: $n_{CaO}{}^*$, $n_{FeO}{}^*$, n_{2CS}, and n_{CS} are, respectively, the number of moles in 100 g of melt of the following species: free CaO (non-combined to form silicate), FeO, Ca_2SiO_4, and $CaSiO_3$. Finally, n_T is the total number of CaO, FeO, 2CS, and CS moles in 100 g of slag according to the molecular model of Schenck. The considered problem has four equations and four unknowns: $n_{CaO}{}^*$, n_{2CS}, n_{CS}, and n_T. The values that are known are:

$$n_{CaO} = 0.6241 \quad (2.107)$$

$$n_{FeO} = n_{FeO}{}^* = 0.5567 \quad (2.108)$$

$$n_{SiO_2} = 0.4160 \quad (2.109)$$

We must express all the variables involved in the equilibrium constant of the reaction (2.95) (n_{2CS}, n_{CS}, and n_T) as a function of $n_{CaO}{}^*$. We obtain the following quadratic equation:

$$a \cdot \left(n_{CaO}{}^*\right)^2 + b \cdot n_{CaO}{}^* + c = 0 \quad (2.110)$$

where:

$$a = 1 + K_{reaction(2.95)} = 1.033 \quad (2.111)$$

$$b = 2 \cdot n_{SiO_2} - n_{CaO} + n_{SiO_2} \cdot K_{reaction(2.95)} + d \cdot K_{reaction(2.95)} - n_{CaO} \cdot K_{reaction(2.95)} = 0.2330 \quad (2.112)$$

$$c = D \cdot n_{SiO_2} \cdot K_{reaction(2.95)} - D \cdot n_{CaO} \cdot K_{reaction(2.95)} = -6.620 \cdot 10^{-3} \quad (2.113)$$

and:

$$D = n_{FeO} + n_{SiO_2} \quad (2.114)$$

The solution to the quadratic equation allows obtaining the rest of the variables of the problem:

$$n_{CaO}{}^* = 2.553 \cdot 10^{-2} \quad (2.115)$$

$$n_{2CS} = 0.1825 \quad (2.116)$$

$$n_{CS} = 0.2335 \quad (2.117)$$

$$n_T = 0.9983 \quad (2.118)$$

Accordingly, the activities of the calcium oxide and iron (II) oxide are identified with the molar fraction of these oxides in the mixture:

$$a_{CaO} = x_{CaO} = \frac{n_{CaO}{}^*}{n_T} = \frac{2.553 \cdot 10^{-3}}{0.9983} = 0.0256 \quad (2.119)$$

$$a_{FeO} = x_{FeO} = \frac{n_{FeO}{}^*}{n_T} = \frac{0.5567}{0.9983} = 0.56 \quad (2.120)$$

Situation B implies that in the melt there is only a complex molecular specie, 2CaO SiO_2, formed by the linear combination of some of the three components-elemental oxides (CaO, FeO, SiO_2) of the system. The ensemble of equations is, in this case, the following one:

$$n_{CaO} = n_{CaO}{}^* + n_{CS} + 2 \cdot n_{2CS} \quad (2.121)$$

$$n_{SiO_2} = n_{SiO_2}{}^* + n_{2CS} \quad (2.122)$$

$$n_{FeO} = n^*_{FeO} \quad (2.123)$$

$$n_T = n_{CaO}{}^* + n_{FeO}{}^* + n_{SiO_2}{}^* + n_{2CS} \quad (2.124)$$

$$K_{reaction(2.97)} = \frac{n_{2CS} \cdot (n_T)^2}{(n_{CaO}{}^*)^2 \cdot n_{SiO_2}{}^*} = \frac{x_{Ca_2SiO_4}}{(x_{CaO})^2 \cdot x_{SiO_2}} = \frac{\left(\frac{n_{2CS}}{n_T}\right)}{\left(\frac{n_{CaO}{}^*}{n_T}\right)^2 \cdot \left(\frac{n_{SiO_2}{}^*}{n_T}\right)} \quad (2.125)$$

where $K_{reaction(2.97)}$ is the equilibrium constant of the reaction of formation of the Ca_2SiO_4, previously considered in the study of the working hypothesis A. The standard free energy and the equilibrium constant at 1650 °C for the reaction (2.97), using the data of Appendix 1, is:

$$\begin{aligned} \Delta_r G^0_{reaction(2.97)}(1650\,°C) &= -141.46\left[kJ \cdot (mol\ Ca_2SiO_4)^{-1}\right] \\ &= -R \cdot T \cdot \ln K_{reaction(2.97)} \rightarrow K_{reaction(2.97)}(1650\,°C) \\ &= 6.957 \cdot 10^3 \end{aligned} \quad (2.126)$$

In this case, we write all the variables involved in the equilibrium constant of the reaction (2.97) as a function of $n_{CaO}{}^*$, and we obtain a cubic equation:

$$a \cdot \left(n_{CaO}{}^{*}\right)^{3} + b \cdot \left(n_{CaO}{}^{*}\right)^{2} + c \cdot n_{CaO}{}^{*} + d = 0 \tag{2.127}$$

where:

$$a = 0.5 \cdot K_{reaction(2.97)} = 3.479 \cdot 10^{3} \tag{2.128}$$

$$b = n_{SiO_2} - 0.5 \cdot n_{CaO} \cdot K_{reaction\ (2.97)} - K_{reaction\ (2.97)} + 0.5 = -9.128 \cdot 10^{3} \tag{2.129}$$

$$c = D - 0.5 \cdot n_{CaO} = 1.2848 \tag{2.130}$$

$$d = D^{2} + 0.5 \cdot n_{CaO} \cdot D = 1.2498 \tag{2.131}$$

The solution of Eq. (2.127) and the values of the rest of the variables involved in the calculations are:

$$n_{CaO}{}^{*} = 1.180 \cdot 10^{-2} \tag{2.132}$$

$$n_{2CS} = 0.3062 \tag{2.133}$$

$$n_{SiO_2}{}^{*} = 0.1099 \tag{2.134}$$

$$n_{FeO}{}^{*} = 0.5567 \tag{2.135}$$

$$n_{T} = 0.9846 \tag{2.136}$$

Consequently, the activities of the calcium oxide and iron (II) oxide are:

$$a_{CaO} = x_{CaO} = \frac{n_{CaO}{}^{*}}{n_{T}} = \frac{1.180 \cdot 10^{-2}}{0.9846} = 0.012 \tag{2.137}$$

$$a_{FeO} = x_{FeO} = \frac{n_{FeO}{}^{*}}{n_{T}} = \frac{0.5567}{0.9846} = 0.57 \tag{2.138}$$

Finally, it is possible to estimate the activity of the silicon oxide using this variant of the molecular model:

$$a_{SiO_2} = x_{SiO_2} = \frac{n_{SiO_2}{}^{*}}{n_{T}} = \frac{0.1099}{0.9846} = 0.11 \tag{2.139}$$

Commentaries: If we take into account the experimental data of the Exercise 2.3, the activity of the CaO is better calculated using the situation A, while in the case of the FeO, better results are obtained using the situation B. However, if we consider the variation of the activities with the temperature, see Springorum (1995, p. 243), the model that better fits to the experimental data would be the situation-model B, which indicates that the most stable silicate at high temperature is Ca_2SiO_4.

Finally, it is necessary to take into account that the data of the standard free energies of formation for both calcium silicates are not significantly accurate. For example, the operation temperature used in the exercise is out of the range of temperatures recommended for the data of the standard free energy of formation. Moreover, the accuracy of the used data for the 2CS of ±18.7 kJ is also great. This everything leads to the following thought: it is necessary to make a greater effort to obtain accurate thermodynamic parameters to study and develop in the future straighter models for the iron and steelmaking slags.

End of the exercise.

Exercise 2.5. Calculate at the temperature of 1500 °C the equilibrium solubility in the following melts:

(a) Fe; (b) Fe-Si (5%); Fe-Si (10%); Fe-Si (15%); (c) Fe–Mn (75%) with different quantities of silicon.

Data: Use as reference the equations that appear in the Appendix 4 for: the carbon solubility in the iron as a function of the temperature, the variation of the self-interaction parameter of the carbon of Henry with the temperature, and the expression to obtain the self-interaction parameter of Raoult from the Henry's value (calculated for a value of the atomic weights of the iron (dissolvent) and carbon (solute) of 55.85 and 12.00 g mol^{-1}, respectively, Sect. 1.3.1).

Solution Starts

One of the advantages of working with Raoult's activities, apart from allowing to know the thermodynamic characteristics of Fe–C melts, would be the possibility of making predictions about the influence of other solutes (for instance, silicon) on the carbon solubility (interaction of other elements over the carbon solubility). This circumstance is basic in the manufacture of ferroalloys using the carbothermal smelting reduction of minerals. Melts resulted from the carbothermal reduction have carbon contents incompatibles with the commercial product. Carbon refinement is related with the carbon solubility losses in the melt due to the presence of other metal that increases the Raoult's activities of the carbon in the solution.

To know the thermodynamic characteristics of Fe–C solutions-melts, at a certain temperature, it would be necessary to detail all the variables involved in the equation that informs about the variation of the coefficient of Raoult's activity of the carbon, γ_C, as a function of both the quantity of dissolved carbon, x_C, the coefficient of activity at infinite dilution, γ_C^*, and the self-interaction parameter of the carbon, ε_C^C.

$$\ln \gamma_C = \ln \gamma_C^* + \varepsilon_C^C \cdot x_C \quad (2.140)$$

On the other hand, at the temperature of 1500 °C, the value of the self-interaction parameter of Henry, e_C^C, Appendix 4, is:

$$e_C^C = \frac{158}{T} + 0.0581 = \frac{158}{1773} + 0.0581 = 0.1472 \quad (2.141)$$

The value of the self-interaction parameter of Raoult, ε_C^C, according to the equation indicated in the Sect. 1.3.1 and Appendix 4 is:

$$\varepsilon_C^C = 49.48 \cdot e_C^C + 0.7851 = 49.48 \cdot 0.172 + 0.7851 = 8.0693 \quad (2.142)$$

By aside, the value of $\gamma_C^=$ for the maximum solubility of the carbon at 1500 °C, which requires to know the equilibrium solubility, $x_C^=$, can be calculated of different manners:

(A) By means of the data provided in the Appendix 4 for the carbon solubility in the iron melt (Fe–C binary solutions):

$$\%C^= = 0.5656 + 2.58 \cdot 10^{-3} \cdot T = 5.14\%C \quad (2.143)$$

We proceed as follows to transform the weight percentage composition of the carbon in molar fraction:

$$x_C^= = \frac{\frac{5.14}{12.01}}{\frac{5.14}{12.01} + \frac{100-5.14}{55.85}} = 0.2013 \quad (2.144)$$

When in the Fe–C binary system the activity of the carbon is one, then we obtain:

$$a_C = 1 = \gamma_C^= \cdot x_C^= \quad (2.145)$$

$$\gamma_C^= = \frac{1}{x_C^=} = \frac{1}{0.2013} = 4.9686 \quad (2.146)$$

Obtained the value of $\gamma_C^=$ using the Eq. (2.140), it is possible to calculate the value of the coefficient of activity at infinite dilution at 1500 °C for the solutions of carbon in the iron melt:

$$\begin{aligned}\ln \gamma_C^* &= \ln \gamma_C - \varepsilon_C^C \cdot x_C \\ &= \ln 4.9686 - 8.0693 \cdot 0.2013 = -0.02121 \rightarrow \gamma_C^* = 0.9790\end{aligned} \quad (2.147)$$

(B) Solving, by successive approximations, the following nonlinear equation in γ_C is used:

$$\ln \gamma_C = \ln \gamma_C^* + \varepsilon_C^C \cdot \left(\frac{1}{\gamma_C}\right) = 0 \quad (2.148)$$

In this case, at the temperature of 1500 °C, the values of γ_C^* and ε_C^C, for the Fe–C melts previously calculated, are 0.9793 and 8.0693, respectively.

Definitely, the nonlinear equation to be solved is:

$$\ln \gamma_C = \ln 0.9793 + 8.0693 \cdot \left(\frac{1}{\gamma_C}\right) = F(\gamma_C) = 0 \quad (2.149)$$

The value of γ_C that makes the function $F(\gamma_C)$ equal to zero is 4.9686, compatible with carbon molar fraction, $x_C^{=}$, of 0.2013 and a percentage, $\%C^{=}$, of 5.14%.

The presence of other metals in the melt can increase or decrease the value of γ_C. Those metals that increase the value of γ_C decrease the carbon solubility in the melt; while those that decrease the value of γ_C increase the carbon solubility in the melt (for instance, Mn or Cr).

Silicon is a metal that produces a decrease of the carbon solubility in the melt (positive interaction with the carbon or interaction parameter ε_C^{Si} positive).

Considering as starting data, the value of the interaction parameter of the silicon on the carbon, e_C^{Si}, at the temperature of 1600 °C (Appendix 4), is 0.080. First, we must correct e_C^{Si} with the effect of the temperature (decreases linearly when the temperature increases):

$$e_C^{Si}(1773K) = 0.08 \cdot \frac{1873}{1773} = 0.085 \quad (2.150)$$

Then, it is necessary to transform the value of the interaction parameter of Henry to Raoult, Eq. (1.9), Sect. 1.3.1 assuming that the atomic weight of the silicon is 28.09 g mol^{-1}:

$$\varepsilon_C^{Si} = \frac{230.3}{55.85} \cdot 28.09 \cdot e_C^{Si} + \frac{55.85 - 28.09}{55.85} = 10.286 \quad (2.151)$$

The nonlinear equation to be solved in the Fe–C–Si system, which allows calculating the value of the coefficient of activity of the carbon in the ternary system when the solubility of the carbon is maximum ($a_C = 1$, and thus, the value obtained for γ_C could be identified with that of $\gamma_C^{=}$) as a function of the silicon composition, expressed in molar fraction, x_{Si}, would be:

$$\ln \gamma_C - \ln \gamma_C^* - \frac{\varepsilon_C^C}{\gamma_C} - \varepsilon_C^{Si} \cdot x_{Si} = F(\gamma_C) = 0 \quad (2.152)$$

$$\ln \gamma_C - \ln 0.9793 - \frac{8.0693}{\gamma_C} - 10.286 \cdot x_{Si} = F(\gamma_C) = 0 \quad (2.153)$$

The criterion that was chosen to calculate the molar fraction of the silicon in the melt is based on the following equation:

$$x_{Si} = \frac{\frac{5.00}{28.09}}{\frac{100}{\overline{PA}}} = \frac{\left(\frac{5.00}{28.09}\right)}{\frac{100}{53.15}} = 0.0946 \quad (2.154)$$

where the average atomic weight of the iron melt, $\overline{PA}$, is estimated for a carbon percentage in the melt of 3.00%C:

$$\overline{PA} = \frac{5.00}{100} \cdot 28.09 + \frac{3.00}{100} \cdot 12.01 + \frac{92.00}{100} \cdot 55.85 = 53.15\ g \cdot mol^{-1} \quad (2.155)$$

We collect in Table 2.3 the values of $F(\gamma_C)$, x_{Si}, γ_C, $x_{\underline{\underline{C}}}$ and $\%C^{=}$.

We see in Table 2.3 that the value of the coefficient of activity of the carbon that makes minimum the value of $F(\gamma_C)$ is 7.55, compatible with a maximum carbon concentration in the ternary melt of 3.33%C.

We collect in Table 2.4 the values of x_{Si}, γ_C, $x_{\underline{\underline{C}}}$ and $\%C^{=}$, consequence of the values obtained for the carbon solubility in melts with different quantities of silicon and values of $\overline{PA}$ at 1500 °C.

The incorporation of a fourth component, manganese, would make that the nonlinear equation to be solved had a new term that quantified the negative effect

Table 2.3 Values of $F(\gamma_C)$, x_{Si}, γ_C, $x_{\underline{\underline{C}}}$ and $\%C^{=}$

$F(\gamma_C)$	x_{Si}	γ_C	$x_{\underline{\underline{C}}}$	$\%C^{=}$
−0.5053	0.0946	6.00	0.1667	4.32
−0.0021	0.0946	7.54	0.1326	3.34
0.0006	0.0946	7.55	0.1325	3.33
0.0033	0.0946	7.56	0.1323	3.33
0.0142	0.0946	7.60	0.1316	3.31

Table 2.4 Values of x_{Si}, γ_C, $x_{\underline{\underline{C}}}$ and $\%C^{=}$ for different melts

Composition of the melt	$\overline{PA}$	x_{Si}	γ_C	$x_{\underline{\underline{C}}}$	$\%C^{=}$
Fe–C-0% Si	53.60	0.000	4.9686	0.2013	5.14
Fe–C-5% Si	53.15	0.0946	7.5500	0.1325	3.33
Fe–C-10% Si	52.64	0.1874	12.7000	0.0787	1.91
Fe–C-15% Si	51.25	0.2237	23.1000	0.0433	1.04

Table 2.5 Values of x_{Si}, γ_C, $x_C^{=}$ and $\%C^{=}$ for different melts also containing manganese

Composition of the melt	$\overline{PA}$	x_{Si}	x_{Mn}	γ_C	$x_C^{=}$	$\%C^{=}$
Fe-75% Mn-0% Si	50.56	0.0000	0.6903	3.0035	0.3329	10.60
Fe-75% Mn-5% Si	50.54	0.9000	0.6899	3.9734	0.2517	7.40
Fe-75% Mn-10% Si	50.22	0.1788	0.6856	5.5870	0.1790	4.96
Fe-75% Mn-15% Si	49.63	0.2650	0.6755	8.4560	0.1183	3.14
Fe-75% Mn-20% Si	48.80	0.3475	0.6662	13.9380	0.0717	1.87
Fe-75% Mn-25% Si	47.83	0.4257	0.6095	26.8100	0.0373	0.96

of the manganese (negative interaction parameter of the manganese over the carbon, ε_C^{Mn}): the presence of manganese in the melt increases the solubility of the carbon in the melt:

$$\begin{aligned} &\ln \gamma_C - \ln \gamma_C^* - \frac{\varepsilon_C^C}{\gamma_C} - \varepsilon_C^{Si} \cdot x_{Si} + \varepsilon_C^{Mn} \cdot x_{Mn} = F(\gamma_C) = 0 \\ &\rightarrow \ln \gamma_C - \ln 0.9793 - \frac{8.0693}{\gamma_C} - 10.286 \cdot x_{Si} + 2.249 \cdot x_{Mn} \\ &= F(\gamma_C) = 0 \end{aligned} \quad (2.156)$$

We show now in Table 2.5 the results for different melts at 1500 °C.

Commentaries: It is possible to proceed in a similar manner with the Fe–Cr-C–Si system, carbon solubility in Fe–Cr solutions is greater due to two reasons: operation temperature is higher (liquidus temperatures of the Fe–Cr system are greater) and negative interaction parameter of the chromium over the carbon (Table 2.1) is greater than in the case of the manganese, see also Rosenqvist (1983, p. 414). However, the same as in the case of the Fe–Mn–C system, the presence of silicon conduces to a loss of carbon solubility in the melt.

Finally, we want to point out that we did not consider the possible interactions-influence that the changes of concentration in the melt could have in the following variables of Eq. (2.156) These variables remained constant and they are not altered by the changes of concentration in the melt:

- Coefficient of activity at infinite dilution of the carbon, γ_C^*.
- Self-interaction parameter of the carbon, ε_C^C.
- Influence of the silicon on the carbon, ε_C^{Si}.
- Influence of the manganese on the carbon, ε_C^{Mn}.

End of the exercise.

Exercise 2.6. Describe in the Si–O-C system all the equilibriums between condensed phases and the gas that can be related with the obtaining of metallic silicon (metallurgical, solar, or electronic quality) and ferrosilicon in electric furnace. The total pressure of the system is considered equal to the atmospheric one.

Solution Starts

Condensed phases-compounds that can be formed as a result of the linear combination of the three components of the Si–O–C system are SiO_2, C, SiC, and Si, while the gaseous compounds are CO and SiO (silicon (II) oxide) (we should also consider the O_2 as gaseous compound but the POS (Potential of Oxygen of the System) is so small that for practical purposes we do not take into account its participation). As the total pressure of the system is equal to one atmosphere and equal to the addition of CO and SiO partial pressures too, the analysis of the system involves three situations:

- System with zero degrees of freedom ($L = 0$). In this situation the equilibrium is reached with four phases in equilibrium: three condensed phases in equilibrium with the gas. Invariant system.
- System with one degree of freedom ($L = 1$). In this case, we would have all those systems with two condensed phases in equilibrium with the gas. To define the system, it is necessary to know one intensive variable: temperature or composition of either SiO or CO in the gas.
- System with two degrees of freedom ($L = 2$). This is the most general case, where the equilibrium would be reached between one of the four possible condensed phases and the gas. The system would be defined if we know two intensive variables.

We start with the study of all those possible monovariant equilibriums ($L = 1$) between two condensed phases and the gas at the total pressure of one atmosphere. The possible monovariant equilibriums of the Si–O–C system are listed below:

$$SiO_2(s) + C(s) \leftrightarrow SiO(g) + CO(g) \quad (2.157)$$

$$2SiO_2(s) + SiC(s) \leftrightarrow 3SiO(g) + CO(g) \quad (2.158)$$

$$SiO_2(s) + Si(l) \leftrightarrow 2SiO(g) \quad (2.159)$$

$$2Si(l) + CO(g) \leftrightarrow SiO(g) + SiC(s) \quad (2.160)$$

$$SiC(s) + CO(g) \leftrightarrow SiO(g) + C(s) \quad (2.161)$$

Using the data of the following reactions of the Appendix 1:

$$Si(l) + O_2(g) \leftrightarrow SiO_2(s) \quad (2.162)$$

$$\Delta_r G^0_{reaction(2.162)}(T) = -945.0 + 0.1980 \cdot T\left[kJ \cdot (mol\ SiO_2)^{-1}\right] \quad (2.163)$$

$$Si(l) + \frac{1}{2}O_2(g) \leftrightarrow SiO(g) \tag{2.164}$$

$$\Delta_r G^0_{reaction(2.164)}(T) = -157.0 - 0.0470 \cdot T\left[kJ \cdot (mol\ SiO)^{-1}\right] \tag{2.165}$$

$$C(s) + \frac{1}{2}O_2(g) \leftrightarrow CO(g) \tag{2.166}$$

$$\Delta_r G^0_{reaction(2.166)}(T) = -118.0 - 0.0840 \cdot T\left[kJ \cdot (mol\ CO)^{-1}\right] \tag{2.167}$$

$$Si(l) + C(s) \leftrightarrow SiC(s) \tag{2.168}$$

$$\Delta_r G^0_{reaction(2.168)}(T) = -113.0 + 0.0360 \cdot T\left[kJ \cdot (mol\ SiC)^{-1}\right] \tag{2.169}$$

we obtained the following results of the standard free energies for the monovariant equilibriums previously indicated:

$$\begin{aligned}\Delta_r G^0_{reaction(2.157)}(T) &= \Delta_r G^0_{reaction(2.164)}(T) + \Delta_r G^0_{reaction(2.166)}(T) - \Delta_r G^0_{reaction(2.162)}(T)\\ &= 670.0 - 0.329 \cdot T\left[kJ \cdot (mol\ SiO_2)^{-1}\right]\end{aligned} \tag{2.170}$$

$$\begin{aligned}\Delta_r G^0_{reaction(2.158)}(T) &= -2 \cdot \Delta_r G^0_{reaction(2.162)}(T) + 3 \cdot \Delta_r G^0_{reaction(2.164)}(T)\\ &\quad + \Delta_r G^0_{reaction(2.166)}(T) - \Delta_r G^0_{reaction(2.168)}(T)\\ &= 1414.0 - 0.657 \cdot T\left[kJ \cdot (mol\ SiC)^{-1}\right]\end{aligned} \tag{2.171}$$

$$\begin{aligned}\Delta_r G^0_{reaction(2.159)}(T) &= 2 \cdot \Delta_r G^0_{reaction(2.164)}(T) - \Delta_r G^0_{reaction(2.162)}(T)\\ &= 631.0 - 0.292 \cdot T\ \left[kJ \cdot (mol\ SiO_2)^{-1}\right]\end{aligned} \tag{2.172}$$

$$\begin{aligned}\Delta_r G^0_{reaction(2.160)}(T) &= \Delta_r G^0_{reaction(2.168)}(T) - \Delta_r G^0_{reaction(2.166)}(T) + \Delta_r G^0_{reaction(2.162)}(T)\\ &= -152.0 + 0.073 \cdot T\left[kJ \cdot (mol\ SiC)^{-1}\right]\end{aligned} \tag{2.173}$$

$$\begin{aligned}\Delta_r G^0_{reaction(2.161)}(T) &= \Delta_r G^0_{reaction(2.164)}(T) - \Delta_r G^0_{reaction(2.166)}(T) - \Delta_r G^0_{reaction(2.168)}(T)\\ &= 74.0 - 0.001 \cdot T\left[kJ \cdot (mol\ SiC)^{-1}\right]\end{aligned} \tag{2.174}$$

In Fig. 2.10 (pressure of SiO in the gas in front of the temperature, P(SiO)–T equilibrium diagram for the Si–O–C system), we represent the straight lines that define the monovariant equilibriums represented by reactions (2.157)–(2.161) previously mentioned. These straight lines would represent the locus along which the two condensed phases would be in equilibrium with the gas. Certainly, even when we draw all the lines, there are some parts of them that would correspond to metastable equilibriums.

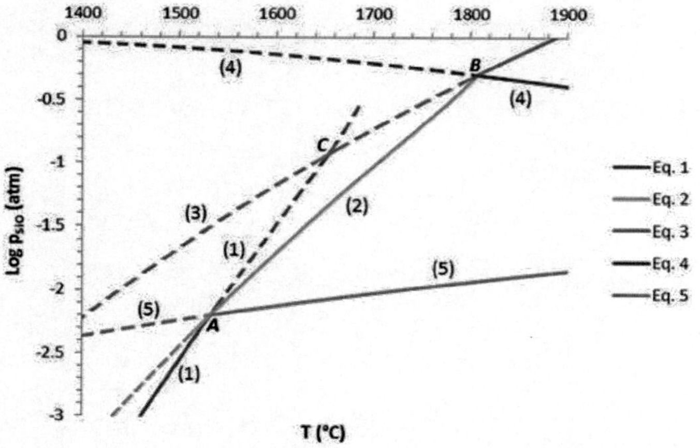

Fig. 2.10 Pressure of SiO in the gas in front of the temperature, P(SiO)-T equilibrium diagram for the Si–O-C system (A (1530, −2.2); B (1807, −0.3); C (1650, –0.94))

The representation of the straight lines of the monovariant equilibriums (P(SiO)-T diagram) allows exploring the possibility of finding invariant points in the system with zero degrees of freedom. For this, it is necessary an equilibrium between three condensed phases and the gas at a constant total pressure of the system, in this case the ambient atmosphere.

We have detected three invariant points: two stable and one metastable. The coordinates of the two detected invariant points, A and B, and the invariant unstable-metastable, C, are:

Point A: equilibrium between condensed phases SiC–SiO_2–C and the gas (SiO and CO) at the temperature of 1530 °C and pressures of SiO and CO of 6.30×10^{-3} atm and 0.9937 atm, respectively. The reaction that would represent the invariant is:

$$3C(s) + SiO_2(s) \leftrightarrow 2CO(g) + SiC(s) \tag{2.175}$$

Point B: equilibrium between condensed phases SiC–SiO_2–Si and the gas (SiO and CO) at the temperature of 1807 °C and pressures of SiO and CO equal to 0.5 atm. The reaction that would represent the invariant is:

$$2SiC(s) + SiO_2(s) \leftrightarrow 3Si(l) + 2CO(g) \tag{2.176}$$

Point C: the unstable phases of SiO_2–C–Si would react with the gas to form SiC at the temperature of 1650 °C and pressures of SiO and CO equal to 0.1138 and 0.8862. The reaction that would represent the unstable invariant would be:

$$SiO_2(s) + 2C(s) \leftrightarrow 2CO(g) + Si(l) \tag{2.177}$$

Using the criterion of minimum free energy, it is possible to define those zones-areas of the diagram with two degrees of freedom, where from the four possible condensed phases, SiO_2, C, SiC, and Si, there is only one that is the most stable. For instance, we are going to consider a temperature of 1700 °C and a P(SiO) of 0.020 atm, log (P(SiO)) = -1.70 to evaluate what would be the most stable condensed phase. Considering the equilibrium (2.161), the free energy of the reaction would be:

$$\Delta_r G^0_{reaction(2.161)}(1973\text{ K}) = 74.0 - 0.001 \cdot 1973 = 72.027\text{ kJ} \cdot (\text{mol SiC})^{-1} \tag{2.178}$$

$$\begin{aligned} \Delta_r G_{reaction(2.161)}(1973\text{ K}) &= \Delta_r G^0_{reaction(2.161)}(1973\text{ K}) + R \cdot T \cdot \ln\left(\frac{a_C \cdot P_{SiO}}{a_{SiC} \cdot P_{CO}}\right) \\ &= 72027\frac{J}{mol} + 8.314\frac{J}{mol \cdot K} \cdot 1973\text{ K} \cdot \ln\left(\frac{1 \cdot 0.02}{1 \cdot 0.98}\right) \\ &= 8.1874\text{ kJ} \cdot (\text{mol SiC})^{-1} \end{aligned} \tag{2.179}$$

Consequently, according to the data of the free energy of the reaction (2.161), the SiC would be more stable than the C.

If we analyze the equilibrium associated to the reaction (2.160):

$$\Delta_r G^0_{reaction(2.160)}(1973\text{ K}) = -152.0 + 0.073 \cdot 1973 = -7.971\text{ kJ} \cdot (\text{mol SiC})^{-1} \tag{2.180}$$

$$\begin{aligned} \Delta_r G_{reaction(2.160)}(1973\text{ K}) &= \Delta_r G^0_{reaction(2.160)}(1973\text{ K}) + R \cdot T \cdot \ln\left(\frac{a_{SiC} \cdot P_{SiO}}{a_{Si}{}^2 \cdot P_{CO}}\right) \\ &= -7917\frac{J}{mol} + 8.314\frac{J}{mol \cdot K} \cdot 1973\text{ K} \cdot \ln\left(\frac{1 \cdot 0.02}{1^2 \cdot 0.98}\right) \\ &= -71.7566\text{ kJ} \cdot (\text{mol SiC})^{-1} \end{aligned} \tag{2.181}$$

Consequently, according to the data of the free energy of the reaction (2.160), the SiC would be more stable than the molten silicon.

Finally, if we analyze the equilibrium associated to the reaction (2.157):

$$\Delta_r G^0_{reaction(2.157)}(1973\text{ K}) = 670.0 - 0.329 \cdot 1973 = 20.883\text{ kJ} \cdot (\text{mol SiO}_2)^{-1} \tag{2.182}$$

$$\begin{aligned}\Delta_r G_{reaction(2.157)}(1973\ K) &= \Delta_r G^0_{reaction(2.157)}(1973\ K) + R \cdot T \cdot \ln\left(\frac{P_{CO} \cdot P_{SiO}}{a_{SiO_2} \cdot a_C}\right)\\ &= 20883\frac{J}{mol} + 8.314\frac{J}{mol \cdot K} \cdot 1973\ K \cdot \ln\left(\frac{0.98 \cdot 0.02}{1 \cdot 1}\right)\\ &= -43.623\ kJ \cdot (mol\ SiO_2)^{-1}\end{aligned} \quad (2.183)$$

SiO_2 would not be a compound stable under the conditions of pressure (CO and SiO) and temperature. Consequently, the phase that is thermodynamically more stable would be the SiO.

Finally, the phase diagram of Fig. 2.10 can be used to study the mechanisms involved in the obtaining of metallic silicon and ferrosilicon in electric furnace. The charge of quartz, coke, and scrap that descends through the shaft of the furnace contacts with a gas rich in SiO. At a temperature higher than 1530 °C, there is a reaction between the SiO and the coke, equilibrium (2.161) to form SiC. The charge of SiC and SiO_2 reaches the hot zone of the furnace, which is at temperatures higher than 1807 °C, and reacts to form metallic silicon and a gas rich in SiO, reaction (2.158) and reaction (2.159). If the coefficient of activity of Raoult of the silicon in the iron is smaller than one, $\gamma_{Si} < 1$, the increase of the silicon concentration in the iron (ferrosilicon) is favored.

Commentaries: The singularity of the carbothermal reduction of the SiO_2 contrasts with that of other oxides of transition metals due to the formation of silicon carbide and silicon monoxide.

The production of silicon carbide, SiC, takes place from quartz and coke in electric furnaces where it is possible to find two zones: one zone of gasification of the SiO at high temperature and, other zone of formation of the SiC located far from the influence of the electric arc.

Solution ends.

2.3 Processes and Reactions in the Electric Furnace

We are going now to analyze in detail the thermodynamic and kinetics fundamentals of the most important reactions that take place in the electric furnace, such as the fusion (smelting), refining of impurities, and reduction.

We make an analysis like that performed for reactions (Sect. 1.3) and kinetic and thermodynamic characteristics (Sect. 1.5) of the BOF process. Moreover, we use the refining and reduction reactions in the electric furnace to study other relevant kinetic aspects.

2.3.1 Smelting (Fusion)

One of the objectives of the electric furnace is to melt the charge of oxides and metallic materials. The enthalpy associated to the process of fusion of the metallic fraction of the charge can be estimated by the rule of mixtures:

$$\Delta_{fusion}H_{metallic\ phase} = \sum x_i \cdot \Delta_{fusion}H_i \tag{2.184}$$

where x_i is the molar fraction of the metal i and $\Delta_{fusion}H_i$ is the enthalpy of fusion associated to the mol of pure metal, which is function of the energy of the metallic bond and, for that reason, of the melting point:

$$\Delta_{fusion}H_i\left(J \cdot mol^{-1}\right) = 10.4 \cdot T_{fusion}(K) \tag{2.185}$$

We use the following equation for the nonmetallic part of the burden (slag) as there is not a well-defined melting point but there is a temperature of formation of the first liquid phases (solidus temperature, $T_{S,i}$) and a liquidus temperature, $T_{L,i}$:

$$\Delta_{fusion}\overline{H} = T_m \cdot \Delta_{fusion}\overline{S} \tag{2.186}$$

where:

$$T_m = 0.5 \cdot \left(T_{S,i} + T_{L,i}\right) \tag{2.187}$$

and $\Delta_{fusion}\overline{S}$ is the entropy associated to the process of fusion/solidification of the nonmetallic burden estimated from the partial molar entropies of each one of the oxides (components) of the system $\Delta_{fusion}\overline{S}_l$ (Table 2.6):

$$\Delta_{fusion}\overline{S} = \sum x_i \cdot \Delta_{fusion}\overline{S}_l \tag{2.188}$$

In this case, the partial molar entropies of each one of the oxides of the slag are smaller than those of the perfectly ordered pure oxides; the energy associated to the fusion process is estimated for solids that, as they are part of the slag, are not similar to the highly ordered pure compound.

2.3.2 Refining

Reactions with oxidizing basic slag of electric furnace are like those described in the Sect. 1.3 for the BOF process. However, the kinetics-rate of the reactions in the

Table 2.6 Partial molar entropies, $\Delta_{fusion}\overline{S}_l$, associated to the fusion of different metallic oxides (Mills and Keene 1987)

Phase	Entropy (J mol^{-1} K^{-1})
Al_2O_3	50.2
CaF_2	18.0
CaO	24.7
Fe	8.4
FeO	9.6
Fe_2O_3	9.6
K_2O	33.9
MgO	24.7
MnO	18.8
Na_2O	33.9
P_2O_5	12.1
SO_3	17.2
SiO_2	4.6
TiO_2	26.4

electric furnace and that in the BOF process cannot be comparable, Sect. 1.5. The kinetic equation considered is very similar to the Eq. (1.127):

$$\frac{d(\%C)}{dt} = k_d \cdot (\%C^0 - \%C^=) \cdot \left(\frac{A}{V}\right) \cdot \left(\frac{65.00}{12.01 \cdot 100}\right) \cdot \left(\frac{1000}{80}\right) \quad (2.189)$$

There is a single substantial change with respect to the Eq. (1.127) and (Eqs. 1.125 and 1.126): the quantity of slag generated, 125 kg to produce 1000 kg of metallic melt in the case of the LD process, is different (it is possible to consider that the requirements of slag in the electric furnaces are in the range 60–100 kg per ton of metallic melt). It is assumed that the average molecular weight of the electric furnace slag is equal to that used in the case of the LD process, even when for the decarburizing in electric furnace, reaction (1.117), the calculation of the coefficient of diffusional mass transfer, k_d, that must be considered in the Eq. (2.189), is performed using the correlation between the Nusselt, Nu, Reynolds, Re, and Schmidt, Sc, numbers (Sect. 1.5):

$$Nu = \frac{k_d \cdot L^*}{D_i} = 0.332 \cdot (Re)^{1/2} \cdot (Sc)^{1/3} \quad (2.190)$$

used in the processes of mass-heat transfer from a flat surface (solid or liquid-molten) to a fluid, nonmiscible with the liquid, that circulates tangentially over it, see Sect. 1.5, where Re is the number of Reynolds and Sc is the number of Schmidt.

Moreover, in Eq. (2.189), the (A/V) ratio in the electric furnace is identified with the height of the metal in the hearth of the furnace (around 40 cm), and, for that reason, the area to volume ratio is 2.5 m^{-1}.

If apart from the diffusion coefficient for the controlling ionic specie in the slag, the O^{2-} anion, is equal to 3.00×10^{-8} m^2 s^{-1} (see Kawai and Shiraishi 1988, p. 196), the difference of carbon content is around 0.40 (it is considered that the refinement capacity of the carbon in the electric furnace is comprised within that of a ferritic steel and that of a pearlitic steel), the viscosity of the slag is 0.50 N s m^{-2} (Pa s), the density is 3000 kg m^{-3}, and the rate of relative movement of the interphases is 3.0×10^{-2} m s^{-1}. The characteristic linear dimension of Reynolds in Eq. (2.190) is identified with the diameter of the furnace (5.0 m). Consequently, the number of Nusselt would be equal to:

$$\begin{aligned}\mathrm{Nu} &= 0.332\cdot\left(\frac{5.0\ \mathrm{m}\cdot 3000\ \mathrm{kg}\cdot\mathrm{m}^{-3}\cdot 3.0\cdot 10^{-2}\mathrm{m}\cdot\mathrm{s}^{-1}}{0.5\ \mathrm{N}\cdot\mathrm{s}\cdot\mathrm{m}^{-2}}\right)^{1/2}\\ &\cdot\left(\frac{0.5\ \mathrm{N}\cdot\mathrm{s}\cdot\mathrm{m}^{-2}}{3\cdot 10^{-8}\mathrm{m}^2\cdot\mathrm{s}^{-1}\cdot 3000\ \mathrm{kg}\cdot\mathrm{m}^{-3}}\right)^{1/3}\\ &= 0.332\cdot(900)^{1/2}\cdot\left(5.56\cdot 10^3\right)^{1/3} = 176.4\end{aligned} \tag{2.191}$$

$$\mathrm{Nu} = 176.4 = \frac{\mathrm{k_d}\cdot \mathrm{L}^*}{\mathrm{D_i}} \tag{2.192}$$

If we identify L^* with δ, thickness of the boundary layer of concentrations, see Sect. 1.5, and taking as characteristic linear dimension for the calculation of the boundary layer in the mass transfer metal-slag, the height of the metal in the electric furnace, 40 cm, we have that δ and, for that reason, L^* reaches the value of 3.993×10^{-3} m, and thus:

$$176.4 = \frac{\mathrm{k_d}\cdot 3.993\cdot 10^{-3}\mathrm{m}}{3.00\cdot 10^{-8}\mathrm{m}^2\cdot\mathrm{s}^{-1}} \rightarrow \mathrm{k_d} = 1.325\cdot 10^{-3}\mathrm{m} = 0.133\ \mathrm{cm}\cdot\mathrm{s}^{-1} \tag{2.193}$$

when the diffusion coefficient is 3.00×10^{-8} m^2 s^{-1}. The value of δ can be calculated with the following equation:

$$\delta = 1.5\cdot \mathrm{D_i}^{1/3}\cdot\left(\frac{\eta}{\rho}\right)^{1/6}\cdot\left(\frac{\mathrm{L}}{\mathrm{v}}\right)^{1/2} \tag{2.194}$$

In this case, the value of the characteristic linear dimension, L, is identified with the height of the metal in the electric furnace (0.40 m):

$$\delta = 1.5 \cdot \left(3.00 \cdot 10^{-8}\mathrm{m}^2 \cdot \mathrm{s}^{-1}\right)^{1/3} \cdot \left(\frac{0.5\ \mathrm{N} \cdot \mathrm{s} \cdot \mathrm{m}^{-2}}{3000\ \mathrm{kg} \cdot \mathrm{m}^{-3}}\right)^{1/6} \cdot \left(\frac{0.4\ \mathrm{m}}{3.0 \cdot 10^{-2}\mathrm{m} \cdot \mathrm{s}^{-1}}\right)^{1/2}$$
$$= 3.993 \cdot 10^{-3}\mathrm{m}\ (3.993\ \mu\mathrm{m}) \tag{2.195}$$

$$176.4 = \frac{\mathrm{k_d} \cdot \mathrm{L}^*}{\mathrm{D_i}} \to \mathrm{k_d} = \frac{176.4 \cdot 3.00 \cdot 10^{-8}\mathrm{m}^2 \cdot \mathrm{s}^{-1}}{3.993 \cdot 10^{-3}\mathrm{m}}$$
$$= 1.325 \cdot 10^{-3}\mathrm{m} \cdot \mathrm{s}^{-1}\left(0.133\ \mathrm{cm} \cdot \mathrm{s}^{-1}\right) \tag{2.196}$$

Definitely, decarburization in electric furnace happens at a rate, according to the kinetic model proposed by Eq. (2.189), of:

$$\left(\frac{x_{\mathrm{FeO}}}{\mathrm{mol\ slag}}\right) \cdot \left(\frac{\mathrm{mol\ C}}{\mathrm{mol\ FeO}}\right) \cdot \left(\frac{12.01\ \mathrm{g\ C}}{\mathrm{mol\ C}}\right) \cdot \left(\frac{\mathrm{mol\ slag}}{65\ \mathrm{g\ slag}}\right) \cdot \left(\frac{80\ \mathrm{g\ slag}}{1000\ \mathrm{g\ melt}}\right) \cdot \left(\frac{100\ \mathrm{g\ melt}}{\%\mathrm{C}}\right)$$
$$\to \frac{\mathrm{d}(\%\mathrm{C})}{\mathrm{dt}}$$
$$= 1.325 \cdot 10^{-3}\mathrm{m} \cdot \mathrm{s}^{-1} \cdot (0.4) \cdot \left(2.5\ \mathrm{m}^{-1}\right) \cdot \left(\frac{65.00}{12.01 \cdot 100}\right) \cdot \left(\frac{1000}{80}\right) \cdot \frac{3600\ \mathrm{s}}{1\ \mathrm{h}}$$
$$= 3.24\%\mathrm{C} \cdot \mathrm{h}^{-1} \tag{2.197}$$

The most significative differences between the decarburization in the LD converter and in the electric furnace can be summarized as follows:

(a) Coefficient of mass transfer: 0.443 cm s^{-1} in the LD and 0.133 cm s^{-1} in the electric furnace.
(b) Driving force of the reaction, difference of concentration: 4.5 in the LD and 0.4 in the electric furnace.

Finally, it is necessary to consider that the order of magnitude of the operating times between the BOF steelworks and the electric steelworks to reach a molten product can be of around 20 min for the BOF and 50 min for the electric steelworks. The possibility of shortening the refinement times in several or tens of minutes might relate to the time required to form the slag: 5–6 min in the BOF and 20 min in the electric furnace. Producing synthetic slags out of the LD or the electric furnace at competitive prices would mean shortening the operation times, decreasing the energy consumption, and increasing the productivity of the steelworks. Concentrated thermal solar energy might be an instrument to be used in the future to produce oxidizing basic slags (to dephosphorize) or reductant basic slags (to remove sulphur) and increase the energy efficiency, reduce the environmental impact and improve the economic benefit of the steelworks (Fernández-González et al. 2018a).

On the other hand, operating versatility of electric furnaces is significant; it is possible to create a reductant environment to achieve the slagging of sulphur or the recovery of metals existing in the slag.

The removal of sulphur dissolved in the liquid is performed by a basic-reductant slag:

$$CaO(dis;\, slag) + S(dis;\, iron) \leftrightarrow CaS(dis;\, slag) + O(dis;\, iron) \quad (2.198)$$

Equilibrium (2.198) is displaced to the right when a basic slag is used in a low oxidizing potential environment. Equally, it is possible to recover oxidized chromium in the electric furnace during the fusion of stainless steel scrap. The addition of ferrosilicon creates a reductant environment that facilitates the re-assimilation of the chromium in the liquid iron:

$$Cr_2O_3(dis;\, slag) + \frac{3}{2}Si(dis;\, iron) \leftrightarrow \frac{3}{2}SiO_2(dis;\, slag) + 2Cr(dis;\, iron) \quad (2.199)$$

The production of stainless steel with 18% Cr and 8% Ni is carried out in electric furnace because of the capacity of this furnace to operate equally with oxidizing and reductant potentials. The growing limitations of carbon content in stainless steels have meant the gradual replacement of the traditional fusion in the electric furnace with double slag by refining in the AOD converter (Fig. 2.11). In this case, the purpose of the electric furnace is to melt the scrap, while the purpose of the AOD converter is recovering chromium and decreasing carbon percentages in the melt.

AOD decarburization takes place because of the displacement to the right of the reaction:

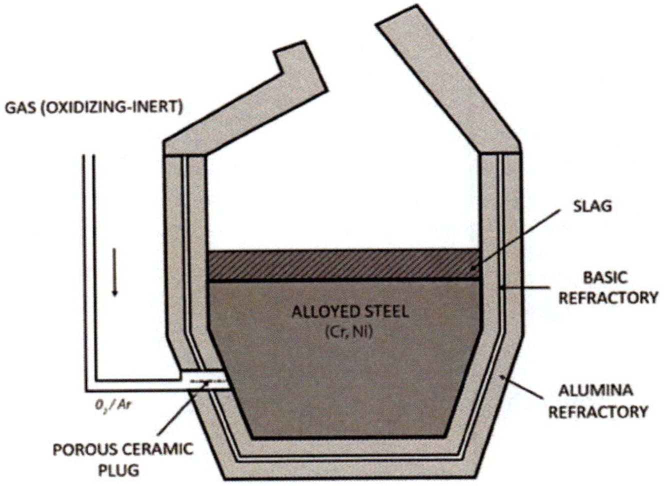

Fig. 2.11 AOD converter—Argon Oxygen Decarburization

$$C(\text{dis; iron}) + O(\text{dis; iron}) \leftrightarrow CO(g) \quad (2.200)$$

as a consequence of the argon as it creates multiple gas-metal interphases, with partial pressures of carbon monoxide very far from the equilibrium values.

2.3.3 *Reduction*

Reduction reactions that can be carried out in the electric furnace are mainly two:

- Carbothermal reduction of raw materials or recycled products to obtain ferroalloys.

$$MeO(\text{dis; slag}) + C(\text{dis; iron}) \leftrightarrow CO(g) + Me(\text{dis; iron}) \quad (2.201)$$

- Metallothermic recovery of valuable metals (using aluminum and silicon) from slags generated during either the reduction or oxidizing thermal refining of metals (Fernández-González et al. 2019a, b):

$$MeO(\text{dis; slag}) + M(\text{dis; metal}) \leftrightarrow MO(\text{dis; slag}) + Me(\text{dis; metal}) \quad (2.202)$$

Table 2.7 (Knacke et al. 1991) completes the thermodynamic data of Appendix 1, the standard free energies of formation associated to the raw materials of the most typical metals of some ferroalloys are indicated in this table.

As it is indicated in the Sect. 2.2.3, one of the disadvantages of the reactions of carbothermal reduction of the transition metals (Mn, Cr, V, Ti, Nb, W, or Mo) is the high carbon contents in the melts. It is necessary to proceed as in the scheme of Fig. 2.12 to obtain carbon contents lower than 100 thousandths in percentage.

Table 2.7 Standard free energies of formation for the manufacture of ferroalloys within a certain range of temperatures

Reaction	$\Delta_f G^0$ (kJ mol^{-1})	Range of temperatures (K)
$Fe(s) + 2Cr(s) + 2O_2(g) = FeCr_2O_4(s)$	$-1453.6 + 0.3255$ T	300–1800
$Fe(s) + W(s) + 2O_2(g) = FeWO_4(s)$	$-1178.7 + 0.3239$ T	300–1300
$Fe(s) + Ti(s) + 3/2O_2(g) = FeTiO_3(s)$	$-1234.5 + 0.2486$ T	300–1673
$Fe(s) + Ti(s) + 3/2O_2(g) = FeTiO_3(s)$	$-1130.0 + 0.857$ T	1673–1800
$Zr(s) + Si(s) + 2O_2(g) = ZrSiO_4(s)$	$-2030.2 + 0.3722$ T	300–1683
$Zr(s) + Si(l) + 2O_2(g) = ZrSiO_4(s)$	$-2056.1 + 0.3866$ T	1683–2000

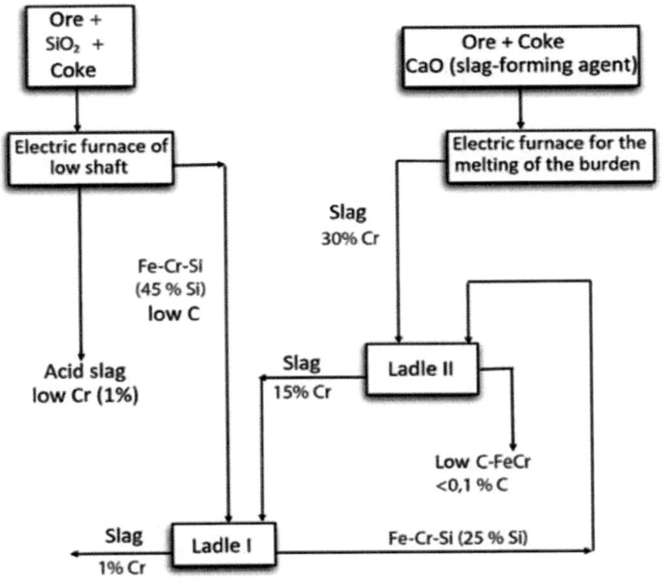

Fig. 2.12 Flow diagram to produce low carbon ferrochromium (silicothermic reduction)

Reactions-mechanisms of reaction that can take place in electric furnaces used to produce pig iron (Sect. 2.2.1) or ferroalloys (Sect. 2.2.3) involve a stoichiometric relation determined by the valence rules between the oxygen atoms combined with the element to be reduced and the oxygen connected to the reductant carbon. This circumstance propitiates a constant relation throughout the stack of the furnace, between the oxygen moles that are transferred from the oxidized charge to the gas. The linear relation between oxygen moles associated to the mineral charge and those associated to the current of reductant carbonaceous gases is translated into the well-known equation of the operating line of the furnace:

$$Y = \mu \cdot X + b \tag{2.203}$$

where:

- Variable Y represents the moles of oxygen combined with the mineral charge for any mol of metal to be reduced. For instance, a charge formed by MnO_2, the value of Y would be equal to 2 mol of oxygen per mol of manganese.
- Variable X is related with the moles of oxygen per mol of carbon in the gas. That is to say, if the gaseous current in contact with the mineral charge of MnO_2 is formed by 100% CO_2, the value of X would be equal to 2 mol of oxygen per mol of carbon.

- The slope of the operating line, μ, refers to the total quantity of carbon-coke (moles of carbon that per mol of metal are required to reduce the oxide). At the same time, the value of the slope has two addends: one to estimate the quantity of coke that is required to reduce the metal, and the other to reflect the quantity of coke that was dissolved in the obtained ferroalloy (carbon losses due to dissolution).
- Finally, constant term, b, is related with the quantity of oxygen moles per mol of metal that were not reduced and that would be part of the slag that, together with the ferroalloy, leaves the furnace. It may be also related with the moles of oxygen per mol of reduced metal different to the main element, iron in a ferroalloy or the metallic and non-metallic (P and S) impurities in the mineral charge. Equally, all contributions of oxygen injected to the furnace in a deliberate way (oxygen or hot blast) or also contributions related with filtrations-entrance of environmental air to the stack of the furnace have influence in the term b.

2.3.4 Solved Exercises

Exercise 2.7. Estimate the minimum energy consumption, in kJ mol^{-1} of Fe and in kWh t^{-1} of Fe, associated to the process of obtaining a melt of low carbon content ($0.02 \leq \%C \leq 0.80$) at 1650 °C from scrap at 25 °C.

We consider that the atomic weight of the iron is 55.85 g mol^{-1}, the mean specific heat of the iron between 25 °C (298 K) and 1536 °C (1809 K, melting point) is 37.66 J mol^{-1} of Fe, the fusion enthalpy of the iron is 13.807 kJ mol^{-1} of Fe and, the variation of the specific heat with the temperature of the molten iron is given by the following equation:

$$c_p(\text{Fe, l}) = 40.878 + 1.647 \cdot 10^{-3} \cdot T\left[J \cdot (\text{mol Fe})^{-1}\right] \tag{2.204}$$

Solution Starts

We work with the hypothesis of identifying the thermal characteristics of a carbon steel with those of the elemental iron. On another note, for compounds or elements that are subjected to allotropic changes, the representation of the specific heat as a function of the temperature shows discontinuities associated to the allotropic changes that can complicate the calculation of the sensitive energy contribution during the heating. It is, for that reason, recommendable to talk about the mean specific heat, $\overline{c_p}$, between two temperatures considering the following equation:

$$\overline{c_p} \cdot (T_2 - T_1) = \int_{T_1}^{T_2} c_p \cdot dT \tag{2.205}$$

The value of the integral can be calculated either by graphical integration or by analytical calculus. In this case, and using the data of the specific heat of the iron provided by the HSC software (function of the temperature), the value obtained for the graphical integration of the data between the temperatures of 1809 K and 298 K is 56,898.3 J mol^{-1} of Fe. Consequently, according to the indications of the enunciation:

$$\overline{c_p} = \frac{56898.3\left[J \cdot (\text{mol Fe})^{-1}\right]}{(1809 - 298)K} = 37.66\ J \cdot (\text{mol Fe})^{-1} \cdot K^{-1} \quad (2.206)$$

After that, we detail the results obtained for the three stages of the calculation process:

Stage 1: sensitive heat required to heat the iron from the room temperature (298 K) to its melting point (1809 K):

$$37.66 \frac{J}{\text{mol Fe} \cdot K} \cdot (1809 - 298)K \cdot \frac{1\ kJ}{10^3 J} = 56.8983 \frac{kJ}{\text{mol Fe}} \cdot \frac{1\ \text{mol Fe}}{55.85\ \text{g Fe}} \cdot \frac{10^6 \text{g Fe}}{1\ \text{t Fe}}$$
$$= 1.019 \cdot 10^6 \frac{kJ}{\text{t Fe}} \cdot \frac{1\ kWh}{3600\ kJ} = 283 \frac{kWh}{\text{t Fe}} \quad (2.207)$$

Stage 2: heat associated to the melting of the iron at 1809 K:

$$13.8070 \frac{kJ}{\text{mol Fe}} \cdot \frac{1\ \text{mol Fe}}{55.85\ \text{g Fe}} \cdot \frac{10^6 \text{g Fe}}{1\ \text{t Fe}} = 2.472 \cdot 10^5 \frac{kJ}{\text{t Fe}} \cdot \frac{1\ kWh}{3600\ kJ} = 69 \frac{kWh}{\text{t Fe}} \quad (2.208)$$

Stage 3: sensitive heat required to heat the iron from the iron melting point (1809 K) to the final temperature of 1650 °C (1923 K):

$$\int_{T_1}^{T_2} c_p \cdot dT = \int_{1809}^{1923} \left(40.878 + \frac{1.646}{1000} \cdot T\right) \cdot dT \quad (2.209)$$

$$\int_{1809}^{1923} c_p \cdot dT = 40.878 \cdot (1923 - 1890) + \frac{1.646}{2 \cdot 1000} \cdot (1923 - 1809)^2$$
$$= 4.671 \cdot 10^3 \frac{J}{\text{mol Fe}} \cdot \frac{1\ kJ}{10^3 J} = 4.671 \frac{kJ}{\text{mol Fe}} \cdot \frac{1\ \text{mol Fe}}{55.85\ \text{g Fe}} \cdot \frac{10^6 \text{g Fe}}{1\ \text{t Fe}}$$
$$= 8.363 \cdot 10^4 \frac{kJ}{\text{t Fe}} \cdot \frac{1\ kWh}{3600\ kJ} = 23 \frac{kWh}{\text{t Fe}} \quad (2.210)$$

The addition of the three above-listed contributions lead to the result of the exercise: minimum energy consumption associated to the obtaining of a low carbon melt at 1650 °C:

$$E = 1.019 \cdot 10^6 \frac{kJ}{tFe} + 2.472 \cdot 10^5 \frac{kJ}{tFe} + 8.363 \cdot 10^4 \frac{kJ}{tFe} = 1.350 \cdot 10^6 \frac{kJ}{tFe} \quad (2.211)$$

$$E = 283 \frac{kWh}{tFe} + 69 \frac{kWh}{tFe} + 23 \frac{kWh}{tFe} = 375 \frac{kWh}{tFe} \quad (2.212)$$

Commentaries: According to Emi (2015, p. 51), the energy consumption of the most efficient electric steelworks is of 345 kWh t^{-1} of steel. Thus, for an estimated energy consumption of 375 kWh t^{-1} of iron, around 75% is the possible contribution of the scrap heating (283 kWh t^{-1} of Fe). It is necessary to make, if we want to increase the energy efficiency of the process, significant savings of electric power, either through the preheating of the scrap with hot gases or by means of the utilization of other primary energy sources (gas, pulverized coal, or fuel oil).

End of the exercise.

Exercise 2.8 Make an estimation about the minimum energy consumption, in kJ mol^{-1} of steel and kWh t^{-1} of steel, associated to the process of formation of the slag during the manufacture of steel in electric furnace. The process of formation of slag takes place from room temperature, 25 °C (298 K), to 1650 °C (1923 K). Consider that the physical–chemical characteristics of the pseudoternary slag are those indicated in the Exercise 2.1. Equally, we should consider the following thermal data:

(a) Mean specific heat of the three components of the slag, from 25 °C and 1460 °C (liquidus temperature of the slag), are: 52.52 J mol^{-1} K^{-1} of CaO, 60.19 J mol^{-1} K^{-1} of FeO, and 67.89 J mol^{-1} K^{-1} of SiO_2, respectively.
(b) The quantity of slag produced is 80.0 kg of slag per ton of steel.
(c) That the enthalpies of formation of the fayalite, Fe_2SiO_4, and of the dicalcium silicate, Ca_2SiO_4, at 1000 °C, are, respectively: −40.039 kJ mol^{-1} of Fe_2SiO_4 and -125.727 kJ mol^{-1} of Ca_2SiO_4. The molecular weights of both phases are: 203.78 g mol^{-1} of Fe_2SiO_4 and 172.24 g mol^{-1} of Ca_2SiO_4.

Solution Starts

The chemical, mineralogical, and physical-thermal characteristics of the slag of three components, Exercise 2.1, considered in the energy simulation of the process of slag-formation are the following ones.

- Composition in weight percentage: 35% CaO, 40% FeO, 25% SiO_2.
- Composition in molar fraction: $x_{CaO} = 0.3908$; $x_{FeO} = 0.3486$; $x_{SiO_2} = 0.2605$.
- Molecular weight of the slag,

$$\begin{aligned}PM_{slag} &= x_{CaO} \cdot PM_{CaO} + x_{FeO} \cdot PM_{FeO} + x_{SiO_2} \cdot PM_{SiO_2} \\ &= 0.3908 \cdot 56.08\ g \cdot mol^{-1} + 0.3486 \cdot 71.85\ g \cdot mol^{-1} \\ &\quad + 0.2605 \cdot 60.09\ g \cdot mol^{-1} = 62.62\ g \cdot mol^{-1}\end{aligned} \quad (2.213)$$

- Specific heat of the molten slag (data in Table 1.13):

$$\begin{aligned}c_p^{slag} &= x_{CaO} \cdot c_p^{CaO} + x_{FeO} \cdot c_p^{FeO} + x_{SiO_2} \cdot c_p^{SiO_2} = 0.3908 \cdot 80.80\ J \cdot mol^{-1} \cdot K^{-1} \\ &\quad + 0.3486 \cdot 76.6\ J \cdot mol^{-1} \cdot K^{-1} + 0.2605 \cdot 87.0\ J \cdot mol^{-1} \cdot K^{-1} \\ &= 80.94 \frac{J}{mol \cdot K} \cdot \frac{mol}{62.62\ g} \cdot \frac{10^3 g}{1\ kg} = 1292.6 \frac{J}{kg \cdot K}\end{aligned} \quad (2.214)$$

- Crystalline phases compatibles with the chemical composition of the slag at room temperature: 53% Ca_2SiO_4, 24% Fe_2SiO_4, 23% FeO.
- Solidus temperature: 1177 °C (1450 K).
- Liquidus temperature: 1460 °C (1773 K).

As in the Exercise 2.7, the minimum energy for the formation of the slag is calculated following three stages:

Stage 1: estimation of the sensitive heat of the three components of the slag (CaO, FeO, SiO_2) from room temperature, 25 °C, to 1460 °C (1773 K). In this case, we work with mean specific heat of the three oxides (between the three above-indicated temperatures):

$$\overline{c_p}(1773 - 298) = \int_{298}^{1773} c_p \cdot dT \quad (2.215)$$

For the CaO, we detail the calculations that is necessary to make:

$$52.52 \frac{J}{mol \cdot K} \cdot (1773 - 298)K = 75367.756 \frac{J}{mol\ CaO} \quad (2.216)$$

$$75.368 \frac{kJ}{mol\ CaO} \cdot \frac{mol\ CaO}{56.08\ g\ CaO} \cdot \frac{35\ g\ CaO}{100\ g\ slag} \cdot \frac{10^3 g}{1\ kg} = 470.377 \frac{kJ}{kg\ slag} \quad (2.217)$$

$$470.377 \frac{kJ}{kg\ slag} \cdot \frac{80\ kg\ slag}{t\ steel} = 3.763 \cdot 10^4 \frac{kJ}{t\ steel} \quad (2.218)$$

For the FeO, we detail the calculations that is necessary to make:

$$60.19 \frac{J}{mol \cdot K} \cdot (1773 - 298)K = 86374.754 \frac{J}{mol\ FeO} \tag{2.219}$$

$$86.375 \frac{kJ}{mol\ FeO} \cdot \frac{mol\ FeO}{71.85\ g\ FeO} \cdot \frac{40\ g\ FeO}{100\ g\ slag} \cdot \frac{10^3 g}{1\ kg} = 480.862 \frac{kJ}{kg\ slag} \tag{2.220}$$

$$480.862 \frac{kJ}{kg\ slag} \cdot \frac{80\ kg\ slag}{t\ steel} = 3.8469 \cdot 10^4 \frac{kJ}{t\ steel} \tag{2.221}$$

For the SiO_2, we detail the calculations that is necessary to make:

$$67.89 \frac{J}{mol \cdot K} \cdot (1773 - 298)K = 97423.82 \frac{J}{mol\ SiO_2} \tag{2.222}$$

$$97.424 \frac{kJ}{mol\ SiO_2} \cdot \frac{mol\ SiO_2}{60.09\ g\ SiO_2} \cdot \frac{25\ g\ SiO_2}{100\ g\ slag} \cdot \frac{10^3 g}{1\ kg} = 405.325 \frac{kJ}{kg\ slag} \tag{2.223}$$

$$405.325 \frac{kJ}{kg\ slag} \cdot \frac{80\ kg\ slag}{t\ steel} = 3.2426 \cdot 10^4 \frac{kJ}{t\ steel} \tag{2.224}$$

Stage 2: heat associated to the melting process of the oxides involved in the slag. We use Eq. (2.186) and the data in Table 2.6:

$$\begin{aligned}
\Delta_{fus}\overline{H} &= T_m \cdot \Delta_{fus}\overline{S} \\
&= \left(\frac{T_S + T_L}{2}\right) \cdot \sum \left(x_{CaO} \cdot \Delta_{fus}\overline{S}_{CaO} + x_{FeO} \cdot \Delta_{fus}\overline{S}_{FeO} + x_{SiO_2} \cdot \Delta_{fus}\overline{S}_{SiO_2}\right) \\
&= \left(\frac{1450 + 1773}{2}\right)K \cdot \left(0.3908 \cdot 24 \frac{J}{mol \cdot K} + 0.3486 \cdot 9.6 \frac{J}{mol \cdot K} + 0.2605 \cdot 4.6 \frac{J}{mol \cdot K}\right) \\
&= 22.5975 \cdot 10^3 \frac{J}{mol\ slag}
\end{aligned} \tag{2.225}$$

$$\Delta_{fus}\overline{H} = 22.5975 \frac{kJ}{mol\ slag} \cdot \frac{mol\ slag}{62.62\ g} \cdot \frac{10^3 g}{1\ kg} = 360.867 \frac{kJ}{kg\ slag} \tag{2.226}$$

$$\Delta_{fus}\overline{H} = 360.867 \frac{kJ}{kg\ slag} \cdot \frac{80\ kg\ slag}{t\ steel} = 2.8869 \cdot 10^4 \frac{kJ}{t\ steel} \tag{2.227}$$

Stage 3: contribution of the sensitive heat of the slag from its liquidus temperature, 1460 °C (1733 K), up to reach 1650 °C (1923 K). Considering that the specific heat of the pseudoternary melt is 1292.6 J kg^{-1} K^{-1}. Consequently:

$$1292.6 \frac{J}{kg \cdot K} \cdot (1923 - 1733)K = 2.46 \cdot 10^5 \frac{J}{kg} \cdot \frac{80\ kg\ slag}{t\ steel} \cdot \frac{1\ kJ}{10^3 J}$$

$$= 1.9648 \cdot 10^4 \frac{\text{kJ}}{\text{t steel}} \tag{2.228}$$

The addition of the three stages gives as a result:

$$\begin{aligned} &3.763 \cdot 10^4 \frac{\text{kJ}}{\text{t steel}} + 3.8469 \cdot 10^4 \frac{\text{kJ}}{\text{t steel}} + 3.2426 \cdot 10^4 \frac{\text{kJ}}{\text{t steel}} \\ &+ 2.8869 \cdot 10^4 \frac{\text{kJ}}{\text{t steel}} + 1.9648 \cdot 10^4 \frac{\text{kJ}}{\text{t steel}} \\ &= 1.5704 \cdot 10^4 \frac{\text{kJ}}{\text{t steel}} \cdot \frac{0.000278 \text{ kWh}}{1 \text{ kJ}} = 43.66 \text{ kWh} \cdot \text{t}^{-1} \end{aligned} \tag{2.229}$$

Finally, we want to make an estimation of the possible contribution, in this case with positive value, referred to the formation of chemical compounds by linear combination of the three components of the system (CaO–FeO–SiO_2). According to the data of the Exercise 2.2, the slag, at room temperature, contains dicalcium silicate, Ca_2SiO_4 (53%), fayalite, Fe_2SiO_4 (24%), and FeO (23%), as chemical compounds compatible with the composition: 35% CaO, 40% FeO, and 25% SiO_2. Considering the data referred to the standard enthalpies of formation at 1000 °C of the HSC5.11 database, the energy supply to the system, as a consequence of the formation of the previously mentioned silicates, would be:

– For the Ca_2SiO_4 (2CS), at 1000 °C, the standard enthalpy, $\Delta_r H^0_{2CS}$, associated to the reaction (2.230):

$$2CaO(s) + SiO_2(s) \leftrightarrow Ca_2SiO_4(s) \tag{2.230}$$

is -125.727 kJ mol^{-1} of 2CS. Consequently, and considering that 80 kg of slag are produced per ton of steel:

$$-125.727 \frac{\text{kJ}}{\text{mol 2CS}} \cdot \frac{\text{mol 2CS}}{172.24 \text{ g}} \cdot \frac{53 \text{ g 2CS}}{100 \text{ g slag}} \cdot \frac{1000 \text{ g}}{1 \text{ t}} = -386.87 \frac{\text{kJ}}{\text{kg slag}} \tag{2.231}$$

The energy supplied to the system by the formation of the 2CS would be:

$$-386.87 \frac{\text{kJ}}{\text{kg slag}} \cdot \frac{80 \text{ kg slag}}{\text{t steel}} = -3.095 \cdot 10^4 \frac{\text{kJ}}{\text{t steel}} \cdot \frac{0.000278 \text{ kWh}}{1 \text{ kJ}} = -8.6 \frac{\text{kWh}}{\text{t steel}} \tag{2.232}$$

– For the Fe_2SiO_4 (2FeS), at 1000 °C, the standard enthalpy, $\Delta_r H^0_{2FeS}$, associated to the reaction (2.233):

$$2FeO(s) + 2SiO_2(s) \leftrightarrow Fe_2SiO_4(s) \tag{2.233}$$

is -40.039 kJ mol^{-1} of 2FeS. Consequently, and considering that 80 kg of slag are produced per ton of steel:

$$-40.039\frac{kJ}{mol\ 2FeS}\cdot\frac{mol\ 2FeS}{203.78\ g}\cdot\frac{24\ g\ 2FeS}{100\ g\ slag}\cdot\frac{1000\ g}{1\ t}=-47.156\frac{kJ}{kg\ slag} \tag{2.234}$$

The energy supplied to the system by the formation of the 2FeS would be:

$$-47.156\frac{kJ}{kg\ slag}\cdot\frac{80\ kg\ slag}{t\ steel}=-3.773\cdot 10^3\frac{kJ}{t\ steel}\cdot\frac{0.000278\ kWh}{1\ kJ}$$
$$=-1.0\frac{kWh}{t\ steel} \tag{2.235}$$

Consequently, the possible reactions of formation of the chemical compounds in the slag would reduce the energy consumption of the furnace in 9.6 kWh per ton of steel.

Commentaries: The process of formation of the slag has also a relevant importance in the energy consumption associated to the production of steel and ferroalloys. It is possible to conclude that around 10% of the total energy consumption required to produce steel is used to generate the suitable slag to complete the refinement operations of the molten metal.

Two types of slag are generated in the electric steelworks: black slag with high content in iron oxides (with variable contents of magnetite) and white slag with low content in iron oxides and high content in CaO. The type of slag considered in the exercise would correspond with the black slag.

End of the exercise.

Exercise 2.9 The first stage in the production of metallurgical quality ferrochromium, Fig. 2.10, consists in the carbothermal reduction of the raw materials, habitually chromite, Cr_2O_3 FeO. Considering that the starting raw material is the chromium (III) oxide, evaluate the energy costs and coke consumption associated to the carbothermal reduction of Cr_2O_3 at temperature of 1910 °C (2183 K). Finally, detail some kinetic considerations of the process, if we consider that in one hour, 500 kg of chromium are obtained from 400 kg of an oxidized melt-slag of high chromium content (68.4% Cr).

Data: Mean specific heats of the chromium (III) oxide and carbon (graphite) obtained from the TAPP database, between 298 and 2183 K, are 120.9259 J (mol of Cr_2O_3)$^{-1}$ K^{-1} and 16.5345 J (mol C)$^{-1}$ K^{-1}, respectively, melting point and enthalpy of fusion of the chromium are 2180 K and 21.014 kJ (mol Cr)$^{-1}$, respectively. Atomic weights of the chromium, oxygen, and carbon are 52 g mol^{-1}, 16 g mol^{-1}, and 12 g mol^{-1}, respectively.

Solution Starts

The global reaction of the carbothermal reduction of the Cr_2O_3 is:

$$Cr_2O_3(s) + 3C(s) \leftrightarrow 3CO(g) + 2Cr(l) \tag{2.236}$$

Considering the data of the Appendix 2, the standard enthalpy of the reaction (2.236) at 2183 K (1910 °C) is the following one:

$$2Cr(s) + \frac{3}{2}O_2(g) \leftrightarrow Cr_2O_3(s) \tag{2.237}$$

$$\Delta_r H^0_{reaction(2.237)} = 1143.0 + 0.0180 \cdot T\left[kJ \cdot (mol\ Cr_2O_3)^{-1}\right] \tag{2.238}$$

$$C(s) + \frac{1}{2}O_2(g) \leftrightarrow CO(g) \tag{2.239}$$

$$\Delta_r H^0_{reaction(2.239)} = -104.0 + 0.0075 \cdot T\left[kJ \cdot (mol\ C)^{-1}\right] \tag{2.240}$$

$$\begin{aligned}\Delta_r H^0_{reaction(2.236)} &= -\Delta_r H^0_{reaction(2.237)} + 3 \cdot \Delta_r H^0_{reaction(2.239)} \\ &= -(1143.0 + 0.0180 \cdot T) + 3 \cdot (-104.0 + 0.0075 \cdot T) \\ &= 831.0 - 0.0405 \cdot T\left[kJ \cdot (mol\ Cr_2O_3)^{-1}\right]\end{aligned} \tag{2.241}$$

Considering that the mol of Cr_2O_3 contains two atoms of chromium and that one kWh is equal to 3.60×10^6 J, the value of the energy consumption in the carbothermal reduction of the chromium (III) oxide at 2183 K (1910 °C) would be:

$$\begin{aligned}&(831.0 - 0.0405 \cdot 2183)\left(\frac{kJ}{mol\ Cr_2O_3}\right) \\ &= 742.5885\frac{kJ}{mol\ Cr_2O_3} \cdot \frac{1\ mol\ Cr_2O_3}{2\ mol\ Cr} \cdot \frac{1\ mol\ Cr}{52.00\ g\ Cr} \cdot \frac{10^6 g}{1\ t} \\ &= 7.1403 \cdot 10^6\frac{kJ}{t\ Cr} \cdot \frac{10^3 J}{1\ kJ} \cdot \frac{1\ kWh}{3.60 \cdot 10^6 J} = 1983\frac{kWh}{t\ Cr}\end{aligned} \tag{2.242}$$

Moreover, as we have seen in the Exercise 2.7, the contributions to be considered in the energy balance of the process are:

(a) The sensitive heat of the reagents of the reaction (2.236), from 298 to 2183 K (1910 °C) would be:

$$\begin{aligned}&120.9259\frac{J}{mol\ Cr_2O_3 \cdot K} \cdot (2183 - 298)K \cdot \frac{1\ mol\ Cr_2O_3}{2\ mol\ Cr} \cdot \frac{1\ mol\ Cr}{52.00\ g\ Cr} \cdot \frac{10^6 g}{1\ t} \\ &+ 16.5345\frac{J}{mol\ C \cdot K} \cdot (2183 - 298)K \cdot \frac{3\ mol\ C}{2\ mol\ Cr} \cdot \frac{1\ mol\ Cr}{52.00\ g\ Cr} \cdot \frac{10^6 g}{1\ t}\end{aligned}$$

$$= 1.1345 \cdot 10^9 \frac{\text{J}}{\text{t Cr}} \cdot \frac{1\ \text{kWh}}{3.60 \cdot 10^6 \text{J}} = 315.152 \frac{\text{kWh}}{\text{t Cr}} \tag{2.243}$$

(b) The energy associated to the melting of the metal at 1907 °C is:

$$21.014 \frac{\text{kJ}}{\text{mol Cr}} \cdot \frac{1\ \text{mol Cr}}{52.00\ \text{g Cr}} \cdot \frac{10^6 \text{g Cr}}{1\ \text{t Cr}} = 4.0412 \cdot 10^5 \frac{\text{kJ}}{\text{t Cr}} \cdot \frac{10^3 \text{J}}{1\ \text{kJ}} \cdot \frac{1\ \text{kWh}}{3.60 \cdot 10^6 \text{J}}$$
$$= 112.2543 \frac{\text{kWh}}{\text{t Cr}} \tag{2.244}$$

In this case, as the final temperature is very close to the melting point of the metal, we did not consider the contribution of the heating of the melt from 1907 °C to 1910 °C. Consequently, the total energy contribution (under the ideal condition of 100% yield and raw material formed by pure Cr_2O_3) attributed to the carbothermal reduction of the chromium is:

$$7.1403 \cdot 10^6 \frac{\text{kJ}}{\text{t Cr}} + 1.1345 \cdot 10^6 \frac{\text{kJ}}{\text{t Cr}} + 4.0412 \cdot 10^5 \frac{\text{kJ}}{\text{t Cr}} = 8.6789 \cdot 10^6 \frac{\text{kJ}}{\text{t Cr}} \tag{2.245}$$

$$1983 \frac{\text{kWh}}{\text{t Cr}} + 315.152 \frac{\text{kWh}}{\text{t Cr}} + 112.2543 \frac{\text{kWh}}{\text{t Cr}} = 2410.41 \frac{\text{kWh}}{\text{t Cr}} \tag{2.246}$$

Coke consumption, identified with the stoichiometric quantity of carbon associated to the reaction (2.236), would be of 1386 kg of coke (formed by 100% carbon) per ton of chromium:

$$\frac{3\ \text{mol C}}{2\ \text{mol Cr}} \cdot \frac{12.01\ \text{g C}}{\text{mol C}} \cdot \frac{1\ \text{mol Cr}}{52.00\ \text{g Cr}} \cdot \frac{1\ \text{kg C}}{1000\ \text{g C}} \cdot \frac{10^6 \text{g}}{\text{t}} = 1386 \frac{\text{kg C}}{\text{t Cr}} \tag{2.247}$$

From the kinetic point of view, it is possible to consider the working hypothesis of the model of permanent regime, see Sect. 1.5, in the slag melt, but instead of considering the slagging of the element (Cr) we consider the assimilation of the element (Cr) to the metallic melt. A mass balance to the chromium of the slag for a time, t, considering that this element exhausts with the time due to the reaction (2.236), would be the following one:

$$(\%\text{Cr})^0_{\text{slag}} \cdot M^0_{\text{slag}} = (\%\text{Cr})_{\text{slag}} \cdot \eta_{\text{m/s}} \cdot \dot{M}_{\text{metal}} \cdot t + (\%\text{Cr})_{\text{slag}} \cdot M^0_{\text{slag}} \tag{2.248}$$

Starting from 400 kg of slag, M^0_{slag}, with an initial chromium content, $(\%\text{Cr})^0_{\text{slag}}$, of 68.4% (which corresponds to the chromium content in the Cr_2O_3) and, a partition coefficient between the metal and the slag of 2.5:

$$\eta_{\text{m/s}} = \frac{(\%\text{Cr})_{\text{metal}}}{(\%\text{Cr})_{\text{slag}}} = 2.5 \tag{2.249}$$

Table 2.8 %Cr as a function of the time

Time, minutes	0.0	1.0	10.0	15	20.0	25.0
%Cr, slag	68.4	65.0	45.0	38.4	33.5	29.7

The kinetic of the reaction would be expressed by the following equation:

$$(\%Cr)_{slag} = \frac{(\%Cr)^0_{slag}}{1 + \eta_{m/s} \cdot \left(\frac{\dot{M}_{metal}}{M^0_{slag}}\right) \cdot t} \quad (2.250)$$

and the results collected in Table 2.8 are obtained when the production of chromium per unit of time, M_{metal}, is of 500 kg per hour or 8.33 kg min^{-1}.

Commentaries: The order of magnitude of the energy consumptions resulted from the production of chromium contrasts with those obtained to produce steel in the electric furnace, Exercise 2.7. While 76% of the energy cost of the manufacture of steel in the electric steelworks was associated to the heating of the scrap, in the production of chromium, the sensitive heat of the raw material, Cr_2O_3, is only 26% while the energy associated to the carbothermal reduction reaches the 70%.

Solution ends.

Exercise 2.10 Study, both from the thermodynamic and the kinetic points of view, the characteristics of the process of the silicothermic reduction of 320 kg of chromite, $FeCr_2O_4$, in a ladle at 1600 °C to obtain 500 kg per hour of a ferrochromium with the following characteristics: 50% Cr, 49% Fe, and 1% of other elements.

Data: the atomic weights of the ferrochromium, chromium, iron, silicon, and oxygen are 53.35, 52.00, 55.85, 28.09, and 16.00 g mol^{-1}, respectively. The activity of the SiO_2, Exercise 2.4, is 0.11.

Solution Starts

Silicothermic reduction of chromite is the following one:

$$FeCr_2O_4(s) + 2Si(l) \leftrightarrow Fe(s) + 2Cr(s) + 2SiO_2(s) \quad (2.251)$$

The standard free energy of the reaction (2.251) can be obtained with the data of the Appendix 1:

$$Fe(s) + 2Cr(s) + 2O_2(g) \leftrightarrow FeCr_2O_4(s) \quad (2.252)$$

$$\Delta_r G^0_{reaction(2.252)}(T) = -1453.6 + 0.3255 \cdot T\left[kJ \cdot (mol\ FeCr_2O_4)^{-1}\right] \quad (2.253)$$

$$Si(l) + O_2(g) \leftrightarrow SiO_2(s) \tag{2.254}$$

$$\Delta_r G^0_{reaction(2.254)}(T) = -945.0 + 0.1980 \cdot T\left[kJ \cdot (mol\ SiO_2)^{-1}\right] \tag{2.255}$$

$$\begin{aligned}\Delta_r G^0_{reaction(2.251)}(T) &= 2 \cdot \Delta_r G^0_{reaction(2.254)}(T) - \Delta_r G^0_{reaction(2.252)}(T)\\ &= 2 \cdot (-945.0 + 0.1980 \cdot T) - (-1453.6 + 0.3255 \cdot T)\\ &= -436.4 + 0.0705 \cdot T\left[kJ \cdot (mol\ FeCr_2O_4)^{-1}\right]\end{aligned} \tag{2.256}$$

$$\begin{aligned}\Delta_r G^0_{reaction(2.251)}(1873\ K) &= -436.4 + 0.0705 \cdot 1873\left[kJ \cdot (mol\ FeCr_2O_4)^{-1}\right]\\ &= -304.3535\left[kJ \cdot (mol\ FeCr_2O_4)^{-1}\right]\end{aligned} \tag{2.257}$$

And the equilibrium constant is:

$$\begin{aligned}\Delta_r G^0_{reaction(2.251)}(1873\ K) &= -R \cdot T \cdot \ln K_{reactinn(2.251)}(1873\ K) \rightarrow K_{reactinn(2.251)}(1873\ K)\\ &= \exp\left(-\frac{\Delta_r G^0_{reaction(2.251)}(1873\ K)}{R \cdot T}\right)\\ &= \exp\left(-\frac{-304353.5\left[J \cdot (mol\ FeCr_2O_4)^{-1}\right]}{8.314\frac{J}{mol \cdot K} \cdot 1873\ K}\right) = 3.0774 \cdot 10^8\end{aligned} \tag{2.258}$$

The homogeneous metallic melt that is formed contains 50% Cr ($x_{Cr} = 0.5130$) and 49% Fe ($x_{Fe} = 0.4680$) and 1% of other elements. However, molar fractions of the metallic components in the melt (Cr and Fe) were estimated assuming that the 1% of other elements indicated in the enunciation was silicon:

$$x_{Cr} = \frac{50\ g\ Cr \cdot \frac{1\ mol\ Cr}{52.00\ g\ Cr}}{50\ g\ Cr \cdot \frac{1\ mol\ Cr}{52.00\ g\ Cr} + 49\ g\ Fe \cdot \frac{1\ mol\ Fe}{55.85\ g\ Fe} + 1\ g\ Si \cdot \frac{1\ mol\ Si}{28.09 g\ Si}} = 0.5130 \tag{2.259}$$

$$x_{Fe} = \frac{49\ g\ Fe \cdot \frac{1\ mol\ Fe}{55.85\ g\ Fe}}{50\ g\ Cr \cdot \frac{1\ mol\ Cr}{52.00\ g\ Cr} + 49\ g\ Fe \cdot \frac{1\ mol\ Fe}{55.85\ g\ Fe} + 1\ g\ Si \cdot \frac{1\ mol\ Si}{28.09 g\ Si}} = 0.4680 \tag{2.260}$$

$$x_{Si} = \frac{1\ g\ Cr \cdot \frac{1\ mol\ Si}{28.09\ g\ Si}}{50\ g\ Cr \cdot \frac{1\ mol\ Cr}{52.00\ g\ Cr} + 49\ g\ Fe \cdot \frac{1\ mol\ Fe}{55.85\ g\ Fe} + 1\ g\ Si \cdot \frac{1\ mol\ Si}{28.09 g\ Si}} = 0.0190 \tag{2.261}$$

Proceeding as in Exercise 2.1, it is possible to calculate the atomic weight of ferrochromium:

$$\begin{aligned}PA_{FeCr} &= x_{Cr} \cdot PA_{Cr} + x_{Fe} \cdot PA_{Fe} + x_{Si} \cdot PA_{Si} \\ &= 0.5130 \cdot 52.00 + 0.4680 \cdot 55.85 + 0.0190 \cdot 28.09 \\ &= 53.35 \text{ g} \cdot \text{mol}^{-1}\end{aligned} \tag{2.262}$$

Liquidus temperature of the melt is around 1600 °C according to the TAPP database. On another note, we assume that the elements chromium and iron have ideal behavior. That is to say, the activity of these elements is identified with their concentration in molar fraction:

$$a_{Cr} = x_{Cr} \tag{2.263}$$

$$a_{Fe} = x_{Fe} \tag{2.264}$$

while the activity of silicon, associated to the 1% of other elements, in the melt would be:

$$a_{Si} = x_{Si} \cdot \gamma^{*}_{Si} \tag{2.265}$$

where the silicon molar fraction, x_{Si}, is corrected by the coefficient of activity at infinite dilution. We show in the data of the Appendix 3 the standard free energy associated to the change of state of reference of the silicon (from the Raoult's activity of the pure metal to the Henry's activity in the diluted solution of the iron melt) solved in the molten iron at 1600 °C:

$$\text{Si(molten, } a_{Si} = 1) \leftrightarrow \text{Si}\big(\text{dis; iron, } a_{H,Si} = 1\big) \tag{2.266}$$

$$\Delta_r G^0_{reaction(2.266)} = -132000 - 17.2 \cdot T\big[\text{J} \cdot (\text{mol Si})^{-1}\big] \tag{2.267}$$

$$\begin{aligned}\Delta_r G^0_{reaction(2.266)}(1873\text{ K}) &= -132000 - 17.2 \cdot 1873\big[\text{J} \cdot (\text{mol Si})^{-1}\big] \\ &= -164215.6\big[\text{J} \cdot (\text{mol Si})^{-1}\big]\end{aligned} \tag{2.268}$$

$$\Delta_r G^0_{reaction\ (2.266)}(1873\text{ K}) = -R \cdot T \cdot \ln\left(\frac{a_{H,Si}}{a_{Si}}\right) \tag{2.269}$$

$$\begin{aligned}-164215.6\frac{\text{J}}{\text{mol Si}} &= -8.314\frac{\text{J}}{\text{mol} \cdot \text{K}} \cdot 1873\text{ K} \cdot \ln\left(\frac{\%\text{Si}}{x_{Si} \cdot \gamma^{*}_{Si}}\right) \rightarrow \exp\left(\frac{-164215.6}{-8.314 \cdot 1873}\right) \\ &= \frac{1\%}{0.019 \cdot \gamma^{*}_{Si}} \rightarrow \gamma^{*}_{Si} \\ &= 1.3848 \cdot 10^{-3}\end{aligned} \tag{2.270}$$

Consequently, the value of the silicon molar fraction, $x_{\overline{\overline{Si}}}$, in equilibrium (minimum silicon that can be obtained in the ferrochromium; this minimum value is defined by the thermodynamics, silicon will be always greater in the ferrochromium) with the compounds and elements of reaction (2.266) is:

$$3.0774 \cdot 10^8 = K_{reactinn(2.251)}(1873\ K) = \frac{x_{Fe} \cdot x_{Cr}^2 \cdot a_{SiO_2}^2}{a_{FeCr_2O_4} \cdot a_{Si}^2} = \frac{x_{Fe} \cdot x_{Cr}^2 \cdot a_{SiO_2}^2}{1 \cdot \left(x_{Si}^{=} \cdot \gamma_{Si}^{*}\right)^2} \tag{2.271}$$

$$x_{Si}^{=} = \sqrt{\frac{0.4680 \cdot 0.5130^2 \cdot 0.11^2}{3.0774 \cdot 10^8 \cdot \left(1.3848 \cdot 10^{-3}\right)^2}} = 1.5891 \cdot 10^{-3} \tag{2.272}$$

The composition of the silicon expressed in weight percentage is:

$$\frac{(\%Si)^{=}}{x_{Si}^{=} \cdot \gamma_{Si}^{*}} = \exp\left(\frac{-164215.6}{-8.314 \cdot 1873}\right) \rightarrow (\%Si)^{=} = x_{Si}^{=} \cdot \gamma_{Si}^{*} \cdot \exp\left(\frac{-164215.6}{-8.314 \cdot 1873}\right) = 0.0836 \tag{2.273}$$

or also calculated as follows:

$$(\%Si)^{=} = 1.5891 \cdot 10^{-3} \frac{\text{mol Si}}{\text{mol FeCr}} \cdot \frac{28.09\ \text{g Si}}{1\ \text{mol Si}} \cdot \frac{1\ \text{mol FeCr}}{53.35\ \text{g FeCr}} \cdot 100 = 0.0837\% \tag{2.274}$$

The obtained results demonstrate that is possible to obtain melts with 50% of chromium and a quantity of silicon dissolved in the ferroalloy of around 0.10%.

As reaction (2.251) has exothermic nature and it does not require energy supply, silicothermic process is usually performed in a ladle, Fig. 2.12. A ladle is an equipment of cylindrical geometry covered by refractory, and it is usually employed in the transportation of metallic melts. In the ladle, it is also possible to perform deoxidation, desulphurizing, and degasification reactions, typical of the secondary metallurgy or ladle metallurgy, Chap. 3.

The recovery of slags with high chromium content is usually performed by means of the contact of the slag with a metallic melt of high silicon content in a ladle, Fig. 2.12. In this case, the partition coefficient between the metal and the slag, considering the data of the $CaO–Cr_2O_3–SiO_2$ ternary system of TAPP, is of around 13.0. Proceeding as in Exercise 2.9, the evolution of the chromium content with the time in the slag would be expressed by the following equation:

$$(\%Cr)_{slag}^{0} = \frac{2\ \text{mol Cr}}{1\ \text{mol}\ FeCr_2O_4} \cdot \frac{1\ \text{mol}\ FeCr_2O_4}{223.85\ \text{g}\ FeCr_2O_4} \cdot \frac{52.00\ \text{g Cr}}{1\ \text{mol Cr}} \cdot 100 = 46.5\% \tag{2.275}$$

$$(\%\mathrm{Cr})_{\mathrm{slag}} = \frac{(\%\mathrm{Cr})^0_{\mathrm{slag}}}{1 + \eta_{\mathrm{m/s}} \cdot \left(\frac{\dot{\mathrm{M}}_{\mathrm{metal}}}{\mathrm{M}^0_{\mathrm{slag}}}\right) \cdot \mathrm{t}} = \frac{46.5}{1 + 13.0 \cdot \left(\frac{500\,\frac{\mathrm{kg}}{\mathrm{h}} \cdot \frac{1\,\mathrm{h}}{60\,\mathrm{min}}}{320\,\mathrm{kg}}\right) \cdot \mathrm{t}} \tag{2.276}$$

For the obtaining of 500 kg h^{-1}, using as starting material 320 kg of slag with 46.5% of chromium, the variation with the time of the chromium content in the slag is collected in Table 2.9.

Commentaries: It is necessary to use two reductants to produce high-quality ferroalloys: coke and an active metal (silicon or aluminum).

End of the exercise.

Exercise 2.11 To produce one ton of ferromanganese in electric furnace with the quality indicated in Table 2.10, we use 1726 kg of a sintered product with the composition indicated in Table 2.11.

As reductant reagent we use 318.50 kg of coke with 90.50% C. We produce 459.65 kg of slag with 40.40% MnO and 609.44 kg of gas with the 45.80% CO and 44.01% CO_2. These gases drag 34.25 kg of powders with 41.41% C and 2.38% MnO. Finally, it is necessary to record the entrance of 58.84 kg of wet ambient air. With the data of the global mass balance, obtain the equation of the operating line of the furnace.

Data: the atomic weights of the Mn, Fe, O, and C are: 54.94, 55.85, 16.00, and 12.01 g mol^{-1}, respectively.

We represent the flows of entrance and exit to the reduction furnace in Fig. 2.13.

Solution Starts

Considering the fundamentals developed in the Sect. 2.3.3 referred to the control of the electric furnaces dedicated to the production of pig iron and ferroalloys using the operating line, we consider that the operation line (Eq. (2.203)) is necessary for the control of the process. For the determination of the operating line, it is necessary to know the coordinates of two points of it or the coordinates of one point and the slope of the line. For instance, it is possible its determination in that plane where the crossing between the mineral charge that enters in the furnace and the gas that leaves the furnace dragging powders takes place.

We are going to consider two situations:

Table 2.9 Chromium content in the slag as a function of the time

% Cr-slag	46.5	34.7	10.6	7.6	6.0	4.9
t (min)	0	1	10	15	20	25

Table 2.10 Composition of the ferromanganese

%Mn	%Fe	%C	%Si
79.00	13.10	7.20	0.40

Table 2.11 Composition of the sintered raw material

%MnO	%Fe_2O_3	%FeO	%SiO_2	%CaO	%MgO	%Al_2O_3	%BaO	%Others
69.16	3.76	6.64	5.63	7.06	2.32	0.96	0.26	4.21

Fig. 2.13 Entrance-exit mass balance

Situation A: In this situation, all the iron oxides available in the sinter (Fe_2O_3 and FeO) can be reduced to sponge iron with the reductant gas (mechanism of reduction with gas) before the entrance to the hearth (indirect reduction with gas in the upper zone of the shaft). In this case, it is necessary to consider the oxygen provided by the iron oxides of the sintered charge to estimate the quantity of oxygen in the charge:

– The number of manganese moles in the charge coming from the 1726 kg of sintered product are:

$$\frac{69.16\text{ g MnO}}{100\text{ g sinter}}\cdot\frac{1\text{ mol MnO}}{70.94\text{ g MnO}}\cdot\frac{1\text{ mol Mn}}{1\text{ mol MnO}}\cdot 1726\cdot 10^3\text{g sinter} = 16826.92\text{ mol Mn} \tag{2.277}$$

– The number of manganese moles coming from the 34.25 kg of powders generated in the reduction process:

$$\frac{2.38\text{ g MnO}}{100\text{ g powder}}\cdot\frac{1\text{ mol MnO}}{70.94\text{ g MnO}}\cdot\frac{1\text{ mol Mn}}{1\text{ mol MnO}}\cdot 34.25\cdot 10^3\text{g powder} = 11.49\text{ mol Mn} \tag{2.278}$$

Thus, the net quantity processed in the furnace is:

$$16826.92\text{ mol Mn} - 11.49\text{ mol Mn} = 16815.43\text{ mol Mn} \tag{2.279}$$

To calculate the number of moles of oxygen, apart from the 16,815.43 mol associated to the MnO, it is necessary to consider those associated to the FeO and Fe_2O_3. That is to say:

- The number of oxygen moles associated to the MnO of the sinter are: 16,826.92 mol.
- The number of oxygen moles associated to the MnO of the powders that leave the furnace are: 11.49 mol.
- The number of oxygen moles associated to the Fe_2O_3 and FeO are calculated as follows:

$$\frac{3.76\ \text{g Fe}_2\text{O}_3}{100\ \text{g sinter}} \cdot \frac{1\ \text{mol Fe}_2\text{O}_3}{159.7\ \text{g Fe}_2\text{O}_3} \cdot \frac{3\ \text{mol O}}{1\ \text{mol Fe}_2\text{O}_3} \cdot 1726 \cdot 10^3\text{g sinter} = 1219.12\ \text{mol O} \tag{2.280}$$

$$\frac{6.64\ \text{g FeO}}{100\ \text{g sinter}} \cdot \frac{1\ \text{mol FeO}}{71.85\ \text{g FeO}} \cdot \frac{1\ \text{mol O}}{1\ \text{mol FeO}} \cdot 1726 \cdot 10^3\text{g sinter} = 1595.07\ \text{mol O} \tag{2.281}$$

This gives as a result that the total quantity of oxygen moles that are charged into the furnace is 19629.62, and for that reason, the relation between the number of oxygen moles and manganese moles is:

$$Y(1) = \frac{\text{moles O}}{\text{moles Mn}} = \frac{19629.62}{16.815.43} = 1.167 \tag{2.282}$$

Manganese sintered product is treated with a gas where the CO/CO_2 ratio is 1.04. Now we calculate the number of oxygen moles in the gas that leaves the furnace:

$$\frac{45.80\ \text{g CO}}{100\ \text{g gas}} \cdot \frac{1\ \text{mol CO}}{28.01\ \text{g CO}} \cdot \frac{1\ \text{mol O}}{1\ \text{mol CO}} \cdot 609.44 \cdot 10^3\text{g gas} = 9965.14\ \text{mol O} \tag{2.283}$$

$$\frac{44.01\ \text{g CO}_2}{100\ \text{g gas}} \cdot \frac{1\ \text{mol CO}_2}{44.01\ \text{g CO}_2} \cdot \frac{2\ \text{mol O}}{1\ \text{mol CO}_2} \cdot 609.44 \cdot 10^3\text{g gas} = 12188.8\ \text{mol O} \tag{2.284}$$

and the number of carbon moles is:

$$\frac{45.80\ \text{g CO}}{100\ \text{g gas}} \cdot \frac{1\ \text{mol CO}}{28.01\ \text{g CO}} \cdot \frac{1\ \text{mol C}}{1\ \text{mol CO}} \cdot 609.44 \cdot 10^3\text{g gas} = 9965.14\ \text{mol C} \tag{2.285}$$

$$\frac{44.01\ \text{g CO}_2}{100\ \text{g gas}} \cdot \frac{1\ \text{mol CO}_2}{44.01\ \text{g CO}_2} \cdot \frac{1\ \text{mol C}}{1\ \text{mol CO}_2} \cdot 609.44 \cdot 10^3\text{g gas} = 6094.4\ \text{mol C} \tag{2.286}$$

Thus:

$$X(1) = \frac{\text{moles O}}{\text{moles C}} = \frac{22153.94}{16059.54} = 1.379 \qquad (2.287)$$

The value of the slope of the operating line is related with the quantity of carbon associated to the reduction processes.

The quantity of carbon in the coke is calculated as follows:

$$\frac{90.50\ \text{g C}}{100\ \text{g coke}} \cdot \frac{1\ \text{mol C}}{12.01\ \text{g C}} \cdot 318.50 \cdot 10^3 \text{g coke} = 24000.21\ \text{mol C} \qquad (2.288)$$

It is necessary to subtract to this quantity both the carbon that is dragged by the ferroalloy and the carbon associated to the powders:

$$\frac{7.20\ \text{g C}}{100\ \text{g ferroalloy}} \cdot \frac{1\ \text{mol C}}{12.01\ \text{g C}} \cdot 1000 \cdot 10^3 \text{g ferroalloy} = 5995\ \text{mol C} \qquad (2.289)$$

$$\frac{41.41\ \text{g C}}{100\ \text{g powder}} \cdot \frac{1\ \text{mol C}}{12.01\ \text{g C}} \cdot 34.25 \cdot 10^3 \text{g gas} = 1180.93\ \text{mol C} \qquad (2.290)$$

Definitely, the quantity of carbon available in the process is:

$$24000.21\ \text{mol C} - 5995\ \text{mol C} - 1180.93\ \text{mol C} = 16824.28\ \text{mol C} \qquad (2.291)$$

to produce 1000 kg of ferroalloy with 79% Mn, or with:

$$\frac{79\ \text{g Mn}}{100\ \text{g ferroalloy}} \cdot \frac{1\ \text{mol Mn}}{54.94\ \text{g Mn}} \cdot 1000 \cdot 10^3 \text{g gas} = 14379.32\ \text{mol Mn} \qquad (2.292)$$

Consequently, the slope, μ, of the operating line is:

$$\mu = \frac{\text{mol C}}{\text{mol Mn}} = \frac{16824.28\ \text{mol C}}{14379.32\ \text{mol Mn}} = 1.170 \qquad (2.293)$$

With the previous data (1.379, 1.167), we calculate the origin ordinate (m) of the operating line:

$$Y = \mu \cdot X + m \rightarrow 1.167 = 1.170 \cdot 1.379 + m \rightarrow m = -0.446 \qquad (2.294)$$

and the equation of the operating line is:

$$Y = 1.170 \cdot X - 0.446 \qquad (2.295)$$

Situation B: All the iron oxides in the sinter (Fe_2O_3 and FeO) are in this case reduced to molten iron by a direct reduction mechanism using coke (direct reduction with coke in the lower part of the shaft).

In this case, the value of the slope of the operating line, μ, and that of the coordinate X(1) are the same that in the **Situation A**. On the contrary, the calculation of the coordinate Y(1) is different. In this situation, as the number of oxygen moles in the mineral charge that enters in the furnace is not increased, the value of Y(1) is equal to one:

$$\mathrm{Y(1)} = \frac{\text{moles O}}{\text{moles Mn}} = \frac{16815.43}{16.815.43} = 1.000 \tag{2.296}$$

As a result, considering the coordinates of the point 1 (1.000, 1.379), we calculate the origin ordinate (m) of the operating line:

$$\mathrm{Y} = \mu \cdot \mathrm{X} + \mathrm{m} \rightarrow 1.000 = 1.170 \cdot 1.379 + \mathrm{m} \rightarrow \mathrm{m} = -0.613 \tag{2.297}$$

and the equation of the operating line is:

$$\mathrm{Y} = 1.170 \cdot \mathrm{X} - 0.613 \tag{2.298}$$

Commentaries: It is not usual to use the operating line in processes of carbothermal reduction in an electric furnace. It is not common to use burden formed by 100% sintered material. However, the concepts and information associated to the operating line can be useful instruments if the functioning of any process of carbothermal reduction of ferroalloys with high carbon content in the final metallic melt is considered.

End of the exercise.

2.4 Design of Electric Furnaces

The relation between the power of the furnace, P, in MW, the specific energy consumption, E, for the manufacture of one ton of product, in kWh t^{-1}, and the time, t, in seconds, required for the obtaining of a certain quantity of product, M, in tons, can be expressed as follows:

$$\mathrm{M} = \mathrm{P} \cdot 10^6 \cdot \mathrm{t} \cdot \left(\frac{\mathrm{kWh}}{3.6 \cdot 10^6 \mathrm{J}}\right) \cdot \left(\frac{\mathrm{ton}}{\mathrm{E\ kWh}}\right) = \frac{\mathrm{P} \cdot \mathrm{t}}{3.6 \cdot \mathrm{E}} \tag{2.299}$$

The variables indicated in Eq. (2.299) may adopt a wide range of values. For example, the specific energy consumption to produce a ton of product can vary from values lower than 400 kWh t^{-1} used in the production of steel, passing through the 4000 kWh t^{-1} required in the case of the ferroalloys, to the 10,000 kWh t^{-1} involved in the production of ceramics derived from the solidification of melts.

Also, the time employed in the manufacture can be of minutes in the case of carbon steel production or hours in the case of ferroalloys and ceramic products. As the variable time has a significant importance in the energetic, economic, and environmental yield of the installation, all those instruments able to reduce the duration of the smelting, as it is the option of the refinement-formation of synthetic slag synthetized out of the process, could play an important role in the increase of the efficiency of these installations (Fernández-González et al. 2018a, b, c, 2019a, b, c).

Finally, as many of the products coming from the electric furnace are related with high electric power consumptions, it would not be strange that, in the future, some of these materials might be considered potential instruments for energy storage (Andresen 1995).

2.4.1 Characteristics and Devices of the Furnaces

One of the first steps to be taken to set the work system that allowed the development of electric furnaces would be defining the most relevant zones and devices. Therefore, we propose to define three zones of interest and two relevant devices in the functioning of the electric furnace as the electrodes and the cooling system are, Fig. 2.14.

The most relevant characteristics of the electric furnace (Fig. 2.14) are indicated as follows:

Zone 1, which is identified with the cover or roof (T) of the furnace. Geometry is similar to that of spherical cap with radius equivalent to that of the crucible of the furnace, R_{HEARTH}, with a height-depth, h_T, that keeps an order of magnitude equivalent to that of refractory materials used in the construction of the lining of the roof. It is possible to find in the roof of some electric furnaces, refrigeration devices based on the circulation of water used to remove heat that, via gases, heats the walls of the roof. Nowadays, it is possible to redirect this situation using designs of the roofs based on dense refractory concrete in contact with the hot gases and with the low thermal conductivity isolating materials that are in contact with the structural steel cover sheet (shell). One of the reasons would be that the dome of the furnaces could be open with the purpose of feeding the furnace. A cover with excessively weight of refractory material and with cooling devices difficulties and impedes these operations.

Zone 2 is associated to the shaft of the furnace in contact with the gas (CG). The geometry of this zone can be identified with a cylinder whose height, h_{CG}, can be in the low-shaft furnaces of the order of magnitude of the radius of the hearth, R_{HEARTH}, as it is possible to see in Fig. 2.14. However, in the high-shaft furnaces, for instance in those used to produce pig iron or high carbon ferroalloys, the height of the shaft can reach 5–10 times the value of the radius of the hearth of the furnace. The height of the shaft with respect to the radius of the hearth is also function of the utility that wants to be given to the CO generated in the reduction with coke. Also in the low-shaft furnaces, the slag melt associated to the Zone 3 (CR1) can be covered with a layer of slag that still did not reached the liquidus temperature that, apart from protecting the

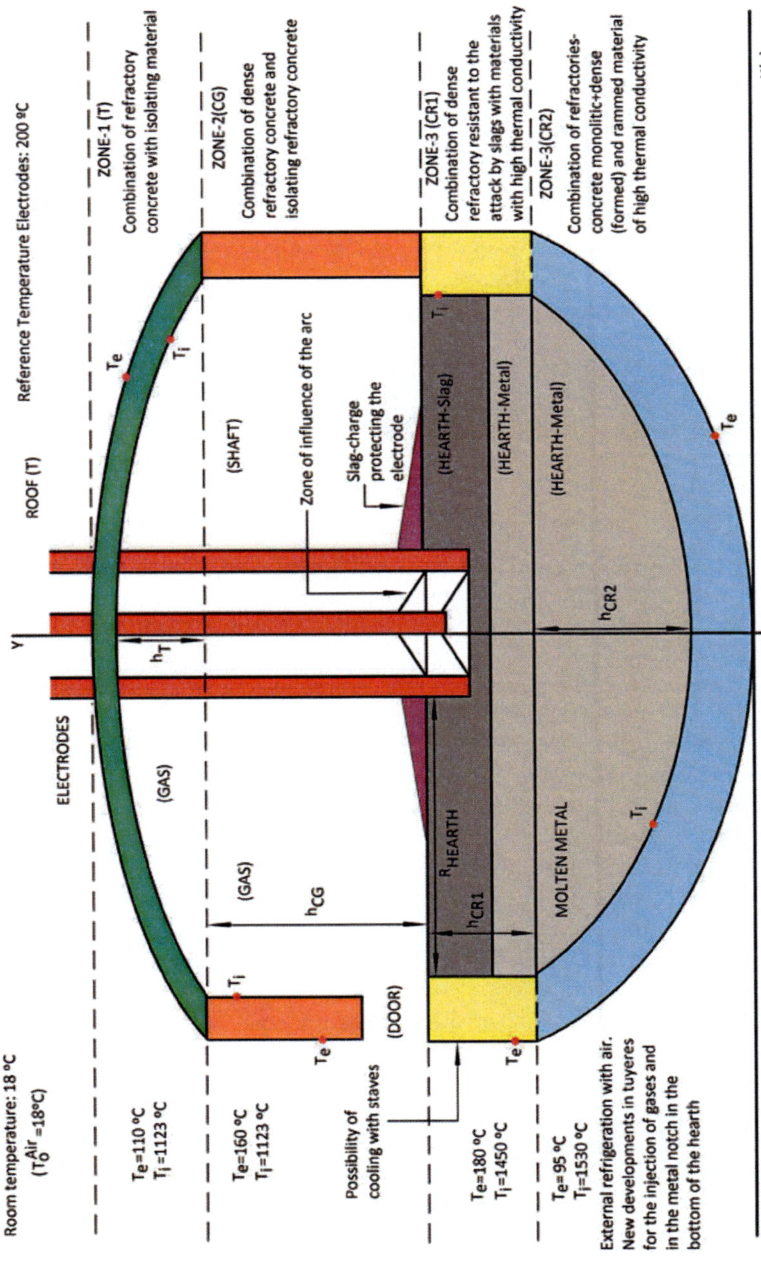

Fig. 2.14 Scheme and boundary conditions of an electric furnace dedicated to the production of an iron melt with a chemical composition equivalent to that of a structural steel

roof from the radiation, acts as chemical reserve in the refinement reactions that can take place. Generally, as the shaft of the furnace is an element that is not susceptible of being removed-displaced from one place to another (it is only possible to rotate it) as in the case of the roof-cover of the furnace, the thickness of the refractory lining usually has a greater thickness than the roof-cover of the furnace, although the main objective of the refractory lining in this zone is to minimize the heat losses.

Zone 3 is associated to the hearth-elaboration zone (CR) of the furnace. However, it is possible to distinguish two parts clearly different in the Zone 3 of the furnace:

- That part situated in the upper zone of the hearth identified as CR1, where slag and metallic melt are in contact with the refractory lining, which can or cannot be externally cooled with water. The geometry of the CR1 zone is identified with that of a cylinder with radius equivalent to that of the furnace, R_{HEARTH}, and with a height referenced as h_{CR1}.
- A lower zone of the hearth identified as CR2, where metallic melt is in contact with a refractory wall formed by three different types of materials that is habitually externally cooled by means of a forced air flow. The geometry of the CR2 zone is identified with that of a spherical cap of radius equivalent to that of the furnace, R_{HEARTH}, and a height identified as h_{CR2}, Fig. 2.14.

The electrodes of the furnace constitute other of the differential elements by means of which the thermal energy, required for the smelting processes and the reduction and refinement reactions of the different elements in the charge, is transferred to the charge. Also the electrodes, due to their special thermal, mechanical, and physical–chemical characteristics, operating in an oxidizing environment, suffer corrosion-wear during the operation of the furnace, and this represents one of the costs to be considered in the production of different materials in electric furnaces, Sect. 2.2.

Even when the three-phase electric arc furnaces are still the most widely used, continuous current electric furnaces are gradually being implemented. This is the case of great size-capacity furnaces. In this situation, the number of electrodes is reduced to one, although its diameter is greater than that of the electrodes used in the three-phase electric arc furnace. Developments in the area are focused on the improvements of the efficiency of the rectification of alternating current systems to generate continuous current, having as main counterpart the noticeable reduction of the acoustic pollution generated by the start-striking—stabilization of the three-phase arc.

Material used in the manufacture of the electrode is coal, with low contents in ashes, and graphite. These raw materials are agglomerated with thermoplastic materials produced in the distilling-coking of coal or produced in the obtaining of light hydrocarbons in petrochemical factories (Verdeja et al. 2014, pp. 325–345). Depending on that the final stage of the electrodes production took place in the own furnace or out of it, we can talk about either Soderberg or prebaked electrodes, respectively. Physical and mechanical characteristics of the electrodes, it does not matter if they are Soderberg or prebaked, are comprised in the following intervals (Verdeja et al. 2014, pp. 325–345):

- Global density: 1640–1750 kg m^{-3}
- Electrical resistance: 4.5–4.6 μΩ m
- Three-point bending test resistance: 10–23 MPa
- Young's modulus: 10–15 GPa
- Coefficient of linear thermal expansion: 1.20 10^{-6}–2.5 10^{-6} °C^{-1}.

New metal-ceramic electrodes have been developed in the last years that allowed increasing the current intensity that can be transferred to the different melts. The design of graphite electrodes with core of 1100 aluminum alloy annealed or alloys of the Al-Si system with aluminum contents greater than that of the eutectic open the field of the novel development of metal-ceramic composites to manufacture electrodes (Pero-Sanz et al. 2017, 2019).

As other auxiliary devices used in electric furnaces, it is possible to mention the injection lances, or collecting shovel-rake used either to remove or homogenize slags in the phosphorus and sulphur refinement during the process employed to produce a metallic melt. Also, recently, technologies aimed at the magnetic stirring of the melt are also included, which might accelerate the kinetics of the smelting and refinement. Improvement of the processes of injection of gases, slags, reagents-deoxidizing products, and desulphurizing agents efficiencies requires the utilization of porous plugs connected to the refractory lining of the bottom of the hearth and also requires the utilization of magnetic stirring systems in those cases where it is possible.

However, electric steelworks are not using systems to open and close hearths similar to those employed in the blast furnace. There are no references about the utilization of staves in the construction of the lining of the Zone 3 of the electric furnace (they are providing good results in blast furnaces) (Verdeja et al. 2020).

Finally, we should also consider the construction design of electric furnaces with hearth of rectangular or circular geometry (see Olsen et al. 2007, p. 63; Gasik 2013, p. 245) and flat bottom like that of the blast furnace, especially in installations with great capacity of production. Isotherms throughout the lining of the furnace hearth (furnace with hearth of spherical cap geometry, also called circular) adapt to the profile of uniform corrosion-wear the hearth experiences. Hearths with flat bottom-cylindrical symmetry experience differential corrosion of the central materials of the hearth different with respect to that of the corners and walls close to the bottom. Habitually, the control of the wear of a hearth of the type of spherical cap, Fig. 2.14, is homogeneous in all the surface of contact melt-refractory and, in this way, the risk of undesirable perforations is reduced (see Verdeja et al. 2014, pp. 175–198).

2.4.2 *Heat Transfer Mechanisms*

In this section, we analyze the heat flows in each one of the most representative zones of the electric furnace considering the scheme of Fig. 2.14. However, the values presented in these pages are consequence of the following working hypothesis:

1. We assume that the stationary regime is reached in the furnace and, the heat flow from the gas to the refractory is the same than that transferred through the refractory lining and that the heat dissipates from the furnace wall to the environment.
2. If the stationary regime is not reached in some zones of the furnace, the fraction of the heat flow that is destined to the energy storage in the metal, the slag, and the refractory is small if compared with the value of the total flow of heat that is being transferred (quasi-stationary regime).

The heat exchange between the furnace walls and the environment can be estimated through the following transfer equation:

$$J_q^{H-air} = h_g \cdot \left(T_i - T_o^{air}\right)\left(W \cdot m^{-2}\right) \tag{2.300}$$

where we consider the global coefficient of heat transfer, h_g, which takes into account the mechanisms of radiation and natural convection at the temperature of the steel shell-sheet of the furnace, T_i, susceptible of being measured and at the air temperature, T_o^{air}.

Even when it is truth that heat flows of the furnace surfaces that connect with the environment are different depending on the spatial orientation, we propose a correlation for the estimation-calculation of the value of h_g expressed in W m^{-2} K^{-1} that is function of the temperature, T_i (K), reached in the shell-sheet of steel:

$$h_g = a \cdot T^3 - b \cdot T^2 + c \cdot T + d\left(W \cdot m^{-2} \cdot K^{-1}\right) \tag{2.301}$$

where $a = 3.89\times10^{-7}$, $b = 3.94\times10^{-4}$, $c = 0.140$, and $d = 0.468$.

This way, if we know the temperature in those most representative surfaces-zones of the furnace, it is possible to obtain a first estimation of the heat losses of the furnace and knowing the weight-contribution-importance of them in the global balance and in the process that is being considered.

2.4.2.1 Conductive Heat Transfer

From the three mechanisms of heat transfer, conduction along the refractory-isolating lining of the furnace has two important characteristics:

(a) the conduction is the slowest heat transfer mechanism, and it controls the heat losses of the furnace.

(b) we assume that the conduction is the single heat transfer mechanism that operates in the different qualities of materials finally selected to build the lining of the furnace.

In stationary state, the heat flow through a refractory lining with three types of materials (first law of Fick), as for instance that used in the design of the Zone 3 (CR2), Fig. 2.14, would be:

$$J_q = \lambda_1 \cdot \left(\frac{T_1 - T_o^{steel}}{\Delta x_1}\right) + \lambda_2 \cdot \left(\frac{T_2 - T_1}{\Delta x_2}\right) + \lambda_3 \cdot \left(\frac{T_3 - T_2}{\Delta x_3}\right) (W \cdot m^{-2}) \tag{2.302}$$

where λ_i represents the thermal conductivity of each one of the products of the lining, Δx_i is the thickness of each layer and ΔT_i are the corresponding temperature differences starting by the temperature reached by the steel shell, T_o^{steel}, and finishing by the temperature of the surface of the refractory in contact with the metallic melt, T_3. It is possible to obtain from Eq. (2.300) three equations that can be used to estimate the temperature drop, in stationary state, in each one of the refractories:

$$T_1 - T_o^{steel} = \frac{J_q}{\left(\frac{\lambda_1}{\Delta x_1}\right)} = \frac{J_q \cdot \Delta x_1}{\lambda_1} \tag{2.303}$$

$$T_2 - T_1 = \frac{J_q}{\left(\frac{\lambda_2}{\Delta x_2}\right)} = \frac{J_q \cdot \Delta x_2}{\lambda_2} \tag{2.304}$$

$$T_3 - T_2 = \frac{J_q}{\left(\frac{\lambda_3}{\Delta x_3}\right)} = \frac{J_q \cdot \Delta x_3}{\lambda_3} \tag{2.305}$$

The addition of Eqs. (2.303), (2.304), and (2.305) gives as a result the following equation:

$$T_3 - T_o^{steel} = \frac{J_q}{\left(\frac{\bar{\lambda}}{\Delta x}\right)} = \frac{J_q \cdot \Delta x}{\bar{\lambda}} \tag{2.306}$$

where the temperature T_3 is 1530 °C, which is reached in the refractory-fluid interphase, Fig. 2.14, and T_o^{steel} is 95 °C, which is the temperature of the steel shell in contact with the current of forced air used as cooling fluid of the hearth in this zone of the furnace.

If we consider that the global coefficient of heat transfer, h_g, in this zone of the hearth cooled by a forced current of air (at a temperature of 18 °C) is 90 W m^{-2} K^{-1}, the heat flow along the surface of the Zone 3 (CR2) of the furnace, Fig. 2.14, would be equal to:

$$J_q^{H-air} = h_g \cdot \left(T_i - T_o^{air}\right) = 90 \frac{W}{m^2 \cdot K}[(273 + 95) - (273 + 18)]K = 6930\ W \cdot m^{-2} \tag{2.307}$$

Consequently, if we transfer the data of the heat flow of Eq. (2.307) to Eq. (2.306), it would be possible to estimate the average conductivity of the materials selected to build the hearth of the furnace in the Zone 3(CR2) of 0.5 m in thickness (500 mm), see Exercise 2.12:

$$\overline{\lambda} = \frac{J_q \cdot \Delta x}{(T_i - T_e)} = \frac{6930\ W \cdot m^{-2} \cdot 0.5\ m}{[(1530 + 273) - (95 + 273)]K} = 2.42\ W \cdot m^{-1} \cdot K^{-1} \tag{2.308}$$

Proceeding as we did for the materials-heat flows in the Zone 3 (CR2), it is possible to analyze the other zones of the electric furnace, Fig. 2.14: Zone 3 (CR1), Zone 2 (CG), and Zone 1 (T).

2.4.2.2 Convective Heat Transfer

Convection heat transfer in the shaft and hearth of the furnace (Zones 2 and 3 in Fig. 2.14) involves the functional relation of the following seven variables:

- Linear characteristic length, L (m).
- Heat transfer coefficient, h_c (W m^{-2} K^{-1}).
- Thermal conductivity, λ (W m^{-1} K^{-1}).
- Viscosity, μ (Pa s).
- Density, ρ (kg m^{-3}).
- Specific heat, c_p (J kg^{-1} K^{-1}).
- Rate, v (m s^{-1}).

The number of dimensionless groups (Nusselt (Nu), Reynolds (Re), and Prandtl (Pr)) that can be derived from the seven above-mentioned variables (dimensional analysis) and the four existing independent variables (mass, length, time, and heat) is determined by means of the Buckingham's π Theorem: the number of dimensionless groups (π groups) is equal to the number of variables involved in the problem (7) minus the corresponding to the independent variables (4), see Guthrie (1989, p. 155). It is possible to deduce from the dimensional analysis the following functional relation:

$$Nu = C \cdot Re^m \cdot Pr^n \tag{2.309}$$

Concretely, to study the process of convective heat transfer from a hot surface-wall-layer toward a fluid (see the similarity with the equation used to obtain the mass transfer coefficient in steel decarburization in electric furnace, see Sect. 2.3.2), the following equation is used:

$$\mathrm{Nu} = 0.664 \cdot \mathrm{Re}^{1/2} \cdot \mathrm{Pr}^{1/3} \tag{2.310}$$

To obtain the value-estimation of the convective heat transfer coefficient, h_c, in the heat transfer from the walls of the shaft of the furnace (L = 1.90 m), Fig. 2.14, at the temperature of 160 °C (433 K) toward the air at 18 °C (291 K) that moves at a rate, v, of 1.0 m s^{-1}, we must consider:

- that the thermal properties of the air are those corresponding to the temperature, $\overline{T}$:

$$\overline{T} = \frac{433 + 291}{2} = 392\ \mathrm{K}(89\,°\mathrm{C}) \tag{2.311}$$

- According to Poirier and Poirier (1992, p. 19), the thermal conductivity, viscosity, density, and specific heat of the air at 89 °C are 0.029945 W m^{-1} K^{-1}, 2.07 10^{-5} Pa s, 1.03 kg m^{-3}, and 1009.9 J kg^{-1} K^{-1}, respectively.

Consequently, the values of Reynolds (Re) and Prandtl (Pr) numbers are:

$$\mathrm{Re} = \frac{v \cdot \rho \cdot L}{\mu} = \frac{1.0\ \mathrm{m} \cdot \mathrm{s}^{-1} \cdot 1.03\ \mathrm{kg} \cdot \mathrm{m}^{-3} \cdot 1.90\ \mathrm{m}}{2.07 \cdot 10^{-5}\mathrm{Pa} \cdot \mathrm{s}} = 9.4720 \cdot 10^{4} \tag{2.312}$$

$$\mathrm{Pr} = \frac{c_p \cdot \mu}{\lambda} = \frac{1009.9\ \mathrm{J} \cdot \mathrm{kg}^{-1} \cdot \mathrm{K}^{-1} \cdot 2.07 \cdot 10^{-5}\mathrm{Pa} \cdot \mathrm{s}}{0.029945\ \mathrm{W} \cdot \mathrm{m}^{-1} \cdot \mathrm{K}^{-1}} = 0.6968 \tag{2.313}$$

The number of Nusselt and the value of h_c, according to Eq. (2.310) are the following ones:

$$\mathrm{Nu} = \frac{h_c \cdot L}{\lambda} \tag{2.315}$$

$$\mathrm{Nu} = 0.664 \cdot \mathrm{Re}^{1/2} \cdot \mathrm{Pr}^{1/3} = 0.664 \cdot \left(9.4720 \cdot 10^{4}\right)^{1/2} \cdot (0.6968)^{1/3} = 1.8117 \cdot 10^{2} \tag{2.316}$$

$$h_c = \frac{\mathrm{Nu} \cdot \lambda}{L} = \frac{1.8117 \cdot 10^{2} \cdot 0.029945\ \mathrm{W} \cdot \mathrm{m}^{-1} \cdot \mathrm{K}^{-1}}{1.90\ \mathrm{m}} = 2.86\ \mathrm{W} \cdot \mathrm{m}^{-2} \cdot \mathrm{K}^{-1} \tag{2.317}$$

If we compare this value with the global coefficient of heat transfer, h_g, from the steel shell-sheet of the furnace toward the environment, Eq. (2.301) (the value of h_g for a wall temperature of 433 K is of 18.80 W m^{-2} K^{-1}), we can conclude that 15% of the heat losses can be attributed to the convective mechanism and the remaining 85% can be attributed to the radiant heat transfer mechanism.

Making a similar operation, it is possible to calculate the convective heat transfer coefficient from the upper wall of the hearth of the furnace (Fig. 2.14, Zone 3 (CR1)) that faces the environment, using the same thermal properties of the air that in the previous calculation, and its value is 5.87 W m^{-2} K^{-1}. In this case, the contribution of the convective mechanism in the heat dissipation of the furnace walls is increased up to the 31% (the value of h_g for a wall temperature of 453 K, Eq. (2.301), is of 19.20 W m^{-2} K^{-1}).

To analyze the behavior of the convective heat transfer mechanism in slags, we have considered a surface of metallic melt (that can act as heat sink or emitting source) that transfers heat to the slag. The physical and thermal properties of the slag, apart from incorporating the data corresponding to the slag of eight components studied in the Exercise 2.1, the estimation of the thermal conductivity coming from the references Kawai and Shiraishi (1988, pp. 241–242) and Springorum (1995, pp. 597–599) is also considered. The thermal conductivity, viscosity, density, and specific heat given to the slag in molten state, independently of the operating temperature, are 0.50 W m^{-1} K^{-1}, 1.40 Pa s, 3204 kg m^{-3}, and 1257.0 J kg^{-1} K^{-1}, respectively. We equally consider that the characteristic linear dimension, L, of the system is identified with that of the inner diameter of the furnace, 3.72 m, and the relative displacement rate of the slag is 1.0 m s^{-1}. The number of Nusselt and the value of h_g, according to Eq. (2.310), would be the following one:

$$\mathrm{Re} = \frac{v \cdot \rho \cdot L}{\mu} = \frac{1.0\ \mathrm{m \cdot s^{-1} \cdot 3204\ kg \cdot m^{-3} \cdot 3.72\ m}}{1.40\ \mathrm{Pa \cdot s}} = 8.5135 \cdot 10^3 \tag{2.318}$$

$$\mathrm{Pr} = \frac{c_p \cdot \mu}{\lambda} = \frac{1257.0\ \mathrm{J \cdot kg^{-1} \cdot K^{-1} \cdot 1.40\ Pa \cdot s}}{0.50\ \mathrm{W \cdot m^{-1} \cdot K^{-1}}} = 3519.6 \tag{2.319}$$

$$\mathrm{Nu} = \frac{h_c \cdot L}{\lambda} \tag{2.320}$$

$$\mathrm{Nu} = 0.664 \cdot \mathrm{Re}^{1/2} \cdot \mathrm{Pr}^{1/3} = 0.664 \cdot \left(8.5135 \cdot 10^3\right)^{1/2} \cdot (3519.6)^{1/3} = 9.3149 \cdot 10^2 \tag{2.321}$$

$$h_c = \frac{\mathrm{Nu} \cdot \lambda}{L} = \frac{9.3149 \cdot 10^2 \cdot 0.50\ \mathrm{W \cdot m^{-1} \cdot K^{-1}}}{3.72\ \mathrm{m}} = 125.3\ \mathrm{W \cdot m^{-2} \cdot K^{-1}} \tag{2.322}$$

It is possible to reach greater values of h_c in slags, for instance of around 450 W m^{-2} K^{-1}, if the thermal conductivity was increased up to 1.50 W m^{-1} K^{-1}, for example, with the addition of calcium fluoride, fluorite, CaF_2.

Finally, Eq. (2.310) should not be applied when $\mathrm{Re} > 10^6$, as is the case of metallic melts. The equation recommended to estimate the number of Nusselt with high Reynolds number would be:

$$\mathrm{Nu} = 0.50 \cdot \mathrm{Re}^{1/2} \cdot \mathrm{Pr}^{1/3} \cdot \left[\frac{0.455}{(\log \mathrm{Re})^{2.584}} - \frac{3300}{\mathrm{Re}}\right] \tag{2.323}$$

In the case of iron melts, and considering, as in the case of the slag previously studied, that the characteristic linear dimension, L, of the system in this case is also the inner diameter of the furnace, 3.72 m (Fig. 2.14), and equally, the value of the relative displacement rate of the slag is 1.0 m s^{-1}. The values of the numbers of Reynolds (Re) and Prandtl (Pr) would be:

$$\mathrm{Re} = \frac{v \cdot \rho \cdot L}{\mu} = \frac{1.0\ \mathrm{m \cdot s^{-1} \cdot 6746\ kg \cdot m^{-3} \cdot 3.72\ m}}{8.930 \cdot 10^{-4}\mathrm{Pa \cdot s}} = 2.810 \cdot 10^{7} \quad (2.324)$$

$$\mathrm{Pr} = \frac{c_p \cdot \mu}{\lambda} = \frac{824.0\ \mathrm{J \cdot kg^{-1} \cdot K^{-1}} \cdot 8.930 \cdot 10^{-4}\mathrm{Pa \cdot s}}{64\ \mathrm{W \cdot m^{-1} \cdot K^{-1}}} = 0.0115 \quad (2.325)$$

The corresponding value of the number of Nusselt (Nu) and that of the convective heat transfer coefficient, h_c, using Eq. (2.323) are:

$$\mathrm{Nu} = 0.50 \cdot \left(2.810 \cdot 10^{7}\right)^{1/2} \cdot (0.0115)^{1/3} \cdot \left[\frac{0.455}{\left(\log\left(2.810 \cdot 10^{7}\right)\right)^{2.584}} - \frac{3300}{2.810 \cdot 10^{7}}\right]$$
$$= 1.4483 \quad (2.326)$$

$$h_c = \frac{\mathrm{Nu} \cdot \lambda}{L} = \frac{1.4483 \cdot 64\ \mathrm{W \cdot m^{-1} \cdot K^{-1}}}{3.72\ \mathrm{m}} = 24.92\ \mathrm{W \cdot m^{-2} \cdot K^{-1}} \quad (2.327)$$

These results lead us to think in low quality of the Nusselt's predictions and in the wrong values of the heat transfer coefficient, h_c, in metallic melts with low values of the Prandtl number (very smaller than the recommended limit of 0.60). We suggest using the correlations previously indicated for fluids with Pr greater than 0.60. If the values of h_c for slags are of around 125 W m^{-2} K^{-1}, it is expected for a metallic melt values of the h_c 10–20 times greater. An option to use these correlations for small numbers of Prandtl is identifying the characteristic linear dimension of Nusselt with the thickness of the thermal boundary layer, δ_T, Fig. 2.15, whose value is usually within 0.50 cm and 5 mm. Consequently, while in slags and specially in gases, if there is not sufficient stirring, there are significant thermal differences, metallic melts are usually considered thermally homogeneous systems.

On another note, for the heat transfer coefficient through static beds of particles (electric furnaces of high shaft), it is possible to apply an equation like Eq. (2.309). However, gas–solid transfer of the Prandtl number of the gases does not exhibit an appreciable variation with the temperature, and the functional dependency of the Nusselt number would be reduced to only the Reynolds number. At the end of the sixties of the last century, Kitaev used the data of Furnas of 1932 to estimate the volumetric heat transfer coefficient, h_v, expressed in kcal m^{-3} h^{-1} °C^{-1} (see Poirier and Geiger 1994, p. 273):

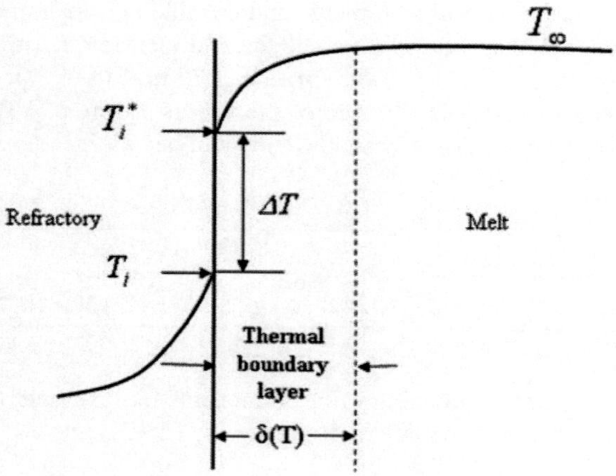

Fig. 2.15 Profile of temperature along the interface solid (refractory or ceramic)-melt

$$h_v = 80 \cdot \left(\frac{v^{0.9} \cdot T^{0.3}}{D_p^{0.75}} \right) \tag{2.328}$$

where v, T, and D_p are the gas speed in m s^{-1}, its temperature in °C, and the size-diameter of the particles of the static bed in meters. To transform the measurement unities of the volumetric heat transfer coefficient, from the kcal m^{-3} h^{-1} °C^{-1} of the equation proposed by Furnas-Kitaev, to W m^{-3} K^{-1}, it is necessary to multiply Eq. (2.328) by the coefficient 1.1622 and, in this way, calories are transformed in Joules and the hours in seconds.

The values of h_v in Eq. (2.328) are big although they have a marked relation with the temperature of the gas. For instance, for a temperature of 1400 °C with a rate of gases circulation of 1.0 m s^{-1} and a diameter of particle of the charge pellets of 25 mm, the value of h_v is:

$$h_v = 80 \cdot \left[\frac{(1\ \text{m} \cdot \text{s}^{-1})^{0.9} \cdot (1400^\circ\text{C})^{0.3}}{(0.025\ \text{m})^{0.75}} \right] \cdot 1.1622 = 12995\ \text{W} \cdot \text{m}^{-3} \cdot \text{K}^{-1} \tag{2.329}$$

while for the same charge and flow conditions (superficial rate of circulation of gases in the furnace) but at 200 °C, the value of h_v is:

$$h_v = 80 \cdot \left[\frac{(1\ \text{m} \cdot \text{s}^{-1})^{0.9} \cdot (200^\circ\text{C})^{0.3}}{(0.025\ \text{m})^{0.75}} \right] \cdot 1.1622 = 7248\ \text{W} \cdot \text{m}^{-3} \cdot \text{K}^{-1} \tag{2.330}$$

However, these results give a quantitative explanation about why the height of the shaft in furnaces of great production does not grow in the same proportion that the diameter of the furnace does. Thus, the productivity (size) of the furnaces is increased by increasing the diameter of the hearth instead of modifying the height of the shaft.

Interface resistances are associated to gradients of speed, temperature, and concentration of fluids and their respective phenomena of movement, mass, and heat transfer, Fig. 2.15. The thickness of the boundary layer associated with each one of the interfacial transfer phenomena would be that of the boundary layer of speeds, δ_v, that of the boundary layer of temperature, δ_T, and that of the boundary layer of concentrations, δ_c. The relation between the thicknesses attributed to the different boundary layers is associated to the dimensionless numbers of Prandtl (Pr) and Schmidt (Sc) where the variables D_i and a are the diffusion coefficient of the specie "i" that diffused and the thermal diffusivity, respectively:

$$\mathrm{Pr} = \frac{\mu}{\rho \cdot a} = \frac{c_p \cdot \mu}{\lambda} \tag{2.331}$$

$$a = \frac{\lambda}{\rho \cdot c_p} \tag{2.332}$$

$$\mathrm{Pr} = \frac{\delta_v}{\delta_T} \tag{2.333}$$

$$\mathrm{Sc} = \frac{\mu}{\rho \cdot D_i} \tag{2.334}$$

$$\mathrm{Sc} = \frac{\delta_v}{\delta_c} \tag{2.335}$$

It is possible to conclude, in a general way, that the value of the thickness of the boundary layers follows the following progression:

$$\delta_v > \delta_T > \delta_c \tag{2.336}$$

Metallic melts are phases especially singulars due to that the thickness of the thermal boundary layer, δ_T, can be nearly 100 times smaller than the thickness of the speed boundary layer, δ_v, with Prandtl numbers between 0.100 and 0.012. On another note, in metallic melts, the interface resistance to the heat transfer is considered in most of the cases very small due to two circumstances:

1. Small values for the thickness of the thermal boundary layer, δ_T, of the order of magnitude of the millimeter.
2. The high thermal conductivity of metallic melts.

However, in the case, for instance, of the continuous solidification of metallic melts in metallic molds, it is industrially promoted a interfacial resistance to the heat transfer (slags of continuous casting, see Sect. 4.3.3); it is tried to not reach the welding-strong joining between metallic mold and melt partially solidified of the ingot that must be continuously removed from the mold.

2.4.2.3 Radiation Heat Transfer.

Reference equation for the calculation of the radiation flow emitted by the black body at a temperature, T (K), is the Stefan-Boltzmann law:

$$J_q = \sigma_{SB} \cdot T^4 \left(W \cdot m^{-2}\right) \tag{2.337}$$

where σ_{SB} is the Stefan-Boltzmann constant that is equal to 5.67051 10^{-8} W m^{-2} K^{-4}.

However, not all the surfaces and black bodies (as carbon black) are able to emit the radiation energy indicated by Eq. (2.337). With the aim of correcting this effect we can talk about grey bodies, whose capacity of emission of radiant energy is equal to that of absorption when a thermal equilibrium between both bodies is reached—radiant surfaces, but in any case, the capacity of emission of grey bodies is always smaller than that of the black body. In this way, the concept of emissivity, ε, is included in the Stefan-Boltzmann equation, and the capacity of radiant energy emission of grey bodies is corrected with respect to those of the black body. The values of emissivity of grey bodies vary within zero and one.

On the other hand, as the radiant energy-electromagnetic radiation does not require a material support to be transferred and it travels longitudinally, we can talk about vision factors of a surface of radiation emission (1) with respect to other surface or reception area for the radiation, Fig. 2.16. The vision factor F_{12} reaches the maximum value of one when the body-radiation element is completely surrounded by the surface-reception element. This situation takes place, for instance, in the electric

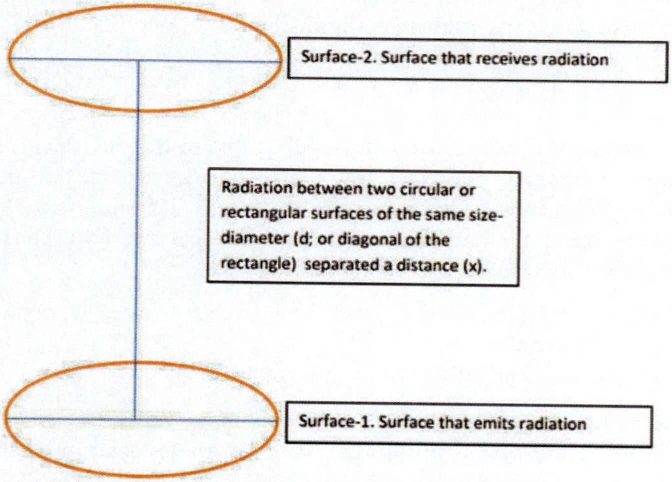

Fig. 2.16 Relation between two surfaces (1 and 2). Two surfaces of the same size separated by a certain distance

furnace, where either the indirect arc or the submerged arc in the slag radiate energy to all the furnace that is in contact with the environment through the structure-shell of ferritic-pearlitic steel. In this case, the Stefan-Boltzmann equation would be:

$$J_q = \sigma_{SB} \cdot \varepsilon \cdot \left(T_1^4 - T_2^4\right)\left(W \cdot m^{-2}\right) \quad (2.338)$$

where T_2 is the temperature of the environment that surrounds the furnace, 18 °C (291 K), Fig. 2.14. For instance, in an electric furnace of 12.67 MW that operates for 45 min (2700 s) with the indicated power, the amount of calorific energy produced by the electric arc generated by the three electrodes would be of 34.25 GJ.

On another note, if we talk about a submerged arc, the emissivity of the source of radiant energy could vary between 0.3 and 0.5 according to the little and few precise data that can be found in the bibliography that was consulted. It is assumed as working hypothesis that the volume-space between the three electrodes of the furnace, inside of which the arc is stroked, is a prism of triangular basis and rectangular sides, Fig. 2.14:

- Length of the sides of the equilateral triangle is 1.375 m (1375 mm).
- Dimensions of the sides of the rectangle of the three lateral faces of the prism: length, 1.375 m; height, 0.175 m (175 mm).
- Surface area of both the bases of the triangular prism and the three rectangular faces (2.36 m^2):

$$\frac{(1.375\ m)^2 \cdot \sqrt{3}}{4} \cdot 2 + 1.375\ m \cdot 0.1753\ m \cdot 3 = 2.36\ m^2 \quad (2.339)$$

It is truth that the two working hypotheses that were established: the emissivity of the radiant energy emitted by the arc and the surface-volume within the electrodes of the furnace where the arc is stroked can vary within a wide range. However, if we maintain the surface that surrounds the radiant energy source in 2.36 m^2 and the emissivity in 0.50, it is possible to estimate using the Eq. (2.338) the temperature that might be reached in the radiant zone between the electrodes compatible with:

- the heat flow of 5.374 MW m^{-2}.
- a generation of heat by square meter during 2700 s of 14.51 GJ m^{-2}.
- and a net quantity of energy of 34.25 GJ.

The value of the temperature would be 3438 °C. Proceeding as in the previous case and keeping constant the surface area of the radiant emission focus in 2.36 m^2, temperatures reached inside of the electric furnace would be 3537 °C, 3650 °C, and 3784 °C if the values of the emissivity of the radiant source would be 0.45, 0.40, and 0.35, respectively.

On another note, the transfer of radiant energy between two surfaces of the same dimensions (for instance between two circular surfaces that might simulate the hearth and the roof of the furnace) separated by a certain distance, Fig. 2.16, leads to a new modification of the Stefan-Boltzmann equation, in this case with the new variable F. The indicated variable, apart from considering the values of the emissivity of the surfaces 1 and 2, also considers the called vision factor between the surfaces F_{12}. The vision factor is not only function of the size (the diameter of a circular surface or the diagonal of a rectangle), it is also function of the relation between the size and the distance, x, that separates both surfaces, Fig. 2.16. When the emitting surface 1 and receiving surface 2 are very close then $F_{12} \rightarrow 1$, while when they are very separated then $F_{12} \rightarrow 0$. Consequently, the radiant flow between these two surfaces is:

$$J_q = \sigma_{SB} \cdot F \cdot \left(T_1^4 - T_2^4\right)\left(W \cdot m^{-2}\right) \quad (2.340)$$

$$F = \frac{1}{\left[\left(\frac{2-2 \cdot F_{12}}{1-F_{12}^2}\right) + \frac{1}{\varepsilon_1} + \frac{1}{\varepsilon_2} - 2\right]} \quad (2.341)$$

$$F_{12} = 1.79 \cdot 10^{-3} \cdot \left(\frac{d}{x}\right)^3 - 0.0337 \cdot \left(\frac{d}{x}\right)^2 + 0.258 \cdot \left(\frac{d}{x}\right) \quad (2.342)$$

where ε_1 and ε_2 are, respectively, the emissivity of the radiant surface 1 and the emissivity of the receiving surface 2. Finally, F_{12} would be the vision factor between two surfaces of the same size, d, separated a distance x.

Finally, it would be useful to know the value of h_r for the mechanism of radiant heat transfer, which is defined by the following equation:

$$J_q = h_r \cdot (T_1 - T_2) = \sigma_{SB} \cdot F \cdot \left(T_1^4 - T_2^4\right)\left(W \cdot m^{-2}\right) \quad (2.343)$$

knowing that:

$$\left(T_1^4 - T_2^4\right) = \left(T_1^2 + T_2^2\right) \cdot (T_1 + T_2) \cdot (T_1 - T_2) \quad (2.344)$$

it is possible to conclude:

$$h_r = \sigma_{SB} \cdot F \cdot \left(T_1^2 + T_2^2\right) \cdot (T_1 + T_2)\left(W \cdot m^{-2} \cdot K^{-1}\right) \quad (2.345)$$

2.4.3 Solved Exercises

Exercise 2.12 Design-detail the most relevant characteristics of an electric furnace, Fig. 2.14. This furnace can produce 25 tons of carbon steel and 2.0 tons of slag (80 kg of slag per ton steel) in 45 min with an energy consumption associated of 380 kWh t^{-1}. The boundary conditions—boundary temperatures fixed in surfaces that are in contact with the air of the environment (18 °C, 291 K) are: Zone 1 (T), 110 °C; Zone 2, 160 °C; Zone 3 (CR1), 180 °C. For the Zone 3 (CR2) we consider an external cooling with environmental air (18 °C) driven by the external fans-blowers that lead to a global heat transfer coefficient of 90 W m^{-2} K^{-1}. The average temperature of the sheet cooled with air is 95 °C.

As general criterion for the assignation of thicknesses to the refractory lining for each one of the zones of the furnace to be considered we have:

- that the zones of greater thickness would be in the Zone 3 (CR1) and Zone 3 (CR2),
- that there is a zone of intermediate thickness, Zone 2 (CG), that is usually associated-structurally connected with the Zone 3, and that does not have capacity of movement-autonomous displacement,
- finally, Zone 1, roof of the furnace, is that zone where the lining has the minimum thickness.

Habitually, the roof of the furnace, together with the electrodes can be vertically displaced with the aim of facilitating the charging operations. We consider that the densities of the metallic melt and slag are 6746 kg m^{-3} and 3200 kg m^{-3}, respectively.

Solution Starts

We consider, with the purpose of detailing those dimensions of the furnace that are more directly related with the production-storage of 27 tons of melts:

- that the internal radius of the furnace (above the hearth and at the height of the furnace's door) is 1.86 m (1860 mm).
- that the external radius, including the thickness of the refractory, expansion joint, and steel sheet, is 2.41 m (2410 mm): 1860 mm + 500 mm (refractory) + 30 mm (expansion joint) + 20 mm (construction steel sheet, density of 7850 kg m^{-3}).

The values of other thicknesses of the lining would be: Zone 1 (T), 150 mm for the roof of the furnace; Zone 2 (CG), 300 mm. We consider that in all the zones of the furnace, as it was previously mentioned, the thickness of the ensemble formed by the expansion joints and the steel sheet are of 50 mm: 30 mm for the expansion joints and 20 mm for the steel. The characteristics of the materials used for the expansion joints are different depending on the zone of the furnace (Pero-Sanz et al. 2019):

- For the roof of the furnace, Zone 1 (T), high density polyethylene (PE) with 950 kg m^{-3} of density and 0.48 W m^{-1} K^{-1} of thermal conductivity.
- For the shaft of the furnace, Zone 2 (CG), expanded clays of low thermal conductivity, glass wool-fiber with densities and thermal conductivities of around 1300 kg m^{-3} and 0.38 W m^{-1} K^{-1}, respectively.
- For the Zone 3 (CR1 and CR2), it is recommended granular material of carbon-silicon carbide of high thermal conductivity and low density: 10 W m^{-1} K^{-1} and 1000 kg m^{-3}, respectively.

To calculate the volume associated to the Zone 3 and to estimate the occupancy of the melts, we consider:

– that the volume of the spherical cap associated to the internal volume of the furnace in the Zone 3 (CR2), Fig. 2.14, with a height, h_{CR2}, of 581.2 mm (with an order of magnitude equivalent to that of the thickness of the lining) and a radius, a, of 1860 mm, is:

$$V_{CR2} = \frac{\pi \cdot h_{CR2}}{6} \cdot \left(3 \cdot a^2 + h_{CR2}{}^2\right)$$
$$= \frac{\pi \cdot 0.5812\ \text{m}}{6} \cdot \left[3 \cdot (1.86\ \text{m})^2 + (0.5812\ \text{m})^2\right] = 3.261\text{m}^2 \quad (2.346)$$

The weight of the molten metal, $M_{\text{molten metal}}$, associated to the volume previously calculated would be:

$$M_{\text{molten metal}} = 6746\frac{\text{kg}}{\text{m}^3} \cdot 3.261\ \text{m}^3 \cdot \frac{\text{t}}{1000\ \text{kg}} = 22.0\ \text{t of metallic melt} \quad (2.347)$$

– the remaining three tons of metallic melt would be in the cylindrical part of the hearth, Zone 3 (CR1), which is designed with a radius of 1860 mm and a height, h_{CR1}, of 450 mm. The melt only occupies the volume of a cylinder of 163.9 mm in height and 1860 mm of radius. This corresponds with an occupancy of 0.445 m^3:

$$V = \pi \cdot r^2 \cdot h = \pi \cdot (1.86\ \text{m})^2 \cdot 0.1639\ \text{m} = 0.445\ \text{m}^3 \quad (2.348)$$

$$M_{\text{molten metal}} = 6746\frac{\text{kg}}{\text{m}^3} \cdot 0.445\ \text{m}^3 \cdot \frac{\text{t}}{1000\ \text{kg}} = 3.0\ \text{t of metallic melt} \quad (2.349)$$

– the 2000 kg of slag would occupy the upper part of the cylinder, which is simulated with the geometry of the Zone 3 (CR1). The height compatible with its density is of 230 mm. That is to say, the volume occupied by the slag would be that of a cylinder with 230 mm in height and 1860 mm of radius:

$$V = \pi \cdot r^2 \cdot h = \pi \cdot (1.86\ \text{m})^2 \cdot 0.230\ \text{m} = 0.625\ \text{m}^3 \quad (2.350)$$

$$M_{\text{molten metal}} = 3200\frac{\text{kg}}{\text{m}^3} \cdot 0.625\ \text{m}^3 \cdot \frac{\text{t}}{1000\ \text{kg}} = 2.0\ \text{t of slag} \quad (2.351)$$

From the energetic point of view, the furnace power, P, in MW (Eq. (2.299), Sect. 2.4) would be:

$$P = \frac{3.6 \cdot M \cdot E}{t} = \frac{3.6 \cdot 25\ t \cdot 380\ kWh \cdot t^{-1}}{2700\ s} = 12.677\ MW \tag{2.352}$$

Consequently, the quantity of energy released by the electric arc during the 45 min that the process lasts would be:

$$12.677 \cdot 10^{6} W \cdot 2700\ s \cdot \left(\frac{GJ}{10^{9} J}\right) = 34.23\ GJ \tag{2.353}$$

The heat losses toward the environment (18 °C), considering the average surface temperatures in the steel sheet in the four zones of the furnace (Zone 1, Zone 2, Zone 3 (CR1), and Zone 3 (CR2)), could be estimated if the next steps are followed:

Step 1: estimation of the heat flow, $W\ m^{-2}$, for each one of the four zones of the furnace (Eqs. (2.300) and (2.301), Sect. 2.4):

– In the Zone 1 (T):

$$\begin{aligned} h_g &= 3.89 \cdot 10^{-7} \cdot 383^3 - 3.94 \cdot 10^{-4} \cdot 383^2 + 0.140 \cdot 383 + 0.468 \\ &= 18.15\ W \cdot m^{-2} \cdot K^{-1} \end{aligned} \tag{2.354}$$

$$J_q^{H-air} = h_g \cdot \left(T_i - T_o^{Air}\right) = 18.15 \frac{W}{m^2 \cdot K} \cdot (383 - 291) K = 1669.55\ W \cdot m^{-2} \tag{2.355}$$

– In the Zone 2 (CG):

$$\begin{aligned} h_g &= 3.89 \cdot 10^{-7} \cdot 433^3 - 3.94 \cdot 10^{-4} \cdot 433^2 + 0.140 \cdot 433 + 0.468 \\ &= 18.80\ W \cdot m^{-2} \cdot K^{-1} \end{aligned} \tag{2.356}$$

$$J_q^{H-air} = h_g \cdot \left(T_i - T_o^{Air}\right) = 18.80 \frac{W}{m^2 \cdot K} \cdot (433 - 291) K = 2669.23\ W \cdot m^{-2} \tag{2.357}$$

– In the Zone 3 (CR1):

$$
\begin{aligned}
h_g &= 3.89 \cdot 10^{-7} \cdot 453^3 - 3.94 \cdot 10^{-4} \cdot 453^2 + 0.140 \cdot 453 + 0.468 \\
&= 19.20\ \mathrm{W \cdot m^{-2} \cdot K^{-1}}
\end{aligned}
\tag{2.358}
$$

$$
J_q^{H-air} = h_g \cdot \left(T_i - T_o^{Air}\right) = 19.20 \frac{W}{m^2 \cdot K} \cdot (453 - 291)K = 3109.91\ \mathrm{W \cdot m^{-2}}
\tag{2.359}
$$

– In the Zone 3 (CR2):

$$
J_q^{H-air} = h_g \cdot \left(T_i - T_o^{Air}\right) = 90 \frac{W}{m^2 \cdot K} \cdot (368 - 291)K = 6930.00\ \mathrm{W \cdot m^{-2}}
\tag{2.360}
$$

Step 2: estimation of the surface area of each one of the four zones of the furnace.

The exchange area in the Zone 1 (T) can be represented by a spherical cap of 2410 mm of radius and 350 mm in height.

2410 mm are the addition of the 1860 mm of inner radius of the furnace, the 500 mm of thickness of the refractory lining and the 50 mm of the ensemble formed by the thickness of the expansion joint and the steel sheet. The height of the cap is 350 mm, and it is equal to the addition of the inner height of the cap, h_T, Fig. 2.14, of 150 mm, 150 mm of the refractory lining, and 50 mm of the ensemble formed by the expansion joint and the steel sheet. The value of the surface area of a spherical cap is function of the radius, a, and its height, h:

$$
A = \pi \cdot \left(a^2 + h^2\right) = \pi \cdot \left[(2.41\ m)^2 + (0.35\ m)^2\right] = 18.632\ m^2
\tag{2.361}
$$

In the Zone 2 (CG), the exchange area is the surface area of a cylinder of 2410 mm of radius and 1900 mm in height, h_{CG}, Fig. 2.14:

$$
A = 2 \cdot \pi \cdot r \cdot h_{CG} = 2 \cdot \pi \cdot 2.41\ m \cdot 1.90\ m = 28.771\ m^2
\tag{2.362}
$$

In the Zone 3 (CR1), the exchange area is the surface area of a cylinder of 2410 mm of radius and 450 mm in height, h_{CR1}, Fig. 2.14:

$$
A = 2 \cdot \pi \cdot r \cdot h_{CR1} = 2 \cdot \pi \cdot 2.41\ m \cdot 0.45\ m = 6.814\ m^2
\tag{2.363}
$$

In the Zone 3 (CR2), the exchange area is the surface area of a cylinder of 2410 mm of radius and 631.2 mm in height, h_{CR2}, Fig. 2.14:

$$A = 2 \cdot \pi \cdot r \cdot h_{CR2} = 2 \cdot \pi \cdot 2.41\ m \cdot 0.6312\ m = 19.498\ m^2 \tag{2.364}$$

Step 3: estimation of the energy losses in Joules through the above-mentioned surfaces in 2700 s.

In the Zone 1 (T):

$$1669.55\ W \cdot m^{-2} \cdot 18.632\ m^2 \cdot 2700\ s \cdot \left(\frac{GJ}{10^9 J}\right) = 0.0840\ GJ \tag{2.365}$$

In the Zone 2 (CG):

$$2669.23\ W \cdot m^{-2} \cdot 28.771\ m^2 \cdot 2700\ s \cdot \left(\frac{GJ}{10^9 J}\right) = 0.2073\ GJ \tag{2.366}$$

In the Zone 3 (CR1):

$$3109.91\ W \cdot m^{-2} \cdot 6.814\ m^2 \cdot 2700\ s \cdot \left(\frac{GJ}{10^9 J}\right) = 0.0572\ GJ \tag{2.367}$$

In the Zone 3 (CR2):

$$6930\ W \cdot m^{-2} \cdot 19.498\ m^2 \cdot 2700\ s \cdot \left(\frac{GJ}{10^9 J}\right) = 0.3648\ GJ \tag{2.368}$$

Consequently, the heat losses toward the environment are 0.7134 GJ. This heat losses represent 2.08% of the 34.23 GJ involved in the manufacture of 25 tons of steel.

Commentaries: In Fig. 2.14, the design that is presented for the Zone 1 (T), roof of the furnace, can lead to two well different interpretations:

1. that the cover of the furnace could not be removed-lift up and that the charge of the furnace was done through the doors located in the lower part of the Zone 2 (CG) that is in contact with the hearth of the furnace. The thickness of the refractory concrete of the roof of the furnace, in this case can even reach the 300–350 mm as in the zone of protection of the shaft of the furnace, Zone 2 (CG).
2. that the cover of the furnace could be removed-lift up, by which, the weight of the cover and the height of the industrial factory acquire a significant relevance. Furnaces with moving cover are specially associated to the smaller sizes, with refractory lining of low density and small thickness. For instance, in the case of the exercise, the thickness of 150 mm is designed with two different qualities:

Quality 1, refractory concrete, in contact with the gas of thermal conductivity of 0.375 W m^{-1} K^{-1} and isolating material in contact with the steel, Quality 2, with thermal conductivity of 0.123 W m^{-1} K^{-1}. The temperature difference inside of the lining, according to Fig. 2.14, is from 1123 °C to 110 °C, ΔT = 1013 °C (or K). The heat flow in the Zone 1 (T) was already calculated, 1669.55 W m^{-2}. Consequently, the value of the thicknesses for each one of the above-mentioned qualities that satisfies the following equation:

$$1013\ \mathrm{K} = 1669.55\ \mathrm{W \cdot m^{-2}} \cdot \left[\frac{(0.150\ \mathrm{m} - \Delta x_2)}{0.375\ \mathrm{W \cdot m^{-1} \cdot K^{-1}}} + \frac{\Delta x_2}{0.123\ \mathrm{W \cdot m^{-1} \cdot K^{-1}}}\right] \tag{2.369}$$

are for the Quality 2, $\Delta x_2 = 37.8$ mm, and for the Quality 1, $\Delta x_1 = 112.2$ mm.

Solution ends.

Exercise 2.13 Study the possibilities of breaking by thermal fatigue (thermal shock) the electrodes of an electric furnace of geometrical characteristics like those shown in Fig. 2.14. We consider that during the 15 min that the process of charge of the furnace (scrap, slag, and fluxes) lasts, the electrodes of the furnace can be subjected to a process-program of cooling that could cause their breaking. The physical, thermal, and mechanical properties of the carbon-graphitized electrodes are:

Global density: 1700 kg m^{-3}; specific heat: 711 J kg^{-1} K^{-1}; thermal conductivity: 50 W m^{-1} K^{-1}; coefficient of linear thermal expansion: 1.50 10^{-6} °C^{-1}; mechanical resistance (three-point bending test): 15 MPa; Young's modulus: 10.5 GPa.

On another note, the thermal boundary conditions of operation of the electrodes and their geometrical characteristics are:

- the maximum temperature during the beginning of the charging process is in the lower extreme of the electrode: 1450 °C, while the minimum temperature, inside of the furnace, is located in that zone of the electrode that is closer to the roof of the furnace: 1123 °C.
- As reference temperature for the heat transfer in the electrodes (heating–cooling) we consider 200 °C, which is the temperature that is reached in the pliers (cooled with water) that hold the electrodes and that are located above the cover of the furnace.
- The height of the electrodes, of cylindrical geometry with a diameter of 350 mm, inside of the furnace is 2.0 m (Zone 1, Zone 2, and a small part of the Zone 3 (CR1), Fig. 2.14).

Solution starts

Supported on the Hooke's and Fourier's laws and on the relation between the linear unitary deformations caused in the materials as a result of the temperature changes, it is possible to obtain an equation that relates heat flow, J_q, and mechanical resistance of the material, σ_{3p}, thermal conductivity, λ, Young's modulus, E, coefficient of linear thermal expansion, α_l, and diameter of the electrode, $d = \Delta x$ (Pero-Sanz et al. 2018, 2019):

$$\sigma_t \simeq \sigma_{3p} = E \cdot \left(\frac{\Delta l}{l_0}\right) \text{Hooke's law} \quad (2.370)$$

$$\left(\frac{\Delta l}{l_0}\right) = \alpha_l \cdot \Delta T \text{ Coefficient of linear thermal expansion} \quad (2.371)$$

$$J_q = \lambda \cdot \left(\frac{\Delta T}{\Delta x}\right) \quad (2.372)$$

$$\sigma_{3p} = E \cdot \alpha_l \cdot \Delta T = E \cdot \alpha_l \cdot \frac{J_q \cdot \Delta x}{\lambda} \quad (2.373)$$

Consequently:

$$J_q = \frac{\sigma_{3p} \cdot \lambda}{E \cdot \alpha_l \cdot \Delta x} \quad (2.374)$$

It is possible to estimate the heat flows in the electrodes before their breaking as a result of the thermal fatigue using Eq. (2.374).

With the data collected in the enunciation, it is possible to estimate the heat flow above which the breaking of the electrode could take place:

$$J_q = \frac{\sigma_{3p} \cdot \lambda}{E \cdot \alpha_l \cdot \Delta x} = \frac{15 \cdot 10^6 \text{Pa} \cdot s \cdot 50\ \text{W} \cdot m^{-1} \cdot K^{-1}}{10.5 \cdot 10^9 \text{Pa} \cdot s \cdot 1.5 \cdot 10^{-6} K^{-1} \cdot 0.350\ m} = 1.36 \cdot 10^5 \text{W} \cdot m^{-2} \quad (2.375)$$

On another note, it would be necessary to justify that during the 15 min that the opening of the furnace cover could take, with the corresponding lifting up of the electrodes, the rate of heat dissipation that can take place in each electrode does not exceed the value previously calculated. Considering each one of the electrodes as a radiant element that emits heat toward the environment (18 °C, 291 K), it is possible to apply the Stefan–Boltzman law, Eq. (2.337), considering that the emissivity of the electrodes is 0.85:

$$J_q = 0.85 \cdot 5.67051 \cdot 10^{-8} \text{W} \cdot m^{-2} \cdot K^{-4} \cdot \left[T_1^4 - (291\ K)^4\right] \quad (2.376)$$

The average value that the temperature can reach in the electrodes, T_1, at the beginning of the period of opening of the cover of the furnace until the closure of the cover could be calculated as the mean between the maximum and the minimum temperatures (that always would be the same during all the process of opening-closing of the furnace, 200 °C) that can be developed inside of the electrode:

Table 2.12 Heat flow as a function of the temperature

Maximum temperature (°C)	Minimum temperature (°C)	Average temperature (°C)	Heat flow (W m^{-2})
1450	200	825	6.97 10^4
1300	200	750	5.24 10^4
1200	200	700	4.29 10^4
1123	200	662	3.64 10^4

$$T_1 = \frac{T_{max} + T_{min}}{2} = \frac{1450 + 200}{2} = 825^{\circ}C \tag{2.377}$$

During the 15 min that the charging process lasts, the maximum temperature could be, as a function of the time that passed, that of 1300 °C, 1200 °C to finally reach, after 15 min, 1123 °C. As a function of these temperatures, in the enclosed Table 2.12, it is possible to find the values of the heat flows calculated using Eq. (2.376).

Other possible alternative to quantify the heat flow released by the electrodes during the process would be through the coefficient of radiant heat transfer, h_r, Eq. (2.345), where F takes the value of 0.85, emissivity of the electrodes:

$$h_r = 5.67051 \cdot 10^{-8} W \cdot m^{-2} \cdot K^{-4} \cdot 0.85 \cdot \left(T_1^2 + T_2^2\right) \cdot (T_1 + T_2)\left(W \cdot m^{-2} \cdot K^{-1}\right) \tag{2.378}$$

$$J_q = h_r \cdot (T_1 - T_2) \tag{2.379}$$

where $T_2 = T_1 - 50$ in °C or $T_2 = T_1 - 323$ in K.

As in the previous estimation, the variation of T_1 during the charging period is between 1450 °C and 1123 °C. We collect in Table 2.13 the values of T_1, T_2, h_r, and J_q.

Finally, the heat flow that can be released by the electrodes during the charging process can be calculated from the heat release of a percentage of accumulated energy (30%) during the heating of the electrode, $Q_{electrode}$:

$$Q_{electrode} = c_e \cdot \rho \cdot \pi \cdot r^2 \cdot (T_i - 200)(J) \tag{2.380}$$

Table 2.13 Values of the coefficient h_r and the heat flow

Temperature T_1 (°C)	Temperature T_2 (°C)	Coefficient h_r (W m^{-2} K^{-1})	Heat flow (W m^{-2})
1450	1400	944.08	4.72 10^4
1300	1250	715.36	3.58 10^4
1200	1150	585.51	2.93 10^4
1123	1073	497.00	2.48 10^4

where $T_i = (1450 + 1123)/2 = 1286.5^\circ C$

$$
\begin{aligned}
J_q = \frac{Q_{electrode}}{\pi \cdot r^2 \cdot t} &= \frac{c_e \cdot \rho \cdot (T_i - 200)}{r \cdot t} \\
&= \frac{711\ J \cdot kg^{-1} \cdot K^{-1} \cdot 1700\ kg \cdot m^{-3} \cdot (1286.5 - 200)K}{0.175\ m \cdot 15\ min \cdot \frac{60\ s}{1\ min}} \cdot 0.30 \\
&= 3.83 \cdot 10^4 W \cdot m^{-2}
\end{aligned} \tag{2.381}
$$

In any of the situations the critical value of 1.36 10^5 W m^{-2} is not reached, value from which the breaking of the electrode could take place. However, either a decrease of both the mechanical resistance and the thermal conductivity or an increase of both the Young's modulus, the coefficient of linear thermal expansion and the diameter of the electrode could lead to the breaking of the electrode in any of the thermal cycles of heating–cooling of the electrode. Finally, we want to point out that the heat flow experienced by the electrodes might be minimized and, consequently, avoiding reaching the critical value of 1.35 10^5 W m^{-2}. In this case, we should consider the heat supplies coming either from the heat emissions of each one of the electrodes over the other two, from the radiant energy coming from the hearth-Zone 3 or also from the walls of the dome-roof of the furnace, if it was designed with a sufficient depth-height, h_T, Fig. 2.14.

Solution ends.

References

Andresen B (1995) Process model for carbothermic production of silicon metal. PhD Thesis (Eng. Dr. Thesis), September 12. University of Trondheim, Norway

Coudurier L, Hopkins DW, Wilkomisky I (1985) Fundamentals of metallurgical processes. Pergamon Press, 2nd edition, Oxford, UK

Emi T (2015) Steelmaking technology for the last 100 years: toward highly efficient mass production systems for high quality steels. ISIJ Int 15(1):36–66

Fernández-González D, Ruiz-Bustinza I, González-Gasca C, Piñuela-Noval J, Mochón-Castaños J, Sancho-Gorostiaga J, Verdeja LF (2018) Concen-trated solar energy applications in materials science and metallurgy. Sol Energy 170(8):520–540

Fernández-González D, Prazuch J, Ruiz-Bustinza I, González-Gasca C, Pi-ñuela-Noval J, Verdeja LF (2018) Iron metallurgy via concentrated solar energy. Metals 8(11):873

Fernández-González D, Prazuch J, Ruiz-Bustinza I, González-Gasca C, Pi-ñuela-Noval J, Verdeja LF (2018) Solar synthesis of calcium alumina-tes. Sol Energy 171(9):658–666

Fernández-González D, Prazuch J, Ruiz-Bustinza I, González-Gasca C, Pi-ñuela-Noval J, Verdeja LF (2019) Transformations in the Si-O-Ca Sys-tem: silicon-calcium via solar energy. Sol Energy 181(3):414–423

Fernández-González D, Prazuch J, Ruiz-Bustinza I, González-Gasca C, Pi-ñuela-Noval J, Verdeja LF (2019) Transformations in the Mn-O-Ca Sys-tem using concentrated solar energy. Sol Energy 184(5):148–152

Fernández-González D, Prazuch J, Ruiz-Bustinza I, González-Gasca C, Pi-ñuela-Noval J, Verdeja LF (2019) The treatment of basic Oxygen Fur-nace (BOF) Slag with concentrated solar energy. Sol Energy 180(3):372–382

Gasik M (2013) Handbook of ferroalloys. Theory and Technology, Elsevier, Oxford, United Kingdom

Guthrie RIL (1989) Engineering in process metallurgy. Clarendon Press-Oxford University Press, New York, USA

Kawai Y, Shiraishi Y (1988) Handbook of physico-chemical properties at high temperatures. The Iron and Steel Institute of Japan-Committee of Japan Society for Promotion of Science, Tokyo, Japan

Knacke O, Kubaschewski O, Hesselmann K (1991) Thermochemical properties of inorganic substances. Springer, 2nd edition, Berlin, Germany

Levin E, Robins CR, McMurdie HF (1969) phase diagrams for ceramists. The American Ceramic Society, Columbus, Ohio, USA (and supplement of figures from the 2067 to the 4149 edited in 1969)

Mills KC, Keene BJ (1987) Physical properties of BOS Slags. Int Mater Rev 32(1–2):1–120

Olsen SE, Tangstad M, Lindstad T (2007) Production of manganese ferroalloys. Tapir Academic Press, Trondheim, Norway

Pero-Sanz JA, Quintana MJ, Verdeja LF (2017) Solidification and solid-state transformations of metals and alloys. Elsevier, 1st edition, Boston, US.

Pero-Sanz JA, Fernández-González D, Verdeja LF (2018) Physcal metallurgy of cast irons. Springer International Publishing, 1 st edition, Cham, Switzerland

Pero-Sanz JA, Fernández-González D, Verdeja LF (2019) Structural materials: properties and selection. Springer International Publishing, 1 st edition, Cham, Switzerland

Poirier DR, Geiger GH (1994) Transport phenomena in materials processing. TMS, Warrendale, Pennsylvania, USA

Poirier DR, Poirier EJ (1992) Heat transfer fundamentals for metal casting. TMS, Pennsylvania, USA

Rosenqvist T (1983) Principles of extractive metallurgy. McGraw-Hill, New York, USA

Sigworth GH, Elliot JM (1974) Thermodynamic of dilute iron alloys. Metals Sci 8:298–310

Springorum D (1995) Slag atlas. Düsseldorf, Germany, Verein Deutscher Eisenhüttenleute (VDEh)

The Japan Society for the Promotion of Science. The 19th Committee on Steelmaking (1988) Steelmaking Data Sourcebook. Gordon and Breach Science Publishers, New York, USA

TAPP Database. Thermochemical and Physical Properties (Software)(1994) E. S. Microwave Inc., Cincinnati, Ohio, USA

Verdeja LF, Ballester A, Sancho JP, González R (2014) Refractory and ceramic materials. Ed. Síntesis, 1st edition, Madrid, Spain

Verdeja JI, Fernández-González D, Verdeja LF (2020) Operations and basic processes in ironmaking. Springer International Publishing, 1 st edition, Cham, Switzerland

Chapter 3
Secondary Metallurgy

Abstract This chapter is dedicated to the group of processes and operations performed in the steelmaking to transform the liquid iron that is obtained either in the converter or in the electric furnace into a melt adjusted in composition and ready for the solidification at the end of the secondary metallurgy. In this chapter, we consider the problem of the sulphur in the metallic melt as well as the degasification process (hydrogen, nitrogen, and oxygen) due to the problems that these elements cause in the steel. The chapter is complemented with several solved exercises.

3.1 Introduction

Secondary metallurgy, also known by the name of ladle metallurgy, is the group of processes and operations performed to transform the liquid iron that comes from the converter or electric furnace (melt produced in the converter or electric furnace will adjust neither from the thermal nor from the compositional points of view) into a melt adjusted in composition and prepared for the solidification at the end of the secondary metallurgy. The aims of the secondary metallurgy are three:

(a) Adjusting the alloying elements: C, Mn, Nb, Ti, etc.
(b) Adjusting the specifications of the nonmetallic (phosphorus, sulphur) and/or gaseous (oxygen, nitrogen, hydrogen) impurities dissolved in the liquid iron to obtain the quality required for the steel.
(c) Adjusting the temperature, as the structure of solidification and the metallurgical properties of the steel depend on the temperature.

In the last years, several technologies of secondary metallurgy have been developed, and in all of them, the most appropriated devices, reagents, and instruments to satisfy the above-mentioned objectives were considered:

- Techniques of powder injection (synthetic slags, alloying elements, REM-Rare Earth Metals, etc.) that, acting via deoxidizing and desulphurizing agents, achieve both the modification of the inclusions and the adjustment of the sulphur and oxygen dissolved in the liquid iron.

L. F. Verdeja González et al., *Operations and Basic Processes in Steelmaking*,
Topics in Mining, Metallurgy and Materials Engineering,
https://doi.org/10.1007/978-3-030-68000-8_3

- Vacuum technologies are designed for the gasification and decarburizing of the liquid iron (Ruhrtahl-Heraeus, RH, and Dortmund-Hoerder, DH, processes and vacuum casting).
- Heating techniques of the metal in the ladle (ladle furnace) to adjust the composition, temperature, and degree of oxidation of the liquid.
- Processes of ferroalloys addition to the refined liquid that comes from the BOF or the electric furnace as the CAS (Composition Adjustment System), which provides an optimal metallurgical performance during the addition of the alloying and deoxidizing elements.

The processes of the secondary metallurgy involve several variables (see points a, b, and c at the beginning of the chapter). It is important not to adjust only two or three metallurgical characteristics of the finished product, it is necessary to adjust most of them in the minimum number of processes or operations. Moreover, this multiplication of the operations before the disposition of the ladle in the casting position has several serious disadvantages:

- Waste of time and productivity of the installations and, especially, the reduction of the temperature of the melt.
- Contamination of the metal by oxidation and nitriding reactions.

The evolution of the secondary metallurgy processes tends to a marked simplification to treat to reach in a simple operation the specifications of the liquid before the casting. Techniques of sulphur and phosphorus refining as well as heating systems to control the temperature are used in the degasification processes that operate under vacuum conditions. The operating costs of applying this type of technologies make it difficult to reach the best qualities.

3.2 Deoxidation

The carbon-oxygen-carbon monoxide equilibrium which was analyzed in the refining of the pig iron (Sect. 1.3) has importance again in all those operations that are carried out during the secondary and tertiary metallurgies. In this way, the influence of the pressure and temperature in the solubility of the elements in the liquid iron is important (Fig. 3.1).

The variation of the standard free energy of oxidation with the temperature (Eq. (1.32)) of the carbon and oxygen dissolved in the liquid iron (Eq. (1.26)) leads to the following constant:

$$\mathrm{C(dis;\,iron)} + \mathrm{O(dis;\,iron)} \rightarrow \mathrm{CO(g)} \tag{3.1}$$

$$\begin{aligned} \Delta G^0_{\text{reaaction}(3.1)} &= -20900 - 39.76 \cdot T(K)\left[J \cdot (\text{mol CO})^{-1}\right] \\ &= -R \cdot T \cdot \ln K_{\text{reaction}\,(3.1)} \rightarrow \ln K_{\text{reaction}\,(3.1)} = 4.78 + \frac{2514}{T} \end{aligned} \tag{3.2}$$

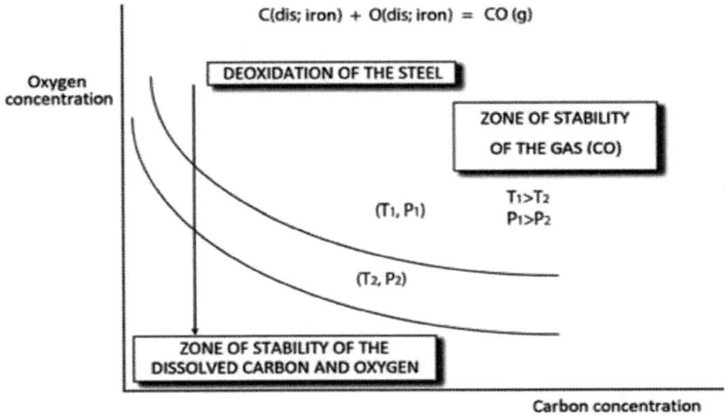

Fig. 3.1 Effect of both temperature and pressure in the solubility of both carbon and oxygen in the liquid iron

$$K_{\text{reaction (3.1)}} = \frac{P_{CO}}{C_O \cdot C_C} \tag{3.3}$$

expressed as a function of either the temperature or pressure and CO, carbon, and oxygen concentrations. An increase of the temperature in the system favors both the decrease of the equilibrium constant and the increase of the quantity of oxygen and carbon dissolved in the liquid iron; equally, an increase of the CO partial pressure, associated to the total pressure of the system, causes an increase in the solubility of both elements. On the contrary, the decrease of the temperature, typical in all process of solidification of metals and alloys, leads, in this case, to a release of carbon monoxide.

Rimmed steel, whose consumption grew in the middle of the twentieth century, is obtained by solidification of the liquid metal in a foundry ingot with a carbon percentage <0.15%. The result is a porous product, without shrinkage (neutralized by the volume of gas trapped in the solidification front and especially interesting for the galvanizing industry (Sect. 6.5)).

Equally, semi-killed steel exists (with carbon percentage in the range 0.15–0.30%) and optimizes the metallurgical yield of the process because it is possible to obtain a solidified material without shrinkage and with a volume equal to that of the ingot (in the case of the rimmed steel, the volume of the steel is greater than that of the ingot).

In any of the previous cases, rimmed or semi-killed steels, a release of carbon monoxide is produced as a consequence of the solubility decrease of the carbon and oxygen dissolved in the liquid. To avoid this gas outburst, metals with greater affinity for oxygen than carbon are added. They are called deoxidizing elements, such as aluminum, ferrosilicon, ferromanganese, calcium silicide, magnesium, and ferroalloys of rare earth elements.

Generally, and with the purpose of avoiding the reoxidation of the melt, deoxidation operations (to avoid the CO outburst) are performed in the last moments of liquid existence. At present, most of the steels are deoxidized with active metals.

The generalized chemical equation of the deoxidizing process is:

$$\text{a Me(dis; iron)} + \text{b O(dis; iron)} \leftrightarrow \text{Me}_a\text{O}_b\text{(dis; slag)} \tag{3.4}$$

and the equilibrium constant of the deoxidation reaction is:

$$K_{\text{reaction (3.4)}} = \frac{a_{Me_aO_b}}{a_{H,Me}{}^{a} \cdot a_{H,O}{}^{b}} \tag{3.5}$$

that, in the best of the cases, can be:

$$a_{H,Me} = C_{Me}(\%) = \%Me \tag{3.6}$$

$$a_{H,O} = C_{O}(\%) = \%O \tag{3.7}$$

when the Henry's coefficients of activity are taken as one. Table 3.1 (carbon steels) and Table 3.2 (alloyed steels) are used for different values of this coefficient. Equally, the activity of the oxide (product of the reaction) might be one on the condition that the action of a slag for trapping inclusions was not considered.

For a certain temperature, the generalized equation to represent the value of the solubility (solubility product) of the Me-O system in the liquid iron is:

$$C_{Me}{}^{a} \cdot C_{O}{}^{b} = \text{constant} \tag{3.8}$$

The efficiency of the deoxidation process involves the following questions:

- Chemical affinity of the metal by the oxygen dissolved in the liquid iron.
- Affinity of the metal to form a solution with the liquid (very low Raoult's activities with negative deviations very far from the ideal behavior).
- Low volatility of the metal at the operation temperature.
- Melting point of the metal and the ferroalloy that is used.
- The greater or smaller possibility of secondary reactions of the following types: (a) active metal-inclusion, (b) active metal-refractory, or (c) oxidation of the active metal.

Ferrosilicon is one of the most widely used metals in the deoxidation of carbon steels. Slags to catch inclusions are used to avoid the presence of metallic oxides (inclusions) in the steel, which negatively influence in the mechanical properties of materials (yield strength, ultimate tensile strength, and toughness). In the case of the silicon oxide, we must use the SiO_2–CaO–Al_2O_3 ternary system of Fig. 3.2 (Springorum 1995, activities of the silicon oxide in the ternary system) to calculate the solubility product of the silicon and oxygen dissolved in the liquid iron (equation similar to Eq. (3.8)).

Table 3.1 Interaction parameters of the Henry's law, e_i^j, for different active metals at 1600 °C according to (a) Sigworth and Elliot (1974) and (b) The Japan Society for the Promotion of Science (1988)

Elements (j)								
Solutes (i)	C	Al	Mg	Ca	Si	Mn	O	S
C	0.14(a) 0.243(b)	0.043(a) 0.043(b)	0.07(a) 0.07(b)	−0.047(a) −0.097(b)	0.08(a) 0.08(b)	−0.012(a) −0.0083(b)	−0.34(a) −0.32(b)	0.046(a) 0.044(b)
Al	0.041(a) 0.091(b)	0.045(a) 0.043(b)	–	−0.047(a) −0.047(b)	0.0056(a) 0.056(b)	–	−6.6(a) −1.98(b)	0.03(a) 0.034(b)
Mg	0.15(a) 0.15(b)	–	–	–	–	–	−3(b)	–
Ca	−0.34(a) −0.34(b)	−0.072(a) −0.072(b)	–	−0.002(a) −0.002(b)	−0.097(a) −0.096(b)	–	−75(a) −580(b)	−140(b)
Si	0.18(a) 0.18(b)	0.058(a) 0.058(b)	–	−0.007(a) −0.066(b)	0.11(a) 0.103(b)	0(a) −0.0146(b)	−0.23(a) −0.119(b)	0.056(a) 0.066(b)
Mn	−0.07(a) −0.0538(b)	–	–	–	0(a) −0.0327(b)	0(a) 0(b)	−0.083(a) −0.083(b)	−0.048(a) −0.047(b)
O	−0.45(a) −0.421(b)	−3.9(a) −1.17(b)	−1.98(a)	–	−0.131(a) −0.066(b)	−0.021(a) −0.021(b)	−0.2(a) −0.17(b)	−0.133(a) −0.133(b)
S	0.11(a) 0.112(b)	0.005(a) 0.042(b)	–	−110(b)	0.063(a) 0.075(b)	−0.026(a) −0.027(b)	−0.27(a) −0.27(b)	−0.028(a) −0.046(b)

Table 3.2 Interaction parameters of the Henry's law, e_i^j, for different active metals at 1600 °C according to (a) Sigworth and Elliot (1974) and (b) The Japan Society for the Promotion of Science (1988)

Elements (j)									
Solutes (i)	Al	Mn	Si	Ti	V	Cr	Ni	O	S
Al	0.045(a) 0.043(b)	–	0.0056(a) 0.056(b)	–	–	–	–	−6.6(a) −1.98(b)	0.03(a) 0.034(b)
Mn	–	0(a) 0(b)	0(a) −0.0327(b)	−0.05(b)	0.0056(b)	−0.0038(b)	−0.0071(b)	−0.083(a) −0.083(b)	−0.048(a) −0.047(b)
Si	0.058(a) 0.058(b)	0(a) −0.0146(b)	0.11(a) 0.103(b)	1.23(b)	0.025(b)	0(b)	0.005(b)	−0.23(a) −0.119(b)	0.056(a) 0.066(b)
Ti	–	−0.043(b)	2.1(b)	0.012(a) 0.042(b)	–	0.055(a)	–	−1.80(a) −3.40(b)	−0.11(a) −0.27(b)
V	–	–	0.042(a) 0.042(b)	–	0.015(a) 0.031(b)	0.0119(b)	–	−0.97(a) −0.46(b)	−0.028(a) −0.033(b)
Cr	–	0.0039(b)	−0.004(a) −0.004(b)	0.059(b)	0.012(b)	−0.0003(a) −0.0003(b)	0.0002(a) 0.0002(b)	−0.14(a) −0.16(b)	−0.02(a) −0.17(b)
Ni	–	−0.008(b)	0.005(a) 0.006(b)	–	–	−0.0003(a) −0.0003(b)	0.0009(a) 0.0007(b)	−0.01(a) −0.01(b)	−0.0037(a) −0.0036(b)
O	−3.9(a) −1.17(b)	−0.0021(a) −0.0021(b)	−0.131(a) −0.066(b)	−0.40(a) −1.12(b)	−0.30(a) −0.14(b)	−0.013(a) −0.055(b)	0.006(a) 0.006(b)	−0.20(a) −0.17(b)	-0.133(a) -0.133(b)
S	0.005(a) 0.042(b)	−0.026(a) −0.027(b)	0.063(a) 0.075(b)	−0.18(b)	−0.019(b)	0.01(a) 0.01(b)	0(a) 0(b)	−0.27(a) −0.27(b)	-0.028(a) -0.046(b)

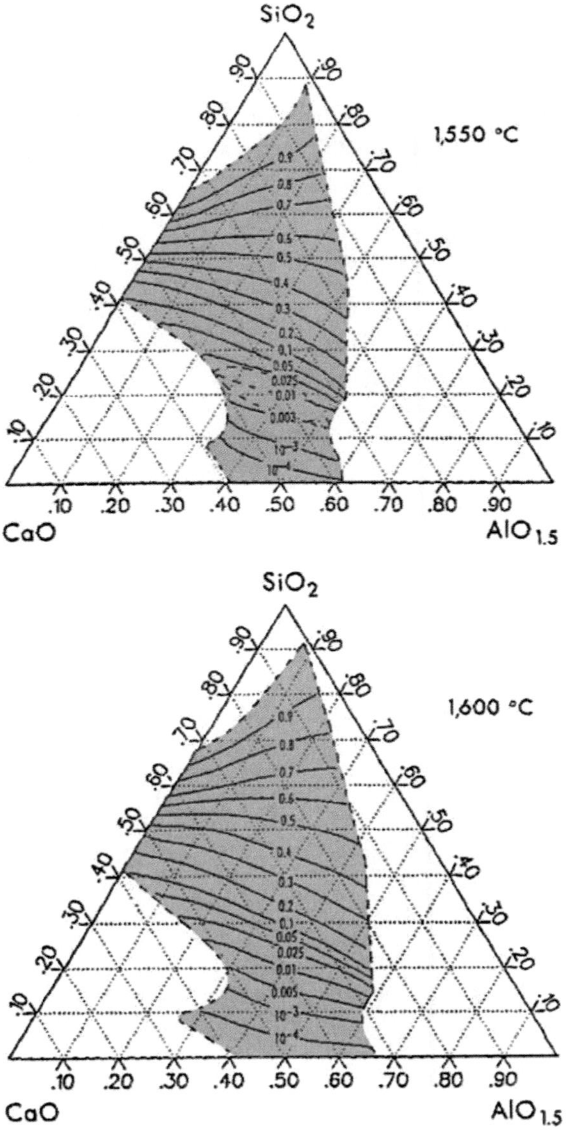

Fig. 3.2 Activity of the SiO_2 in the SiO_2–CaO–Al_2O_3 ternary system, at 1550 and 1600 °C

3.2.1 Solved Exercises

Exercise 3.1 In an iron melt with 0.0050% of oxygen (5 ppm) we must add vanadium until reaching 0.80% at 1560 °C. Analyze, when adding vanadium as ferrovanadium (Fe-30% V), if it is possible to lose this element in the slag as vanadium (II) oxide.

Solution starts:

Reaction that must be considered, if it is thermodynamically possible, is:

$$V(dis;\, a_{H,V} = 1) + O(dis;\, a_{H,O} = 1) \leftrightarrow VO(s) \quad (3.9)$$

We use the data of the Appendices 1 and 3 to calculate the standard free energy associated to the reaction (3.9):

$$V(s) + \frac{1}{2}O_2(g) \leftrightarrow VO(s) \quad (3.10)$$

$$\Delta_r G^0_{reaction\,(3.10)}(T) = -425.0 + 0.0810 \cdot T(K)\,\left[kJ \cdot (mol\,VO)^{-1}\right] \quad (3.11)$$

$$\frac{1}{2}O_2(g) \leftrightarrow O(dis;\, a_{H,O} = 1) \quad (3.12)$$

$$\Delta_r G^0_{reaction\,(3.12)}(T) = -117.0 + 0.0314 \cdot T(K)\,\left[kJ \cdot (mol\,O)^{-1}\right] \quad (3.13)$$

$$V(s) \leftrightarrow V(dis;\, a_{H,V} = 1) \quad (3.14)$$

$$\Delta_r G^0_{reaction\,(3.14)}(T) = -20.7 + 0.0456 \cdot T(K)\,\left[kJ \cdot (mol\,V)^{-1}\right] \quad (3.15)$$

and:

$$\Delta_r G^0_{reaction\,(3.9)}(T) = \Delta_r G^0_{reaction\,(3.10)}(T) - \Delta_r G^0_{reaction\,(3.12)}(T) - \Delta_r G^0_{reaction\,(3.14)}(T)$$
$$= -287.3 + 0.1297 \cdot T(K)\,\left[kJ \cdot (mol\,VO)^{-1}\right] \quad (3.16)$$

At the temperature of 1560 °C (1823 K), the values of both the standard free energy and the equilibrium constant are:

$$\Delta_r G^0_{reaction\,(3.9)}(1560\,°C) = -287.3 + 0.1297 \cdot 1833 = -49.4866\,\left[kJ \cdot (mol\,VO)^{-1}\right] \quad (3.17)$$

$$\Delta_r G^0_{reaction\,(3.9)}(1560\,°C) = -R \cdot T \cdot \ln K_{reaction\,(3.9)} \rightarrow K_{reaction\,(3.9)}$$

$$= \exp\left(-\frac{\Delta_r G^0_{\text{reaction (3.9)}}(1560\,^\circ\text{C})}{R \cdot T}\right)$$

$$= \exp\left(-\frac{-49486.6\left[J \cdot (\text{mol VO})^{-1}\right]}{8.314\left[J \cdot \text{mol}^{-1} \cdot K^{-1}\right] \cdot 1833\,K}\right) = 25.72 \tag{3.18}$$

The thermodynamic possibility that the vanadium oxidation could be carried out, according to the mechanism proposed by the reaction (3.9), would be determined by the free energy associated to the process:

$$\begin{aligned}\Delta_r G_{\text{reaction (3.9)}}(1560\,^\circ\text{C}) &= \Delta_r G^0_{\text{reaction (3.9)}}(1560\,^\circ\text{C}) + R \cdot T \cdot \ln\left(\frac{1}{\%V \cdot \%O}\right)\\ &= -49486.6\left[J \cdot (\text{mol VO})^{-1}\right]\\ &\quad + 8.314\left[J \cdot \text{mol}^{-1} \cdot K^{-1}\right] \cdot 1833K \cdot \ln\left(\frac{1}{0.80 \cdot 0.0050}\right)\\ &= 34658.05\left[J \cdot (\text{mol VO})^{-1}\right]\end{aligned} \tag{3.19}$$

Thus, vanadium oxidation, when its concentration is 0.80%, in a deoxidized iron melt (50 ppm O) at 1560 °C, does not take place because the free energy is positive. However, if we calculate the free energy of the process when the degree of oxidation of the melt is greater (500 ppm), for instance at the end of the fusion process in an electric furnace, the value is negative, and the reaction is favorable:

$$\begin{aligned}\Delta_r G_{\text{reaction (3.9)}}(1560\,^\circ\text{C}) &= \Delta_r G^0_{\text{reaction (3.9)}}(1560\,^\circ\text{C}) + R \cdot T \cdot \ln\left(\frac{1}{\%V \cdot \%O}\right)\\ &= -49486.6\left[J \cdot (\text{mol VO})^{-1}\right]\\ &\quad + 8.314\left[J \cdot \text{mol}^{-1} \cdot K^{-1}\right] \cdot 1833K \cdot \ln\left(\frac{1}{0.80 \cdot 0.050}\right)\\ &= -432.34\left[J \cdot (\text{mol VO})^{-1}\right]\end{aligned} \tag{3.20}$$

Finally, it is possible to quantify the deviations resulted from considering that the Henry's coefficients of activity of both vanadium, f_V, and oxygen, f_O, are different than one:

$$\begin{aligned}\log f_V &= e_V^V \cdot \%V + e_V^O \cdot \%O = 0.015 \cdot \left(\frac{1873}{1833}\right) \cdot 0.8 - 0.97 \cdot \left(\frac{1833}{1873}\right) \cdot 0.005\\ &= 7.5154 \cdot 10^{-3} \rightarrow f_V = 1.0175\end{aligned} \tag{3.21}$$

$$\log f_O = e_O^O \cdot \%O + e_O^V \cdot \%V = -0.20 \cdot \left(\frac{1833}{1873}\right) \cdot 0.005 - 0.30 \cdot \left(\frac{1833}{1873}\right) \cdot 0.80$$

$$= -0.2359 \rightarrow f_O = 0.5810 \tag{3.22}$$

Considering the data of the Appendix 3 at 1600 °C (1873 K) and correcting them with the effect of the temperature, Sect. 1.3.1, we obtain for the vanadium, as it was previously calculated, a value of $f_V = 1.0175$ while for the oxygen the value is $f_O = 0.5810$. The contribution of the oxidation process to the free energy is:

$$\begin{aligned} -R \cdot T \cdot \ln(f_V \cdot f_O) &= -8.314\left[J \cdot mol^{-1} \cdot K^{-1}\right] \cdot 1873\,K \cdot \ln(1.0175 \cdot 0.5810) \\ &= 8185.58\left[J \cdot (mol\,VO)^{-1}\right] \end{aligned} \tag{3.23}$$

The result previously obtained justifies that, in the following, it is not given excessive importance to the Henry's coefficients of activity and their possible influence in metallurgically directing a process, except in highly oxidized melts and with elements dissolved in quantities greater than 1.0%.

Commentaries: The solubility of compounds that form the slag in the melt are usually expressed in the format known as solubility product. In our case:

$$\%V \cdot \%O = \frac{1}{K_{reaction\,(3.9)}} = \frac{1}{25.72} = 3.89 \cdot 10^{-2} \tag{3.24}$$

The representation of the percentage of oxygen in front of the vanadium percentage, as in Fig. 3.1 for the C–O system, shows a hyperbolic curve. Above the equilibrium curve, the stability corresponds to the vanadium (II) oxide, while below it is shown a zone of stability of the elements-solutes dissolved in the iron melt (solvent).

Finally, we want to indicate that the criterium chosen to modify the interaction parameters linearly with the temperature is supported by the idea that any increase of temperature leads to an increase in the solubility. Consequently, for positive interactions, a decrease of temperature causes an increase of the positive interaction parameters, while a decrease of the temperature would mean reaching fewer negative values (more positive) in those negative interactions.

Solution ends.

Exercise 3.2 During the treatment with calcium of an iron melt at 1600 °C, the quantity of oxygen dissolved can reach 2 ppm. Analyze the thermodynamic possibility of contaminating the melt with silicon coming from the matrix of silica-alumina (60% SiO_2–40% Al_2O_3).

Data: Assume that the SiO_2 activity in the refractory matrix is equal to 10^{-3}.

Solution starts:

The equilibrium between the refractory matrix and the melt is the following one:

$$Si(dis;\ a_{H,Si} = 1) + 2O(dis;\ a_{H,O} = 1) \leftrightarrow SiO_2(s) \tag{3.25}$$

We use the data of the Appendices 1 and 3 to calculate the standard free energy, the equilibrium constant and the solubility product of the SiO_2 in the melt for reaction (3.25):

$$Si(l) + O_2(g) \leftrightarrow SiO_2(s) \tag{3.26}$$

$$\Delta_r G^0_{reaction\,(3.26)}(T) = -945.0 + 0.1980 \cdot T(K)\left[kJ \cdot (mol\,SiO_2)^{-1}\right] \tag{3.27}$$

$$\frac{1}{2}O_2(g) \leftrightarrow O\left(dis;\, a_{H,O} = 1\right) \tag{3.28}$$

$$\Delta_r G^0_{reaction\,(3.28)}(T) = -117.0 - 0.00314 \cdot T(K)\left[kJ \cdot (mol\,O)^{-1}\right] \tag{3.29}$$

$$Si(l) \leftrightarrow Si\left(dis;\, a_{H,Si} = 1\right) \tag{3.30}$$

$$\Delta_r G^0_{reaction\,(3.30)}(T) = -132.0 - 0.0172 \cdot T(K)\left[kJ \cdot (mol\,Si)^{-1}\right] \tag{3.31}$$

and:

$$\begin{aligned}\Delta_r G^0_{reaction\,(3.25)}(T) &= \Delta_r G^0_{reaction\,(3.26)}(T) - 2 \cdot \Delta_r G^0_{reaction\,(3.28)}(T) \\ &\quad - \Delta_r G^0_{reaction\,(3.30)}(T) \\ &= -579 + 0.22148 \cdot T(K)\left[kJ \cdot (mol\,SiO_2)^{-1}\right]\end{aligned} \tag{3.32}$$

At 1600 °C:

$$\begin{aligned}\Delta_r G^0_{reaction\,(3.25)}(1873K) &= -579 + 0.22148 \cdot 1873 \\ &= -164.1680\left[kJ \cdot (mol\,SiO_2)^{-1}\right]\end{aligned} \tag{3.33}$$

$$\begin{aligned}\Delta_r G^0_{reaction\,(3.25)}(1600\,°C) &= -R \cdot T \cdot \ln K_{reaction\,(3.25)} \rightarrow K_{reaction\,(3.25)} \\ &= \exp\left(-\frac{\Delta_r G^0_{reaction\,(3.25)}(1600\,°C)}{R \cdot T}\right) \\ &= \exp\left(-\frac{-164168\left[J \cdot (mol\,SiO_2)^{-1}\right]}{8.314\left[J \cdot mol^{-1} \cdot K^{-1}\right] \cdot 1873K}\right) \\ &= 37889.41\end{aligned} \tag{3.34}$$

$$K_{reaction\,(3.25)} = \frac{a_{SiO_2}}{(\%Si) \cdot (\%O)^2} \rightarrow 37889.41 = \frac{10^{-3}}{(\%Si) \cdot (\%O)^2} \tag{3.35}$$

Table 3.3 Maximum quantities of silicon compatibles with a certain concentration of oxygen

%Si (maximum)	0.6598	0.0066	0.0003	<0.001
O (ppm)	2.0	20.0	100.0	200.0

As it is possible to deduce from Eq. (3.35), the possibilities of silicon contamination in the melt depend on the quantity of oxygen dissolved in the melt. We collect in Table 3.3 the maximum quantities of silicon compatibles with a certain concentration of oxygen.

Commentaries: It is possible to check that, only in very deoxidized melts, with small quantities of oxygen dissolved or low potential of oxygen, exists the possibility of dissolution for the silicon oxide available in the refractory matrix of the lining, being possible an alteration of the silicon chemical specification in the melt.

Finally, it is necessary to consider that the duration of the refractory lining (corrosion) is usually studied taking into account the kinetics of the wear of those elements of the refractory that are thermodynamically susceptible of being attacked (see Verdeja et al. 2014, p. 177).

Solution ends.

Exercise 3.3 A ladle of 250 tons of iron melt at 1600 °C contains 0.030% of oxygen (300 ppm) after the fusion-refining processes in the electric furnace. Calculate the minimum quantity of aluminum required to reduce the oxygen below 10 ppm. We assume:

(a) That the Henry's law is verified for all the elements dissolved in the melt.
(b) That the conditions for the thermodynamic equilibrium are reached, and that the atomic weights of aluminum and oxygen are 26.98 and 16.00 g·mol^{-1}, respectively.

Solution starts:

Aluminum is one of the most widely used deoxidizing elements in the iron and steelmaking processes. Reactions that should be considered are:

$$2\mathrm{Al}\left(\mathrm{dis};\ a_{\mathrm{H,Al}}=1\right)+3\mathrm{O}\left(\mathrm{dis};\ a_{\mathrm{H,O}}=1\right)\leftrightarrow \mathrm{Al_2O_3(s)} \tag{3.36}$$

We use the data of the Appendices 1 and 3 to calculate the standard free energy, the equilibrium constant, and the solubility product of the SiO_2 in the melt for reaction (3.36):

$$2\mathrm{Al(l)}+\frac{3}{2}\mathrm{O_2(g)}\leftrightarrow \mathrm{Al_2O_3(s)} \tag{3.37}$$

$$\Delta_r G^0_{\text{reaction (3.37)}}(T)=-1680.0+0.3240\cdot T(K)\left[\mathrm{kJ}\cdot(\mathrm{mol\ Al_2O_3})^{-1}\right] \tag{3.38}$$

$$\frac{1}{2}O_2(g) \leftrightarrow O\big(dis;\ a_{H,O} = 1\big) \tag{3.39}$$

$$\Delta_r G^0_{reaction\,(3.39)}(T) = -117.0 - 0.00314 \cdot T(K)\left[kJ \cdot (mol\,O)^{-1}\right] \tag{3.40}$$

$$Al(l) \leftrightarrow Al\big(dis;\ a_{H,Al} = 1\big) \tag{3.41}$$

$$\Delta_r G^0_{reaction\,(3.41)}(T) = -67.1 - 0.0237 \cdot T(K)\left[kJ \cdot (mol\,Al)^{-1}\right] \tag{3.42}$$

and:

$$\begin{aligned}\Delta_r G^0_{reaction\,(3.36)}(T) &= \Delta_r G^0_{reaction\,(3.37)}(T) - 3 \cdot \Delta_r G^0_{reaction\,(3.39)}(T) \\ &\quad - 2 \cdot \Delta_r G^0_{reaction\,(3.41)}(T) \\ &= -1194.8 + 0.38092 \cdot T(K)\left[kJ \cdot (mol\,Al_2O_3)^{-1}\right]\end{aligned} \tag{3.43}$$

At 1600 °C:

$$\begin{aligned}\Delta_r G^0_{reaction\,(3.36)}(1873K) &= -1194.8 + 0.38092 \cdot 1873 \\ &= -481.3368\left[kJ \cdot (mol\,Al_2O_3)^{-1}\right]\end{aligned} \tag{3.44}$$

$$\begin{aligned}\Delta_r G^0_{reaction\,(3.36)}(1600\,°C) &= -R \cdot T \cdot \ln K_{reaction\,(3.36)} \rightarrow K_{reaction\,(3.36)} \\ &= \exp\left(-\frac{\Delta_r G^0_{reaction\,(3.36)}(1600\,°C)}{R \cdot T}\right) \\ &= \exp\left(-\frac{-481336.8\left[J \cdot (mol\,Al_2O_3)^{-1}\right]}{8.314\left[J \cdot mol^{-1} \cdot K^{-1}\right] \cdot 1873K}\right) \\ &= 2.66 \cdot 10^{13}\end{aligned} \tag{3.45}$$

$$K_{reaction\,(3.36)} = \frac{a_{Al_2O_3}}{(\%Al)^2 \cdot (\%O)^3} \rightarrow 2.66 \cdot 10^{13} = \frac{1}{(\%Al)^2 \cdot (\%O)^3} \tag{3.46}$$

The quantity of aluminum dissolved in equilibrium with 10 ppm of oxygen at 1600 °C is:

$$(\%Al) = \sqrt{\frac{1}{2.66 \cdot 10^{13} \cdot (0.001)^3}} = 0.0061\%\,Al(61\,ppm\,Al) \tag{3.47}$$

The quantity of aluminum in 250 tons of melt is:

$$250\,\text{t melt} \cdot \frac{0.0061\,\text{t Al}}{100\,\text{t melt}} \cdot \frac{1000\,\text{kg Al}}{1\,\text{t Al}} = 15.26\,\text{kg Al} \tag{3.48}$$

On the other hand, it is necessary to consider the quantity of aluminum that reacts with the oxygen to reduce its concentration from 300 to 10 ppm:

$$250\,\text{t melt} \cdot \frac{0.0029\,\text{t O}}{100\,\text{t melt}} \cdot \frac{1000\,\text{kg O}}{1\,\text{t O}} \cdot \frac{1\,\text{kmol O}}{16.00\,\text{kg O}} \cdot \frac{2\,\text{kmol Al}}{3\,\text{kmol O}} \cdot \frac{26.98\,\text{kg Al}}{1\,\text{kmol Al}} = 81.50\,\text{kg Al} \tag{3.49}$$

Definitely, the minimum quantity of aluminum would be 96.76 kg (15.26 kg + 81.50 kg). It is also necessary to consider that the efficiency of the deoxidation is not complete, there are losses due to the oxidation of the aluminum that basically depend on the technology of addition that is used.

Commentaries: In the deoxidation processes of iron melts using aluminum, the final quantity of aluminum is usually greater than 61 ppm (it is usually up to 240 ppm (0.024%)). In this case, the quantity of aluminum dissolved in the melt would be:

$$250\,\text{t melt} \cdot \frac{0.024\,\text{t Al}}{100\,\text{t melt}} \cdot \frac{1000\,\text{kg Al}}{1\,\text{t Al}} = 60.0\,\text{kg Al} \tag{3.50}$$

Under this consideration, the minimum quantity of aluminum would reach 141.50 kg (81.50 kg + 60 kg). Apart from the removal of the dissolved oxygen as Al_2O_3 to avoid the effervescence of the CO, Fig. 3.1, there are also other reasons that justify the presence of aluminum as for instance the capture of free nitrogen.

Solution ends.

3.3 Desulphurizing

Sulphur dissolved in the liquid iron can be removed using one of the following mechanisms:

- using elements with great affinity for the sulphur, which behave according to a reaction of the type:

$$c\text{Me(dis; iron)} + d\text{S(dis; iron)} \leftrightarrow \text{Me}_c\text{S}_d\text{(dis; slag)} \tag{3.51}$$

- by means of a slag with great ability to pick up sulphur according to the following ionic reaction:

$$\text{S}^{2-}\text{(dis; iron)} + \text{O(dis; slag)} \leftrightarrow \text{S(dis; slag)} + \text{O}^{2-}\text{(dis; iron)} \tag{3.52}$$

or, considering the corresponding molecular reaction:

$$S(\text{dis; iron}) + MeO(\text{dis; slag}) \leftrightarrow MeS(\text{dis; slag}) + O(\text{dis; iron}) \quad (3.53)$$

In the production of mild steels, especially for low and medium added value products, desulphurizing agents are those with the lowest price: ferromanganese, calcium carbide, synthetic slags with great activity in calcium oxide, etc. In alloyed steel or in the greatest qualities of mild steel, it is justified the use of calcium silicide, magnesium, or rare earth elements.

The capacity of one slag to pick up sulphur, $C_{S^{2-},slag/g}$, can be defined by the following slag-gas equilibrium:

$$\frac{1}{2}S_2(g) + O^{2-}(\text{dis; slag}) \leftrightarrow S^{2-}(\text{dis; slag}) + \frac{1}{2}O_2(g) \quad (3.54)$$

and, therefore:

$$C_{S^{2-},slag/g} = C_{S^{2-},slag} \cdot \left(\frac{P_{O_2}}{P_{S_2}}\right)^{1/2} = K_{reaction\,(3.54)} \cdot C_{O^{2-},slag} \quad (3.55)$$

or it might be also defined by the metal-slag equilibrium, $C^{*}_{S^{2-},slag/m}$ (capacity of sulphur removal):

$$S(\text{dis; iron}) + O^{2-}(\text{dis; slag}) \leftrightarrow S^{2-}(\text{dis; slag}) + O(\text{dis; iron}) \quad (3.56)$$

$$C^{*}_{S^{2-},slag/m} = C_{S^{2-},slag} \cdot \left(\frac{a_{H,O}}{a_{H,S}}\right) = K_{reaction\,(3.56)} \cdot C_{O^{2-},slag} \quad (3.57)$$

where $a_{H,O}$ and $a_{H,S}$ are the activities of Henry of the oxygen and sulphur in the metal, which are habitually identified with the weight percentage and $C^{*}_{S^{2-},slag/m}$ (concentration of sulphur in the slag that is usually expressed in percentage).

If we consider the following equilibrium, where the elements S and O dissolved in the iron melt are identified with the activities of Henry:

$$S(\text{dis; iron}) + \frac{1}{2}O_2(g) \leftrightarrow O(\text{dis; iron}) + \frac{1}{2}S_2(g) \quad (3.58)$$

where K_{O_2/S_2} is the equilibrium constant associated to the gas-metal (reaction (3.58)), it is possible to check that the addition of standard free energies associated to the reactions (3.54) and (3.58) leads to that of the reaction (3.56):

$$\Delta_r G^0_{reaction\,(3.56)}(T) = \Delta_r G^0_{reaction\,(3.54)}(T) + \Delta_r G^0_{reaction\,(3.58)}(T) \quad (3.59)$$

As a consequence of this, the capacity of picking up sulphur defined according to the metal-slag equilibrium, $C^{*}_{S^{2-},slag/m}$, Eq. (3.56), and the capacity of picking up sulphur according to the slag-gas equilibrium, $C_{S^{2-},slag/g}$, are related by the equilibrium constant of the reaction (3.58), K_{O_2/S_2}. The following functional relation between both capacities of removal is obtained:

$$C^*_{S^{2-},slag/m} = C_{S^{2-},slag} \cdot \left(\frac{a_{H,O}}{a_{H,S}}\right) = \left(\frac{P_{S_2}}{P_{O_2}}\right)^{1/2} \cdot C_{S^{2-},slag} \cdot \left(\frac{a_{H,O}}{a_{H,S}}\right) \cdot \left(\frac{P_{O_2}}{P_{S_2}}\right)^{1/2}$$
$$= K_{O_2/S_2} \cdot C_{S^{2-},slag/g} \qquad (3.60)$$

Finally, the capacity of sulphur removal, $C^*_{S^{2-},slag/m}$, is related with a variable commonly used in the desulphurizing of iron melts, the sulphur partition coefficient between the slag and the metal. If we identify the activities of Henry with their composition in percentage, we obtain:

$$C^*_{S^{2-},slag/m} = C_{S^{2-},slag} \cdot \left(\frac{(\%O)_{metal}}{(\%S)_{metal}}\right) \rightarrow \frac{C_{S^{2-},slag}}{(\%S)_{metal}} = \frac{C^*_{S^{2-},slag/m}}{(\%O)_{metal}} \qquad (3.61)$$

The more basic the slag is, the greater the quantity of anions that it provides and, thus, this facilitates the displacement to the right of the equilibrium (3.54). In the SiO_2–CaO–Al_2O_3 ternary system, the activity of the calcium oxide indicates the most suitable slags to remove the sulphur dissolved in the liquid iron (Fig. 3.3, Springorum 1995).

Tables 3.4 and 3.5 collect the values of the activity of calcium oxide in two systems of slags especially important in the sulphur refining:

- CaF_2–CaO–SiO_2 system.
- CaF_2–CaO–Al_2O_3 system (refining synthetic slags by vacuum re-melting).

The relation between the capacity of removing sulphur from the slag, $C_{S^{2-},slag/g}$, and the optical basicity (Springorum 1995, p. 11) is:

$$\log C_{S^{2-},slag/g} = \frac{22690 - 54640 \cdot \Lambda_{slag}}{T} + 43.6 \cdot \Lambda_{slag} - 25.2 \qquad (3.62)$$

where Λ_{slag} is the optical basicity of the slag.

The desulphurizing reaction is normally controlled by the diffusional mass transfer through the metal-slag interphase. From the first law of Fick of the stationary diffusional flow of matter, see Sect. 1.5, for the transportation of sulphur in the melt, whose concentration is usually expressed in percentage, we have:

$$\frac{dC_S}{dt} = \left(\frac{D_i}{\Delta x}\right) \cdot \left(C^0_S - C^=_S\right) \qquad (3.63)$$

If we integrate the Eq. (3.63) within the following boundary conditions:

- boundary condition 1: when $t = 0$, $C_S = C^0_S$.

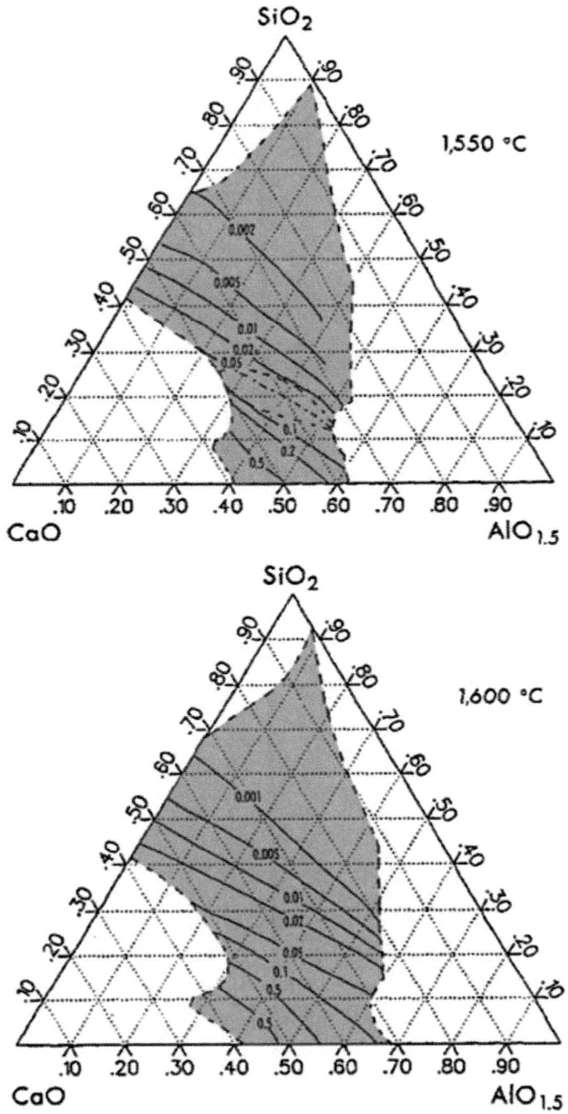

Fig. 3.3 Activities of the calcium oxide in the SiO_2–CaO–Al_2O_3 system at 1500 and 1600 °C

- boundary condition 2: for a time of reaction, t, sulphur concentration takes a value C_S.

We obtain the following equation that expresses the variation of the sulphur concentration with the time:

Table 3.4 Activity of the calcium oxide at 1400 °C in the CaF_2–CaO–SiO_2 system (Springorum 1995)

x_{CaO}	a_{CaO}	x_{CaO}/x_{SiO_2}
0.62	0.110	3.30
0.52	0.170	3.02
0.37	0.180	3.10
0.22	0.130	3.40
0.51	0.013	2.00
0.45	0.019	2.00
0.38	0.098	2.80
0.26	0.030	2.90

Table 3.5 Activity of the calcium oxide at 1400 °C in the CaF_2–CaO–Al_2O_3 system (Springorum 1995)

x_{CaO}	a_{CaO}	$x_{CaO}/a_{Al_2O_3}$
0.403	0.05	3.05
0.470	0.09	3.14
0.520	0.09	3.08
0.560	0.10	3.10
0.220	0.08	3.90
0.380	0.17	4.99
0.410	0.18	6.27
0.320	0.22	7.02
0.260	0.24	7.67

$$\ln\left(\frac{C_S - C_S^{=}}{C_S^0 - C_S^{=}}\right) = -k_d \cdot t \tag{3.64}$$

where C_S^0 is the sulphur concentration at the beginning of the treatment, $C_S^{=}$ is the sulphur concentration in equilibrium with the slag of the ladle, and k_d is the kinetics constant of the diffusional process of mass transfer in unities of speed (length divided by time):

$$k_d = \left(\frac{D_i}{\Delta x}\right) = \frac{D_i}{\delta_i^0} \tag{3.65}$$

where δ_i^0 is the thickness of fluid, tight boundary layer, where the only possible transfer mechanism is diffusion.

When the concentration of sulphur in equilibrium tends to zero, $C_S^{=} \rightarrow 0$, Eq. (3.64) can be expressed as follows:

$$\ln\left(\frac{C_S}{C_S^0}\right) = -k_d \cdot t \rightarrow C_S = C_S^0 \cdot \exp(-k_d \cdot t) \tag{3.66}$$

On another note, the kinetic constant of the diffusional process, expressed in unities of time raised to -1:

$$k = k_d \cdot \left(\frac{A}{V}\right) \tag{3.67}$$

is the function of the rate of argon injection used as inert gas for the transportation of active materials:

$$k\left(\min^{-1}\right) = 8.0 \cdot 10^{-6} \cdot (\tilde{\varepsilon})^{2.1} \tag{3.68}$$

where $\tilde{\varepsilon}$ is the mixing power per ton of metal ($W{\cdot}t^{-1}$), which is equal to:

$$\tilde{\varepsilon} = 14.23 \cdot \frac{V_g \cdot T}{M_{metal}} \cdot \log\left(1 + \frac{H}{1.48 \cdot P_0}\right) \tag{3.69}$$

where V_g is the gas flow in $Nm^3{\cdot}min^{-1}$ and H is the depth in meters reached by the injection measured from the slag-gas interface. If the injection of gas takes place via a porous plug, H is equal to the height of the melt in the ladle, P_0 is the pressure of the gas in the surface of the melt (in atmospheres), M_{metal} is the quantity of steel that was treated, in tons, and T is the temperature in K.

If the controlling mechanism of the desulphurizing is the chemical reaction in the metal-slag interface (of first order), the variation of the sulphur concentration (per mol) with the time of treatment is equivalent to that obtained with the Eq. (3.64). The method of finding the real controlling mechanism is analyzing the variation of the kinetics constant of the process, k, with the temperature, as an equation of Arrhenius type:

$$k = A \cdot \exp\left(-\frac{\Delta E}{R \cdot T}\right) \tag{3.70}$$

The activation energy of the process, ΔE, indicates whether the control of the desulphurizing process is the diffusion (low values) or the chemistry (ΔE of around the energy of the chemical bond). If the kinetics of the sulphur removal is studied considering the chemical characteristics of the slag during the treatment, there are two possibilities (Fig. 3.4):

1. Sulphur concentration in the slag remains constant during all the treatment: model of transient regime.
2. Sulphur concentration in the slag varies with the time until reaching the value of the equilibrium concentration: model of permanent regime.

The equation of the model of transient regime applied to the liquid iron desulphurizing (the same criterion can be applied to reactions of steel killing) is the following one:

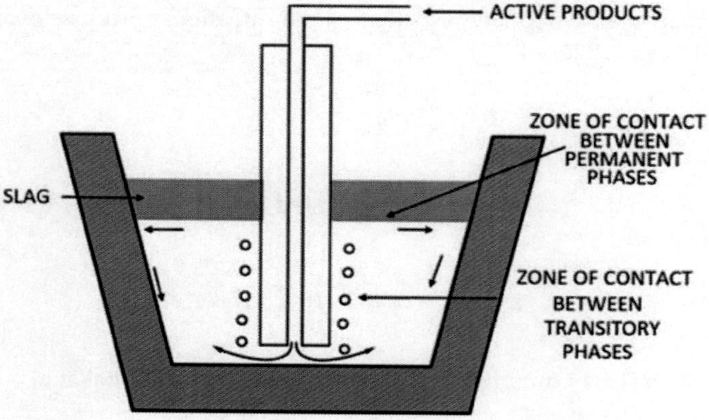

Fig. 3.4 Kinetics models for processes of treatment in the ladle. Processes: TN (Thyssen-Niederrhein) and SL (Scandinavian Lancers, developed by the Scandinavian Steelmakers Society)

$$\frac{C_{S,metal}}{C^0_{S,metal}} = \frac{\%S_{metal}}{\%S^0_{metal}} = \exp(-\eta \cdot Y) \tag{3.71}$$

where $C_{S,metal}$ is the sulphur concentration in the metal at the end of the treatment, $C^0_{S,metal}$ is the initial sulphur concentration in the metal and η is the sulphur partition coefficient between the slag and the metal at the beginning of the treatment, which can be expressed as follows:

$$\eta = \frac{C_{S,slag}}{C_{S,metal}} = \frac{\%S_{slag}}{\%S_{metal}} \tag{3.72}$$

In the model of transient regime, sulphur concentration in the synthetic slag injected in the melt varies during the time that lasts the transient contact of the slag with the melt.

Finally, Y is a dimensionless variable that expresses the relation between the quantity of slag injected per unit of time, $\dot{V}_{slag}$, the quantity of metal that is processed, M_{metal}, and the time, t.

For a certain time of injection, t:

$$\frac{C_S}{C^0_S} = \exp\left(-\eta \cdot \frac{\dot{V}_{slag}}{M_{metal}} \cdot t\right) \tag{3.73}$$

That is to say, the functional relation between Y and the quantity of slag injected per unit of time, $\dot{V}_{slag}$, the quantity of metal that is processed, M_{metal}, and the time, t, is:

$$Y = \frac{\dot{V}_{slag}}{M_{metal}} \cdot t \tag{3.74}$$

If we compare the kinetic equations of the transient regime and the derivative of the mechanism of diffusional transfer for the sulphur removal in an iron melt, Eq. (3.64), it is possible to obtain the following relation between kinetic variables:

$$k = k_d \cdot \left(\frac{A}{V}\right) = \left(\frac{D_S}{\delta_c}\right) \cdot \left(\frac{A}{V}\right) = \eta \cdot \frac{\dot{V}_{slag}}{M_{metal}} = \eta \cdot \left(\frac{Y}{t}\right) \tag{3.75}$$

where D_S is the diffusion coefficient of the sulphur in the melt and δ_c is the thickness of the tight boundary layer of sulphur concentration (thickness of fluid in contact with the slag where the only mass transfer mechanism is the diffusion) and the value of A/V is the relation area to volume associated to the metal/slag contact.

On the other hand, the application of the model of permanent regime to the liquid iron desulphurizing is developed when the slag with high capacity to remove sulphur is stationary contact with an iron melt. In this case, the sulphur concentration in the slag does not change with the time. The kinetic expression of the permanent regime model to the sulphur slagging leads to the following equation, see Sect. 1.5:

$$\frac{C_S^0}{C_S} = 1 + \eta \cdot Y \tag{3.76}$$

If we compare Eq. (3.73) and Eq. (3.76), it is possible to deduce that, operating with the same quantity of slag, the model of transient regime gives a faster sulphur slagging rate than in the case of the model of permanent regime.

The model of transient regime is adjusted to the experimental data when the quantity of sulphur dissolved in the liquid iron is small. On another note, when the sulphur concentration is high, the desulphurizing reaction is preferentially identified with the model of permanent regime.

3.3.1 Solved Exercises

Exercise 3.4 Calculate the sulphur partition coefficient for a melt at 1600 °C with 40 ppm (0.004%) of oxygen that is treated with a slag of the CaO–SiO_2–CaF_2 ternary system (Table 3.4). The chemical characteristics of the slag correspond to a molar fraction of calcium oxide of 0.62 (see Table 3.4). Assume that the activity of the calcium oxide does not change with the temperature. The atomic weights of the CaO, SiO_2, and CaF_2 are 56.08, 60.09, and 78.08 g·mol^{-1} and the CaS exhibits ideal behavior in the slag.

Solution starts:

From the data of the enunciation and those collected in Table 3.4, it is possible to express the composition of the slag in molar fraction:

$$x_{CaO} = 0.62 \quad (3.77)$$

$$\frac{x_{CaO}}{x_{SiO_2}} = 3.30 \rightarrow x_{SiO_2} = \frac{0.62}{3.3} = 0.1879 \quad (3.78)$$

$$x_{CaF_2} = 1 - x_{CaO} - x_{SiO_2} = 1 - 0.62 - 0.1879 = 0.1921 \quad (3.79)$$

On the other hand, other chemical characteristics of the slag are:

- Composition in percentage:

$$CaO = 0.62\,\text{mol CaO} \cdot \frac{56.08\,\text{g CaO}}{1\,\text{mol CaO}} = 34.7696\,\text{g CaO} \rightarrow \%CaO = \frac{34.7696}{61.0597} \cdot 100 = 56.94\% \quad (3.80)$$

$$SiO_2 = 0.1879\,\text{mol}\,SiO_2 \cdot \frac{60.09\,\text{g}\,SiO_2}{1\,\text{mol}\,SiO_2} = 11.2909\,\text{g}\,SiO_2 \rightarrow \%SiO_2 = \frac{11.2909}{61.0597} \cdot 100 = 18.49\% \quad (3.81)$$

$$CaF_2 = 0.1921\,\text{mol}\,CaF_2 \cdot \frac{78.08\,\text{g}\,CaF_2}{1\,\text{mol}\,CaF_2} = 14.9992\,\text{g}\,CaF_2 \rightarrow \%CaF_2 = \frac{14.9992}{61.0597} \cdot 100 = 24.57\% \quad (3.82)$$

The molecular weight of the slag is:

$$PM_{slag} = x_{CaO} \cdot PM_{CaO} + x_{CaF_2} \cdot PM_{CaF_2} + x_{SiO_2} \cdot PM_{SiO_2} = 0.62 \cdot 56.08 + 0.1921 \cdot 78.08 + 0.1879 \cdot 60.09 = 61.06\,\text{g} \cdot \text{mol}^{-1} \quad (3.83)$$

Now, we calculate the optical basicity of the slag:

1. Number of moles in 100 g of slag:

$$\text{Moles} = \frac{56.94}{56.08} + \frac{18.49}{60.09} + \frac{24.57}{78.08} = 1.6377\,\text{mol in 100 g of slag} \quad (3.84)$$

2. Composition expressed in molar fraction, of each one of the oxides of the slag. We have the molar fractions at the beginning of the exercise.

3. Summation of the product of the molar fraction of each oxide of the slag, x_{MeO}, and the number of oxygen atoms in each oxide, n(O) (the contribution of the CaF_2 to the optical basicity is the same as that of the CaO):

$$\sum x_{MeO} \cdot n(O) = 0.6200 \cdot 1 + 0.1879 \cdot 2 + 0.1921 \cdot 1 = 1.1879 \tag{3.85}$$

4. We calculate X_{MeO}:

$$X_{CaO} = \frac{x_{CaO} \cdot n(O)}{\sum x_{MeO} \cdot n(O)} = \frac{0.62 \cdot 1}{1.1879} = 0.5219 \tag{3.86}$$

$$X_{SiO_2} = \frac{x_{SiO_2} \cdot n(O)}{\sum x_{MeO} \cdot n(O)} = \frac{0.1879 \cdot 2}{1.1879} = 0.3164 \tag{3.87}$$

$$X_{CaF_2} = \frac{x_{CaF_2} \cdot n(O)}{\sum x_{MeO} \cdot n(O)} = \frac{0.1921 \cdot 1}{1.1879} = 0.1617 \tag{3.88}$$

5. Finally, we calculate the optical basicity of the slag, Λ_{slag};

$$\begin{aligned}\Lambda_{slag} &= X_{CaO} \cdot \Lambda_{CaO} + X_{SiO_2} \cdot \Lambda_{SiO_2} + X_{CaF_2} \cdot \Lambda_{CaF_2} \\ &= 0.5219 \cdot 1 + 0.3164 \cdot 0.48 + 0.1617 \cdot 1 = 0.8355\end{aligned} \tag{3.89}$$

The value of the standard free energy associated to the equilibrium of the sulphur distribution between the metal and the slag is calculated from the equation in the ionic form, Eq. (3.56), using the data of the Appendices 1 and 3:

$$Ca(l) + \frac{1}{2}O_2(g) \leftrightarrow CaO(s) \tag{3.90}$$

$$\Delta_r G^0_{reaction\ (3.90)}(T) = -788.0 + 0.193 \cdot T(K)\left[kJ \cdot (mol\,CaO)^{-1}\right] \tag{3.91}$$

$$Ca(l) + \frac{1}{2}S_2(g) \leftrightarrow CaS(s) \tag{3.92}$$

$$\Delta_r G^0_{reaction\ (3.92)}(T) = -696.0 + 0.189 \cdot T(K)\left[kJ \cdot (mol\,CaS)^{-1}\right] \tag{3.93}$$

$$\frac{1}{2}O_2(g) \leftrightarrow O\left(dis;\ a_{H,O} = 1\right) \tag{3.94}$$

$$\Delta_r G^0_{reaction\ (3.94)}(T) = -117.0 - 3.14 \cdot 10^{-3} \cdot T(K)\left[kJ \cdot (mol\,O)^{-1}\right] \tag{3.95}$$

$$\frac{1}{2}S_2(g) \leftrightarrow S\left(dis;\ a_{H,S} = 1\right) \tag{3.96}$$

$$\Delta_r G^0_{\text{reaction (3.96)}}(T) = -130.0 + 2.10 \cdot 10^{-2} \cdot T(K) \left[kJ \cdot (\text{mol S})^{-1}\right] \quad (3.97)$$

The global reaction is:

$$CaO(\text{dis; slag}) + S\left(\text{dis; } a_{H,S} = 1\right) \leftrightarrow CaS(\text{dis; slag}) + O\left(\text{dis; } a_{H,O} = 1\right) \quad (3.98)$$

and the value of the $\Delta_r G^0_{\text{reaction (3.98)}}(T)$ is:

$$\begin{aligned} \Delta_r G^0_{\text{reaction (3.98)}}(T) &= -\Delta_r G^0_{\text{reaction (3.90)}}(T) + \Delta_r G^0_{\text{reaction (3.92)}}(T) + \Delta_r G^0_{\text{reaction (3.94)}}(T) \\ &\quad - \Delta_r G^0_{\text{reaction (3.96)}}(T) \\ &= 105.0 - 2.184 \cdot 10^{-2} \cdot T(K) \left[kJ \cdot (\text{mol CaO})^{-1}\right] \end{aligned} \quad (3.99)$$

At 1600 °C (1873 K):

$$\begin{aligned} \Delta_r G^0_{\text{reaction (3.98)}}(1873\,K) &= 105.0 - 2.184 \cdot 10^{-2} \cdot 1873 \\ &= 52.2938\,kJ \cdot (\text{mol CaO})^{-1} \end{aligned} \quad (3.100)$$

And the equilibrium constant is:

$$\begin{aligned} \Delta_r G^0_{\text{reaction (3.98)}}(1873\,K) &= -R \cdot T \cdot \ln K_{\text{reaction (3.98)}} \rightarrow K_{\text{reaction (3.98)}} \\ &= \exp\left(-\frac{\Delta_r G^0_{\text{reaction (3.98)}}(1873\,K)}{R \cdot T}\right) \\ &= \exp\left(-\frac{52293.8\,kJ \cdot (\text{mol CaO})^{-1}}{8.314\,J \cdot mol^{-1} \cdot K^{-1} \cdot 1873\,K}\right) \\ &= 0.0348 \end{aligned} \quad (3.101)$$

$$K_{\text{reaction (3.98)}} = \frac{a_{H,O} \cdot a_{CaS}}{a_{H,S} \cdot a_{CaO}} \quad (3.102)$$

As:

$$a_{H,O} = \%O \quad (3.103)$$

$$a_{H,S} = (\%S)_{\text{metal}} \quad (3.104)$$

and as the CaS has ideal behavior in the slag:

$$a_{CaS} = x_{CaS} \quad (3.105)$$

then:

$$\frac{x_{CaS}}{(\%S)_{metal}} = K_{reaction\,(3.98)} \cdot \left(\frac{a_{CaO}}{\%O}\right) \rightarrow (\%S)_{metal} = \frac{x_{CaS} \cdot \%O}{K_{reaction\,(3.98)} \cdot a_{CaO}} \tag{3.106}$$

$$x_{CaS} = \frac{(\%S)_{slag}}{100} \cdot \frac{PM_{slag}}{32.06} \rightarrow (\%S)_{slag} = \frac{x_{CaS} \cdot 100 \cdot 32.06}{PM_{slag}} \tag{3.107}$$

$$\eta = \frac{(\%S)_{slag}}{(\%S)_{metal}} = \frac{\frac{x_{CaS} \cdot 100 \cdot 32.06}{PM_{slag}}}{\frac{x_{CaS} \cdot \%O}{K_{reaction\,(3.98)} \cdot a_{CaO}}} = \frac{\frac{100 \cdot 32.06}{61.06}}{\frac{\%O}{0.0348 \cdot a_{CaO}}} \tag{3.108}$$

All the terms of the equation of the partition coefficient are known: the activity of the calcium oxide in the slag and the oxygen concentration in the metallic melt. Consequently, the value of the partition coefficient of the considered slag is (a_{CaO} = 0.11, see Table 3.4):

$$\eta = \frac{(\%S)_{slag}}{\left(\vec{\%S}\right)_{metal}} = \frac{\frac{100 \cdot 32.06}{61.06}}{\frac{0.004}{0.0348 \cdot 0.11}} = 50.25 \tag{3.109}$$

Commentaries: The sulphur partition coefficient between the slag and the metal is very sensitive to the variations of the activity of the calcium oxide in the slag and the degree of deoxidation reached in the deoxidation of the melt.

If we would have used the $CaO–Al_2O_3–CaF_2$ ternary system, Table 3.5, the chemical characteristics of the slag corresponding to a CaO molar fraction of 0.26 are described in the following lines.

We express the composition of the slag in molar fraction:

$$x_{CaO} = 0.26 \tag{3.110}$$

$$\frac{x_{CaO}}{x_{Al_2O_3}} = 7.67 \rightarrow x_{Al_2O_3} = \frac{0.26}{7.67} = 0.0334 \tag{3.111}$$

$$x_{CaF_2} = 1 - x_{CaO} - x_{Al_2O_3} = 1 - 0.26 - 0.0334 = 0.7066 \tag{3.112}$$

On the other hand, other chemical characteristics of the slag are:

– Composition in mass percentage:

$$CaO = 0.26\,\text{mol CaO} \cdot \frac{56.08\,\text{g CaO}}{1\,\text{mol CaO}} = 14.5808\,\text{g CaO} \rightarrow \%CaO = \frac{14.5808}{73.1576} \cdot 100 = 19.93\% \tag{3.113}$$

$$Al_2O_3 = 0.0334\,\text{mol}\,Al_2O_3 \cdot \frac{101.96\,\text{g}\,Al_2O_3}{1\,\text{mol}\,Al_2O_3} = 3.4055\,\text{g}\,Al_2O_3 \rightarrow \%Al_2O_3$$

$$= \frac{3.4055}{73.1576} \cdot 100 = 4.66\% \tag{3.114}$$

$$CaF_2 = 0.7066\,\text{mol}\,CaF_2 \cdot \frac{78.08\,\text{g}\,CaF_2}{1\,\text{mol}\,CaF_2} = 55.1713\,\text{g}\,CaF_2 \rightarrow \%CaF_2$$

$$= \frac{55.1713}{73.1576} \cdot 100 = 75.41\% \tag{3.115}$$

The molecular weight of the slag is:

$$\begin{aligned} PM_{slag} &= x_{CaO} \cdot PM_{CaO} + x_{CaF_2} \cdot PM_{CaF_2} + x_{Al_2O_3} \cdot PM_{Al_2O_3} \\ &= 0.26 \cdot 56.08 + 0.7066 \cdot 78.08 + 0.0334 \cdot 101.96 = 73.16\,\text{g} \cdot \text{mol}^{-1} \end{aligned} \tag{3.116}$$

Now, we calculate the optical basicity of the slag:

1. Number of moles in 100 g of slag:

$$\text{Moles} = \frac{19.93}{56.08} + \frac{4.66}{101.96} + \frac{75.41}{78.08} = 1.3669\,\text{mol in 100 g of slag} \tag{3.117}$$

2. Composition expressed in molar fraction, of each one of the oxides of the slag. We have the molar fractions at the beginning of this section of the exercise.
3. Summation of the product of the molar fraction of each oxide of the slag, x_{MeO}, and the number of oxygen atoms in each oxide, n(O) (the contribution of the CaF_2 to the optical basicity is the same as that of the CaO):

$$\sum x_{MeO} \cdot n(O) = 0.26 \cdot 1 + 0.0334 \cdot 3 + 0.7066 \cdot 1 = 1.0668 \tag{3.118}$$

4. We calculate X_{MeO}:

$$X_{CaO} = \frac{x_{CaO} \cdot n(O)}{\sum x_{MeO} \cdot n(O)} = \frac{0.26 \cdot 1}{1.0668} = 0.2437 \tag{3.119}$$

$$X_{Al_2O_3} = \frac{x_{Al_2O_3} \cdot n(O)}{\sum x_{MeO} \cdot n(O)} = \frac{0.0334 \cdot 3}{1.0668} = 0.0939 \tag{3.120}$$

$$X_{CaF_2} = \frac{x_{CaF_2} \cdot n(O)}{\sum x_{MeO} \cdot n(O)} = \frac{0.7066 \cdot 1}{1.0668} = 0.6624 \tag{3.121}$$

5. Finally, we calculate the optical basicity of the slag, Λ_{slag}:

$$\begin{aligned} \Lambda_{slag} &= X_{CaO} \cdot \Lambda_{CaO} + X_{Al_2O_3} \cdot \Lambda_{Al_2O_3} + X_{CaF_2} \cdot \Lambda_{CaF_2} \\ &= 0.2437 \cdot 1 + 0.0939 \cdot 0.61 + 0.6624 \cdot 1 = 0.9634 \end{aligned} \tag{3.122}$$

All the terms of the equation of the partition coefficient are known: the activity of the calcium oxide in the slag and the oxygen concentration in the metallic melt. Consequently, the value of the partition coefficient of the considered slag is (a_{CaO} = 0.24, see Table 3.5):

$$\eta = \frac{(\%S)_{slag}}{(\%S)_{metal}} = \frac{\frac{100 \cdot 32.06}{73.16}}{\frac{0.004}{0.0348 \cdot 0.24}} = 91.50 \tag{3.123}$$

However, from the two studied systems, Tables 3.4 and 3.5, the obtaining of high activities for the calcium oxide in the slag are associated to the $CaO–Al_2O_3–CaF_2$ system with great CaF_2 contents. High CaF_2 contents, apart from accelerating the corrosion of the refractory lining, can have associated environmental problems.

Finally, we want to point out, as it is expressed in the Appendix 4 and in the physical-chemical fundamentals of the optical basicity of the oxides that constitute the slag, see Deo and Boom (1993) (pp. 77–78), that the value given to the optical basicity of an slag with CaF_2 is the same than that given to the combination of the calcium with the oxygen: $\Lambda_{CaO} = \Lambda_{CaF_2}$. This criterion is also applicable to the combinations of the fluorine and chlorine with the calcium, sodium, and potassium, that is to say: $\Lambda_{CaO} = \Lambda_{CaF_2}$, $\Lambda_{CaO} = \Lambda_{CaCl_2}$ and $\Lambda_{Na_2O} = \Lambda_{NaF} = \Lambda_{NaCl}$.

Solution ends.

Exercise 3.5 Calculate the sulphur partition coefficient at 1600 °C between a slag and an iron melt with 40 ppm of oxygen when the characteristics of the slag, free of CaF_2, correspond to the $CaO–MgO–SiO_2$ system: 40% CaO, 20% MgO, and 40% SiO_2. The molecular weights of the components of the slag are: 56.08, 40.31, and 60.09 g·mol^{-1} for the CaO, MgO, and SiO_2, respectively.

Solution starts:

In this case, and as opposed to the Exercise 3.4, the calculation of the sulphur partition coefficient between the slag and the metal is associated to the capacity of picking up sulphur by the slag. On the other hand, the capacity of a slag to pick up sulphur is function of the optical basicity, Eq. (3.62). The optical basicity of the slag is calculated as follows:

Now, we calculate the optical basicity of the slag:

1. Number of moles in 100 g of slag:

$$\text{Moles} = \frac{40}{56.08} + \frac{20}{40.31} + \frac{40}{60.09} = 1.8751 \text{ mol in 100 g of slag} \tag{3.124}$$

2. Composition expressed in molar fraction, of each one of the oxides of the slag. We calculate the molar fractions:

$$x_{CaO} = \frac{40\,g\,CaO \cdot \frac{1\,mol\,CaO}{56.08\,g\,CaO}}{40\,g\,CaO \cdot \frac{1\,mol\,CaO}{56.08\,g\,CaO} + 20\,g\,MgO \cdot \frac{1\,mol\,MgO}{40.31\,g\,MgO} + 40\,g\,SiO_2 \cdot \frac{1\,mol\,SiO_2}{60.09\,g\,SiO_2}} = 0.3804 \quad (3.125)$$

$$x_{MgO} = \frac{20\,g\,MgO \cdot \frac{1\,mol\,MgO}{40.31\,g\,MgO}}{40\,g\,CaO \cdot \frac{1\,mol\,CaO}{56.08\,g\,CaO} + 20\,g\,MgO \cdot \frac{1\,mol\,MgO}{40.31\,g\,MgO} + 40\,g\,SiO_2 \cdot \frac{1\,mol\,SiO_2}{60.09\,g\,SiO_2}} = 0.2646 \quad (3.126)$$

$$x_{SiO_2} = \frac{40\,g\,SiO_2 \cdot \frac{1\,mol\,SiO_2}{60.09\,g\,SiO_2}}{40\,g\,CaO \cdot \frac{1\,mol\,CaO}{56.08\,g\,CaO} + 20\,g\,MgO \cdot \frac{1\,mol\,MgO}{40.31\,g\,MgO} + 40\,g\,SiO_2 \cdot \frac{1\,mol\,SiO_2}{60.09\,g\,SiO_2}} = 0.3550 \quad (3.127)$$

3. Summation of the product of the molar fraction of each oxide of the slag, x_{MeO}, and the number of oxygen atoms in each oxide, n(O):

$$\sum x_{MeO} \cdot n(O) = 0.3804 \cdot 1 + 0.2646 \cdot 1 + 0.3550 \cdot 2 = 1.355 \quad (3.128)$$

4. We calculate X_{MeO}:

$$X_{CaO} = \frac{x_{CaO} \cdot n(O)}{\sum x_{MeO} \cdot n(O)} = \frac{0.3804 \cdot 1}{1.355} = 0.2807 \quad (3.129)$$

$$X_{MgO} = \frac{x_{MgO} \cdot n(O)}{\sum x_{MeO} \cdot n(O)} = \frac{0.2646 \cdot 1}{1.355} = 0.1953 \quad (3.130)$$

$$X_{SiO_2} = \frac{x_{SiO_2} \cdot n(O)}{\sum x_{MeO} \cdot n(O)} = \frac{0.3550 \cdot 2}{1.355} = 0.5240 \quad (3.131)$$

5. Finally, we calculate the optical basicity of the slag, Λ_{slag}:

$$\Lambda_{slag} = X_{CaO} \cdot \Lambda_{CaO} + X_{MgO} \cdot \Lambda_{MgO} + X_{SiO_2} \cdot \Lambda_{SiO_2} = 0.2807 \cdot 1 + 0.1953 \cdot 0.87 + 0.5240 \cdot 0.48 = 0.7021 \quad (3.132)$$

If we apply Eq. (3.62) at the temperature of 1873 K, we obtain:

$$\log C_{S^{2-},slag/g} = \frac{22690 - 54640 \cdot \Lambda_{slag}}{T} + 43.6 \cdot \Lambda_{slag} - 25.2 = \frac{22690 - 54640 \cdot 0.7021}{1873} + 43.6 \cdot 0.7021 - 25.2 = -2.9562 \rightarrow C_{S^{2-},slag/g} = 1.1062 \cdot 10^{-3} \quad (3.133)$$

As it was explained in Sect. 3.3, the relation between the capacity of picking up sulphur considering the slag-gas equilibrium and the corresponding equilibrium of the slag with the metal is calculated by means of the equilibrium constant of the following reaction:

$$S(\text{dis}; a_{H,S} = 1) + \frac{1}{2}O_2(g) \leftrightarrow O(\text{dis}; a_{H,O} = 1) + \frac{1}{2}S_2(g) \quad (3.134)$$

With the data of the Appendix 3, we can calculate the values of the standard free energy and the equilibrium constant of the reaction (3.134):

$$\frac{1}{2}O_2(g) \leftrightarrow O(\text{dis}; a_{H,O} = 1) \quad (3.135)$$

$$\Delta_r G^0_{\text{reaction (3.135)}}(T) = -117.0 - 3.14 \cdot 10^{-3} \cdot T(K)\left[kJ \cdot (\text{mol O})^{-1}\right] \quad (3.136)$$

$$\frac{1}{2}S_2(g) \leftrightarrow S(\text{dis}; a_{H,S} = 1) \quad (3.137)$$

$$\Delta_r G^0_{\text{reaction (3.137)}}(T) = -130.0 + 2.1 \cdot 10^{-2} \cdot T(K)\left[kJ \cdot (\text{mol S})^{-1}\right] \quad (3.138)$$

$$\begin{aligned}\Delta_r G^0_{\text{reaction (3.134)}}(T) &= \Delta_r G^0_{\text{reaction (3.135)}}(T) - \Delta_r G^0_{\text{reaction (3.137)}}(T)\\ &= 13.0 - 0.02414 \cdot T(K)\left[kJ \cdot (\text{mol S})^{-1}\right]\end{aligned} \quad (3.139)$$

At 1600 °C (1873 K):

$$\Delta_r G^0_{\text{reaction (3.134)}}(1873) = 13.0 - 0.02414 \cdot 1873 = -32.2142\,kJ \cdot (\text{mol S})^{-1} \quad (3.140)$$

$$\begin{aligned}\Delta_r G^0_{\text{reaction (3.134)}}(1873\,K) &= -R \cdot T \cdot \ln K_{\text{reaction (3.134)}} \rightarrow K_{\text{reaction (3.134)}}\\ &= \exp\left(-\frac{\Delta_r G^0_{\text{reaction (3.134)}}(1873K)}{R \cdot T}\right)\\ &= \exp\left(-\frac{-32214.2\,kJ \cdot (\text{mol S})^{-1}}{8.314\,J \cdot mol^{-1} \cdot K^{-1} \cdot 1873\,K}\right) = 7.9146\end{aligned} \quad (3.141)$$

Consequently, and identifying the value of the equilibrium constant of the reaction (3.134) with the value given to K_{O_2/S_2}, see Sect. 3.3, that is to say:

$$K_{\text{reaction (3.134)}} = K_{O_2/S_2} \quad (3.142)$$

$$C^*_{S^{2-},slag/m} = K_{O_2/S_2} \cdot C_{S^{2-},slag/g} = 7.9146 \cdot 1.1062 \cdot 10^{-3} = 8.7551 \cdot 10^{-3} \tag{3.143}$$

Finally, the value of the partition coefficient, we assume that $a_{H,O} = (\%O)_{metal}$ and $a_{H,S} = (\%S)_{metal}$, is calculated as follows:

$$C^*_{S^{2-},slag/m} = C_{S^{2-},slag} \cdot \frac{(\%O)_{metal}}{(\%S)_{metal}} = \left[\frac{(\%S)_{slag}}{(\%S)_{metal}}\right] \cdot (\%O)_{metal} = \eta \cdot (\%O)_{metal} \tag{3.144}$$

$$\eta = \frac{(\%S)_{slag}}{(\%S)_{metal}} = \frac{C^*_{S^{2-},slag/m}}{(\%O)_{metal}} = \frac{8.7551 \cdot 10^{-3}}{(\%O)_{metal}} = \frac{8.7551 \cdot 10^{-3}}{4.0 \cdot 10^{-3}} = 2.1888 \tag{3.145}$$

Commentaries: In this case, as opposed to the Exercise 3.4, it is only necessary to know the chemical characteristics (optical basicity) of the slag and the degree of oxidation of the melt (POS, Potential of Oxygen of the System, expressed by the concentration of oxygen atoms in the melt) to know the sulphur partition coefficient between the slag and the metal.

On the other hand, with an optical basicity of the slag of 0.7021, and a basicity measured through the relation between the calcium oxide and the silicon oxide, $I_B(I) = 1.0$, we have a capacity of picking up sulphur for this slag (40% CaO, 20% MgO, and 40% SiO_2) noticeably lower than that obtained in the Exercise 3.4, where the optical basicities were 0.8355 and 0.9634 with partition coefficients clearly greater, of 50.25 and 91.50, respectively.

Solution ends.

Exercise 3.6 Assuming that the model of Temkin can be applied to the slags of the CaO–Al_2O_3–CaF_2 ternary system, Table 3.5, determine the activity of the calcium slag in a slag, where the molar fraction of the calcium oxide is 0.520 and the relation of molar fractions CaO/Al_2O_3 is 3.08 at 1400 °C. Use the molecular weights of the Exercise 3.4.

Solution starts:

The composition of the slag expressed in molar fraction, derived from the data of the enunciation, is the following one:

$$x_{CaO} = 0.5200 \tag{3.146}$$

$$\frac{x_{CaO}}{x_{Al_2O_3}} = 3.08 \rightarrow x_{Al_2O_3} = \frac{0.520}{3.08} = 0.1688 \tag{3.147}$$

$$x_{CaF_2} = 1 - x_{CaO} - x_{Al_2O_3} = 1 - 0.5200 - 0.1688 = 0.3112 \tag{3.148}$$

On the one hand, the composition of the slag expressed in mass percentage:

$$\mathrm{CaO} = 0.5200\,\mathrm{mol\,CaO} \cdot \frac{56.08\,\mathrm{g\,CaO}}{1\,\mathrm{mol\,CaO}} = 29.1616\,\mathrm{g\,CaO} \rightarrow \%\mathrm{CaO}$$
$$= \frac{29.1616}{70.0485} \cdot 100 = 41.63\% \quad (3.149)$$

$$\mathrm{Al_2O_3} = 0.1688\,\mathrm{mol\,Al_2O_3} \cdot \frac{101.96\,\mathrm{g\,Al_2O_3}}{1\,\mathrm{mol\,Al_2O_3}} = 17.2108\,\mathrm{g\,Al_2O_3} \rightarrow \%\mathrm{Al_2O_3}$$
$$= \frac{17.2108}{70.0485} \cdot 100 = 24.57\% \quad (3.150)$$

$$\mathrm{CaF_2} = 0.3112\,\mathrm{mol\,CaF_2} \cdot \frac{78.08\,\mathrm{g\,CaF_2}}{1\,\mathrm{mol\,CaF_2}} = 23.6761\,\mathrm{g\,CaF_2} \rightarrow \%\mathrm{CaF_2}$$
$$= \frac{23.6761}{70.0485} \cdot 100 = 33.80\% \quad (3.151)$$

On the other hand, we have in 100 g of slag:

$$\mathrm{CaO} : \frac{41.63}{56.08} = 0.7423 \quad (3.152)$$

$$\mathrm{Al_2O_3} : \frac{24.57}{101.96} = 0.2410 \quad (3.153)$$

$$\mathrm{CaF_2} : \frac{33.80}{78.08} = 0.4329 \quad (3.154)$$

As we have seen in Sects. 1.3.1 and 1.6, in the ionic model of Temkin, the activity of one oxide in the melt at any temperature is equal to the product of the equivalent fractions of the oxygen anion in the ideal mixture of anions, and the product of the equivalent fractions of the cations in the ideal mixture of cations:

$$a_{\mathrm{CaO}} = N_{e,\mathrm{Ca^{2+}}} \cdot N_{e,\mathrm{O^{2-}}} \quad (3.155)$$

It is necessary to make the balance of ions in the ideal mixture of anions and cations that, according to Temkin, is the model of thermodynamic behavior of the slag. Thus, in a first step of the calculus process, it is necessary to define the working hypothesis, to identify what are the cations and anions that constitute the melt. The hypothesis that we are going to consider in the exercise is that the melt is formed by the following ionic species:

$$\text{Anions} : \mathrm{Al_2O_6^{6-}}, \mathrm{O^{2-}} \text{ and } \mathrm{F^-} \quad (3.156)$$

$$\text{Cations} : \mathrm{Ca^{2+}} \quad (3.157)$$

On the other hand, the acid–base reactions where these ionic species are involved are the following ones:

$$Al_2O_3(s) + 3O^{2-} \rightarrow Al_2O_6^{6-}(\text{acid oxide}) \tag{3.158}$$

$$CaO(s) \rightarrow Ca^{2+} + O^{2-}(\text{basic oxide}) \tag{3.159}$$

$$CaF_2(l) \rightarrow Ca^{2+} + 2F^{-}(\text{flux}) \tag{3.160}$$

Considering both the number of moles in 100 g of slag (0.7423 mol of CaO, 0.2410 mol of Al_2O_3 and 0.4329 mol of CaF_2) and that all reactions previously defined are completely displaced to the formation of the products (total ionization), we obtain the following results:

$$\text{Gram} - \text{equivalents of } Al_2O_6^{6-} \text{anions} : 0.2410 \cdot 6 = 1.446\,\text{mol} \tag{3.161}$$

$$\text{Gram} - \text{equivalents of } O^{2-} \text{anions provided by the CaO} : 0.7423 \cdot 2 = 1.4846\,\text{mol} \tag{3.162}$$

$$\text{Gram} - \text{equivalents of } O^{2-} \text{anions that react with the } Al_2O_3 : 0.2410 \cdot 3 \cdot 2 = 1.446\,\text{mol} \tag{3.163}$$

$$\text{Gram} - \text{equivalents of } F^{-} \text{anions} : 0.4329 \cdot 2 = 0.8658 \tag{3.164}$$

$$\begin{gathered}\text{Total gram} - \text{equivalents in the mixture ofanions}: \\ 1.446 + 1.4846 - 1.446 + 0.8658 = 2.3504\end{gathered} \tag{3.165}$$

For the mixture of cations, as calcium is the only ionic species that is possible, the total quantity of calcium equivalents will be 2.3504. The equivalent fraction of calcium cations is equal to one. Consequently:

$$a_{CaO} = 1 \cdot N_{e,O^{2-}} = \frac{1.4846 - 1.446}{2.3504} = 0.014 \tag{3.166}$$

If we compare the result of Temkin with the data of Table 3.5, $a_{CaO} = 0.09$, the quality of the Temkin's prediction might be acceptable regarding the order of magnitude, but exhibits low accuracy.

Commentaries: If we have considered as working hypothesis that one where the distribution of anions and cations in the slag was the following:

$$\text{Anions} : Al_2O_4^{2-}, O^{2-} \text{ and } F^{-} \tag{3.167}$$

$$\text{Cations} : Ca^{2+} \tag{3.168}$$

On the other hand, the acid–base reactions where these ionic species are involved are the following ones:

$$Al_2O_3(s) + O^{2-} \rightarrow Al_2O_4^{2-}(\text{acid oxide}) \tag{3.169}$$

$$CaO(s) \rightarrow Ca^{2+} + O^{2-}(\text{basic oxide}) \tag{3.170}$$

$$CaF_2(l) \rightarrow Ca^{2+} + 2F^{-}(\text{flux}) \tag{3.171}$$

Considering both the number of moles in 100 g of slag (0.7423 mol of CaO, 0.2410 mol of Al_2O_3 and 0.4329 mol of CaF_2) and that all reactions previously defined are completely displaced to the formation of the products (total ionization), we obtain the following results:

$$\text{Gram} - \text{equivalents of } Al_2O_4^{2-}\text{anions} : 0.2410 \cdot 2 = 0.482\,\text{mol} \tag{3.172}$$

$$\text{Gram} - \text{equivalents of } O^{2-}\text{anions provided by the CaO} : 0.7423 \cdot 2 = 1.4846\,\text{mol} \tag{3.173}$$

$$\text{Gram} - \text{equivalents of } O^{2-}\text{anions that react with the } Al_2O_3 : 0.2410 \cdot 2 = 0.482\,\text{mol} \tag{3.174}$$

$$\text{Gram} - \text{equivalents of } F^{-}\text{anions} : 0.4329 \cdot 2 = 0.8658 \tag{3.175}$$

$$\begin{gathered}\text{Total gram} - \text{equivalents in the mixture of anions} : \\ 0.482 + 1.4846 - 0.482 + 0.8658 = 2.3504\end{gathered} \tag{3.176}$$

For the mixture of cations, as calcium is the only ionic species that is possible, the total quantity of calcium equivalents will be 2.3504. The equivalent fraction of calcium cations is equal to one. Consequently:

$$a_{CaO} = 1 \cdot N_{e,O^{2-}} = \frac{1.4846 - 0.482}{2.3504} = 0.4266 \tag{3.177}$$

The activity of Temkin of the CaO would be 0.4266, very far from the experimental results collected in Table 3.5 (a_{CaO}=0.09).

We should remember that Temkin's estimations for the chemical activity of the ionic pairs do not depend on the temperature at which the prediction was made because at any temperature that would be considered, the ionization reactions do not allow partial ionizations.

End of the exercise.

Exercise 3.7 The kinetics of the desulphurization of 220 tons of iron melt at 1600 °C, with 0.030% S, in the ladle is described by the following equation:

$$(\%S) = \left(\%S^0 - \%S^=\right) \cdot \exp(-\beta \cdot t) + \%S^= \tag{3.178}$$

where the time t is expressed in minutes. The injection of a slag able to pick up sulphur is performed by means of a ceramic lance. This slag belongs to the $CaO–Al_2O_3–CaF_2$ ternary system, which has a sulphur partition coefficient of 145. Calculate, assuming that is possible to apply the kinetic model of transient regime, the quantity of slag that is necessary to inject to reach 0.010% S in 15 min. Finally, if the mean size of the particles of injected slag is 5.0 mm and the thickness of the tight boundary layer is of 2.5 mm, calculate the diffusion coefficient of the sulphur.

Data: Sulphur percentage, $\%S^=$, dissolved in the melt, in equilibrium with the slag, is $2.0 \cdot 10^{-4}$ (2.0 ppm).

Solution starts:

It is possible to obtain the value of the constant β (min^{-1}) from the kinetics equation of the enunciation because the rest of the variables are known.

The initial sulphur percentage is: $\%S^0 = 0.030$; the percentage that is reached in the equilibrium is: $\%S^= = 2.0 \cdot 10^{-4}$; when the time, $t = 15\,min$; the sulphur concentration in the melt is: $\%S = 0.010\%$. Consequently:

$$\beta = \ln\left(\frac{\%S - \%S^=}{\%S^0 - \%S^=}\right) \cdot \left(\frac{1}{t}\right) = \ln\left(\frac{0.010 - 2.0 \cdot 10^{-4}}{0.030 - 2.0 \cdot 10^{-4}}\right) \cdot \left(\frac{1}{15\,min}\right)$$
$$= 0.07414\,min^{-1} \tag{3.179}$$

If we identify the kinetics equation of the enunciation with the equations of Sect. 3.3 that develop the model of transient regime, we obtain the rate of injection of the slag:

$$\beta = \frac{\eta \cdot \dot{V}_{slag}}{M_{metal}} = \frac{145}{220\,t} \cdot \dot{V}_{slag} \tag{3.180}$$

$$\dot{V}_{slag} = \frac{220\,t \cdot 0.07414\,min^{-1}}{145} \cdot \frac{1000\,kg}{1\,t} = 112.49\,kg \cdot min^{-1} \tag{3.181}$$

Definitely, the quantity of slag used in the process is:

$$M_{metal} = \dot{V}_{slag} \cdot 15\,min = 112.49\,kg \cdot min^{-1} \cdot 15\,min = 1687\,kg \text{ of slag} \tag{3.182}$$

Finally, if we identify the constant β with the value of k of the Eq. (3.75) of Sect. 3.3, assuming that the area to volume ratio (A/V) for the spherical particles of the injected synthetic slag is:

$$\frac{A}{V} = \frac{4 \cdot \pi \cdot r^2}{\frac{4}{3} \cdot \pi \cdot r^3} = \frac{6}{D_p} \tag{3.183}$$

$$k = \left(\frac{D_S}{\delta_c}\right) \cdot \left(\frac{A}{V}\right) \rightarrow k \cdot \delta_c = D_S \cdot \frac{6}{\overline{D_p}} \rightarrow D_S = \frac{k \cdot \delta_c \cdot \overline{D_p}}{6} \tag{3.184}$$

$$D_S = \frac{0.07414\,\text{min}^{-1} \cdot 2.5\,\text{mm} \cdot 5\,\text{mm}}{6} \cdot \frac{1\,\text{cm}^2}{10^2\,\text{mm}^2} \cdot \frac{1\,\text{min}}{60\,\text{s}} = 2.754 \cdot 10^{-5}\,\text{cm}^2 \cdot \text{s}^{-1} \tag{3.185}$$

Commentaries: If we have used this same quantity of slag but we have applied the metal-slag contact according to the model of permanent regime, Eq. (3.76), Sect. 3.3, the value of sulphur concentration after 15 min would have been:

$$\frac{\%S^0}{\%S} = 1 + \beta \cdot t \rightarrow \%S = \frac{\%S^0}{1 + \beta \cdot t} \tag{3.186}$$

$$\%S = \frac{\%S^0}{1 + \beta \cdot t} = \frac{0.03\%}{1 + 0.07414\,\text{min}^{-1} \cdot 15\,\text{min}} = 0.014\% \tag{3.187}$$

Thus, if the desulphurizing mechanism follows the model of permanent regime, the quantity of sulphur in the melt would be 0.004% greater.

End of the exercise.

3.4 Vacuum Treatments. Degasification

One of the most important problems in the production of high-quality steel is controlling the gasses dissolved in elemental or atomic state:

- Hydrogen is a gas particularly harmful, responsible for cracks, porosity, brittleness, and loss of ductility in the finished product.
- Nitrogen, if it is in solid solution in the ferrite, causes losses of toughness in ferritic-pearlitic steels. However, if nitrogen is combined with aluminum, vanadium, titanium, or niobium, it can cooperate in the structural hardening of the steel due to the formation of submicroscopic precipitates of metallic nitrides and carbonitrides.
- Oxygen, as main responsible for the thermal refining of the iron, plays a determining role in the composition and properties of the steel. On the other hand, as the elimination of gases dissolved in the liquid iron is not strictly controlled by physical mechanisms (difference of solubility with the pressure and temperature), there are chemical equilibriums, as the oxidation of the carbon in the liquid iron, where the dissolved oxygen plays a fundamental role.

The pressure required to form stable bubbles of gas dissolved in the liquid iron is equal to:

$$P_T = P_{slag/gas} + \rho \cdot g \cdot h + 2 \cdot \frac{\gamma_{metal/gas}}{r} \quad (3.188)$$

where $P_{slag/gas}$ is the pressure on the surface of the melt (slag-gas interface), $\rho \cdot g \cdot h$ is the metallostatic pressure of the liquid (ρ is the density of the metal, g is the gravitational constant and h is the height of the molten metal above the bubble), and finally, the third of the addends represents the contribution of the interface metal-gas forces in the stabilization of the bubbles ($\gamma_{metal/gas}$ is the interface energy and r is the radius of the bubble).

According to Eq. (3.188), very high pressures are required to form, by homogeneous clustering, a stable bubble of small diameter. It is possible to happen, although it is not frequent, that the controlling mechanism of the degasification of the liquid iron will progress by the homogeneous clustering rate of the stable bubble of gas.

From the thermodynamics point of view, the dissolution of the diatomic gases (G_2, where G can be H, N, F or Cl) in molten metals, can be expressed by means of the following equilibrium:

$$G_2(g) \leftrightarrow 2G(dis; melt) \quad (3.189)$$

and:

$$K_{reaction\,(3.189)} = \frac{x_G^2}{P_{G_2}} \quad (3.190)$$

where P_{G_2} would be the partial pressure of the gas G_2 and x_G is the concentration of the gas dissolved in the melt expressed in molar fraction. Nevertheless, the gas concentration in the melt is expressed either in weight percentage (parts-per-million, ppm) or in the volume of gas measured in normal conditions, per unit of weight or volume of melt. In any case, reaction (3.189) represents the Sievert's law: the capacity of dissolution of gases in molten metals is directly proportional to the square root of the partial pressure of the gas molecule. Consequently, from the Sievert's law it is possible to derive that the gas dissolution mechanism in the metallic melt is through the atoms and not through the molecular species.

Industrial operations of vacuum degasification started in 1950, and they can proceed via the following basic technologies:

- Degasification in the ladle:

The casting ladle is located in the vacuum chamber where the pressure can reach 10^{-4} atm (Fig. 3.5).

The pressure drop causes the violent elimination of the carbon monoxide, accompanied by significant quantities of hydrogen and, in small proportion, of nitrogen. A better efficiency in the degasification of the metal can be achieved by electromagnetic stirring or bubble up of argon in the melt.

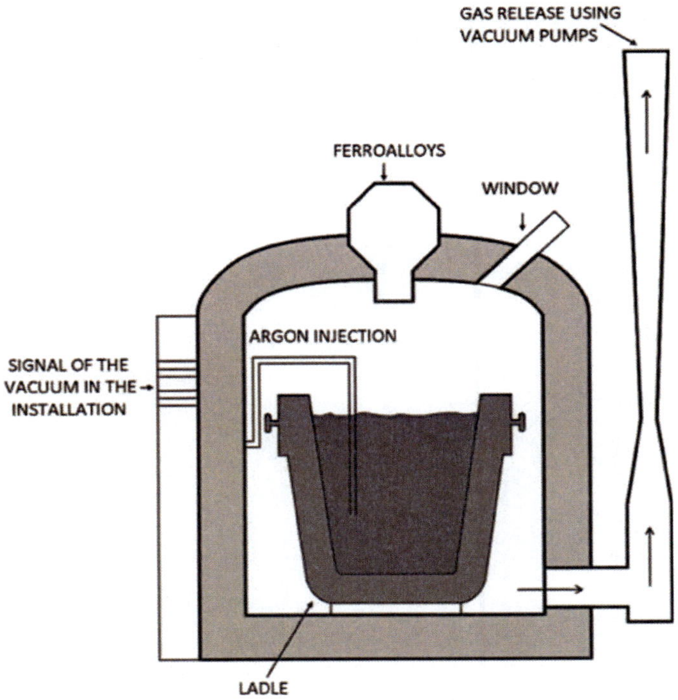

Fig. 3.5 Vacuum installation in a ladle. General characteristics

– Degasification during the tapping:

The melt of the ladle is casted in an ingot that is kept inside of a vacuum chamber (vacuum ingot casting), as it is indicated in Fig. 3.6. The metal stream is violently disintegrated into drops once bubbles of gas are formed due to the vacuum conditions. The efficiency in the elimination of the hydrogen, oxygen, and nitrogen dissolved in the melt is very high as a consequence of the great surface of reaction even with very short degasification times.

– DH (Dortmund-Hoerder) process of intermittent suction:

In this process, the ascending movement of the ladle forces the metal to enter in a vacuum chamber where the degasification takes place. The cycle is completed when the ladle descends to the initial height and forces the degasified liquid to homogenize with the rest of the melt charged in the ladle (Fig. 3.7). The concentration of the gas in the metal has the following exponential law of behavior:

$$C_{gas} = C^0_{gas} \cdot (1 - x)^n \quad (3.191)$$

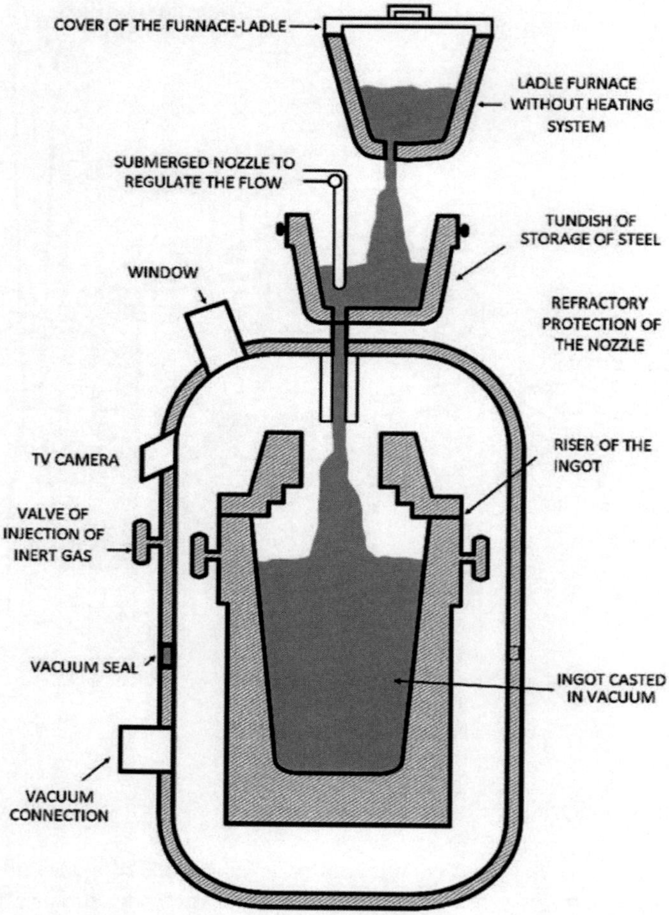

Fig. 3.6 Degasification during the tapping process in the ingot

where C^0_{gas} is the concentration of the gas in the metal at the beginning of the treatment, C_{gas} is the concentration of the gas after the cycle n of suction and x is the fraction of metal that, in each cycle, enters in the vacuum chamber.

– RH (Ruhrtahl-Heraeus) process of continuous circulation:

In this case, a vacuum chamber is connected to the ladle by means of two tuyeres. In one of them, the ascending one, argon is injected to locally diminish the density of the metal and facilitate the circulation between the vacuum chamber and the ladle. The circulation of the metal activates the kinetics of the reactions of gases elimination increasing, with respect to the DH, the efficiency of the process (Fig. 3.8).

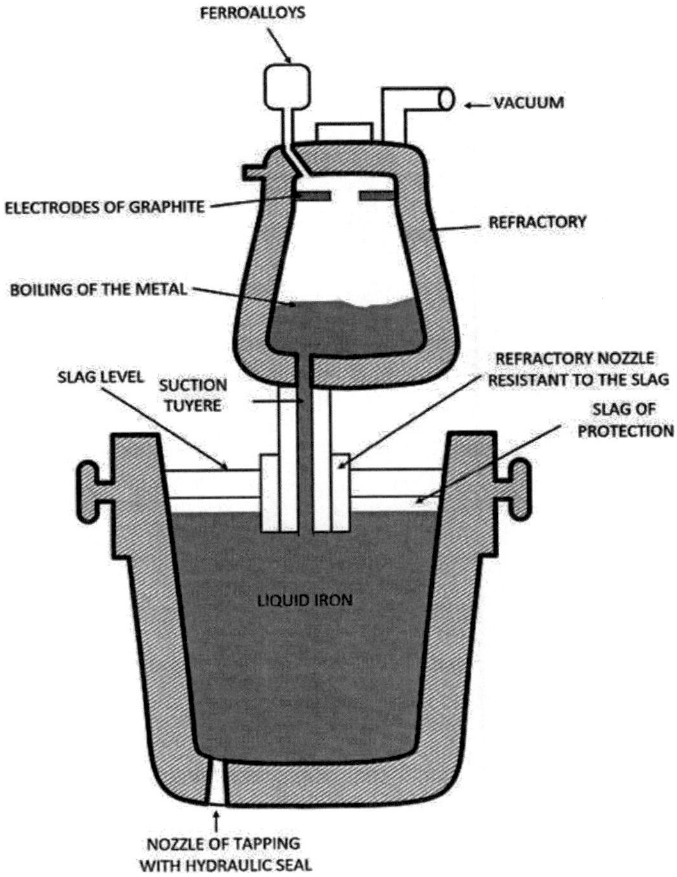

Fig. 3.7 DH process (Dortmund-Hoerder)

– RH-OB (Nippon Steel Corporation) continuous circulation process.

All the processes of degasification with vacuum, or by means of injection of noble gases, have a disadvantage that the metal losses heat during the treatment, which requires:

- An important overheating of the melt in the BOF process or in the electric furnace.
- A limited duration of the process: from 20 to 30 min.

The RH-OB technology can accept different heating system:

- Injection of oxygen.
- Oxidation of the aluminum.
- Indirect heating by means of electric arc.

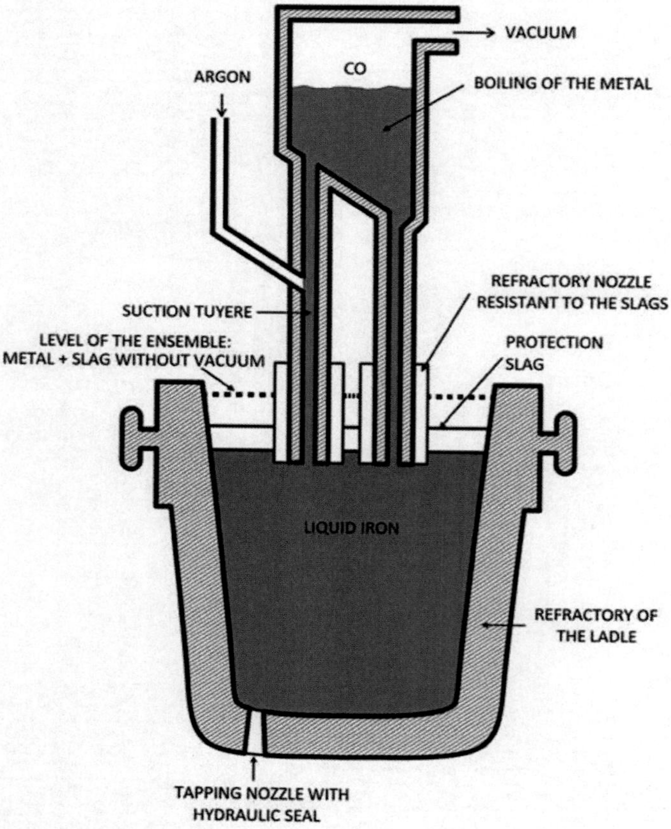

Fig. 3.8 RH (Ruhrtahl-Heraeus) process

- Direct heating by means of gas burners.

DH, RH, and RH-OB technologies are able of processing big quantities of steel (250 tons) in short periods of time with the additional advantage of being possible to adjust the composition and temperature (RH-OB) at the end of the tapping thanks to the good mixing conditions.

The kinetics of the carbon oxidation process (decarburization), when the carbon percentage is >0.10%, is controlled by the diffusional transfer of the reagents in the metallic phase. Thus, the variation of the carbon concentration with the time, C_C, in percentage, follows the next equation:

$$\ln\left(\frac{C_C}{C_C^0}\right) = -k_{A/V} \cdot t \tag{3.192}$$

where C_C^0 is the carbon concentration at the beginning of the process and $k_{A/V}$ is the coefficient of mass transfer of the system, in min^{-1}, defined by:

$$k_{A/V} = \left(\frac{Q}{V}\right) \cdot \left(\frac{k_A}{k_A + Q}\right) \tag{3.193}$$

$$k_{A/V} = k \cdot \left(\frac{A}{V}\right) \tag{3.194}$$

where Q is the volumetric flow of circulation of metal in the RH or RH-OB process in $m^3 \cdot min^{-1}$, V is the volume of steel in the ladle in m^3, k_A is the coefficient of mass transfer in $m^3 \cdot min^{-1}$ and k is the same coefficient of mass transfer that the k_d (coefficient of diffusional mass transfer), expressed in $m \cdot min^{-1}$, and finally it is verified that $k_A = k \cdot A$, where A is the surface involved in the mass transfer process. On the other hand, the kinetics of the hydrogen elimination is controlled by the diffusion of the monoatomic gas in the melt:

$$\ln\left(\frac{C_H}{C_H^0}\right) = -k_{A/V} \cdot t \tag{3.195}$$

where C_H is the concentration of hydrogen in ppm at the end of the process, which can be even smaller than 2.0 ppm, $k_{A/V}$ is the coefficient of mass transfer, in min^{-1}, of the hydrogen refining process.

With the purpose of simplifying the operations of the secondary metallurgy, several initiatives were developed to incorporate in the vacuum equipment devices to adjust the final composition of the liquid in those elements that are not susceptible of vacuum refining. The addition of active elements and the incorporation of an oxygen lance or sources of heating are examples of simplifying trends in the new iron and steelmaking technologies.

Even though the adjustment of the sulphur and phosphorus is performed in the CAS, in the BOF or in the electric furnace, vacuum technologies provide a last opportunity for the refining of non-gaseous impurities. Slags that incorporate sodium compounds are excellent for the desulphurizing and dephosphorizing of the liquid iron. We show the activities of the sodium oxide in the SiO_2–FeO–Na_2O and P_2O_5–SiO_2–Na_2O ternary systems in Fig. 3.9 (edited from Kawai and Shiraishi 1988) and Fig. 3.10 (edited from Kawai and Shiraishi 1988).

As a complement to the correlations of the Appendix 1, we collect in Table 3.6 the free energies of formation of some of the most stable metallic phosphates that can produce alternative phases in the dephosphorizing of the liquid iron.

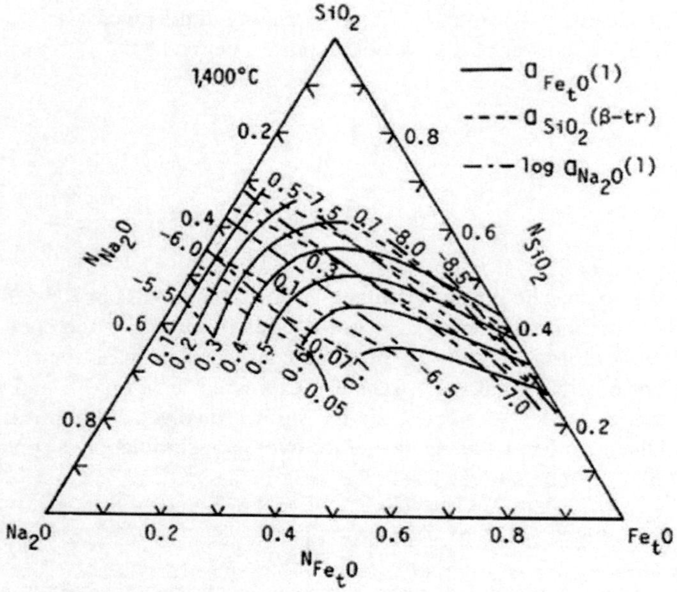

Fig. 3.9 Activities of the FeO, Na_2O and SiO_2 for the ternary SiO_2–FeO–Na_2O at 1400 °C

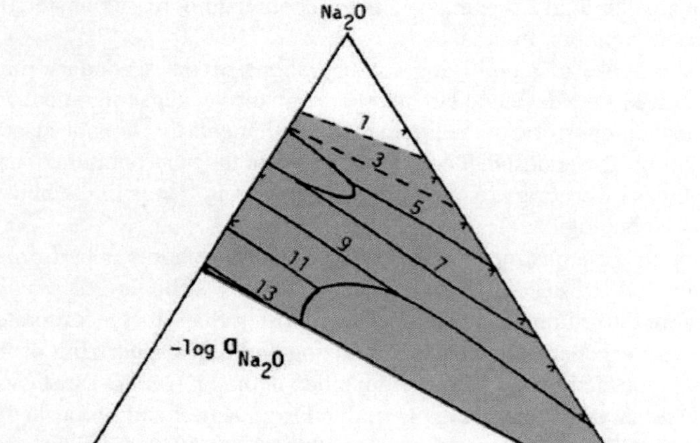

Fig. 3.10 Activity of the Na_2O in the P_2O_5–SiO_2–Na_2O ternary system at 1300 °C

Table 3.6 Standard free energies of formation of metallic phosphates in the range of above-mentioned temperatures (Knacke et al. 1991)

Reaction	$\Delta_f G^0(T)\ [kJ \cdot (mol)^{-1}]$	Range of T(K)
$6Na(l) + 4O_2(g) + P_2(g) = P_2O_8Na_6(s)$	$-3989.3 + 1.0318 \cdot T$	371–1000
$3K(l) + 2O_2(g) + 1/2P_2(g) = PO_4K_3(s)$	$-2062.2 + 0.5096 \cdot T$	336–1000
$Al(s) + 2O_2(g) + 1/2P_2(g) = PO_4Al(s)$	$-1801.8 + 0.4488 \cdot T$	300–933
$Al(l) + 2O_2(g) + 1/2P_2(g) = PO_4Al(s)$	$-1778.5 + 0.4246 \cdot T$	933–2000

3.5 Kinetics of the Refining Processes with Interface Resistances

In the refining of metals, it can happen that the controlling stage of the process is neither the diffusional mass transfer nor the own chemical reaction, but the absorption of the reagents in the metal-slag or metal-gas interface. In the iron refining, there are two examples that, due to their importance, are analyzed in this section: the adsorption or elimination of nitrogen from the liquid iron and the iron decarburizing with carbon dioxide.

The absorption of a solute (gaseous molecule or atomic species dissolved in a liquid) over an interface to form an ideal monolayer (there is no interaction between the atoms of the absorbed specie) follows the Langmuir isotherm, which for the specific case of the absorption of gas in a liquid is equal to:

$$\theta = \frac{K_a \cdot P}{1 + K_a \cdot P} \tag{3.196}$$

where K_a is the equilibrium constant of absorption, P is the partial pressure of the gas and θ is the fraction of the surface of the gas-liquid interface occupied by the atoms of the absorbed species. On some occasion, the absorption of the solute is preceded by the dissociation of the gas molecule, thus it is verified that:

$$1 - \theta = \frac{1}{1 + \left(K_a^* \cdot P\right)^{0.5}} \tag{3.197}$$

where is K_a^* is:

$$K_a^* = K_a \cdot K_d^{0.5} \tag{3.198}$$

K_d is the equilibrium constant for the dissociation process, see Exercise 3.14 for further explanations.

Especially, the desorption mechanism (absorption) of the nitrogen dissolved in the liquid iron has the following stages:

1. Diffusion of nitrogen atoms from inside of the metal to the metal-gas interface:

$$N^{O}\text{(atoms of nitrogen inside of the metal)} \leftrightarrow N^{i}\text{(interfacial position)} \tag{3.199}$$

2. Absorption on the surface:

$$N^{O} + v\text{(vacancy positions)} \leftrightarrow N^{a} \tag{3.201}$$

3. Chemical reaction of the atomic nitrogen in the interface to form the N_2 molecule (two possibilities):

$$N^{a} + N^{i} \leftrightarrow N_2^{a} \tag{3.202}$$

$$N^{a} + N^{a} \leftrightarrow N_2^{a} \tag{3.203}$$

4. Desorption of nitrogen molecules:

$$N_2^{a} \leftrightarrow N_2^{i} + v \tag{3.204}$$

5. Diffusion in the gaseous phase:

$$N_2^{i} \leftrightarrow N_2(g) \tag{3.205}$$

If the controlling stage of the process is the chemical reaction in the interface (stage 3), of second order and molecularity equal to two (two nitrogen atoms are involved in the reaction), the rate of the desorption (absorption) process is equal to:

$$\frac{dC_N}{dt} = k_q \cdot C_N^2 \cdot \left(\frac{A}{V}\right) \tag{3.206}$$

where C_N is the nitrogen concentration in percentage and k_q, expressed by $k_q = k_q^* \cdot (1 - \theta)$, is the kinetic constant of the process, which includes the fraction of the surface that is not occupied by active atoms $(1 - \theta)$ in cm·s^{-1}. On the other hand:

$$k_q = \frac{f_N \cdot (A + B)}{C_N} \tag{3.207}$$

where:

$$A = \frac{3.15}{1 + 360 \cdot a_{H,O} + 120 \cdot a_{H,S}} \tag{3.208}$$

$$B = \frac{25 \cdot \Phi}{1 + 360 \cdot a_{H,O} + 120 \cdot a_{H,S}} \qquad (3.209)$$

where f_N is the coefficient of activity of Henry for the nitrogen in the melt, $a_{H,O}$ and $a_{H,S}$ are the activities of Henry of the oxygen and sulphur in the liquid iron and, finally, when the process of desorption of the nitrogen takes place at environmental pressure, Φ in Eq. (3.207) is zero, while it takes values of one when the degasification takes place under vacuum conditions: DH, RH, or RH-OB processes.

In the decarburizing with carbon dioxide, the controlling stage of the process can be the chemisorption rate of the CO_2 molecule in the metal-gas interphase for the reaction:

$$CO_2(g) + C(\text{dis; iron}) \leftrightarrow 2CO(g) \qquad (3.210)$$

with kinetics equation:

$$\frac{dC_C}{dt} = -k_q^* \cdot (1 - \theta) \cdot P_{CO} \qquad (3.211)$$

where:

$$\log k_q^* = -\frac{5080}{T} + 2.65 \qquad (3.212)$$

where k_q^* is the kinetics constant of the process in g·cm^{-2}·min^{-1}·atm^{-1} and C_C is the interfacial concentration (superficial) of the carbon in g·cm^{-2}.

3.6 Solved Exercises: Vacuum Treatments. Degasification. Kinetics of the Refining Process with Interface Resistances

Exercise 3.8 Calculate the free energy associated to the hydrogen dissolution in the liquid iron at 1600 °C (1873 K) when the hydrogen pressure in the gas is equal to one atmosphere and the concentration that the hydrogen can reach in the melt is equal to 100 ppm.

Data: The standard free energy associated to the dissociation of the hydrogen molecule (data of the TAPP database (1994)) is:

$$\Delta_d G^0(T) = 453.0 - 0.1186 \cdot T(K) \left[kJ \cdot (mol\ H_2)^{-1} \right] \qquad (3.213)$$

The coefficient of activity of the hydrogen in the iron solution at 1600 °C is 0.435, and atomic weight of the hydrogen is 1.01 g·mol^{-1}.

Solution starts:

The dissolution of hydrogen in the iron melt can be represented by means of the following reaction:

$$H_2(g)\left(P_{H_2} = 1\,atm\right) \leftrightarrow 2H(dis;\, a_H = 1) \tag{3.214}$$

In reaction (3.214), the state of reference of activity one is the monoatomic hydrogen at the pressure of one atmosphere. On another note, the dissolution of the hydrogen can follow the next stages (dissolution mechanisms):

(a) Dissociation of the hydrogen molecule:

$$H_2(g)\left(P_{H_2} = 1\,atm\right) \leftrightarrow 2H(P_H = 1\,atm) \tag{3.215}$$

(b) Dissolution of the hydrogen atom in the iron:

$$2H(g)(P_H = 1\,atm) \leftrightarrow 2H(dis;\, a_H = 1) \tag{3.216}$$

To obtain the value of the free energy of the reaction (3.214):

$$\Delta_r G_{reaction\,(3.214)}(T) = \Delta_d G^0_{reaction\,(3.215)}(T) + \Delta_r G_{reaction\,(3.216)}(T) \tag{3.217}$$

where:

$$\Delta_d G^0_{reaction\,(3.215)}(T) = 453.0 - 0.1186 \cdot T(K)\left[kJ \cdot (mol\,H_2)^{-1}\right] \tag{3.218}$$

$$\Delta_r G_{reaction\,(3.216)}(T) = R \cdot T \cdot \ln\left[(\gamma_H \cdot x_H)^2\right]\left[kJ \cdot (mol\,H_2)^{-1}\right] \tag{3.219}$$

Definitely, the value of the free energy associated to the reaction (3.214), when:

$$\gamma_H = 0.435 \tag{3.220}$$

$$x_H = \frac{\frac{0.01\,g\,H}{1.01\,g\,H\cdot mol^{-1}}}{\frac{0.01\,g\,H}{1.01\,g\,H\cdot mol^{-1}} + \frac{99.99\,g\,Fe}{55.85\,g\,Fe\cdot mol^{-1}}} = 5.5 \cdot 10^{-3} \tag{3.221}$$

is:

$$\Delta_r G_{reaction\,(3.214)}(T) = 453.0 - 0.1186 \cdot T(K) + R \cdot T \cdot \ln\left[(\gamma_H \cdot x_H)^2\right]\left[kJ \cdot (mol\,H_2)^{-1}\right] \tag{3.222}$$

at 1600 °C (1873 K):

$$\Delta_r G_{\text{reaction (3.214)}}(1873\,\text{K}) = 453000 - 118.6 \cdot 1873 + 8.314 \cdot 1873 \cdot \ln\left[\left(0.435 \cdot 5.5 \cdot 10^{-3}\right)^2\right]\left[\text{J} \cdot (\text{mol H}_2)^{-1}\right] = 42893.72\left[\text{J} \cdot (\text{mol H}_2)^{-1}\right] \tag{3.223}$$

Thus, the process of hydrogen dissolution in the iron melt when the pressure of hydrogen, H_2, in the gas is equal to one atmosphere and the concentration in the liquid is 100 ppm H, is not thermodynamically possible (positive variation of the free energy). The reason that explains this situation is the quantity of hydrogen dissolved at 1600 °C in the liquid, which is excessively high. The maximum concentration, which corresponds to the thermodynamic equilibrium, is:

$$\Delta_r G_{\text{reaction (3.214)}}(1873\,\text{K}) = 0 = 453000 - 118.6 \cdot 1873 + 8.314 \cdot 1873 \cdot \ln\left[(0.435 \cdot x_H)^2\right]\left[\text{J} \cdot (\text{mol H}_2)^{-1}\right] \rightarrow x_H = \frac{\sqrt{\exp\left(\frac{-453000+118.6 \cdot 1873}{8.314 \cdot 1873}\right)}}{0.435} = 1.387 \cdot 10^{-3} \tag{3.224}$$

If the concentration is expressed in mass, we have:

$$H = 1.387 \cdot 10^{-3}\,\text{mol H} \cdot \frac{1.01\,\text{g H}}{1\,\text{mol H}} = 1.4009 \cdot 10^{-3}\,\text{g H} \rightarrow \%H = \frac{1.4009 \cdot 10^{-3}}{55.7739} \cdot 100 = 2.5117 \cdot 10^{-3}\%(25.117\,\text{ppm}) \tag{3.225}$$

$$\text{Fe} = 0.998613\,\text{mol Fe} \cdot \frac{55.85\,\text{g Fe}}{1\,\text{mol Fe}} = 55.7725\,\text{g Fe} \rightarrow \%\text{Fe} = \frac{55.7725}{55.7739} \cdot 100 = 99.9974883\% \tag{3.226}$$

Commentaries: The dissolution of gases in liquid metals takes place through the atoms and not the molecules. As the process of dissociation of the gas molecules, habitually stable, is associated to a standard free energy negative, the quantity of gas dissolved in the melt is usually small.

Solution ends.

Exercise 3.9 The degasification of an iron melt takes place by means of a process of intermittent suction at 1600 °C, see Fig. 3.7. If it is necessary to reduce the 8 ppm of hydrogen dissolved in the melt, calculate:

(a) The minimum quantity of hydrogen that can be reached if the vacuum in the chamber is 10^{-3} atm.

(b) The number of cycles that is necessary to apply to reach a concentration of 3 ppm of hydrogen in the melt. Assume that only 10% of the melt is suctioned from the ladle to the vacuum chamber.

Solution starts:

The variation of the standard free energy associated to the change of state of reference of the hydrogen, Appendix 3:

$$\frac{1}{2}H_2(g)\left(P_{H_2};\ atm\right) \leftrightarrow H\left(dis; a_{H,H} = 1\right) \tag{3.227}$$

can be applied to calculate the solubility of the hydrogen in the iron melt under different conditions of pressure and temperature. In this case, it is necessary to calculate the minimum hydrogen concentration in the metal compatible with a pressure in the gas of 10^{-3} atmospheres at 1600 °C. First, we calculate the equilibrium constant associated to the reaction (3.227):

$$\Delta_r G^0_{reaction\ (3.227)}(T) = 36500 + 30.5 \cdot T(K)\left[J \cdot (mol\ H)^{-1}\right] \tag{3.228}$$

$$\Delta_r G^0_{reaction\ (3.227)}(T) = -R \cdot T \cdot \ln K_{reaction\ (3.227)} \rightarrow K_{reaction\ (3.227)} = \exp\left(-\frac{\Delta_r G^0_{reaction\ (3.227)}(T)}{R \cdot T}\right) \tag{3.229}$$

at 1600 °C (1873 K):

$$K_{reaction\ (3.227)} = \exp\left[-\frac{36500 + 30.5 \cdot 1873\ K\left(J \cdot (mol\ H)^{-1}\right)}{8.314\ J \cdot mol^{-1} \cdot K^{-1} \cdot 1873\ K}\right] = 2.448 \cdot 10^{-3} \tag{3.230}$$

The equilibrium constant is:

$$K_{reaction\ (3.227)} = \frac{\%H}{\sqrt{P_{H_2}}} \rightarrow \%H = K_{reaction\ (3.227)} \cdot \sqrt{P_{H_2}} \rightarrow \%H = 2.448 \cdot 10^{-3} \cdot \sqrt{10^{-3}\ atm} = 7.74 \cdot 10^{-5}\%(0.77\ ppm) \tag{3.231}$$

To calculate the number, n, of suctions-discharges of the melt that is necessary to perform in the DH process, Fig. 3.7, it is necessary to use the Eq. (3.191):

$$C_{gas} = C^0_{gas} \cdot (1 - x)^n \rightarrow \frac{C_{gas}}{C^0_{gas}} = (1 - x)^n \rightarrow \ln\left(\frac{C_{gas}}{C^0_{gas}}\right)$$

$$= \ln(1-x)^n \rightarrow \ln\left(\frac{C_{gas}}{C^0_{gas}}\right) = n \cdot \ln(1-x) \rightarrow n = \frac{\ln\left(\frac{C_{gas}}{C^0_{gas}}\right)}{\ln(1-x)} \quad (3.232)$$

$$n = \frac{\ln\left(\frac{C_{gas}}{C^0_{gas}}\right)}{\ln(1-x)} = \frac{\ln\left(\frac{3}{8}\right)}{\ln(1-0.1)} = 9.3 \quad (3.233)$$

Consequently, to ensure that the melt could reach the specification of 3 ppm of H, it would be necessary to perform 10 cycles to obtain that value at the end of the process:

$$C_{gas} = C^0_{gas} \cdot (1-x)^n = 8\,\text{ppm} \cdot (1-0.1)^{10} = 2.79\,\text{ppm H} \quad (3.234)$$

Finally, we want to mention that the constant of the law of Sievert, see Sect. 3.4, for this temperature would take the value 2.448×10^{-3} when the concentration in the melt is expressed in percentage and the pressure of H_2 is indicated in atmospheres as it is indicated in Eq. (3.230).

Commentaries: If the hydrogen concentration dissolved in the melt is high, the consequence is that microporosity in the finished product would be frequent. Nevertheless, due to the great capacity of hydrogen diffusion, it is not easy to analytically detect its presence in the finished products. Elements as manganese, chromium, nickel, titanium, vanadium, niobium, and oxygen, see data of the Appendix 3, make the coefficient of activity of Henry smaller than one and, for that reason, the solubility of hydrogen in iron melts is high.

Finally, we indicate that the thermodynamic characteristics of the binary dissolutions of molten iron-hydrogen can be defined by means of the coefficient of activity of the hydrogen at infinite dilution, γ^*_H, in the iron melts:

$$\Delta_d G^0_{reaction\,(3.227)}(T) = R \cdot T \cdot \ln\left(\frac{\gamma^*_H \cdot 55.85}{100 \cdot 1.01}\right) \rightarrow \gamma^*_H$$
$$= \exp\left(\frac{\Delta_d G^0_{reaction\,(3.227)}(T)}{R \cdot T}\right) \cdot 100 \cdot 1.01 \cdot \frac{1}{55.85} \rightarrow \gamma^*_H$$
$$= \exp\left[-\frac{36500 + 30.5 \cdot 1873\,K\left(J \cdot (mol\,H)^{-1}\right)}{8.314\,J \cdot mol^{-1} \cdot K^{-1} \cdot 1873\,K}\right] \cdot$$
$$100 \cdot 1.01 \cdot \frac{1}{55.85} \rightarrow \gamma^*_H = 738.5 \quad (3.235)$$

value that exhibits a strong positive deviation of the ideal behavior of Raoult of the iron-hydrogen solutions at 1600 °C.

End of the exercise.

Exercise 3.10 Calculate the maximum concentration of nitrogen in the liquid iron at 1600 °C in equilibrium with an atmosphere where the pressure of nitrogen is 0.79 atm. Moreover, determine the Henry's and Raoult's activities for the dissolved nitrogen. Data: The atomic weight of the nitrogen is 14.01 g·mol^{-1}.

Solution starts:

The reaction associated to the change of state of reference of the nitrogen, Appendix 3, is:

$$\frac{1}{2}N_2(g)\left(P_{N_2};\ atm\right) \leftrightarrow N\left(dis;\ a_{H,N} = 1\right) \tag{3.236}$$

The standard free energy and the equilibrium constant at 1873 K are the following ones:

$$\Delta_r G^0_{reaction\ (3.236)}(T) = 6757 + 22 \cdot T(K)\ \left[J \cdot (mol\ H)^{-1}\right] \tag{3.237}$$

$$\Delta_r G^0_{reaction\ (3.236)}(T) = -R \cdot T \cdot \ln K_{reaction\ (3.236)} \rightarrow K_{reaction\ (3.236)} = \exp\left(-\frac{\Delta_r G^0_{reaction\ (3.236)}(T)}{R \cdot T}\right) \tag{3.238}$$

at 1600 °C (1873 K):

$$K_{reaction\ (3.236)} = \exp\left[-\frac{6757 + 22 \cdot 1873\ K\left(J \cdot (mol\ H)^{-1}\right)}{8.314\ J \cdot mol^{-1} \cdot K^{-1} \cdot 1873\ K}\right] = 4.596 \cdot 10^{-2} \tag{3.239}$$

The maximum solubility of the nitrogen, expressed in mass percentage or in molar fraction, is:

$$K_{reaction\ (3.236)} = \frac{\%N}{\sqrt{P_{N_2}}} \rightarrow \%N = K_{reaction\ (3.236)} \cdot \sqrt{P_{N_2}} \rightarrow \%N = 4.596 \cdot 10^{-2} \cdot \sqrt{0.79\ atm} = 4.085 \cdot 10^{-2}\%(408.5\ ppm) \tag{3.240}$$

$$x_N = \frac{4.085 \cdot 10^{-2} g\, N \frac{1\ mol\ N}{14.04\ g\ N}}{4.085 \cdot 10^{-2} g\, N \frac{1\ mol\ N}{14.04\ g\ N} + \left(100 - 4.085 \cdot 10^{-2} g\, N\right) gFe \cdot \frac{1\ mol\ Fe}{55.85\ g\ Fe}} = 1.627 \cdot 10^{-3} \tag{3.241}$$

On the other hand, the value of the standard free energy of the reaction (3.236) is identified with the coefficient of activity of Raoult at infinite dilution of the nitrogen in the iron melt, γ_N^*:

$$\Delta_d G^0_{\text{reaction (3.236)}}(T) = R \cdot T \cdot \ln\left(\frac{\gamma_N^* \cdot 55.85}{100 \cdot 1.01}\right) \rightarrow \gamma_N^*$$

$$= \exp\left(\frac{\Delta_d G^0_{\text{reaction (3.236)}}(T)}{R \cdot T}\right) \cdot 100 \cdot 14.01 \cdot \frac{1}{55.85} \rightarrow \gamma_N^*$$

$$= \exp\left[\frac{6757 + 22 \cdot 1873\text{K}\left(\text{J} \cdot (\text{mol H})^{-1}\right)}{8.314\,\text{J} \cdot \text{mol}^{-1} \cdot \text{K}^{-1} \cdot 1873\,\text{K}}\right].$$

$$100 \cdot 14.01 \cdot \frac{1}{55.85} \rightarrow \gamma_N^* = 545.84 \quad (3.242)$$

Consequently, if the self-interaction parameter of the nitrogen is zero (the same happens with the hydrogen), the activity of Raoult is:

$$a_N = \gamma_N^* \cdot x_N = 545.84 \cdot 1.627 \cdot 10^{-3} = 0.8881 \quad (3.243)$$

while the activity of Henry, as the value of f_N is assumed to be one, is:

$$a_{H,N} = f_N \cdot (\%N) = 1 \cdot 4.085 \cdot 10^{-2} = 4.085 \cdot 10^{-2} \quad (3.244)$$

Finally, the value of the constant of the law of Sievert, Sect. 3.4, at 1600 °C, and when the pressure is expressed in atmospheres and the concentration is expressed in mass percentage is $4.596 \cdot 10^{-2}$.

Commentaries: The value of the activity of Raoult for the nitrogen would be equal to one when the pressure of nitrogen in the gas is equal to one atmosphere. The coefficient of activity at infinite dilution of the hydrogen, γ_H^* (Exercise 3.9) at 1600 °C ($\gamma_H^* = 738.5$) is greater than that previously calculated for the nitrogen ($\gamma_N^* = 545.84$). Resulting from that, the solubility of the hydrogen in the melt is much lower. Both hydrogen and nitrogen are two of the gases that have more influence in the quality of the finished product (steels).

End of the exercise.

Exercise 3.11 Develop the equation that allows determining the minimum quantity of argon that is necessary to inject (ceramic lance or porous plug) in an iron melt to reduce the hydrogen content dissolved in the melt. Assume that the bubble of argon that leaves the melt reaches the Sievert's equilibrium and that the atomic weight of the hydrogen is 1.01 g·mol^{-1}.

Solution starts:

In each bubble of argon, we consider that there is a quantity of gas, δn_{Ar}, expressed in moles of argon per kilogram of melt that, during its ascending path picks up δn_H of hydrogen dissolved in the iron until reaching the saturation, thermodynamic equilibrium, Sievert's law, see Exercise 3.9. The gas that leaves the melt, apart from argon, contains hydrogen, in a degasification operation that is performed at room pressure:

$$K_{Sievert,H} = \frac{\%H}{\sqrt{P_{H_2}}} \tag{3.245}$$

$$P_{H_2} = \frac{(\%H)^2}{(K_{Sievert,H})^2} = \frac{\delta n_H}{\delta n_H + \delta n_{Ar}} \tag{3.246}$$

Reordering the terms of the previous equation, we obtain:

$$\delta n_{Ar} \cdot \left[\frac{(\%H)^2}{(K_{Sievert,H})^2}\right] = \left[1 - \frac{(\%H)^2}{(K_{Sievert,H})^2}\right] \cdot \delta n_H \tag{3.247}$$

$$\delta n_{Ar} = \left[\frac{(K_{Sievert,H})^2}{(\%H)^2} - 1\right] \cdot \delta n_H \tag{3.248}$$

When δn_H are removed from the melt as H_2 in the bubble of gas, the change that is produced in the composition of the hydrogen in the melt expressed in percentage, $\delta(\%H)$, for 1000 g of melt would be equal to:

$$\delta(\%H) = \delta n_H \cdot \left(\frac{1.01\,\text{g}\,\text{H} \cdot 2}{1000\,\text{g}\,\text{melt}}\right) \cdot 100 \simeq -\frac{\delta n_H}{5} \tag{3.249}$$

$$\delta(\%H) = \frac{\delta n_{Ar}}{5 \cdot X} \tag{3.250}$$

where:

$$X = \left[1 - \frac{(K_{Sievert,H})^2}{(\%H)^2}\right] \tag{3.251}$$

If we integrate for all the period of argon injection (to remove the hydrogen from the melt):

$$\frac{1}{5} \cdot \int_{\%H^0}^{\%H} \delta n_{Ar} = \frac{N_{Ar}}{5} = \int_{\%H^0}^{\%H} \left[1 - \frac{(K_{Sievert,H})^2}{(\%H)^2}\right] \cdot \delta(\%H) \tag{3.252}$$

$$V_{Ar}^{STP} = 22.4 \cdot 5 \cdot \left[\int_{\%H^0}^{\%H} \delta(\%H) - \int_{\%H^0}^{\%H} \left[\frac{(K_{Sievert,H})^2}{(\%H)^2}\right] \cdot \delta(\%H)\right] \tag{3.253}$$

where N_{Ar} is the minimum quantity of argon that is required, expressed in moles. As a mol of gas in standard conditions of pressure and temperature (273 K and 1 atm) occupies 22.4 L, the volume of argon per kilogram of melt is:

$$V_{Ar}^{STP} = 22.4 \cdot 5 \cdot \left[(\%H - \%H^0) + (K_{Sievert,H})^2 \cdot \left(\frac{1}{\%H} - \frac{1}{\%H^0} \right) \right] \quad (3.254)$$

$$V_{Ar}^{STP} = 112 \cdot \left[(K_{Sievert,H})^2 \cdot \left(\frac{1}{\%H} - \frac{1}{\%H^0} \right) - (\%H^0 - \%H) \right] \quad (3.255)$$

Comment: If the initial and final quantities of hydrogen, Exercise 3.9, are 8.0 and 3.0 ppm, respectively and using the same constant for the law of Sievert, $2.448 \cdot 10^{-3}$, the volume of argon per kilogram of melt, measured under standard conditions of pressure and temperature is:

$$V_{Ar}^{STP} = 112 \cdot \left[\left(2.448 \cdot 10^{-3}\right)^2 \cdot \left(\frac{1}{3.0 \cdot 10^{-4}} - \frac{1}{8.0 \cdot 10^{-4}} \right) - \left(8.0 \cdot 10^{-4} - 3.0 \cdot 10^{-4}\right) \right]$$
$$= 1.34\,\text{L of Ar per kilogram of melt} \quad (3.256)$$

End of the exercise.

Exercise 3.12 Apart from favoring the degasification of the melt, the injection of argon (or the injection of any other inert gas) can contribute to the stirring and thermal homogenization of the melt. The bubble of argon, when expands, forces the fluid that surrounds it to move. The expansion mechanism of this bubble of argon can be produced by:

(a) a change of temperature.
(b) a variation of pressure.

Assuming applicable the equation of the state for an ideal gas and considering one mol of gas, find the equation that estimates the work and power associated to the injection of argon.

Solution starts:

The energy derived from the thermal expansion (at constant pressure), W_{te}, of an argon bubble would be:

$$W_{te} = \int_{T_1}^{T_2} P \cdot dV = \int_{T_1}^{T_2} \frac{P \cdot R}{P} \cdot dT = \int_{T_1}^{T_2} R \cdot dT \quad (3.257)$$

$$W_{te} = R \cdot (T_2 - T_1) = R \cdot T_2 \cdot \left(1 - \frac{T_1}{T_2}\right) \quad (3.258)$$

where T_2 is the temperature that reaches the bubble of gas in contact with the melt and T_1 is the temperature of the gas injected by the ceramic lance or the porous plug.

On the other hand, the work–energy supplied by a change of pressure would be equally equivalent to the product of the pressure and variation of the volume, as it is expressed in Eq. (3.257). Nevertheless, in this case there is not only the variation of volume, V, but also of pressure, P. If the variable that remains constant is the temperature, T_2, we have for one mole of gas:

$$P \cdot V = R \cdot T_2 \tag{3.259}$$

$$P \cdot dV + V \cdot dP = 0 \tag{3.260}$$

Consequently, the work due to the change in the pressure, W_{cp}, of the bubble of gas is:

$$W_{cp} = -\int_{P_1}^{P_0} V \cdot dP = -R \cdot T_2 \int_{P_1}^{P_0} \frac{dP}{P} = R \cdot T_2 \cdot \ln\left(\frac{P_1}{P_0}\right) \tag{3.261}$$

where P_1 is the pressure of injection of the gas and P_0 is the pressure of the ambient where the operation is carried out (the usual situation is to have the atmospheric pressure, but it is also possible to operate under vacuum conditions).

Finally, the value of the energy, W_t, provided by the injection of argon, would be:

$$W_t = W_{te} + W_{cp} = R \cdot T_2 \cdot \left[1 - \frac{T_1}{T_2} + \ln\left(\frac{P_1}{P_0}\right)\right]\left[J \cdot (\text{mol Ar})^{-1}\right] \tag{3.262}$$

$$W_t = \frac{R \cdot T_2}{22.4} \cdot \left[1 - \frac{T_1}{T_2} + \ln\left(\frac{P_1}{P_0}\right)\right]\left[J \cdot (\text{liter Ar})^{-1}\right] \tag{3.263}$$

$$W_t = 44.64 \cdot R \cdot T_2 \cdot \left[1 - \frac{T_1}{T_2} + \ln\left(\frac{P_1}{P_0}\right)\right]\left[J \cdot \left(\text{Nm}^3\ \text{Ar}\right)^{-1}\right] \tag{3.264}$$

If we know the flow of argon injected in $\text{Nm}^3 \cdot \text{s}^{-1}$, V_g, it is possible to know the power provided by the injection, $\varepsilon(W)$, expressed in watts:

$$\varepsilon(W) = 44.64 \cdot R \cdot T_2 \cdot V_g \cdot \left[1 - \frac{T_1}{T_2} + \ln\left(\frac{P_1}{P_0}\right)\right] \tag{3.265}$$

Commentaries: If we identify the previous equation with that that gives the power of mixture in watts per ton of metal, $\tilde{\varepsilon}\left(W \cdot t^{-1}\right)$, Eq. (3.69):

$$\tilde{\varepsilon} = 14.23 \cdot \frac{V_g \cdot T}{M_{metal}} \cdot \log\left[1 + \left(\frac{H}{1.48 \cdot P_0}\right)\right] \tag{3.266}$$

it is possible to have a more detailed knowledge about the following thermal variables: T_1, T_2, and T or the position H of the point of injection of the gas in the ladle or at the bottom/hearth of the furnace.

Metallostatic pressure is that produced by the column of molten metal measured from the gas-melt interface. For the iron melts, it is assumed that for each 1.53 m, there is an increase of pressure in 1 atm or 0.101 MPa.

End of the exercise.

Exercise 3.13 Analyze the possibility of performing the deoxidation of an iron melt with 0.30%C and 0.01% O (100 ppm O) at 1600 °C, by means of the injection of argon with the aim of obtaining an oxygen content in the melt of 10 ppm. The atomic weight of the oxygen is 16 g·mol^{-1}.

Solution starts:

If we assume that during the time that the bubble of gas remains in contact with the melt, the following equilibrium between the oxygen and the carbon dissolved in the iron with the CO of the gas is established:

$$C\left(\text{dis; } a_{H,C} = 1\right) + O\left(\text{dis; } a_{H,O} = 1\right) \leftrightarrow CO(g) \tag{3.267}$$

We use the data of the Appendices 1 and 3 to obtain the standard free energy associated to the reaction (3.267):

$$C(s) + \frac{1}{2}O_2(g) \leftrightarrow CO(g) \tag{3.268}$$

$$\Delta_r G^0_{\text{reaction (3.268)}}(T) = -118.0 - 0.0840 \cdot T(K)\left[kJ \cdot (\text{mol CO})^{-1}\right] \tag{3.269}$$

$$\frac{1}{2}O_2(g) \leftrightarrow O\left(\text{dis; } a_{H,O} = 1\right) \tag{3.270}$$

$$\Delta_r G^0_{\text{reaction (3.270)}}(T) = -117.0 - 0.00314 \cdot T(K)\left[kJ \cdot (\text{mol O})^{-1}\right] \tag{3.271}$$

$$C(s) \leftrightarrow C\left(\text{dis; } a_{H,C} = 1\right) \tag{3.272}$$

$$\Delta_r G^0_{\text{reaction (3.272)}}(T) = 19.9 - 0.0411 \cdot T(K)\left[kJ \cdot (\text{mol C})^{-1}\right] \tag{3.273}$$

$$\begin{aligned}\Delta_r G^0_{\text{reaction (3.267)}}(T) &= \Delta_r G^0_{\text{reaction (3.268)}}(T) - \Delta_r G^0_{\text{reaction (3.270)}}(T) \\ &\quad - \Delta_r G^0_{\text{reaction (3.272)}}(T) \\ &= -20.9 - 0.03976 \cdot T(K)\left[kJ \cdot (\text{mol CO})^{-1}\right]\end{aligned} \tag{3.274}$$

at 1600 °C (1873 K):

$$\Delta_r G^0_{\text{reaction (3.267)}}(1873\,\text{K}) = -20.9 - 0.03976 \cdot 1873 = -95.3705\,\text{kJ} \cdot (\text{mol CO})^{-1} \tag{3.275}$$

The value of the equilibrium constant as a function of the temperature for the reaction (3.267) was already calculated in the Eq. (3.2) and, at the temperature of 1600 °C, Eq. (3.267), reaches the value of 457:

$$K_{\text{reaction (3.267)}} = \frac{P_{CO}}{a_{H,C} \cdot a_{H,O}} = 457 \rightarrow P_{CO} = 457 \cdot (\%C) \cdot (\%O) \tag{3.276}$$

In a bubble of argon of volume, dV, oxygen accumulates in the form of CO. The differential variation of oxygen, dn_O, expressed in moles or in grams, dm_O, are:

$$dn_O = \frac{P_{CO} \cdot dV}{R \cdot T} \tag{3.277}$$

$$dm_O = 16 \cdot \frac{P_{CO} \cdot dV}{R \cdot T} \tag{3.278}$$

The differential quantity of oxygen, expressed in grams, dm_O, that disappears from the melt would be equal to:

$$dm_O = -\frac{1000 \cdot M_{metal}}{100} \cdot d(\%O) \tag{3.279}$$

where M_{metal} is the quantity of melt to be deoxidized in kilograms. Regrouping terms and replacing the value of the CO pressure in the argon, equilibrium (3.267), if the carbon concentration in the melt remains constant, we obtain the following differential equation:

$$dV = -10 \cdot \frac{R \cdot T \cdot M_{metal} \cdot d(\%O)}{16 \cdot 457 \cdot (\%C) \cdot (\%O)} = -10 \cdot \frac{R \cdot T \cdot M_{metal}}{16 \cdot 457 \cdot 0.30} \cdot \left[\frac{d(\%O)}{(\%O)}\right] \tag{3.280}$$

We integrate Eq. (3.280) with the following boundary conditions:

Boundary condition 1: for $t = 0$, $V = 0$ and $(\%O) = 0.01\%(100\,\text{ppm})$.

Boundary condition 2: for a time t, $V = V$ and $(\%O) = 0.001\%(10\,\text{ppm})$.

When the thermal and chemical equilibrium in the injected gas is reached, the volume, V, in liters of injected gas would be equal to:

$$V(l) = 10 \cdot \frac{R \cdot T \cdot M_{metal}}{16 \cdot 457 \cdot 0.30} \cdot \ln\left(\frac{0.01}{0.001}\right) = 1.61 \cdot M_{metal} \tag{3.281}$$

The variation of the carbon concentration, considering the stoichiometry of the reaction (3.267), would be equal to

$$(0.01 - 0.001) \cdot \left(\frac{12.01}{16.00}\right) = \Delta C(\%) = 6.75 \cdot 10^{-3}\% \quad (3.282)$$

that is to say, the carbon concentration in the melt decreases only 67.5 ppm with respect to the 3000 ppm of carbon that melt had at the beginning of the treatment.

Commentaries: As we have seen in the Exercise 3.11, the argon injection can be considered a basic process for the hydrogen removal from an iron melt. In this case, we have used a medium with high-carbon percentage (3000 ppm) for the deoxidizing of the melt. We could also think in the decarburizing of a stainless steel with initial carbon percentages within 400 and 300 ppm, to values of around 100 ppm when, in this case, the initial oxygen content was around 3000 ppm.

End of the exercise.

Exercise. 3.14 One of the reasons that explains why the degasification of metallic melts in general, and iron melts in particular, is especially problematic is caused by the existence of processes of superficial absorption. Derivate Eqs. (3.196) and (3.197), which are consequence of the different mechanisms of solutes absorption in the liquid-gas interface.

Solution starts:

The mechanism of absorption of a gas molecule, G, on the surface of a melt that is in contact with the gas can be analyzed by the following stages:

(a) Absorption reaction:

$$G + V_v \rightarrow G - V \quad (3.283)$$

where G represents the absorbed gas and V_v is a lattice vacancy (lattice defect, position without other atoms) where the gas molecule can be located-adsorbed. The corresponding rate of the absorption process, v_{ab}, reaction (3.283), is equal to:

$$v_{ab} = k_1 \cdot P \cdot (1 - \theta) \quad (3.284)$$

where k_1 is the kinetic constant of the absorption process, P is the partial pressure of the gas G and θ is the fraction of the melt's surface occupied by atoms of the solute superficially actives.

(b) Desorption reaction:

$$G - V \rightarrow G + V_v \quad (3.285)$$

where the corresponding rate of the desorption process, v_{de}, is:

$$v_{de} = k_2 \cdot \theta \tag{3.286}$$

When the rates of the adsorption-desorption stages are the same (thermodynamic equilibrium):

$$k_1 \cdot P \cdot (1 - \theta) = k_2 \cdot \theta \tag{3.287}$$

$$k_1 \cdot P - k_1 \cdot P \cdot \theta = k_2 \cdot \theta \tag{3.288}$$

we obtain the following equation if we reorder terms:

$$\frac{k_1}{k_2} \cdot P = \theta + \left(\frac{k_1}{k_2}\right) \cdot P \cdot \theta \tag{3.289}$$

$$K_a \cdot P = \theta + K_a \cdot P \cdot \theta \tag{3.290}$$

where the relation between the kinetic constants of the absorption-desorption stages are identified with the equilibrium constant of the process, K_a (Eqs. (3.283) and (3.285)):

$$\theta = \frac{K_a \cdot P}{1 + K_a \cdot P} \tag{3.291}$$

$$1 - \theta = \frac{1}{1 + K_a \cdot P} \tag{3.292}$$

If the dissociation of the gas molecule, A_2, takes place before the adsorption stage:

$$A_2(g) \leftrightarrow 2A(g) \tag{3.293}$$

the equilibrium constant of the dissociation process, K_d, is:

$$K_d = \frac{P_A^2}{P_{A_2}} \tag{3.294}$$

$$P_A = \sqrt{K_d \cdot P_{A_2}} \tag{3.295}$$

Consequently, Eq. (3.291) (or Eq. (3.196)) is transformed into Eq. (3.197):

$$1 - \theta = \frac{1}{1 + K_a \cdot \sqrt{K_d \cdot P_{A_2}}} \tag{3.296}$$

$$K_a^* = K_a \cdot K_d^{0.5} \tag{3.297}$$

$$P = P_{A_2} \tag{3.298}$$

Commentaries: The difficulties that can be found in the removal of nitrogen and hydrogen from the iron melt can be increased due to occupation of the surface positions by sulphur and oxygen atoms. Consequently, it is difficult the dehydrogenization and de-nitriding of melts with high sulphur and oxygen concentrations.

End of the exercise.

References

Deo B, Boom R (1993) Fundamentals of steelmaking metallurgy. Prentice Hall International, Hertfordshire, UK

Kawai Y, Shiraishi Y (1988) Handbook of physico-chemical properties at high temperatures. The Iron and Steel Institute of Japan-Committee of Japan Society for Promotion of Science, Tokyo, Japan

Knacke O, Kubaschewski O, Hesselmann K (1991) Thermochemical proper-ties of inorganic substances, 2nd edn. Springer-Verlag, Berlin, Germany

Sigworth GH, Elliot JM (1974) Thermodynamic of dilute iron alloys. Metals Sci 8:298–310

Springorum D (1995) Slag Atlas. Düsseldorf, Germany, Verein Deutscher Eisenhüttenleute (VDEh)

The Japan Society for the Promotion of Science. The 19th Committee on Steelmak-ing (1988) Steelmaking data sourcebook. Gordon and Breach Science Publish-ers, New York, USA

TAPP Database (1994) Thermochemical and physical properties (Software). E. S. Microwave Inc., Cincinnati, Ohio, USA

Verdeja LF, Ballester A, Sancho JP, González R (2014) Refractory and ceramic materials, 1st edn. Ed. Síntesis, Madrid, Spain

Chapter 4
Solidification of the Steel

Abstract Iron metallurgy in liquid phase finishes with the processes of forming by casting, molding, or production of powders. We study in this chapter the two most relevant processes of solidification: the molding and the continuous casting. Molding practice is described from the obtaining of the mold to the study of the solidification from the thermal point of view. Continuous casting is also described as the most recent technology for the iron melt solidification, where we indicate the recent advances that are mainly focused on obtaining the near-net-shape. Additionally, the problem of inclusions is also indicated as it is important for the solidified product. Finally, we include several solved exercises where we emphasize on the theoretical aspects.

4.1 Introduction

From the conceptual point of view, the iron metallurgy in liquid phase (iron and steelmaking) finishes with the processes of forming by casting, molding, or production of powders. After that, the operations of the iron metallurgy in solid phase are studied by the metal engineering (Fig. 4.1), which includes the structural metallurgy, the physical metallurgy, and the mechanics metallurgy.

Casting is the usual procedure for the forming of semifinished or finished steel products (Flinn 1963). Less than one percent of the world's production of steel is formed by molding or by powder metallurgy (Table 4.1). On the contrary, cast irons are the other big family of structural materials based on the iron (Fe–C alloys with around 3.5% C, Pero-Sanz et al. 2018), which are exclusively formed by molding.

The control of the variables involved in the casting process is basic to achieve the desired structure in the applications planned during the design. In the last decades, continuous casting has imposed as the most efficient process of forming for mild steels and for those special materials with high demand (for instance, certain qualities of stainless steels). However, in some regions and countries of the developed world and with advanced iron and steelmaking technology, it is observed that the quantity of steel that solidifies by continuous casting reaches a maximum of 90–95% of steel

L. F. Verdeja González et al., *Operations and Basic Processes in Steelmaking*,
Topics in Mining, Metallurgy and Materials Engineering,
https://doi.org/10.1007/978-3-030-68000-8_4

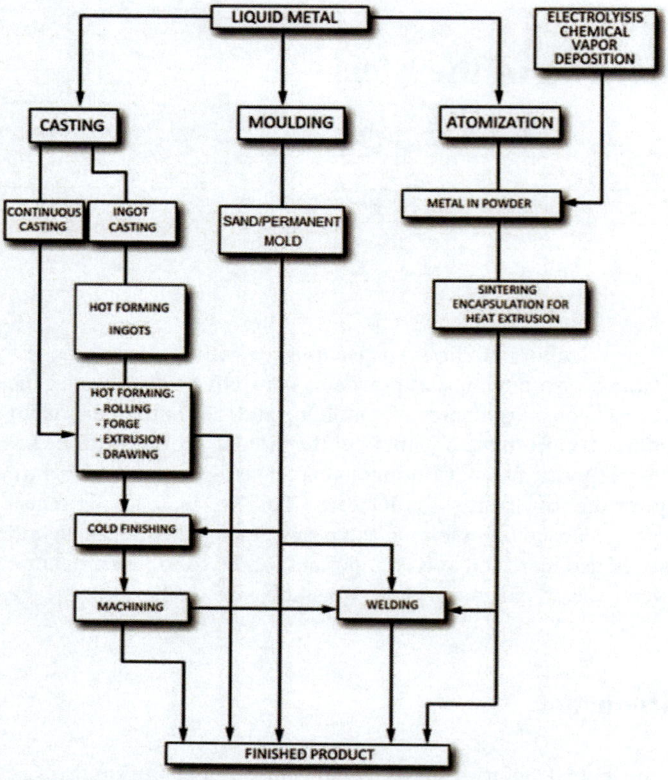

Fig. 4.1 Alternatives to the forming of either the liquid iron or the non-ferrous metals

production, depending on the greater or smaller installed capacity for special steels production in those regions or countries.

In the conventional casting, liquid iron is poured into molds with simple geometric shape called ingot molds (Pero-Sanz 2006; pp. 162–164; Pero-Sanz et al. 2017, 2018, 2019), which usually have tapered shape to facilitate the removal from the mold and square, rectangular or with special shape cross section. The filling of the ingot molds can be done as follows:

A. Direct casting through the upper zone by locating the ladle over the ingot mold and making the melt to flow into it (Fig. 4.2). It is necessary to pay attention to the shape and center position of the liquid stream in this kind of operation to avoid splashes (projections) on the walls that could mean defects in the semifinished product.
B. Bottom casting via filling the ingot mold by the bottom because the ladle is situated on a central flue from which several horizontal channels of distribution start. This practice is very recommendable when it is necessary to reach a

Table 4.1 World steel production from 1950 to 2015

Year	Production (Mt)
1950	189
1960	347
1970	595
1980	717
1985	719
1990	770
1995	753
2000	850
2003	971
2006	1250
2008	1343
2010	1433
2012	1559
2014	1662
2015	1620
2019	1870

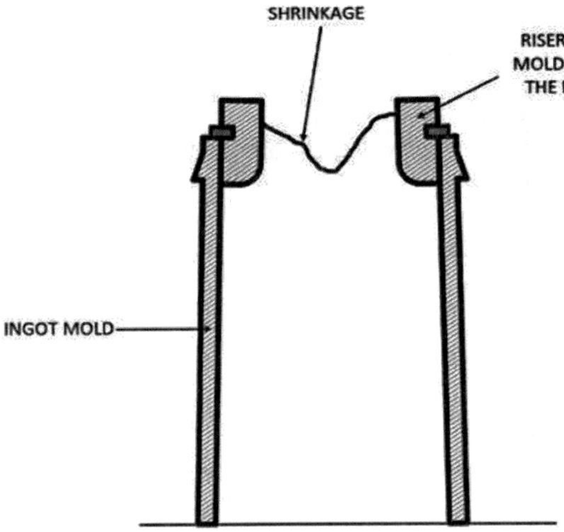

Fig. 4.2 Ingot mold for the conventional casting of the liquid iron

maximum surface and internal quality in the ingot, and it has been widely used in the production of special steels.

If the gasification of the carbon is aborted, either partially or totally, rough ingots obtained by ingot casting might have capped, semi-killed, killed, or rimmed qualities

(Fig. 3.1). However, it is necessary to consider that the forming technology of the liquid by continuous casting requires deoxidizing of steel.

In the first developments and tests at the industrial scale of the continuous casting, the quality of the continuous casting steel was lower, regarding mechanical specifications (yield strength, ultimate tensile strength, ductility, and toughness), than that obtained by means of the traditional casting. Continuous casting has imposed as the forming technology of liquid iron only after solving the problem of the inclusions removal (Kiessling and Lange 1978), through the deoxidizing and refining of the melt.

The tertiary metallurgy of the steel has its origin in the operations performed on the ready to cast liquid metal, before the solidification (Flemings 1974), with the aim of keeping and, in some points, improving the specifications established by the secondary metallurgy. That is to say, the objective is to keep the specifications referred to the alloy elements added to reach the desired quality and to improve the specifications referred to the cleaning of the metal (volumetric fraction of inclusions), all of this with respect to the melt that was obtained in the secondary metallurgy.

4.2 Molding

In the molding processes, liquid iron, normally with an equivalent carbon quantity close to that of the Fe–C system eutectic point (it is also possible to process liquids with low carbon percentages, that is to say, steels), is poured into an empty mold whose cavity reproduces the shape of the part. The most important reason to use this forming technique, facing either the conventional or continuous casting, is economical. There are situations where the geometrical complexity, size, and number of parts allow obtaining by molding the parts in a more economical way than using forge processes (hot plastic deformation) or that allow obtaining different shapes than those obtained by ingot casting.

The process of manufacture of a molded part has the following stages:

1. Manufacture and preparation of the model (pattern).
2. Manufacture of the mold from the model.
3. Production of the alloyed liquid iron.
4. Pouring the melt into the mold.
5. Removal of parts from the mold.
6. Cleaning and finishing the parts.
7. Heat treatment of the parts (stress relieving heat treatment and others).

The model (pattern) of the part, which is manufactured in wood, metal, thermoplastic, or wax, is used to obtain the mold and reproduces the geometrical characteristics of the part. It may be permanent or disposable (cope-and-drag sand casting, rejectable, one single casting operation, for instance, when thermoplastics or wax are used). The model is covered in all the situations with a refractory material (silica sand, zirconium sand, or zirconia sand), which by means of chemical, organic, or ceramic agglomeration adjusts to the geometrical characteristics of the part to be manufactured.

When a permanent model is used, manufactured in wood or in metal, the mold must allow removing the model. Normally, the recovery of the model is achieved by making the mold separable into two parts.

We present in Fig. 4.3 the foundry process of a part using sand mold and polystyrene model, which is burnt during the molding process. Gases of the polystyrene combustion escape through the open porosity of the mold and the heat of the polymer combustion improves both the adhesion of the metal to the sand and the surface characteristics of the part.

The refractory mold (sand molding) has as main advantages:

- That can adapt to produce series of parts with a very low number of items.
- That is adequate for parts with complex geometry.
- That can adapt to the manufacture of both small models and big parts.

The refractory mold, due to its thermal properties (low conductivity), can be only used once, although it is possible to recycle the foundry sand that was not affected by the liquid.

The other possibility is the solidification of the melt in a permanent metallic mold that is normally cooled in water (especially metallic melts with low liquidus temperature) (Winegard 1974). With respect to the sand casting, the solidification in permanent mold improves the metallurgical characteristics of the parts in the following aspects:

- Smaller grain sizes.
- Better surface characteristics.
- Reduction of the porosity.

However, solidification in permanent mold has limitations regarding the size and shape of the parts. The casting in permanent metallic mold can be also carried out under pressure, by injection of the liquid at enough pressure to avoid several physical heterogeneities that are inherent to any conventional process of solidification, concretely certain types of shrinkage and porosity due to the gas release (Pero-Sanz 2006, pp. 169–175). On the other hand, the die casting, or liquid forge, allows reaching high rates in the production of parts with the following characteristics:

- Fine-grained structures due to the high cooling rates.
- Isotropic mechanical properties very similar to those of the forged parts and greater than those obtained by conventional molding processes.

We show in Fig. 4.4 the characteristics of an L-shaped steel part with infinite thickness (z dimension). Likewise, the dimensions of the ceramic or metallic mold are detailed.

The thickness of the part is sufficiently great to achieve that the flow of heat in the direction z was small in comparison with the ordinates or abscissas. The variation of the temperature with the time, in any point of the part, is analyzed using the following differential equation (two-dimension Fourier equation):

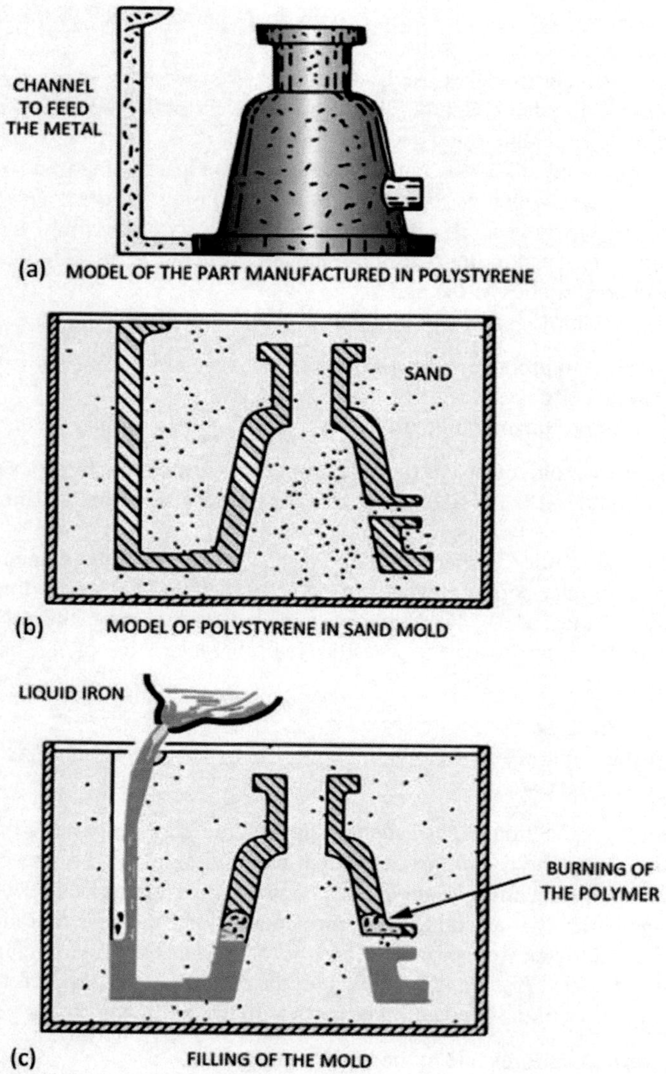

Fig. 4.3 Scheme of a sand molding process with a non-recoverable model of polystyrene

$$\left(\frac{\mathrm{d}T}{\mathrm{d}t}\right)_{x,y} = \alpha \cdot \left[\left(\frac{\mathrm{d}T}{\mathrm{d}x}\right)_{t,y} + \left(\frac{\mathrm{d}T}{\mathrm{d}y}\right)_{t,x}\right] \tag{4.1}$$

where α is the thermal diffusivity in $\mathrm{m}^2 \cdot \mathrm{s}^{-1}$, which is defined as follows:

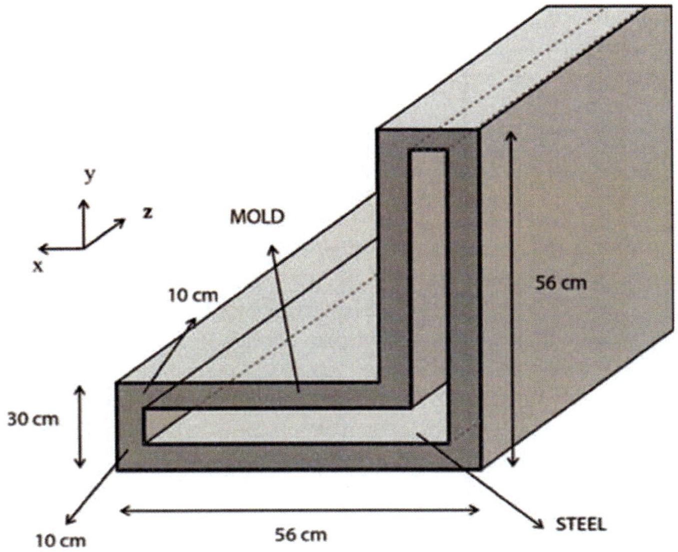

Fig. 4.4 Dimensional characteristics of both a part and a mold (ceramic or metallic, with infinite thickness)

$$\alpha = \frac{\lambda}{\rho \cdot c_p} \quad (4.2)$$

where λ is the thermal conductivity of the metal, ρ is the density and c_p is the specific heat at constant pressure.

Differential Eq. (4.1) is solved using the finite element software ANSYS and considering the following boundary conditions:

- Initial temperature of the steel: 1580 °C.
- Initial temperature of the mold: 25 °C.
- Variation of both the enthalpy and thermal conductivity of the steel with the temperature are calculated taking as reference the values of Table 4.2.

The calculation analyzes the variation of the temperature with the time in one point of the periphery of the part, coordinate (15, 15) in Fig. 4.4, with respect to the other point located in the center of the part, coordinate (26, 15), in two situations:

Table 4.2 Variation of both the thermal conductivity and thermal enthalpy of the steel with the temperature

Temperature (°C)	0	1450	1500	1580
Thermal conductivity ($W \cdot m^{-1} \cdot K^{-1}$)	29.8	31.9	25.3	25.3
Enthalpy ($J \cdot m^{-3}$)	0	825×10^7	1054×10^7	1121×10^7

(a) With a sand refractory mold (quartzite, SiO_2) with 0.519 $W \cdot m^{-1} \cdot K^{-1}$ of thermal conductivity (sand molding).

(b) With a copper metallic mold of high thermal conductivity (392 $W \cdot m^{-1} \cdot K^{-1}$).

We want, with these two situations, to simulate the steel solidification conditions in sand mold in front of those of the solidification in a permanent metallic mold (Davies 1973). We show in Figs. 4.5 and 4.6 the change of temperature with the time in the points (15, 15) and (26, 15), for the solidification in sand or in copper mold. The difference of temperatures between the periphery and the nucleus is significant in the case of the sand molding, and the heterogeneities that the parts have as a consequence of the different cooling rates between both points makes necessary the heat treatment as mandatory final stage in the operations of sand molding.

The microstructure that is obtained, as a consequence of the cooling in metallic mold at pressure (where the pressure increases the coefficient of heat transfer in the metal), is fine (small grain size) and does not usually require heterogeneous nucleation agents. It is possible to act either on the temperature of the melt or on the temperature of the mold to control the microstructure of the material. The CastCAE software, as others that are similar and used in foundries, allows:

- Determining the productivity of the process by solving the Navier–Stokes equations that estimate the rate of the mold filling (Pero-Sanz et al. 2018).
- Verifying the heat transfer properties, and performing the thermal simulation of the process, using the data of the temperature incorporated to the mold of the part.

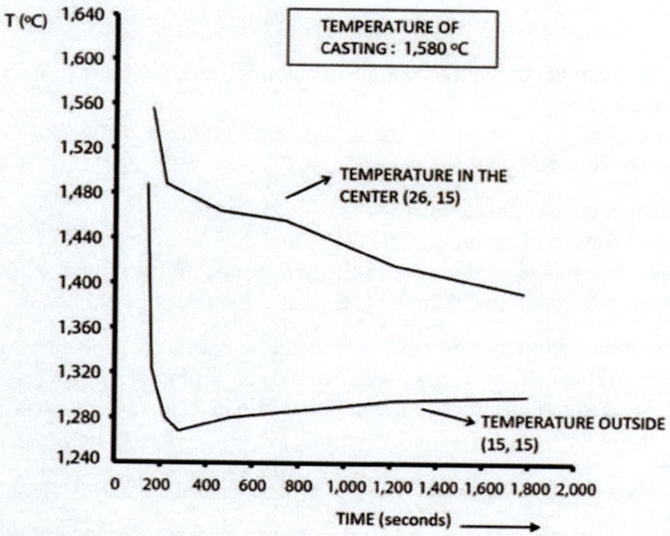

Fig. 4.5 Variation of the temperature with the time in the coordinates (15, 15) and (26, 15) of Fig. 4.4, in sand mold. Properties of the sand: thermal conductivity, 0.519 $W \cdot m^{-1} \cdot K^{-1}$; density, 1494 $kg \cdot m^{-3}$; and, specific heat, 241 $J \cdot kg^{-1} \cdot K^{-1}$

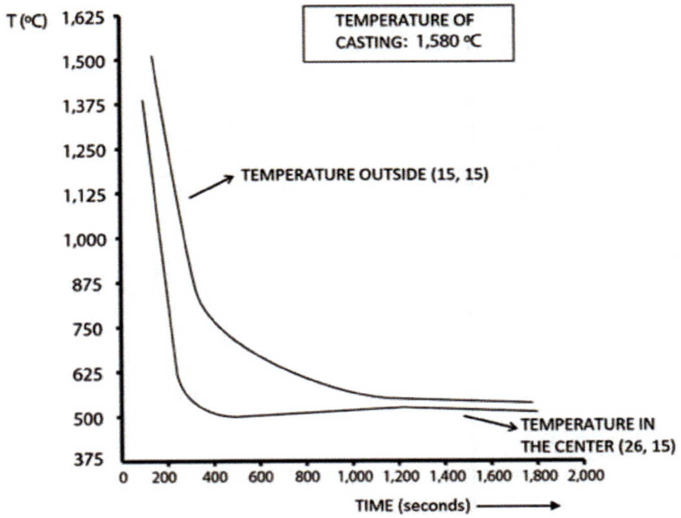

Fig. 4.6 Variation of the temperature with the time, in the coordinates (15, 15) and (26, 15) of Fig. 4.4, in copper mold. Properties of the copper: thermal conductivity, 392 $W \cdot m^{-1} \cdot K^{-1}$; density, 8960 $kg \cdot m^{-3}$; and, specific heat, 385 $J \cdot kg^{-1} \cdot K^{-1}$

- In gray cast irons, forecasting both the hardness and volumetric fraction of graphite nodules.

This everything provides the following economic advantages to the industry of molding cast iron and steel parts:

- Energy savings in the range 3–6%.
- Improvement in the quality of the product.
- Decrease in the number of parts that must be either remelted or machined before being considered as finished products.
- Optimization of the idle times in the process, resulting in an increase of the productivity.

4.2.1 *Solved Exercises*

Exercise 4.1 It is possible to initially consider solidification of the pure solvent, the iron, to calculate the times required for the solidification of steels and cast irons. Analyzing the solidification of 250 kg of iron at 1750 °C in a ceramic mold of cylindric geometry with 0.40 m of diameter (see Fig. 4.7), estimate:

(a) The time required to reduce the overheating of the melt.
(b) The time required to solidify, at 1538 °C, 250 kg of iron.
(c) The rate of solidification in centimeters per second.

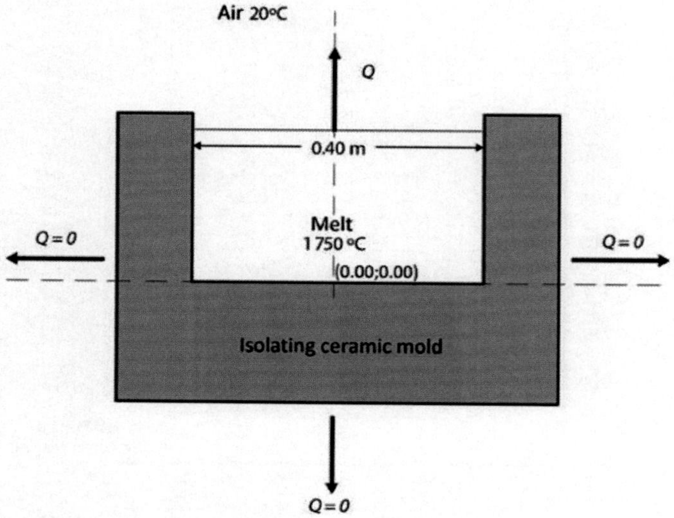

Fig. 4.7 Solidification of the pure metal in isolating ceramic mold (ideal)

Data from the HSC5.1: density of the liquid iron, 6746 kg · m^{-3}; average specific heat of the metallic melt, 46.02 J· mol^{-1} · K^{-1}; latent heat of fusion of the iron, 13,807 J · mol^{-1}; iron melting point, 1538 °C; atomic weight of the iron, 55.85 g · mol^{-1}.

Finally, assume that the global coefficient of heat transfer, h_g, from the melt to the air is 130 W · m^{-2} · K^{-1}; room temperature, T_0, 20 °C.

Solution starts:

The time required to dissipate the overheating of the melt is calculated from the following equation:

$$h_g \cdot A \cdot (T_1 - T_0) \cdot t = \bar{c}_p \cdot M_{metal} \cdot \Delta T \tag{4.3}$$

where:

$$A = \pi \cdot r^2 \tag{4.4}$$

and, thus:

$$t = \frac{\bar{c}_p \cdot M_{metal} \cdot \Delta T}{h_g \cdot \pi \cdot r^2 \cdot (T_1 - T_0)} \tag{4.5}$$

$$t = \frac{46.02\ \mathrm{J \cdot mol^{-1} \cdot K^{-1}} \cdot 250\ \mathrm{kg} \cdot (2023 - 1811)\mathrm{K}}{130\mathrm{W \cdot m^{-2} \cdot K^{-1}} \cdot \pi \cdot \left(\frac{0.4}{2}\ \mathrm{m}\right)^2 \cdot (2023 - 293)\mathrm{K}} \cdot \frac{1000\ \mathrm{g}}{1\ \mathrm{kg}} \cdot \frac{1\ \mathrm{mol}}{55.85\ \mathrm{g\ Fe}} = 1545\ \mathrm{s} = 25.75\ \mathrm{min} \tag{4.6}$$

On the other hand, the time required to achieve that the iron melt contained in the cylindric mold could dissipate the heat associated with the solidification of the iron is:

$$h_g \cdot \mathrm{A} \cdot (\mathrm{T_M} - \mathrm{T_0}) \cdot \mathrm{t} = \Delta \mathrm{H}_f \cdot \mathrm{M_{metal}} \tag{4.7}$$

$$\begin{aligned} t &= \frac{\mathrm{H}_f \cdot \mathrm{M_{metal}}}{h_g \cdot \pi \cdot r^2 \cdot (\mathrm{T_1} - \mathrm{T_0})} \\ &= \frac{13807\ \mathrm{J \cdot mol^{-1}} \cdot 250\ \mathrm{kg}}{130\mathrm{W \cdot m^{-2} \cdot K^{-1}} \cdot \pi \cdot \left(\frac{0.4}{2}\ \mathrm{m}\right)^2 \cdot (1811 - 293)\mathrm{K}} \cdot \frac{1000\ \mathrm{g}}{1\ \mathrm{kg}} \\ &\quad \cdot \frac{1\ \mathrm{mol}}{55.85\ \mathrm{g\ Fe}} = 2492\ \mathrm{s} = 41.5\ \mathrm{min} \end{aligned} \tag{4.8}$$

Finally, the solidification rate can be deduced from the total time involved in the solidification:

$$\mathrm{t} = 25.75 + 41.5 = 67.25\ \mathrm{min}(4037\ \mathrm{s}) \tag{4.9}$$

The special characteristics of the ceramic mold lead to a flat solidification front. We use the density of the iron to calculate the height (h) of the cylindric mold where the melt solidifies:

$$h = \frac{250\ \mathrm{kg}}{6746\ \mathrm{kg \cdot m^{-3}}} \times \frac{1}{\pi \cdot r^2} = 0.295\ \mathrm{m} \tag{4.10}$$

Therefore, the solidification rate of the melt, $\mathrm{v_{sol}}$, is:

$$\mathrm{v_{sol}} = \frac{0.295\ \mathrm{m}}{4037\ \mathrm{s}} \frac{100\ \mathrm{cm}}{1\ \mathrm{m}} = 7.3 \times 10^{-3}\ \mathrm{cm \cdot s^{-1}} \tag{4.11}$$

Commentaries: In steels and cast irons, the temperature at which the solidification begins, and finishes is not usually the same (the alloy with eutectic composition would be the exception).

Likewise, it is not probable that the walls of the ceramic mold were perfectly isolating. Normally, there is a heat transfer through the walls of the ceramic mold and the time required to reduce the overheating and carrying out the solidification would be shorter.

Finally, we proceed as follows to explain the value of h_g ($130\ W \cdot m^{-2} \cdot K^{-1}$, from the melt to the air) proposed in the enunciation of the exercise:

The value of the coefficient of radiant transfer, h_r, is equal to:

$$h_r = \sigma_{SB} \cdot F \cdot \left(T_1{}^2 + T_2{}^2\right) \cdot (T_1 + T_2)\left(W \cdot m^{-2} \cdot K^{-1}\right) \tag{4.12}$$

where σ_{SB} is the Stefan–Boltzmann parameter into the equation, 5.67051×10^{-8} $W \cdot m^{-2} \cdot K^{-4}$; F is the emissivity of the iron melt, 0.28 (see Poirier and Poirier 1992, p. 28); $T_2 = T_0$ is the absolute temperature of the air, 293 K; and, there are two options for the estimation of the temperature T_1, temperature in the melt-air interface:

- *Hypothesis A*: by means of the following equation that, at the beginning of the cooling process, reaches the following value:

$$T_1 = \frac{1750 + 1538}{2} = 1644\,°C(1823\,K) \tag{4.13}$$

- *Hypothesis B*: assuming that the temperature at the beginning of the solidification process in the melt-air interface is only 12 °C greater than the iron melting point (1538 °C): 1550 °C.

The variation of the temperature with the time, as the solidification process progresses, could be, in the case of *Hypothesis A*: 1700, 1650, or 1630 °C. The values of the corresponding interfacial temperatures, T_1, would be, respectively, Eq. (4.13): 1616 °C (1889 K), 1596 °C (1869 K), and 1584 °C (1857 K). The values of h_r estimated with the Eq. (4.12), considering the values of T_1 previously indicated, would be, respectively:

$$h_r = 5.67051 \times 10^{-8} \cdot W \cdot m^{-2} \cdot K^{-4} \cdot 0.28 \cdot \left[(1889\ K)^2 + (293\ K)^2\right] \cdot\ [(1889K) + (293\ K)] = 126.6\ W \cdot m^{-2} \cdot K^{-1} \tag{4.14}$$

$$h_r = 5.67051 \times 10^{-8} \cdot W \cdot m^{-2} \cdot K^{-4} \cdot 0.28 \cdot \left[(1869\ K)^2 + (293\ K)^2\right] \cdot\ [(1869K) + (293\ K)] = 122.9\ W \cdot m^{-2} \cdot K^{-1} \tag{4.15}$$

$$h_r = 5.67051 \times 10^{-8} \cdot W \cdot m^{-2} \cdot K^{-4} \cdot 0.28 \cdot \left[(1857\ K)^2 + (293\ K)^2\right] \cdot\ [(1857K) + (293\ K)] = 120.6\ W \cdot m^{-2} \cdot K^{-1} \tag{4.16}$$

The variation with the time in the case of applying the *Hypothesis B*, as the only surface available to dissipate the heat of the solidification is through the melt-air interface, leads to that the temperature could increase with the time: 1560 °C (1833 K), 1580 °C (1853 K), and 1584 °C (1857 K). The values of h_r estimated with the Eq. (4.12), considering the values of T_1 previously indicated, would be, respectively:

$$h_r = 5.67051 \times 10^{-8} \cdot \mathrm{W} \cdot \mathrm{m}^{-2} \cdot \mathrm{K}^{-4} \cdot 0.28 \cdot \left[(1833\,\mathrm{K})^2 + (293\,\mathrm{K})^2\right] \cdot [(1833\,\mathrm{K}) + (293\,\mathrm{K})] = 116.3\mathrm{W} \cdot \mathrm{m}^{-2} \cdot \mathrm{K}^{-1} \tag{4.17}$$

$$\mathrm{h_r} = 5.67051 \times 10^{-8} \cdot \mathrm{W} \cdot \mathrm{m}^{-2} \cdot \mathrm{K}^{-4} \cdot 0.28 \cdot \left[(1853\,\mathrm{K})^2 + (293\,\mathrm{K})^2\right] \cdot [(1853\,\mathrm{K}) + (293\,\mathrm{K})] = 119.9\mathrm{W} \cdot \mathrm{m}^{-2} \cdot \mathrm{K}^{-1} \tag{4.18}$$

$$\mathrm{h_r} = 5.67051 \times 10^{-8} \cdot \mathrm{W} \cdot \mathrm{m}^{-2} \cdot \mathrm{K}^{-4} \cdot 0.28 \cdot \left[(1857\,\mathrm{K})^2 + (293\,\mathrm{K})^2\right] \cdot [(1857\,\mathrm{K}) + (293\,\mathrm{K})] = 120.6\mathrm{W} \cdot \mathrm{m}^{-2} \cdot \mathrm{K}^{-1} \tag{4.19}$$

As it is possible to check, the values of h_r, estimated by means of the two above-indicated hypotheses, converge in the interfacial temperature of 1584 °C (1857 K) and the value of h_r of 120.6 $\mathrm{W} \cdot \mathrm{m}^{-2} \cdot \mathrm{K}^{-1}$. The difference between this value and the 130 $\mathrm{W} \cdot \mathrm{m}^{-2} \cdot \mathrm{K}^{-1}$ indicated in the enunciation are attributed to the convective mechanism of heat transfer in the air.

End of the exercise.

Exercise 4.2 Calculate the time that is necessary for the solidification of a carbon steel part of 2.5 cm in thickness in a mold of zirconium silicate sand.

We assume that the temperature is the only function of both the coordinate, x, and the time, t, T(x, t), during the solidification of the melt and the corresponding heat transfer.

Data (zircon mold): thermal conductivity, λ, 1.04 $\mathrm{W} \cdot \mathrm{m}^{-1} \cdot \mathrm{K}^{-1}$; average specific heat, $\bar{c}_p$, 837 $\mathrm{J} \cdot \mathrm{kg}^{-1} \cdot \mathrm{K}^{-1}$; density, ρ, 2723 $\mathrm{kg} \cdot \mathrm{m}^{-3}$.

Data (Fe-melt): density, ρ_{fus}, 6746 $\mathrm{kg} \cdot \mathrm{m}^{-3}$; melting point, T_M, 1811 K; enthalpy of fusion of the Fe, ΔH_{fus}, 277 $\mathrm{kJ} \cdot \mathrm{kg}^{-1}$. The physical and thermal properties of the iron melt with a carbon percentage < 0.80% are identified with those of a melt of pure iron (see Fine and Geiger 1993, p. 431).

Solution starts:

Metal is poured exactly at its melting point against a flat mold. Thus, the mold surface is suddenly heated to $\mathrm{T_M}$ in a time t = 0. This is a transient one-dimensional heat-flow problem, and the solution must conform with the Fourier partial differential equation:

$$\frac{\partial T}{\partial t} = \alpha_{\mathrm{m}} \cdot \frac{\partial^2 T}{\partial x^2} \tag{4.20}$$

The situation of the origin of coordinates for the 2D–3D solidification models is in the central zone of the melt to be solidified, Fig. 4.7. The temperature in this point, $x = 0$, for any time, is identified with the liquidus temperature of the melt, if there is not overheating, or with that of melting point (solidification) of the melt, $\mathrm{T_M}$, if it is pure metal.

The variation of the temperature from the origin of coordinates toward the walls of the ceramic mold that is in contact with an environment at constant temperature, T_0 (infinite heat sink), is calculated as follows:

$$T(x, t) = A + B \cdot \operatorname{erf}(z) \tag{4.21}$$

When $x = 0 \rightarrow T(0, t) = T_M \rightarrow \operatorname{erf}(z) = 0$
When $x = \infty \rightarrow T(\infty, t) = T_0 \rightarrow \operatorname{erf}(z) = 1$
We know that:

$$\operatorname{erf}(z) = \frac{2}{\sqrt{\pi}} \cdot \int_0^z e^{-u^2} \cdot du \tag{4.22}$$

$$z = \pm \frac{x}{2 \cdot \sqrt{\alpha_m \cdot t}} \tag{4.23}$$

$$\alpha_m = \frac{\lambda_m}{\rho_m \cdot c_m} \tag{4.24}$$

For that reason, under these boundary conditions:

$$A = T_M \tag{4.25}$$

$$B = T_0 - T_M \tag{4.26}$$

Thus:

$$T(x, t) = T_M + (T_0 - T_M) \cdot \operatorname{erf}(z) = T_M - (T_M - T_0) \cdot \operatorname{erf}(z) \tag{4.27}$$

where:

$$\frac{T - T_M}{T_0 - T_M} = \operatorname{erf}(z) \tag{4.28}$$

It is necessary to calculate the heat flow, J_q, through the ceramic mold-melt interface (First Law of Fourier) to consider that:

$$J_q = -\lambda \cdot \left(\frac{dT}{dx}\right) \tag{4.29}$$

$$-\left(\frac{dT}{dx}\right) = \frac{2}{\sqrt{\pi}} \cdot (T_M - T_0) \cdot \exp(z^2) \cdot \left(\frac{dz}{dx}\right) \tag{4.30}$$

The derivative with respect to x of the variable z:

$$\left(\frac{dz}{dx}\right) = \frac{1}{\sqrt{4 \cdot \alpha \cdot t}} \tag{4.31}$$

where the value of the variable z is equal to:

$$z = \frac{x}{\sqrt{4 \cdot \alpha \cdot t}} \tag{4.32}$$

Thus, the thermal gradient, for $x = 0$ where α is the thermal diffusivity, would be:

$$\left(\frac{dT}{dx}\right)_{x=0} = \frac{2}{\sqrt{\pi}} \cdot (T_M - T_0) \cdot \frac{1}{\sqrt{4 \cdot \alpha \cdot t}} \tag{4.33}$$

which leads to a value of the heat flow, J_q, in $W \cdot m^{-2}$ of:

$$J_q = \frac{\lambda}{\sqrt{\pi \cdot \alpha \cdot t}} \cdot (T_M - T_0) = \frac{1}{\sqrt{\pi}} \cdot (T_M - T_0) \cdot \sqrt{\lambda \cdot c_p \cdot \rho} \cdot \left(t^{-0.5}\right) \tag{4.34}$$

The heat rate, ϕ, in W, through the area, A, of the solid-melt interface is:

$$\phi = J_q \cdot A = \frac{A}{\sqrt{\pi}} \cdot (T_M - T_0) \cdot \sqrt{\lambda \cdot c_p \cdot \rho} \cdot \left(t^{-0.5}\right) \tag{4.35}$$

On the other hand, the rate associated with the solidification of the melt, when the variation of the thickness of solidified material with the time is (de/dt):

$$\Delta H_{fus} \cdot \rho_{fus} \cdot A \cdot \left(\frac{de}{dt}\right) \tag{4.36}$$

If we identify the last two equations, and solving for the variation of the thickness of solidified material with the time, we have:

$$\left(\frac{de}{dt}\right) = \frac{\sqrt{\lambda \cdot c_p \cdot \rho}}{\sqrt{\pi} \cdot \Delta H_{fus} \cdot \rho_{fus}} \cdot (T_M - T_0) \cdot \left(t^{-0.5}\right) \tag{4.37}$$

We integrate the last differential equation, and we obtain the functional relation between time and thickness of melt that has solidified:

$$e = \frac{2}{\sqrt{\pi}} \cdot \frac{\sqrt{\lambda \cdot c_p \cdot \rho}}{\Delta H_{fus} \cdot \rho_{fus}} \cdot (T_M - T_0) \cdot \left(t^{0.5}\right) \tag{4.38}$$

Using the data of the enunciation and considering the thickness of solidified material $e = 2.5\,\text{cm} \cdot 0.5 = 1.25\text{cm}(1.25 \times 10^{-2}\,\text{m})$ and the room temperature, $T_0 = 298\,\text{K}(25\,°\text{C})$, we obtain a time required for the solidification of the carbon steel part of:

$$t = \left[\frac{\sqrt{\pi}}{2} \cdot \frac{\Delta H_{fus} \cdot \rho_{fus}}{\sqrt{\lambda \cdot c_p \cdot \rho}} \cdot \frac{e}{(T_M - T_0)}\right]^2 \tag{4.39}$$

$$t = \left[\frac{\sqrt{\pi}}{2} \cdot \frac{277000\,\text{kJ} \cdot \text{kg}^{-1} 6746\,\text{kg} \cdot \text{m}^{-3}}{\sqrt{1.04\text{W} \cdot \text{m}^{-1} \cdot \text{K}^{-1} 837\,\text{J} \cdot \text{kg}^{-1} \cdot \text{K}^{-1} 2723\,\text{kg} \cdot \text{m}^{-3}}} \cdot \frac{1.25 \times 10^{-2}\,\text{m}}{(1811 - 298)\,\text{K}}\right]^2 = 79\,\text{s} \tag{4.40}$$

The rate of advance of the solidification front, v_{sol}, would be:

$$v_{sol} = \frac{1.25 \times 10^{-2}\,\text{m}}{79} \times \frac{100\,\text{cm}}{1\,\text{m}} = 1.58 \times 10^{-2}\,\text{cm} \cdot \text{s}^{-1} \tag{4.41}$$

Commentaries: the time required for the solidification of a carbon steel part was calculated considering the following restrictions:

(a) The used ceramic mold has the typical characteristics of a semifinite solid: the only heat transfer inside of the mold takes place in the direction x. The temperature in the face of the mold that is in contact with the environment is always 25 °C (298 K), that is to say, the ambient acts as an infinite heat sink.
(b) The resistance to the heat transfer between the melt and the ceramic mold is considered negligible. The interfacial resistance is very small as the metallic melt has great thermal conductivity, λ, with a small thickness of the thermal boundary layer, δ_T, see Sect. 2.4.2.2.
(c) The thickness of solidified metal is considered sufficiently small to think that it does not represent an appreciable resistance to the heat transfer from the melt to the mold.
(d) We consider that the temperature of the melt is corresponding to that of the melting point: there is not overheating. If the melt reaches a temperature $T > T_M$, the heat that must be removed to complete the solidification is calculated as follows:

$$\Delta H_f = \Delta H_{fus} + \bar{c}_{p/l} \cdot (T - T_M) \tag{4.42}$$

End of the exercise.

Exercise 4.3 Develop the equation that relates the time required to reach the solidification of a melt with semi-thickness, e, of the flat, smooth part when, apart from the flat, smooth mold wall, we consider that there is an interfacial resistance to the heat transfer, h_g. That is to say:

$$R_{tc} = \frac{1}{A \cdot h_g} + \frac{\sqrt{\pi \cdot t}}{A \cdot \sqrt{\lambda \cdot c_p \cdot \rho}} \tag{4.43}$$

where R_{tc} is the value of the total interfacial resistance to the heat transfer.

Solution starts:

We have developed in Exercise 4.2 the equation that relates the solidified thickness, e, and the time, t, considering that the only resistance to the heat transfer was in the mold. The rate of heat transfer, ϕ, is:

$$\phi = \frac{A}{\sqrt{\pi}} \cdot (T_M - T_0) \cdot \sqrt{\lambda \cdot c_p \cdot \rho} \cdot \left(t^{-0.5}\right) = \frac{T_M - T_0}{R_{tc}} \tag{4.44}$$

where the resistance to the heat transfer is identified as follows, in this case:

$$R_{tc} = \frac{\sqrt{\pi \cdot t}}{A \cdot \sqrt{\lambda \cdot c_p \cdot \rho}} \tag{4.45}$$

The enunciation of this exercise incorporates a new interfacial resistance between the melt and the solidified product represented by h_g (global coefficient of heat transfer). The influence of h_g is greater when the solidified thickness or when corrosion processes derived from the contact between the melt and the mold could take place. The heat that the system must dissipate through the interfacial and mold resistances is equal to:

$$\Delta H_f = \Delta H_{fus} + \bar{c}_{p/l} \cdot (T - T_M) \tag{4.46}$$

where we assume that the melt is not at the melting temperature (as in the case of Exercise 4.2), but there is a certain overheating, T. In this case, $\bar{c}_{p/l}$ is the average specific heat of the melt.

The differential equation that results from the identification of the heat dissipation rate, consequence of the cooling and solidification of the melt, with the rate that is verified through the ensemble formed by solid-melt interface and mold, is the following one:

$$\frac{de}{dt} = \frac{T_M - T_0}{\Delta H_f \cdot \rho_{melt}} \cdot \left(\frac{1}{h_g} + \frac{\sqrt{\pi \cdot t}}{\sqrt{\lambda \cdot c_p \cdot \rho}}\right)^{-1} \tag{4.47}$$

It is necessary to consider the following change of variable for the integration of this differential equation:

$$x = \frac{h_g \cdot \sqrt{\pi \cdot t}}{\sqrt{\lambda \cdot c_p \cdot \rho}} \tag{4.48}$$

Thus, the relationship between dt and dx is equal to:

$$\mathrm{d}t = \frac{2 \cdot \lambda \cdot c_p \cdot \rho}{h_g^2 \cdot \pi} \cdot x \cdot \mathrm{d}x \tag{4.49}$$

and the value of de is equal to:

$$\mathrm{d}e = A \cdot \left(\frac{2 \cdot \lambda \cdot c_p \cdot \rho}{h_g \cdot \pi}\right) \cdot \left(\frac{x \cdot \mathrm{d}x}{1 + x}\right) \tag{4.50}$$

where:

$$A = \frac{T_M - T_0}{\Delta H_f \cdot \rho_{melt}} \tag{4.51}$$

If we know that the value of the integral is:

$$\int \frac{x \cdot \mathrm{d}x}{c \cdot x + \mathrm{d}} = \frac{x}{c} - \frac{d}{c^2} \cdot \ln(1 + x) \tag{4.52}$$

and for $t = 0$, $e = 0$, where $c = d = 1$, and the value of the thickness, e, is:

$$e = A \cdot \frac{2 \cdot \lambda \cdot c_p \cdot \rho}{h_g \cdot \pi} \cdot [x - \ln(1 + x)] \tag{4.53}$$

If we undo the change of variables, and regrouping terms we obtain the following equation:

$$e = \left(\frac{2}{\sqrt{\pi}} \cdot \frac{T_M - T_0}{\Delta H_f \cdot \rho_{melt}} \cdot \sqrt{\lambda \cdot c_p \cdot \rho \cdot t^{0.5}}\right) \cdot f_{int} \tag{4.54}$$

If we compare the result with that obtained in Exercise 4.2, it is possible to conclude that if we take into account a resistance in the solid-melt interface, the solidified thickness is affected by a dimensionless coefficient, f_{int}, that is equal to:

$$f_{int} = \left[1 - \frac{\sqrt{\lambda \cdot c_p \cdot \rho}}{h_g \cdot \sqrt{\pi \cdot t}} \cdot \ln\left(1 + \frac{h_g \cdot \sqrt{\pi \cdot t}}{\sqrt{\lambda \cdot c_p \cdot \rho}}\right)\right] \tag{4.55}$$

Commentaries: In the solidification of parts in molds with different geometrical characteristics, it is usually interesting to identify the thickness, e, with the relation area, A, to volume, V, of the mold/melt system:

$$e = \frac{V}{A} \tag{4.56}$$

This relation, as we have seen, can be related to the solidification rate by means of a constant, C:

$$\frac{V}{A} \simeq C \cdot t^{0.5} \tag{4.57}$$

where the value of C is calculated as follows:

$$C = \frac{2}{\sqrt{\pi}} \cdot \frac{T_M - T_0}{\Delta H_f \cdot \rho_{melt}} \cdot \sqrt{\lambda \cdot c_p \cdot \rho} \cdot f_{int} \tag{4.58}$$

that depends on the thermal characteristics of both the melt and the mold, as well as on the geometrical characteristics. Parts of different sizes solidified in molds of the same thermal characteristics have the same value of C when their geometry exhibits a certain similarity (the relation between the volume of melt, its corresponding area of heat transfer, and the time required for the solidification of the melt is known with the name of Chvorinov law, Pero-Sanz et al. (2017, Chap. 2, pp. 63–65).

The following dimensionless variables are defined to try to quantify the effects of the different geometries of solidification (layers of defined thickness and infinite length, cylinders of infinite length or spheres):

$$\beta = \left(\frac{V}{A}\right) \cdot \frac{1}{\sqrt{\alpha \cdot t}} \tag{4.59}$$

$$\gamma = f_{int} \cdot \frac{T_M - T_0}{\Delta H_f \cdot \rho_{melt}} \cdot c_p \cdot \rho \tag{4.60}$$

Thus, the time that three basic formats (flat, cylindrical, and spherical) previously mentioned will require to solidify is given by:

$$\beta = \gamma \cdot \left(\frac{2}{\sqrt{2}} + \frac{1}{b \cdot \beta}\right) \tag{4.61}$$

where b is equal to infinite, ∞, for sheets of infinite length; $b = 4$ for cylindrical molds and, finally, $b = 3$ in the case of spheres.

For instance, to solidify a sheet of iron melt of infinite length and 2.5 cm in thickness (use the same thermal data of Exercise 4.2 for the metal and ceramic mold), the time and the solidification rate, as a function of h_g, are given in Table 4.3.

Table 4.3 h_g, solidification time and solidification rate

$h_g(\mathrm{W \cdot m^{-2} \cdot K^{-1}})$	Solidification time (s)	Solidification rate $(\mathrm{cm \cdot s^{-1}})$
30	1106	1.131×10^{-3}
60	742	1.686×10^{-3}
80	644	1.941×10^{-3}
130	527	2.373×10^{-3}
600	366	3.418×10^{-3}

We proceed as follows to obtain the solidification rate of a melt in the form of sheet with infinite length and 2.5 cm in thickness for different values of interfacial resistances, h_g, previously indicated:

1. Starting from a value of the time required to reach the solidification of the melt, we calculate the corresponding solidification rate using the Eq. (4.47).
2. With the value of the solidification rate previously calculated, we multiply this rate by the time that initially was considered until obtaining a value of solidified thickness of 1.25 cm.

Finally, it is convenient to remember that the dendritic growing rates associated with the solidification of the pure metal, see Pero-Sanz (2006, pp. 96–100), are around 2 cm per second (Pero-Sanz et al. 2017, 2018, 2019).

End of the exercise.

Exercise 4.4 When we work with molds of finite dimensions, it is possible that part of the solidification heat could not be transferred through the mold because of being this saturated of heat. In this case, the time required for the solidification of the melt is greater and the restrictions and equations used in Exercises 4.2 and 4.3 are different. Describe the dimensionless variables that are used to calculate the solidification times in the melt under these conditions.

Solution starts:

The dimensionless numbers that appear in these situations are the following:

- The geometrical characteristic:
  - for the part:

$$\left(\frac{\mathrm{V}}{\mathrm{A}}\right)_{\mathrm{part}} \tag{4.62}$$

  - for the mold:

$$\left(\frac{1}{\mathrm{L}}\right)_{\mathrm{mold}} \tag{4.63}$$

- Biot's number:

$$\mathrm{Bi} = \frac{h_g \cdot \mathrm{L}}{\lambda} \tag{4.64}$$

- Fourier's number:

$$\mathrm{Fo} = \frac{\alpha \cdot \mathrm{t}}{\mathrm{L}^2} \tag{4.65}$$

where:

$$\alpha = \frac{\lambda}{\mathrm{c_p} \cdot \rho} \tag{4.66}$$

where V and A make reference to the volume and cross section of the part-material that solidifies; L is the thickness of the mold used in the experiments; h_g is the coefficient of global resistance to the heat transfer between melt and mold; α and λ are, respectively, the thermal diffusivity and thermal conductivity of the mold; and, t, is the time required for the solidification of the metal.

In these cases, the value of γ^* (to distinguish it from the γ of Exercise 4.3) is identified with the following equation:

$$\gamma^* = \left(\frac{\mathrm{T_{fus}} - \mathrm{T_0}}{\Delta\,\mathrm{H_f} \cdot \rho_{\mathrm{melt}}}\right)_{\mathrm{melt}} \cdot \left(\rho \cdot \mathrm{c_p}\right)_{\mathrm{mold}} \tag{4.67}$$

The relation between the dimensionless numbers previously defined facilitates the calculation of the solidification times:

$$\frac{\mathrm{V}}{\mathrm{A} \cdot \mathrm{L}} = \gamma^* \cdot \left(\frac{\mathrm{Bi}}{1 + \mathrm{Bi}}\right) \cdot \mathrm{Fo} + \mathrm{C} \rightarrow \mathrm{Fo} = \frac{\frac{\mathrm{V}}{\mathrm{A \cdot L}} - \mathrm{C}}{\gamma^* \cdot \left(\frac{\mathrm{Bi}}{1+\mathrm{Bi}}\right)} \tag{4.68}$$

The value of the constant C for short solidification times (parts of small size) and using a preheated mold is 0.25, while in the case of long solidification times (parts of great dimensions) its value might be considered zero.

For instance, we will see the case of solidifying a cylindrical part of 250 kg of iron, Exercise 4.1, in a ceramic zirconium silicate mold of 10 cm in thickness, data of the Exercise 4.1, with a liquidus temperature of 2023 K (1750 °C) and a room temperature of 293 K (20 °C). In this case:

$$\begin{aligned} H_f &= H_{fus} + \bar{c}_{p/l} \cdot (T - T_{fus}) = 13807 \frac{J}{mol\ Fe} \cdot \frac{1\ mol\ Fe}{55.85\ g\ Fe} \\ &\cdot \frac{1000\ g\ Fe}{1\ kg\ Fe} + 46.02 \frac{J}{mol \cdot K} \cdot \frac{1\ mol\ Fe}{55.85\ g\ Fe} \cdot \frac{1000\ g\ Fe}{1\ kg\ Fe} \\ &\cdot (2023 - 1811)K = 421826 \frac{J}{kg\ Fe} \end{aligned} \tag{4.69}$$

The value of γ^* is:

$$\begin{aligned} \gamma^* &= \left(\frac{T_{fus} - T_0}{\Delta H_f \cdot \rho_{melt}}\right)_{melt} \cdot (\rho \cdot c_p)_{mold} \\ &= \left[\frac{(2023 - 293)\ K}{421826 \frac{J}{kg\ Fe} \cdot 6746 \frac{kg}{m^3}}\right]_{melt} \cdot \left(2723 \frac{kg}{m^3} \cdot 837 \frac{J}{kg \cdot K}\right)_{mold} \\ &= 1.3856 \end{aligned} \tag{4.70}$$

The relation volume to area of the part to be obtained, which has cylindrical geometry, is:

$$\left(\frac{V}{A}\right)_{part} = \frac{\pi \cdot r^2 \cdot h}{\pi \cdot r^2} = h = 0.295\,m \tag{4.71}$$

With these values, as a function of interfacial resistance, h_g, and thickness of the mold, L, for a solidified thickness of 17.37 cm: mean value of the diameter (in meters) and height of the part (expressed in meters). The calculus of the solidified thickness, for the estimation of the rate of advance of the solidification front, is identified with the summation of the semi-length of the diameter and the semi-height (of the height) of the cylindrical mold divided by two), data in Exercises 4.1 and 4.2:

$$\text{Solidified thickness} = \left(\frac{\frac{0.40\,m}{2} + \frac{0.295\,m}{2}}{2}\right) \frac{100\,cm}{1\,m} = 17.38\,cm \tag{4.72}$$

The time and solidification rate are indicated in Table 4.4 (properties of the mold: $\lambda = 1.04\ W \cdot m^{-1} \cdot K^{-1}$; $c_p = 837\ J \cdot kg^{-1} \cdot K^{-1}$; $\rho = 2723\ kg \cdot m^{-3}$; $\alpha = 4.5631 \times 10^{-7}\ m^2 \cdot s^{-1}$), for different thickness of mold, L, 0.10, 0.08, and 0.05 m and different values of h_g, 130 and 600 $W \cdot m^{-2} \cdot K^{-1}$, C = 0 (part of great dimensions). Biot's number is calculated from Eq. (4.64). Fourier's number is calculated using Eq. (4.68). Solidification times are calculated by solving for t the Eq. (4.65). Finally, the solidification rate is calculated by dividing the solidified thickness (Eq. 4.72) by the solidification time.

Commentaries: We have used the thermal properties of the melt of two different bibliographic sources in the previous exercises, Exercises 4.1, 4.2, and 4.3: Fine and Geiger (1993, p. 431) and the database of the HSC-5.1. The most noticeable

Table 4.4 Solidification times and rates

$h_g(\mathrm{W \cdot m^{-2} \cdot K^{-1}})$	Thickness of the mold, L (m)	Biot's number	Fourier's number	Solidification time (s)	Solidification rate $(\mathrm{cm \cdot s^{-1}})$
130	0.10	12.5	2.2990	50,382	3.444×10^{-4}
130	0.08	10.0	2.9270	41,052	4.226×10^{-4}
130	0.05	6.25	4.9386	27,057	6.412×10^{-4}
600	0.10	57.69	2.1656	47,455	3.656×10^{-4}
600	0.08	46.15	2.7185	38,129	4.550×10^{-4}
600	0.05	28.85	4.4050	24,134	7.189×10^{-4}

differences are observed in the enthalpy of fusion of the iron. We use in this example the data of enthalpy of fusion of the HSC5.1 database: $247{,}216\ \mathrm{J \cdot kg^{-1}}$.

End of the exercise.

Exercise 4.5 Calculate the time required for the solidification of one kilogram of liquid iron, with an overheating of 40 °C, in a ceramic mold of zirconium silicate (zirconium sand) of 10 cm in thickness. The part is a cylinder of 5.0 cm in diameter. The thermal properties of the iron melt can be taken from Exercise 4.1, while those of the zirconium silicate mold can be found in Exercise 4.2.

Finally, assume that the heat transfer coefficient of the melt toward the mold is $130\ \mathrm{W \cdot m^{-2} \cdot K^{-1}}$ and the room temperature is 20 °C.

Solution starts:

As it is a solidification process in a cylindric mold of 5.0 cm in diameter and 0.0755 m in height, the height is estimated by the volume occupied by one kilogram of liquid iron with a density of $6764\ \mathrm{kg \cdot m^{-3}}$:

$$\frac{1\ \mathrm{kg}}{\pi \cdot r^2 \cdot h} = \rho \rightarrow h = \frac{1\ \mathrm{kg}}{\pi \cdot r^2 \cdot \rho}$$
$$= \frac{1\ \mathrm{kg}}{\pi \times (0.025\ \mathrm{m})^2 \times 6764\ \mathrm{kg \cdot m^3}} = 0.0753\ \mathrm{m} \tag{4.73}$$

In this case, the solidified thickness is:

$$\text{Solidified thickness} = \left(\frac{0.050\ \mathrm{m} + 0.753\ \mathrm{m}}{4}\right) \times \frac{100\ \mathrm{cm}}{1\ \mathrm{m}} = 3.14\ \mathrm{cm} \tag{4.74}$$

We are going to calculate the resistances to the removal of heat from the melt during its solidification. Consequently, considering the dimensionless numbers and the correlations of Exercise 4.4, we obtain:

(a) The dimensionless number related with the geometries of the part and the mold is:

$$\left(\frac{V}{A}\right)_{\text{part}} \cdot \left(\frac{1}{L}\right)_{\text{mold}} = \frac{V}{A \cdot L} = \frac{h}{L} = \frac{0.0753\,\text{m}}{0.10\,\text{m}} = 0.753 \tag{4.75}$$

(b) The quantity of heat that is necessary to dissipate to solidify the iron melt (data of Exercise 4.1) is:

$$\begin{aligned} H_f &= H_{\text{fus}} + \bar{c}_{p/l} \cdot (T - T_M) \\ &= 13807 \frac{J}{\text{mol Fe}} \cdot \frac{1\text{ mol Fe}}{55.85\text{ g Fe}} \cdot \frac{1000\text{ g Fe}}{1\text{ kg Fe}} \\ &\quad + 46.02 \frac{J}{\text{mol} \cdot K} \cdot \frac{1\text{ mol Fe}}{55.85\text{ g Fe}} \cdot \frac{1000\text{ g Fe}}{1\text{ kg Fe}} \\ &\quad \cdot [(1538 + 40 + 273) - (1538 + 273)]K \\ &= 280175 \frac{J}{\text{kg Fe}} \end{aligned} \tag{4.76}$$

(c) We also obtain the dimensionless number γ^*

$$\begin{aligned} \gamma^* &= \left(\frac{T_{M,\text{ overheating}} - T_0}{\Delta H_f \cdot \rho_{\text{melt}}}\right)_{\text{melt}} \cdot (\rho \cdot c_p)_{\text{mold}} \\ &= \left[\frac{(1851 - 293)K}{280175 \frac{J}{\text{kg Fe}} \cdot 6746 \frac{\text{kg}}{m^3}}\right]_{\text{melt}} \cdot \left(2723 \frac{\text{kg}}{m^3} \cdot 837 \frac{J}{\text{kg} \cdot K}\right)_{\text{mold}} \\ &= 1.8787 \end{aligned} \tag{4.77}$$

(d) Finally, the value of the Biot's number is:

$$\text{Bi} = \frac{h_g \cdot L}{\lambda} = \frac{130\text{W} \cdot m^{-2} \cdot K^{-1} \cdot 0.10\,\text{m}}{1.04\text{W} \cdot m^{-1} \cdot K^{-1}} = 12.5 \tag{4.78}$$

Consequently, the value of the Fourier's number, Fo, when the value of the constant C of Exercise 4.4 is equal to 0.25 (parts of small dimensions) is:

$$\text{Fo} = \frac{\frac{V}{A \cdot L} - C}{\gamma^* \cdot \left(\frac{\text{Bi}}{1+\text{Bi}}\right)} \tag{4.79}$$

With this value of Fo, the solidification time is:

$$\text{Fo} = \frac{\alpha \cdot t}{L^2} \rightarrow t = \frac{\text{Fo} \cdot \alpha}{L^2} \tag{4.80}$$

Table 4.5 Solidification times and rates in Exercise 4.5

h_g ($W \cdot m^{-2} \cdot K^{-1}$)	Thickness of the mold, L (m)	Biot's number	Fourier's number	Solidification time (s)	Solidification rate ($cm \cdot s^{-1}$)
130	0.10	12.5	0.2903	6361	4.932×10^{-4}
130	0.08	10.0	0.4062	5697	5.507×10^{-4}
130	0.05	6.25	0.7779	4262	7.361×10^{-4}
600	0.10	57.69	0.2734	5992	5.236×10^{-4}
600	0.08	46.15	0.3772	5291	5.930×10^{-4}
600	0.05	28.85	0.6939	3802	8.253×10^{-4}

and the speed of the solidification front is:

$$\text{Solidification rate} = \frac{3.14\,\text{cm}}{\text{solidified time}} \tag{4.81}$$

when the thickness of solidified material is 3.14 cm (sum of the semi-length of the diameter and the semi-length of the height divided by two).

The time and the solidification rate are indicated in Table 4.5 (properties of the mold: $\lambda = 1.04\ W \cdot m^{-1} \cdot K^{-1}$; $c_p = 837\ J \cdot kg^{-1} \cdot K^{-1}$; $\rho = 2723\ kg \cdot m^{-3}$; $\alpha = 4.5631 \times 10^{-7}\ m^2 \cdot s^{-1}$), for different thickness of mold, L, 0.10, 0.08, and 0.05 m and different values of h_g, 130 and 600 $W \cdot m^{-2} \cdot K^{-1}$, C = 0.25 (part of a small size).

Commentaries: the difference between the densities of the melt and the solidified product determines that the kilograms of available liquid are always greater than the mass of the finished product. The metallurgical yield and economic efficiency of the molding are determined by the suitable design of the system used to fill the mold, Fig. 4.3. Between 70 and 80% of the available quantity of metal is used to manufacture the designed part.

End of the exercise.

4.3 Continuous Casting

4.3.1 Introduction

Continuous casting is a process of forming metals (Fig. 4.1) where the melt is poured into a bottomless mold whose cross section has the geometrical shape of the semi-finished product that we want to manufacture (Fig. 4.8).

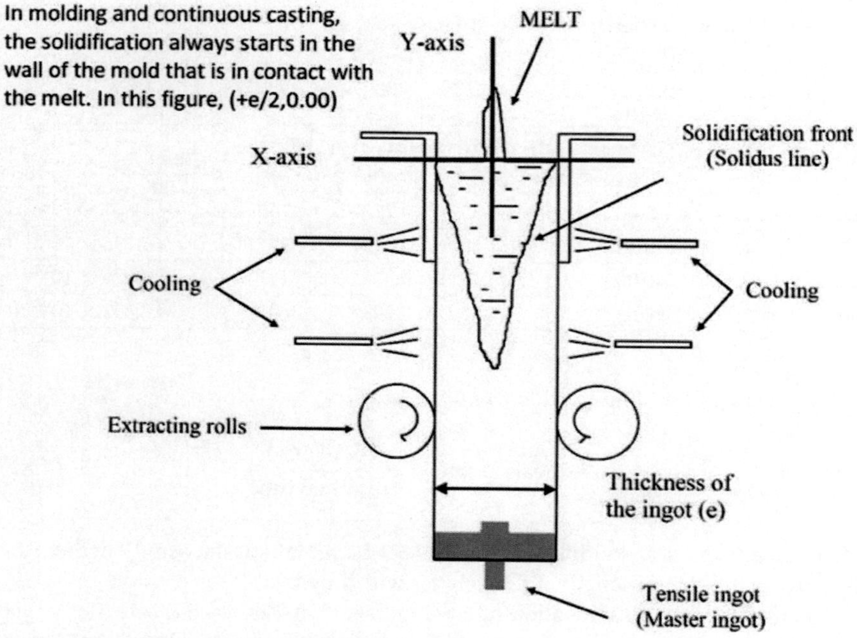

Fig. 4.8 Detail of a mold for a continuous casting machine

In the iron and steelmaking industry, the most common formats of continuous casting are:

- Rectangular long products of 150–300 mm in thickness (slabs).
- Rectangular flat products of medium thickness: 80–100 mm (medium-sized slabs, Sumitomo or Control technologies).
- Rectangular flat products of thin thickness: 40–60 mm (thin slabs, Siemag CSP, Mannesmann Demag ISP, or Danieli FTSR technologies).
- Squared or cylindrical long products (blooms).
- Long products with double T cross section (beam blank).
- Billet.

The continuous casting has competed for many years with the rimmed and semi-killed qualities (Fig. 3.1), where it demonstrated that it is possible to obtain a material with equal or better mechanical properties. On the other hand, the development of continuous casting has taken place in parallel to the improvements in secondary and tertiary metallurgies of the steel. Once achieved the metallurgical specifications required for the semi-finished product, it is possible to set out a compared study of the two forming processes by casting the liquid steel, the continuous process, and the process in the ingot mold (Fig. 4.1). The best qualities of the continuous process are summarized in the following points.

- The continuous casting has allowed eliminating the rolling-forming processes from the ingots to an intermediate rough product (roughing machine) from which the slabs (slabbing train), the blooms (blooming train), and the billets (billet mill) were obtained by hot forming, as it is shown in Fig. 4.9.
- There was a reduction of the workforce as a consequence of the automation of operations with respect to the ingot casting.
- There was an improvement in the metallurgical yield as the crops (head and end) were avoided in each one of the hot rough rolling processes of the ingot.
- There was a remarkable reduction in the processing times required to obtain the semi-finished product.
- Expensive operations were eliminated (especially the ingot molds).
- There was a significant reduction in the consumption of primary energy (mixed gas and fuel) in reheating furnaces before the rolling.

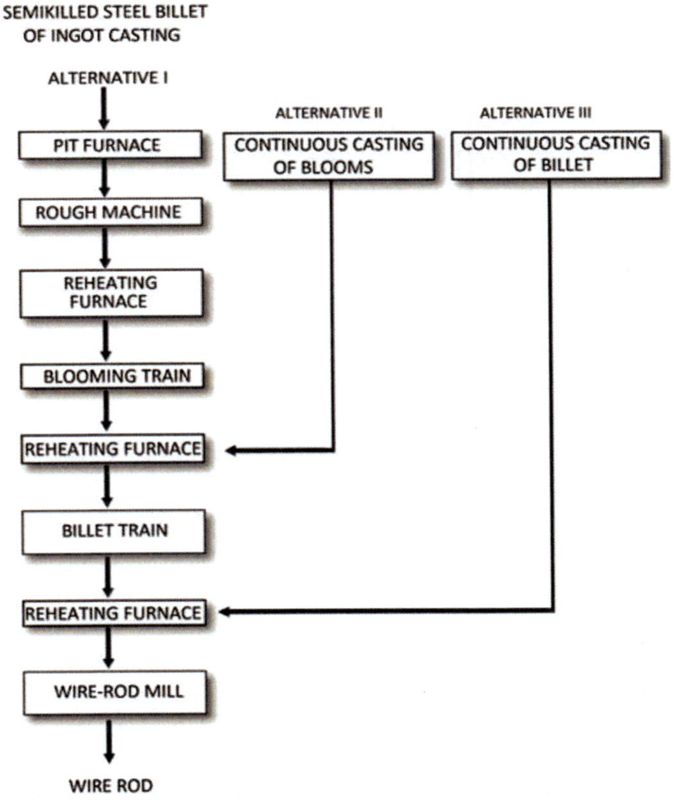

Fig. 4.9 Alternatives to produce wire rod: ingot casting, continuous casting of blooms, and continuous casting of billet

- It was also achieved an improvement in the yield of other equipment and facilities of the iron and steelmaking industry, as the ladle.
- Gradually, it was produced a significant reduction of the specific consumption of refractory in the ladles.
- The surface finishing that provides the continuous casting has noticeably reduced the costs of recovering by scarfing of the steel ingots.

Nowadays, it is possible to state that the only commercial alternatives to the continuous casting for the forming of structural steel products are the molding, the powder metallurgy, and the hot forging. However, during the R&D stages of tune-up, development or introduction in the market of novel and at the vanguard steel qualities, as the TWIP (Twinning Induced Plasticity) steels (Pero-Sanz 2004), with small productions, the alternative of the ingot casting is still a suitable alternative for the solidification of those preforms that will be hot forged (plastic deformation of the steel) to obtain the final geometry for the finishing.

4.3.2 *The Continuous Casting Process*

Continuous casting machines have developed in their construction from the first machines completely vertical (Fig. 4.10) to those of curved geometry. Thus, it is not possible to reduce more the height of the installations as the horizontal continuous casting was recently developed.

Horizontal continuous casting machines successfully operate from 1966 when Dawy-Loewy Ltd. installed the first machine of this type in Great Britain. The main advantages are:

- Less height of the installations, facilitates the design engineering of the steelworks, and the handling of the ladle.
- Reduction of the ferrostatic pressure on the product.

Independently of the height that the continuous casting installations could reach, from the level of maximum height (where the ladle is located) to the lower zone (where the product is discharged) there are the following elements in common:

- A planetary platform where the ladle coming from either the BOF, the electric furnace or the secondary metallurgy installations locates.
- An element distributor and regulator of the liquid (tundish) between the different lines of the machine.
- Equally, it is important the function of the tundish in the pick up and flotation of the inclusions (tertiary metallurgy of the steel).
- The ingot mold or copper (or copper-nickel alloys) mold which, with the shape of the final product, allows solidifying the first proportions of liquid (casting skin). The molds can be vertical (Fig. 4.11) or curved (Fig. 4.12).

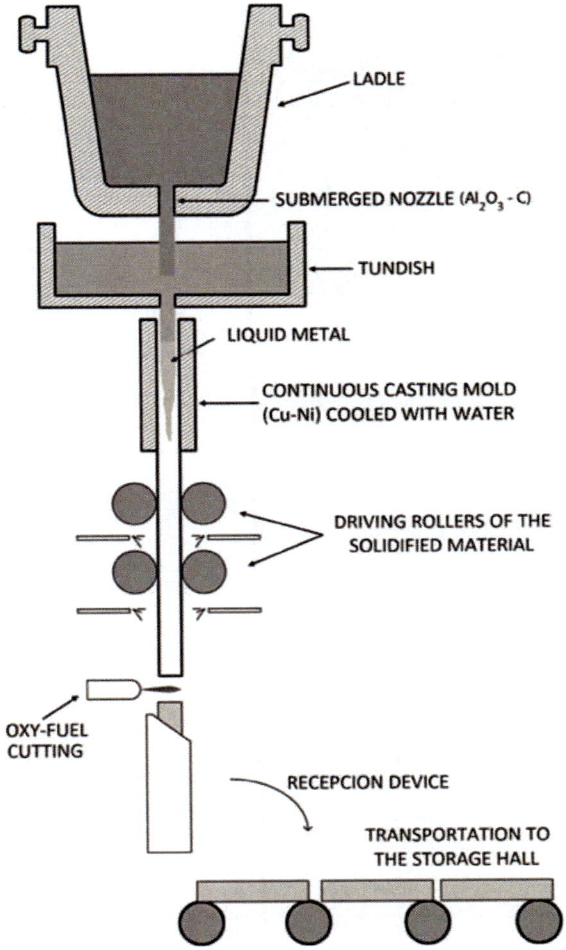

Fig. 4.10 Vertical continuous casting

As happens in the tundish, it is important to emphasize (in the mold) the nature and role that synthetic slags (casting powders) play, which protects the metal from the contact with the gases (O_2, N_2, H_2) and catch the inclusions from the liquid.

- The oscillator, which allows the ingot mold to move alternatively up (negative vibrations) or down (positive vibrations).
  During the period of negative vibrations, the metal can detach from the wall of the mold. The period of positive vibration is in resonance with the tensile impulses of the chain that pulls the master ingot and regulates the casting rate of the machine (Fig. 4.8).
- Secondary cooling system that allows the heat release through the steel-water interphase to complete the total solidification of the product.

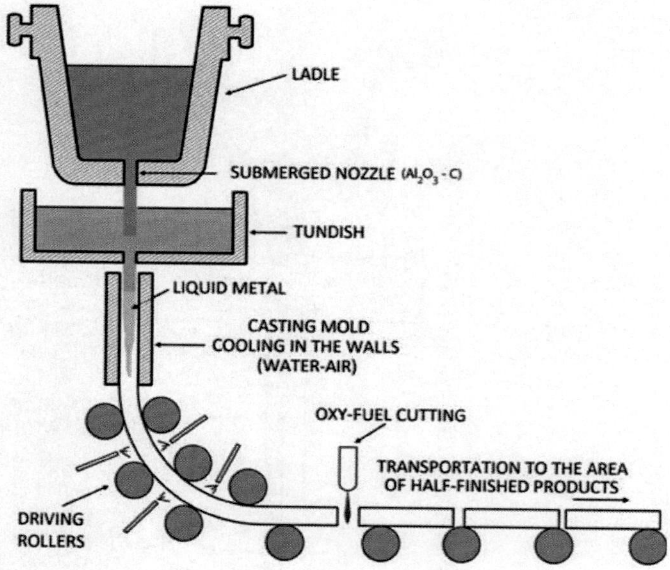

Fig. 4.11 Curve continuous casting with vertical mold

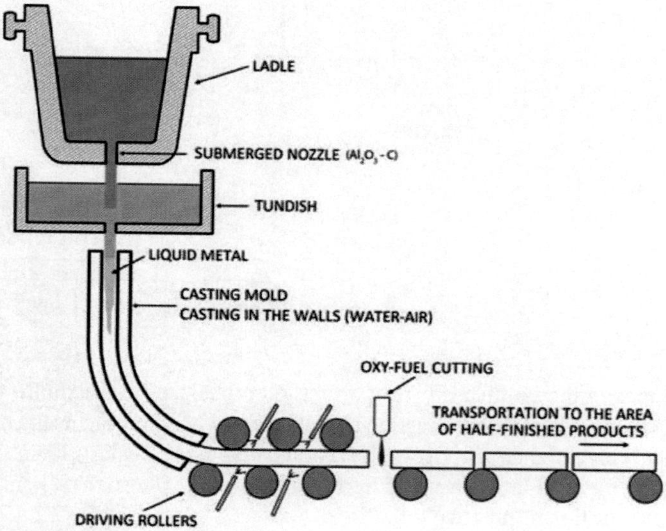

Fig. 4.12 Continuous casting with curved mold

Equally, continuous casting machines have a system of magnetic stirring that, apart from guaranteeing the uniformity in the composition of the liquid and influencing the heat transfer rate of the liquid to the walls, drive the metal to occupy inter-dendritic spaces to minimize the problems derived from both the micro-shrinkage and the macro-segregation.

– Cutting mechanisms that produce sections of metal with the desired length.

4.3.2.1 Solved Exercises

Exercise 4.6 Continuous casting to produce ferritic slabs with 0.10% C (Figs. 4.1 and 4.2) of $250 \times 1600\ \text{mm}^2$ (cross section) is performed in a machine (Fig. 4.11) whose mold has a height of 1 m. If the rate of the process (rate of the ingot removal from the mold) is $2.0\ \text{m} \cdot \text{min}^{-1}$ and the overheating of the melt is 50 °C. Determine:

(a) The liquidus and solidus temperatures of the melt with 0.10% C.
(b) The position of the solidus temperature along the height of the mold.
(c) The time required for the complete solidification of the melt.

Considering a 2D model for the solidification of the melt with infinite sump (sink) of heat removal, the temperature of the walls of the ingot that is in contact with the mold, in the primary cooling is 200 and 280 °C in the secondary cooling, when the walls of the slag contact directly with the cooling medium. The values of the thermal diffusivity, α, for ferritic steel with primary or secondary refrigeration are, respectively: 9.07×10^{-6} and $5.60 \times 10^{-6}\ \text{m}^2 \cdot \text{s}^{-1}$. The density of the slab is $7800\ \text{kg} \cdot \text{m}^{-3}$.

Solution starts:

The liquidus temperature, T_L, of a melt with 0.10% C can be estimated by means of the data of temperature and composition given by the metastable Fe–C diagram, see Pero-Sanz (2006, p. 267), Pero-Sanz et al. (2017, 2018):

$$T_L = (1538 - 1495)\,°\text{C} \cdot \frac{0.43}{0.53} + 1495\,°\text{C} = 1530\,°\text{C} \tag{4.82}$$

On the other hand, the solidus temperature, T_S, of the solid corresponds to the peritectic reaction of the Fe–C metastable diagram, 1495 °C. However, the presence of alloying elements, when interacting with the carbon, slightly modifies the values previously estimated. There is specialized software to calculate both temperatures as the Thermo-Calc or the FactSage for metallic or oxidized multicomponent systems.

The generalized 2D equation for sources and infinite sumps (sinks) of heat is:

$$T(x, t) = A + B \cdot \text{erf}(z) \tag{4.83}$$

where:

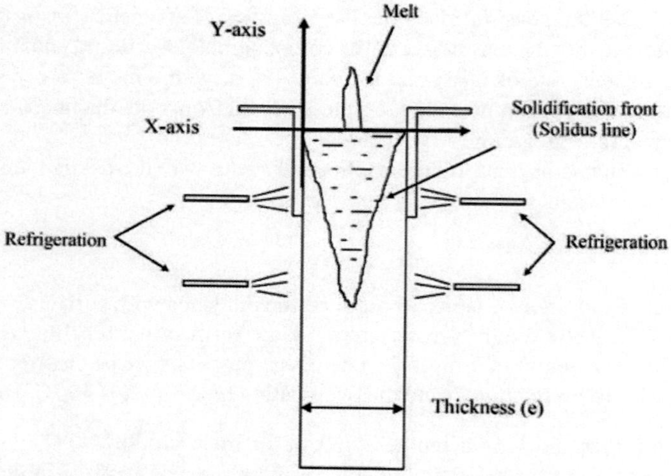

Fig. 4.13 Molding, definition of the axis, and origin of coordinates. We point out again that in molding and continuous casting the solidification always begins in the wall of the mold that is in contact with the melt

$$z = \frac{x}{\sqrt{4 \cdot \alpha \cdot t}} \tag{4.84}$$

Putting the origin of coordinates in the upper left corner of the ingot that is in contact with the mold (Fig. 4.13), anyway, we point out again that in molding and continuous casting the solidification always begins in the wall of the mold that is in contact with the melt. The calculation of the constants A and B to reach the value of the temperature of the solidification front, as a function of the position and the temperature, is as follows:

- Along the position $x = 0$, the origin of coordinates, the ingot wall enters in contact either with the cooled mold (primary cooling) or directly with the coolant (secondary cooling), and, for any time, its temperature is assumed to be constant. In this case, the value of the error function is zero and consequently, the value of A is 200 °C during the process of primary cooling when the slab is inside of the mold, or 280 °C during the time when the secondary cooling is operating (see error function values in Appendix 7).
- In the center of the casting mold of 250 mm in thickness ($x = 125$ mm) and for any time, the temperature of the melt is constant and equal to the sum of the liquidus temperature, T_L, and the corresponding overheating, ΔT, of the melt. When the variable z reaches the value of 2.114, the polynomic function which approximates erf(z) takes the value of 1.0000 (see error function values in Appendix 7). The value of the constant B is:

$$B = \frac{T_L + \Delta T - A}{1.0000} = \frac{1530\,^{\circ}C + 50\,^{\circ}C - 200\,^{\circ}C}{1.0000} = 1380\,^{\circ}C \quad (4.85)$$

For zones of primary cooling, the model of thermal distribution defined for this exercise would be given by the following equation:

$$T(x, t) = 200 + 1380 \cdot \text{erf}(z) \quad (4.86)$$

as a function of the time, t, the position, x, of the solidus temperature (T_S) in the slab, when it is inside of the mold, would be given by the following equation:

$$T_S = 200 + 1380 \cdot \text{erf}(z) \rightarrow \text{erf}(z) = \frac{1495 - 200}{1380}$$
$$= 0.9384 \rightarrow z = 1.3361 \quad (4.87)$$

$$z = \frac{x}{\sqrt{4 \cdot \alpha \cdot t}} \rightarrow x = z \cdot \sqrt{4 \cdot \alpha \cdot t}$$
$$= 1.3361 \times \sqrt{4 \times 9.07 \times 10^{-6}\ m^2 \cdot s^{-1} \cdot t} \quad (4.88)$$

We collect in Table 4.6 the thickness of the metal that has solidified, x, in millimeters or in percentage (from the origin of coordinates to the center, as the machine is symmetric, the same thickness solidifies from the opposite extreme of the machine, Fig. 4.13) as a function of the time (s) and the height from the origin of coordinates of the mold (m). The height of the mold is calculated as follows as a function of the speed of extraction, v_{ext}, for different times being the height of the mold, y, 1 m:

$$v_{ext} = \frac{y}{t} \rightarrow y = v_{ext} \cdot t \cdot \frac{1\ \text{min}}{60\ \text{s}} \quad (4.89)$$

while the solidified percentage is:

Table 4.6 Solidified thickness as a function of the height of the mold and the time

Time (s)	Height of the mold, y (m)	Solidified thickness, x (mm)	% Solidified
1	0.03	8.0	6.40
5	0.17	18.0	14.40
10	0.33	25.4	20.32
15	0.50	31.2	24.96
20	0.67	36.0	28.80
25	0.83	40.2	32.16
30	1.00	44.1	35.28

Table 4.7 Solidified thickness as a function of the length of the slab and the time

Time (s)	Length of the slab, y (m)	Solidified thickness, x (mm)	% Solidified
95	3.17	51.0	40.80
235	7.83	80.3	64.24
570	19.00	125.0	100.00

$$\%\text{solidified} = \frac{x}{125\,\text{mm}} \times 100 \tag{4.90}$$

where x is the position of the T_S, and the semi-thickness of the slab is 125 mm (250 mm/2).

As it is possible to appreciate, the ingot that leaves the mold of the casting machine still contains liquid phase without solidifying inside. The symmetry of the solidification system that was analyzed leads to that the solidified thickness, x, that would be necessary to reach the complete solidification of the slab is of 125 mm (250 mm is the total thickness of the slab).

When the ingot leaves the mold and enters in contact with the secondary refrigeration, the position of the solidus temperature, inside of the slab in the secondary refrigeration, would be determined by the following equations:

$$T_S = 280 + 1380 \cdot \text{erf}(z) \rightarrow \text{erf}(z) = \frac{1495 - 280}{1380} = 0.8804 \rightarrow z = 1.1057 \tag{4.91}$$

$$z = \frac{x}{\sqrt{4 \cdot \alpha \cdot \text{t}}} \rightarrow x = z \cdot \sqrt{4 \cdot \alpha \cdot \text{t}} = 1.1057 \times \sqrt{4 \times 5.61 \times 10^{-6}\text{m}^2 \cdot \text{s}^{-1} \cdot \text{t}} \tag{4.92}$$

We collect in Table 4.7 the thickness of the slab that has solidified, x, in millimeters or in percentage (from the origin of coordinates to the center, as the machine is symmetric, the same thickness solidifies from the opposite extreme of the machine, Fig. 4.13) as a function of the time (s) and the height from the origin of coordinates of the mold (m). The height of the mold is calculated as follows as a function of speed of extraction, v_{ext}, for different times being the height of the mold, y, 1 m:

$$v_{\text{ext}} = \frac{y}{t} \rightarrow y = v_{\text{ext}} \cdot t \cdot \frac{1\,\text{min}}{60\,\text{s}} \tag{4.93}$$

while the solidified percentage is:

$$\%\text{solidified} = \frac{x}{125\,\text{mm}} \times 100 \tag{4.94}$$

where x is the position of the T_S, and the semi-thickness of the slab is 125 mm (250 mm/2).

After 570 s from the beginning of the solidification process in the ingot mold head, it is possible to confirm that the solidification of the melt in the central zone of the slab has finished.

Commentaries: it is possible to estimate the hourly production, P, in tons per hour, of the casting line, for a density of the product of 7800 kg $\cdot$ m^{-3}, from the production of the installation in cubic meters of metal obtained per minute (slabs of 250 mm in thickness and 1600 mm in width):

$$\mathrm{P} = 0.250\,\mathrm{m} \times 1.6\,\mathrm{m} \times \frac{2.0\,\mathrm{m}}{\mathrm{min}} = 0.800\frac{\mathrm{m}^3}{\mathrm{min}} \tag{4.95}$$

$$\mathrm{P} = 0.800\frac{\mathrm{m}^3}{\mathrm{min}} \times \frac{1\,\mathrm{min}}{60\,\mathrm{s}} \times \frac{7800\,\mathrm{kg}}{\mathrm{m}^3} \times \frac{1\,\mathrm{t}}{1000\,\mathrm{kg}} = 374\,\mathrm{t}\cdot\mathrm{h}^{-1} \tag{4.96}$$

A possible improvement to approach to the real operation of the installation would be considering the heat sumps, although it is possible to consider that a constant temperature is reached (typical from an infinite sump of heat). This temperature might not be the same along the path of the material through the primary cooling (200 °C) or secondary cooling (280 °C), that is to say, it can change with the time according to a law of linear or parabolic behavior.

End of the exercise.

Exercise 4.7 The process of continuous casting for the obtaining of thin ferritic slabs of 50 × 1600 mm^2 (cross section) with 0.10% C is performed in a continuous casting machine whose mold has one meter in height (Exercise 4.6 , and Fig. 4.11). If the process rate (that corresponds to the rate of removal of the ingot from the mold) is 6.0 m $\cdot$ min^{-1} and the overheating of the melt is 50 °C. Determine:

(a) The time required for the complete solidification of the melt.
(b) The cooling rate at one millimeter from mold wall when the slab leaves the ingot mold.

Consider the same working hypothesis that in Exercise 4.6.

Solution starts:

The function that estimates the distribution of temperatures as a function of the coordinates "x" and "y" is the same that was used in Exercise 4.6: A = 200 °C and B = 1380 °C, for the zone of primary cooling. We also remember the liquidus (1530 °C) and solidus (1495 °C) temperatures of the melt of chemical composition equivalent to that of a ferritic steel, the overheating (50 °C) as well as the temperature of the infinite sump of heat, which is able to maintain in 200 °C the temperature in the wall of the ingot, being the origin of coordinates $x = 0$ and $y = 0$ indicated in the Fig. 4.13.

Equally, the values of A = 200 °C and B = 1380 °C, as in Exercise 4.6, lead to the following equation:

$$T(x, t) = 200 + 1380 \cdot \mathrm{erf}(z) \tag{4.97}$$

that allows calculating the position along the height (y) of the mold of the continuous casting when the solidus temperature (1495 °C) is reached:

$$\mathrm{erf}(z) = \frac{T_S - 200}{1380} = \frac{1495 - 200}{1380} = 0.9384 \tag{4.98}$$

On the other hand, as it was previously indicated, the value of the error function is 0.9384, which gives a value of z equal to 1.3357 (see Appendix 7).

That is to say, along the primary cooling of the continuous casting of thin slabs, the position of solidification front (solidus temperature of the melt) and the time (y coordinate, which is the function of the ingot extraction rate from the mold) are related as follows:

$$x = z \cdot \sqrt{4 \cdot \alpha \cdot t} = 1.3357 \times \sqrt{4 \times 9.07 \times 10^{-6} \mathrm{m}^2 \cdot \mathrm{s}^{-1} \cdot \mathrm{t}} \tag{4.99}$$

We collect in Table 4.8 the thickness of the slab that has solidified, x, in millimeters or in percentage (from the origin of coordinates to the center, as the machine is symmetric, the same thickness solidifies from the opposite extreme of the machine, Fig. 4.13) as a function of the time (s) and the height from the origin of coordinates of the mold (m). The height of the mold is calculated as follows as a function of speed of extraction, v_{ext} (6.0 m · min^{-1}), for different times being the height of the mold, y, 1 m:

$$v_{ext} = \frac{y}{t} \rightarrow y = v_{ext} \cdot t \cdot \frac{1\ \mathrm{min}}{60\ \mathrm{s}} \tag{4.100}$$

while the solidified percentage is:

Table 4.8 Solidified thickness as a function of the height of the mold and the time (Exercise 4.7)

Time (s)	Height of the mold, y (m)	Solidified thickness, x (mm)	% Solidified
1.0	0.10	8.0	32.0
3.0	0.30	13.9	55.6
5.0	0.50	18.0	72.0
7.0	0.70	21.3	85.2
8.0	0.80	22.7	90.8
9.0	0.90	24.1	96.4
9.7	0.97	25.0	100

$$\%\text{solidified} = \frac{x}{25\text{ mm}} \times 100 \tag{4.101}$$

where x is the position of the T_S, and the semi-thickness of the slab is 25 mm (50 mm/2).

In this case, the solidification of the slab takes place inside of the ingot mold where the primary cooling acts. When the slab leaves the mold all the melt has already solidified; the time required for the operation does not exceed 10 s.

One of the advantages of being possible to analytically express the variation of the temperature with the variables coordinate (x) and time (t) is to determine the derivative of the temperature with the position for a defined time. In this case, to know the derivative of this function:

$$\mathrm{T}(x, t) = 200 + 1380 \cdot \mathrm{erf}(z) \tag{4.102}$$

it is necessary to know the derivative of the error function:

$$\frac{\mathrm{derf}(z)}{\mathrm{d}t} = \left(\frac{\mathrm{derf}(z)}{\mathrm{d}z}\right) \cdot \left(\frac{\mathrm{d}z}{\mathrm{d}t}\right) = \exp(-z^2) \cdot \left(-\frac{x}{\sqrt{4 \cdot \pi \cdot \alpha \cdot t^3}}\right) \tag{4.103}$$

Consequently, the derivative of the temperature with respect to the position, x, is:

$$\left(\frac{\mathrm{d}T}{\mathrm{d}t}\right)_x = \frac{\mathrm{B}}{\sqrt{\pi}} \cdot \exp(-z^2) \cdot \left(\frac{x}{\sqrt{4 \cdot \alpha \cdot t^3}}\right) \tag{4.104}$$

$$\left(\frac{\mathrm{d}T}{\mathrm{d}t}\right)_{t=10\text{ s}} = \frac{1380}{\sqrt{\pi}} \cdot \exp\left[-\frac{(10^{-3}\text{ m})^2}{4 \times 9.07 \times 10^{-6}\text{ m}^2 \cdot \text{s}^{-1} \cdot 10\text{ s}}\right] \cdot \left[\frac{10^{-3}\text{ m}}{\sqrt{4 \times 9.07 \times 10^{-6}\text{m}^2 \cdot \text{s}^{-1} \cdot (10\text{ s})^3}}\right] = 4.08\ {}^\circ\text{C} \cdot \text{s}^{-1} \tag{4.105}$$

Commentaries: Thermal gradients, consequence of different cooling rates of the material along the zones close to the walls of the slab, can cause the appearance of cross or lengthwise cracks (Pero-Sanz et al. 2017, pp. 206–208). The gradients of temperature are greater in the zones close to the head of the ingot that is solidifying. In the position expressed in millimeters (1.0; 100), the cooling rate would be equal to 126 °C · s^{-1}. If we keep constant the abscissa (x), 1.0 mm, and increasing the value of the ordinate, y, the cooling rate is increasingly smaller. For instance, for the position (1.0; 1000), the value that is obtained is 4.08 °C · s^{-1}, and the possibility of finding cracks is smaller.

On the other hand, the decreasing of the thickness of the continuous casting slabs can be so significant that the direct solidification of the steel strip can take place, Direct Strip Casting (DSC). Table 4.9 collects a comparative analysis of the

Table 4.9 Comparative analysis of the continuous casting of conventional, thin, and DSC slabs

Continuous casting type	Thickness (mm)	Rate ($m \cdot min^{-1}$)	Heat flow ($MW \cdot m^{-2}$)	Solidification time (s)	Cooling rate ($°C \cdot s^{-1}$)
Traditional casting	250	2.0	0.50	570	15
Thin slabs	50	6.0	0.20	10	50
DSC	1.0	80.0	14.0	0.15	1700

continuous casting of conventional, thin, and DSC slabs. The value of the heat flow is calculated as follows:

$$J_q = \frac{2 \cdot B \cdot \lambda}{\sqrt{\pi}} \cdot \exp(-z^2) \cdot \left(\frac{1}{\sqrt{4 \cdot \alpha \cdot t}}\right) \quad (4.106)$$

Finally, it is necessary to take into account that both the cooling rate and the heat flow that are given in Table 4.9 represent average values associated to the zone of primary cooling of the slabs inside of the casting mold.

End of the exercise.

Exercise 4.8 The continuous casting to obtain pearlitic billet with 0.70% C and 150 mm × 150 mm is carried out in a continuous casting machine whose mold has a height of one meter. If the process rate (speed of removal of the ingot from the mold) is 3.0 $m \cdot min^{-1}$ (50.0 $mm \cdot s^{-1}$) and the overheating of the melt is 50 °C, calculate:

(a) The time required for the complete solidification of the melt.
(b) The heat flow, through the solidification front of the melt inside of the mold, for a position equivalent to the semi-height of the mold, y = 0.50 m.
(c) Analyze if it is possible to use a refrigeration different from either water or air.

Assume that it is possible to apply the same working hypothesis that in Exercise 4.6 to solve the questions of the enunciation. The thermal conductivity of the solidified product is estimated at 46 $W \cdot m^{-1} \cdot K^{-1}$.

Solution starts:

In the case of a pearlitic steel, the liquidus and solidus temperatures are noticeably different from those obtained in Exercises 4.6 and 4.7 for the ferritic product. Considering the data of the Fe–C metastable diagram presented in Pero-Sanz (2006, p. 267, 2017, 2018, 2019):

$$T_L = \left(\frac{4.30 - 0.70}{4.30}\right) \times (1538 - 1148) + 1148 = 1475\,°C \quad (4.107)$$

$$T_S = \left(\frac{2.11 - 0.70}{2.11 - 0.17}\right) \times (1495 - 1148) + 1148 = 1400\,°C \quad (4.108)$$

The equation that must be used to determine the position of the profile of solidification of the melt in the mold of continuous casting of 150 mm × 150 mm of section would be:

$$T(x,t) = A + B \cdot \operatorname{erf}(z) \tag{4.109}$$

where the value of the constant A would be the same that in Exercises 4.6 and 4.7: 200. On the contrary, the value of the constant B would be equal to:

$$B = \frac{T_L + \Delta T - A}{1.0000} = \frac{1475\,^\circ C + 50\,^\circ C - 200\,^\circ C}{1.0000} = 1324\,^\circ C \tag{4.110}$$

and Eq. (4.109) is transformed into:

$$T(x,t) = 200 + 1324 \cdot \operatorname{erf}(z) \tag{4.111}$$

The calculation of the solidified thickness in the mold of the primary cooling would be obtained by means of the following equations (see Appendix 7):

$$\operatorname{erf}(z) = \frac{T_S - 200}{1324} = \frac{1400 - 200}{1324} = 0.9063 \rightarrow z = 1.1915 \tag{4.112}$$

$$x = z \cdot \sqrt{4 \cdot \alpha \cdot t} = 1.1915 \times \sqrt{4 \times 9.07 \times 10^{-6} m^2 \cdot s^{-1} \cdot t} \tag{4.113}$$

We collect in Table 4.10 the evolution of the solidified thickness, x, in millimeters or percentage (from the origin of coordinates to the center, as the machine is symmetric, the same thickness solidifies from the opposite extreme of the machine, Fig. 4.13) as a function of the time (s) and the height from the origin of coordinates of the mold (m), inside of the mold (primary cooling). The height of the mold is calculated as follows as a function of speed of extraction, v_{ext} (3.0 m · min^{-1}), for different times being the height of the mold, y, 1 m (total height):

Table 4.10 Solidified thickness as a function of the height of the mold and the time (Exercise 4.8)

Time (s)	Height of the mold, y (m)	Solidified thickness, x (mm)	% Solidified
1	0.05	7.2	9.6
5	0.25	16.0	21.3
10	0.50	22.7	30.3
15	0.75	27.8	37.1
17	0.85	29.6	39.5
18	0.90	30.4	40.5
20	1.00	32.1	42.8

$$v_{ext} = \frac{y}{t} \rightarrow y = v_{ext} \cdot t \cdot \frac{1\ \text{min}}{60\ \text{s}} \quad (4.114)$$

while the solidified percentage is:

$$\%\text{solidified} = \frac{x}{75\ \text{mm}} \times 100 \quad (4.115)$$

where x is the position of the T_S, and the semi-thickness of the billet is 75 mm (150 mm/2).

The solidification of the billet takes place out of the mold of primary cooling of the continuous casting machine. In this case, as in Exercise 4.6, the equations that should be considered are:

$$T(x, t) = 280 + 1324 \cdot \text{erf}(z) \quad (4.116)$$

The calculation of the solidified thickness in the mold of the primary cooling would be obtained by means of the following equations:

$$\text{erf}(z) = \frac{T_S - 280}{1324} = \frac{1400 - 280}{1324} = 0.8459 \rightarrow z = 1.0801 \quad (4.117)$$

$$x = z \cdot \sqrt{4 \cdot \alpha \cdot t} = 1.0801 \times \sqrt{4 \times 9.07 \times 10^{-6}\text{m}^2 \cdot \text{s}^{-1} \cdot \text{t}} \quad (4.118)$$

In this case, as opposed to Exercise 4.6, we have not considered differences in the thermal diffusivity of the metal in the primary cooling or secondary cooling. The reason for this consideration is that the time required for the complete solidification of the melt is too short. We indicate in Table 4.11 the evolution of the profile of solidification along the secondary cooling of the continuous casting machine. We collect the values of the thickness that has solidified, x, in millimeters or percentage (from the origin of coordinates to the center, as the machine is symmetric, the same thickness solidifies from the opposite extreme of the machine, Fig. 4.13) as a function of the time (s) and the height from the origin of coordinates of the mold (m). The height of the mold is calculated as a function of speed of extraction, v_{ext}, as follows for different times being the height of the mold, y, 1 m:

Table 4.11 Solidified thickness as a function of the length of the billet and the time

Time (s)	Length of the billet, y (m)	Solidified thickness, x (mm)	% Solidified
50	2.5	43.0	57.3
100	5.0	60.9	81.2
152	7.6	75.0	100

$$v_{ext} = \frac{y}{t} \rightarrow y = v_{ext} \cdot \mathrm{t} \cdot \frac{1\,\mathrm{min}}{60\,\mathrm{s}} \tag{4.119}$$

while the solidified percentage is:

$$\%\mathrm{solidied} = \frac{x}{75\,\mathrm{mm}} \times 100 \tag{4.120}$$

where x is the position of the $\mathrm{T_S}$, and the semi-thickness of the billet is 75 mm (150 mm/2).

It is possible to make an estimation of the heat flow along the solidification front of the melt to estimate the heat flow, $\mathrm{J_q}$, that is necessary to remove from the mold of primary cooling of a continuous casting process. As we have the equation that gives the distribution of the temperature in the mold, and considering the First Law of Fourier:

$$\mathrm{J_q} = -\lambda \cdot \left(\frac{\mathrm{d}T}{\mathrm{d}x}\right)_i \tag{4.121}$$

$$\left(\frac{\mathrm{d}T}{\mathrm{d}x}\right)_i = \frac{\mathrm{B}}{\sqrt{\pi}} \cdot \exp\left(-z^2\right) \cdot \left(\frac{1}{\sqrt{\alpha \cdot t}}\right) \tag{4.122}$$

the values of $\mathrm{J_q}$ are high in the upper zone of the mold and decrease as we descend in the mold (the height of the mold is one meter). A primary estimation of the average value of the heat flow that would be necessary to remove to reach the solidification of the melt might be calculated in the solid-melt interface for a height of 0.5 m (half the height of the mold), position that is reached when the casting time is 10 s (casting rate of 3 $\mathrm{m \cdot min^{-1}}$, see Table 4.10):

$$\begin{aligned}
\mathrm{J_q} = {} & \frac{2 \times 1324\,\mathrm{K} \times 46\,\mathrm{W \cdot m^{-1} \cdot K^{-1}}}{\sqrt{\pi}} \cdot \exp\left[-\left(\frac{x}{\sqrt{4 \cdot \alpha \cdot \mathrm{t}}}\right)^2\right] \\
& \cdot \left(\frac{1}{\sqrt{4 \times 9.07 \times 10^{-6}\,\mathrm{m^2 \cdot s^{-1}} \cdot 10\,\mathrm{s}}}\right) \\
= {} & \frac{2 \times 1324\,\mathrm{K} \times 46\,\mathrm{W \cdot m^{-1} \cdot K^{-1}}}{\sqrt{\pi}} \cdot \exp\left[-\left(\frac{0.0227\,\mathrm{m}}{\sqrt{4 \times 9.07 \times 10^{-6}\,\mathrm{m^2 \cdot s^{-1}} \cdot 10\,\mathrm{s}}}\right)^2\right] \\
& \cdot \left(\frac{1}{\sqrt{4 \times 9.07 \times 10^{-6}\,\mathrm{m^2 \cdot s^{-1}} \cdot 10\,\mathrm{s}}}\right) = 8.72 \times 10^5\,\mathrm{W \cdot m^{-2}}
\end{aligned} \tag{4.123}$$

Even though the water and the pressurized air are the coolants habitually used, it is possible to evaluate the liquid oxygen that is used in the conversion of the pig iron (Chap. 1). Although it is a complex operation from the logistics point of view, it is necessary to take into account that the oxygen plant annexed to the LD steelworks

Table 4.12 Thermal properties of the O_2

Mean specific heat O_2-liquid ($J \cdot mol^{-1} \cdot K^{-1}$)	Boiling point of the O_2 (K)	Vaporization enthalpy ($J \cdot mol^{-1}$)	Mean specific heat O_2-gas ($J \cdot mol^{-1} \cdot K^{-1}$)
53.7985	90.2	7857	26.348

has a fluid at 54.4 K, of molecular weight of 32.0 $g \cdot mol^{-1}$ and whose thermal properties are collected in Table 4.12.

If the final temperature of the oxygen-gas is 373 K (100 °C), the cooling capacity of the O_2 will be:

$$C_{e,l} \cdot \left(T_{Boiling} - T_{fusion}\right) + H_{vap} + C_{e,g} \cdot \left(T - T_{Boiling}\right)$$
$$= 53.7985 \frac{J}{mol \cdot K} \cdot (90.2 - 54.4)K + 7857 \frac{J}{mol} + 26.348 \frac{J}{mol \cdot K}(373 - 90.2)K$$
$$= 17234.2 \frac{J}{mol\ O_2} \cdot \frac{mol\ O_2}{32g\ O_2} \cdot \frac{1000g\ O_2}{1kg\ O_2} = 538568 \frac{J}{kg\ O_2} \tag{4.124}$$

If the oxygen consumption of the LD steelworks is 80 kg $O_2 \cdot t^{-1}$ of steel, in a factory that brought about 3.0×10^6 t of steel every year (8219 daily t), the cooling potential of the oxygen would be:

$$538568 \frac{J}{kgO_2} \cdot \frac{80\,kgO_2}{t\ steel} \cdot \frac{8219\,t\ steel}{day} = 3.541 \times 10^{11} \frac{J}{day} \tag{4.125}$$

Commentaries: we have previously made an estimation of the heat flow to be removed in the continuous casting mold for the billet of 150 mm × 150 mm. Using the same criterion of calculus, it is possible to estimate the heat flow that would be required to be removed in the case of continuous casting of thick slabs (250 mm × 1600 mm) or thin slabs (50 mm × 1600 mm), Exercises 4.6 and 4.7. We collect in Table 4.13 the values of the heat flow in different semi-finished products, the heat transfer area in each case, and the heat that should be removed in each one of the lines every day (the height of the mold is one meter in all the cases). Finally, we should take into

Table 4.13 Heat removal requirements in different products of the continuous casting

Format of solidification	Heat flow ($W \cdot m^{-2}$)	Area of the heat transfer (m^2)	Cooling requirements ($J \cdot day^{-1}$)
250 mm × 1600 mm	9.341×10^5	3.7 (0.25 m · 1 m· 2 + 1.6 m · 1 m · 2)	2.986×10^{10}
50 mm × 1600 mm	2.161×10^5	3.3 (0.05 m · 1 m· 2 + 1.6 m · 1 m · 2)	6.161×10^{10}
150 mm × 150 mm	8.72×10^5	0.6 (0.150 m · 1 m· 4)	4.525×10^{10}

account that the utilization of this cooling system might have consequences on the microstructure of the final product.

End of the exercise.

4.3.3 Tertiary Metallurgy: Inclusions

Tertiary metallurgy, also called tundish metallurgy, has as objectives:

- Reducing the number of inclusions in the liquid (enhancing the steel cleanliness).
- Avoiding the oxidation and nitriding of the metal.
- Controlling that the deviations in the percentages of the alloying elements adjusted by the secondary metallurgy operations are not produced.
- Performing a tracking and if it is required, adjusting the temperature of the melt.

The following elements are necessaries to accomplish the above-mentioned objectives:

- Gases, as argon, to avoid the oxidation and nitriding of the liquid and to activate the mechanism of inclusions flotation.
- Devices of heating by induction or plasma to control the temperature.
- Mechanism of pneumatic and electromagnetic stirring that could ensure the thermal and chemical homogeneity of the melt.
- Ceramic filtration systems to retain the inclusions of the smallest size.

Conceptually, an inclusion is a non-metallic particle that is non-soluble both in liquid iron and in steel and that might be of two types:

- Exogenous inclusions, which come from phases or mediums in contact with the liquid iron, slag, refractory, components of the burden or components of the channels and casting pipes.
- Endogenous inclusions, which come from chemical reactions or solubility losses of the elements inside of the liquid iron.

Even though the liquid iron is in contact with both potential sources of endogenous (refining reactions of the sulfur and oxygen) and exogenous (slag and refractories) inclusions during all the process of manufacture of steel, it is equally truth that, simultaneously, removal mechanisms take place, as for instance:

- Boiling of the metal resulting in the CO evolution.
- Catching by means of the slag and refractory.
- Coalescence of inclusions.
- Flotation.

Inclusions are produced as a consequence of:

- The deoxidizing and desulfurizing of the steel.

Table 4.14 Chemical composition, in weight %, of the continuous casting synthetic slags (tundish-semi-product or ingot mold)

Phase	Slag Type I	Slag Type II
SiO_2	29	27
CaO	29	17
MgO	–	9
Al_2O_3	5	4
Fe_2O_3	1	1
Na_2O	6	16
CaF_2	10	–
B_2O_3	–	4
C	13	15
CO_2	7	7

- Reactions of oxidation and nitriding of the ready to cast liquid iron during the processes of the secondary metallurgy.
- Tribology wear or chemical attack of the refractory lining by the active elements of the metal.
- In less frequent situations, emulsion of the slag in the metal.

Inclusions are mainly produced as a consequence of reactions of deoxidizing and refining of sulphur in the secondary metallurgy and during the time that liquid iron is in contact with the materials of the tundish and the mold of the continuous casting machine.

It is a habitual practice, in the degasification of the liquid iron, the use of synthetic slags of the CaF_2–CaO–Al_2O_3 system to perfectly cover the interphase of the liquid in contact with the air. In the tundish and in the mold of the continuous casting machine synthetic slags of high fluidity (Table 4.14) are used, which ensure perfect isolation of the liquid. They are classified in:

- Slag type I, with fluorspar, CaF_2, in their composition.
- Slag type II, where fluorspar is replaced by sodium oxide.

We suggest using the following equivalence for its transformation into a pseudoternary product and, this way, studying the continuous casting synthetic slag, using the SiO_2–CaO–Na_2O ternary diagram of Fig. 4.14 (Levin et al. 1969):

$$\%SiO_2 = SiO_2 + Al_2O_3 + Fe_2O_3 + C + CO_2 \tag{4.126}$$

$$\%CaO = CaO + MgO - CaO \cdot CaMg(CO_3)_2 \tag{4.127}$$

$$\%Na_2O = Na_2O + CaF_2 \tag{4.128}$$

The resulting pseudoternary composition for the slag Type I is: 72% SiO_2, 12% CaO, and 16% Na_2O, using the above-mentioned equivalences.

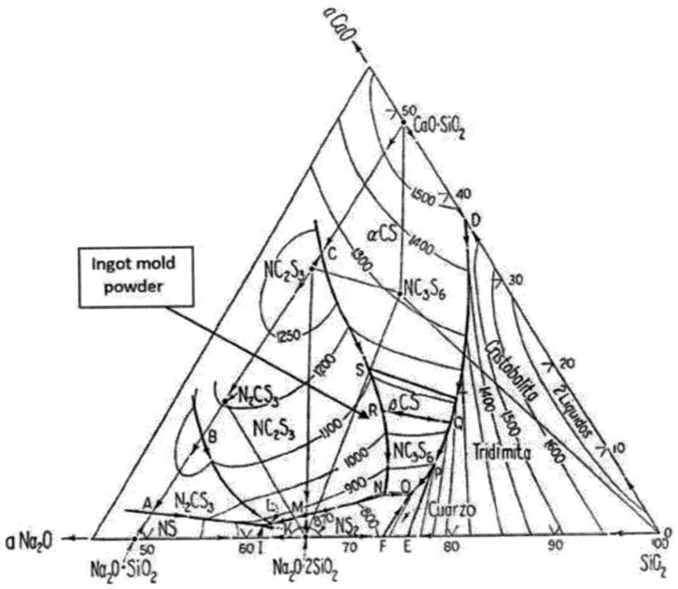

Fig. 4.14 SiO_2–CaO–Na_2O ternary diagram

It is possible to use the following mechanisms to eliminate the inclusions that are present in the liquid iron and to have clean steel:

- Flotation (natural or forced).
- Growing of the particles.
- Catching by a slag or by the refractory.
- Modification and coalescence.

Stable inclusions, which had a bigger size than the critical one (clustering mechanism), make, inside of the liquid iron, a heterogeneous phase that displaces the liquid–gas interface due to the flotation forces. The dynamic equilibrium of forces on a spherical inclusion located inside of liquid iron is equal to:

$$F_f - F_s - F_g = 0 \tag{4.129}$$

where F_f is the flotation component, F_g is the gravitational component and F_s is the Stokes frictional force. These forces are calculated as follows:

$$F_f = \frac{4 \cdot \pi \cdot r^3 \cdot \rho_l \cdot g}{3} \tag{4.130}$$

$$F_g = \frac{4 \cdot \pi \cdot r^3 \cdot \rho_i \cdot g}{3} \tag{4.131}$$

$$F_s = 6 \cdot \pi \cdot r \cdot \eta \cdot v \quad (4.132)$$

where r is the radius of the inclusion, ρ_l is the density of the liquid metal, g is the gravitational acceleration, ρ_i is the density of the inclusion, η is the viscosity and, finally, v is the flotation rate of the spherical inclusions, where this last term is calculated as follows (Stoke's law):

$$v = \frac{2}{9} \cdot \left(\frac{g \cdot r^2}{\eta}\right) \cdot (\rho_l - \rho_i) \quad (4.133)$$

Endogenous inclusions, formed by the elimination reaction of the dissolved sulfur and oxygen, may have a size so small that the flotation mechanisms cannot explain their disappearance from the melt. Deoxidation with aluminum, for instance, produces alumina, Al_2O_3, inclusions that, once reached their critical size, can be substrates for the growth of new particles of aluminum oxide during the process of steel deoxidation. The controlling stage of the process is the growth of the alumina particles until they have a size that the flotation rate does not achieve their elimination from the melt.

On the other hand, an alumina inclusion, apart from leaving the metal by means of growing and flotation mechanisms, may also disappear if it is trapped by the refractory of the ladle, the tundish, or the slag emulsified inside of the melt.

The driving force of the inclusions catching process is the chemical potential. Alumina, of activity equal to one, is captured by a slag in which the activity of the aluminum oxide is lower than one. According to the SiO_2–Al_2O_3–CaO ternary diagram of Fig. 4.15 (Levin et al. 1969), it is possible to estimate the activity of the alumina along the boundary lines of the following fields of primary crystallization: $2CaO \cdot SiO_2$–$12CaO \cdot 7Al_2O_3$, $12CaO \cdot 7Al_2O_3$–$3CaO \cdot Al_2O_3$, and $12CaO \cdot 7Al_2O_3$–$CaO \cdot Al_2O_3$.

Slags used by the secondary and tertiary metallurgies are characterized by both the low liquidus temperature and the high basicity. The chemical composition of the suitable slags to satisfy these objectives are those that are close to the eutectic of 1335 °C:

$$12CaO \cdot 7Al_2O_3(s) + 2CaO \cdot SiO_2(s) + CaO \cdot Al_2O_3(s) \leftrightarrow \text{liquid} \quad (4.134)$$

Other alternative is to use slags of the CaF_2-Al_2O_3-CaO system, which give better performance with respect to the values of the calcium oxide in the melt. Along the line that delimitates the fields of primary crystallization of the CaO–$5CaO{\cdot}3Al_2O_3$ and the CaO–CaF_2, the activity of the calcium oxide in the liquid is equal to one (Ballester et al. 2000, p. 500).

Dispersed constituents and the matrix of the refractories of the ladle and the tundish can be active centers able to catch or dissolve the inclusions of the melt. The kinetics of catching inclusions by these refractories is promoted when the number of Reynolds of the metal is increased.

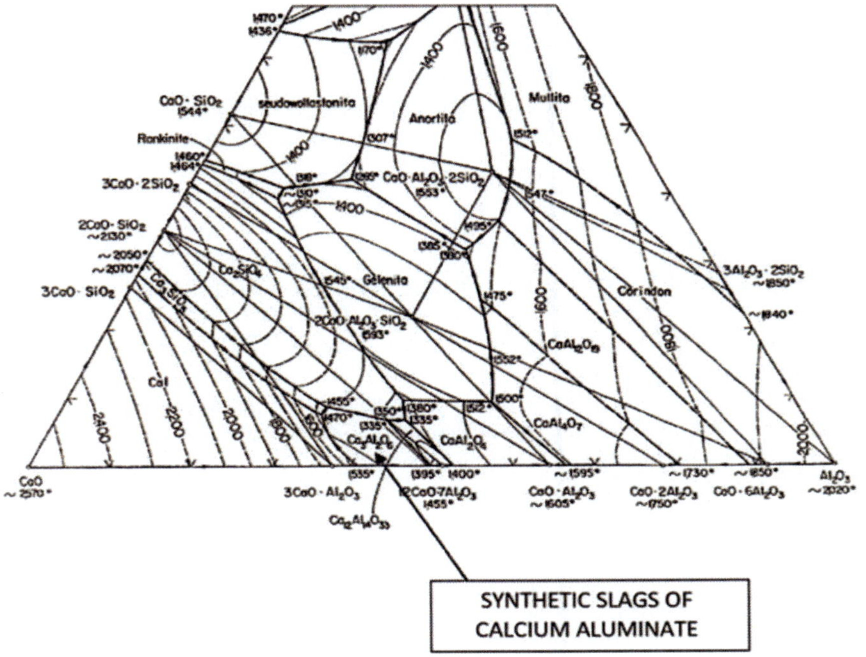

Fig. 4.15 SiO_2–Al_2O_3–CaO ternary diagram. Slags for the capture of the alumina inclusions

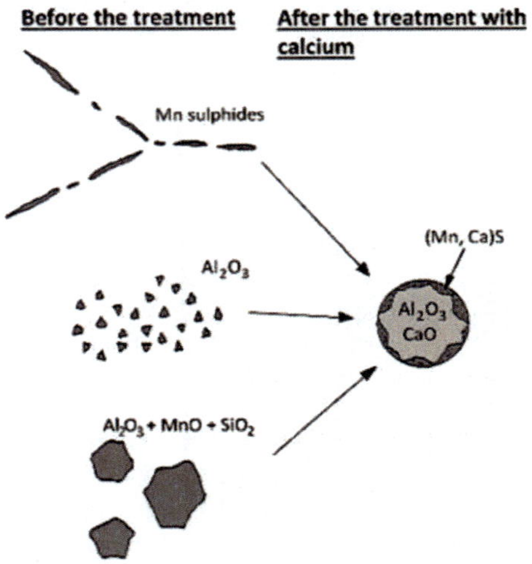

Fig. 4.16 Characteristics of the inclusions before and after the treatment with calcium

The modification and coalescence of inclusions is other of the alternatives existing in the obtaining of clean steel. We show in Fig. 4.16 a scheme of the modification of the inclusions of manganese sulfide, alumina, manganese (II) oxide, and silicon oxide.

The time, t, required to achieve the coalescence of the inclusions dispersed in an iron melt is related with the size-diameter of the particles, D_p, the viscosity of the melt, η, and the solid–liquid surface energy of the inclusions, γ_{SL}. The dimensional analysis can demonstrate that the functional relation that exists between the coalescence time and the size, viscosity and surface energy is related by the following equation:

$$t = \frac{D_p \cdot \eta}{\gamma_{SL}} \tag{4.135}$$

If the values of the variables viscosity and surface energy are in the range 5–20 Pa · s and 0.30–0.60 J · m^{-2}, respectively, the coalescence times for inclusions of small size, 5.0 μm, are: 3.33×10^{-4} s (when the viscosity is 20.0 Pa · s and the surface energy is 0.30 J · m^{-2}) or 4.17×10^{-5} s (when the viscosity is 5.0 Pa · s and the surface energy is 0.60 J · m^{-2}), which indicates instantaneous coalescence. Times of coalescence of seconds are obtained when the size of the inclusions is 15 mm (when the viscosity is 20.0 Pa · s and the surface energy is 0.30 J · m^{-2}) or 120 mm (when the viscosity is 5.0 Pa · s and the surface energy is 0.60 J · m^{-2}). Consequently, the coalescence of the inclusions of any size is a process kinetically favorable always and when it is possible to find the conditions that propitiate or favor the contact between the inclusions that exist in the melt.

The presence of inclusions in a metallic matrix produces frequent lack of cohesion (decohesion) during the hot forming of the steel. However, if we admit that the presence of inclusions is unavoidable, the non-favorable interferences during the hot plastic deformation will decrease if:

- the lengthened morphology (or acicular) is replaced by the spherical morphology.
- the hot plasticity of the inclusions is increased.

We achieve by means of the treatment with calcium (Fig. 4.16 adapted from Deo and Boom 1993):

- a reduction of the size and a modification of the shape of the manganese sulfide inclusions.
- the transformation of a hard and brittle inclusion of alumina into a calcium aluminate with variable quantities of dissolved sulfur and calcium, which have more plasticity.
- the disappearance of the inclusions of manganese oxide and silicon oxide by dissolution in the liquid iron.

Calcium is an active metal of low melting point (850 °C) and volatile (boiling point, 1487 °C). The yield in the dissolution of the metal in the liquid iron does not reach the 10% as great part of the calcium is released as vapor. So, calcium is usually employed alloyed with silicon (calcium silicide, CaSi) to increase the yield of the treatment, and there are two methods of adding it:

- injecting, normally in the ladle, the powders of CaSi (0.60–1.0 mm) with ceramic lance and driven for a carrying gas at a pressure of 5–15 atm.

– CaSi powders encapsulated inside of a wrap of steel with cylindrical shape (steel wire) that can reach any point inside of the liquid. The injection machine of the wire, with the encapsulated material, can adapt to any place of the steelworks by introducing the active products in the ladle or the tundish at the considered rate.

0.50–0.70 kg of CaSi per ton of steel are required to modify the geometry of the manganese sulfide. If it is necessary to modify the inclusions of alumina, it is necessary to use either calcium quantities greater than 0.40 kg per ton of steel or Ca/Al ratios, in the liquid iron, >0.14. There are risks that solid aluminates of low plasticity could be formed (Fig. 4.15) when quantities greater than 1.0 kg of calcium per t of steel are employed.

The injection of an inert gas is a usual operation in the tertiary metallurgy due to the following aspects:

– enhances the flotation of the inclusions.
– impedes the formation of chemical and thermal segregations in the liquid.
– finalizes the operations of dissolved gases refining that were performed during the secondary metallurgy.

Normally, the kinetics control of the degasification, which can be performed in the continuous casting tundish, depends on the rate of diffusional transfer of the hydrogen, nitrogen, and oxygen in the liquid iron. The recommended correlation for the calculation of the mass transfer coefficient between a liquid and the ascending bubbles of argon is:

$$\mathrm{St}^* = 1.13 \times \mathrm{Re}^{0.5} \times \mathrm{Sc}^{-0.5} \tag{4.136}$$

where Re is Reynolds's number, Sc is the Schmidt's number, and St* is Stanton's number for the mass transfer, which is defined by:

$$\mathrm{St}^* = \frac{k_d}{v} \tag{4.137}$$

where v is the average ascending rate of the argon bubbles and k_d is the coefficient of mass transfer.

4.3.3.1 Solved Exercises

Exercise 4.9 In a rod for prestressing (rebar, reinforced bar) concrete, inclusions of alumina, Al_2O_3, of 4.0 μm in diameter are detected. The time that passes between the deoxidation of the melt and the solidification of the billet in a continuous casting machine is of two hours and fifteen minutes. Analyze:

(a) Can be these inclusions eliminated by flotation?

(b) Regarding the origin of the inclusions, can they be associated to the fact that 1 kg of alumina could be emulsified in an iron melt creating particles of 4.0 μm in diameter when it is dropped from a certain height?

Data: density of the liquid iron: 7000 kg · m^{-3}; density of the alumina: 3950 kg · m^{-3}; viscosity of the liquid iron: 4.0 × 10^{-3} Pa · s; surface energy of the alumina: 740 mJ · m^{-2}.

Solution starts:

We calculate, using the Eq. (4.133), the flotation rate of the inclusions of alumina:

$$v = \frac{2}{9} \cdot \left(\frac{g \cdot r^2}{\eta}\right) \cdot (\rho_l - \rho_i) = \frac{2}{9} \cdot \left[\frac{9.81\ \mathrm{m \cdot s^{-2}} \cdot \left(2.0 \times 10^{-6}\ \mathrm{m}\right)^2}{4.0 \times 10^{-3}\ \mathrm{Pa \cdot s}}\right]$$
$$\cdot \left(7000\ \mathrm{kg \cdot m^{-3}} - 3950\ \mathrm{kg \cdot m^{-3}}\right) = 6.649 \times 10^{-6}\ \mathrm{m \cdot s^{-1}} \quad (4.138)$$

For a time of two hours and fifteen minutes (8100 s), the distance covered by the inclusion is of 0.0539 m (5.4 cm). Consequently, considering these values, only around six centimeters of melt, from its contact with the protecting slag, are free of alumina inclusions of these characteristics.

To analyze if it is possible the conversion-disaggregation of one kilogram of alumina into particles of 2.0 μm in radius, it is necessary to calculate the number of alumina particles of 4.0 μm in diameter. The volume associated to one particle of 2.0 μm in radius is:

$$V = \frac{4}{3} \cdot \pi \cdot \left(2.0 \times 10^{-6}\ \mathrm{m}\right)^3 = 3.35 \times 10^{-17}\mathrm{m^3 \cdot particle^{-1}} \quad (4.139)$$

The mass, M, associated to each particle and the number of particles, N, per kg are:

$$\begin{aligned} M &= 3.35 \times 10^{-17}\ \mathrm{m^3 \cdot particle^{-1}} \times 3950\ \mathrm{kg \cdot m^{-3}} \\ &= 1.324 \times 10^{-13}\ \mathrm{kg \cdot particle^{-1}} \end{aligned} \quad (4.140)$$

$$\begin{aligned} N &= \frac{1}{M} = \frac{1}{1.324 \times 10^{-13}\ \mathrm{kg \cdot particle^{-1}}} \\ &= 7.56 \times 10^{12}\ \mathrm{particles \cdot kg^{-1}} \end{aligned} \quad (4.141)$$

Similarly, the area associated to a particle of 2.0 μm in radius, A_p, and the area corresponding to the ensemble of alumina particles that exist in 1 kg, specific surface, $S_{e/M}$ are:

$$\begin{aligned} A_p &= 4 \cdot \pi \cdot r^2 = 4 \cdot \pi \cdot \left(2.0 \times 10^{-6}\ \mathrm{m}\right)^2 \\ &= 5.027 \times 10^{-11}\ \mathrm{m^2 \cdot particle^{-1}} \end{aligned} \quad (4.142)$$

$$S_{e/M} = 5.027 \times 10^{-11}\ m^2 \cdot particle^{-1} \cdot 7.56 \times 10^{12}\ particles \cdot kg^{-1}$$
$$= 379.7\ m^2 \cdot kg^{-1} \quad (4.143)$$

The surface energy, G_s, associated to one kilogram of alumina inclusions of 2.0 μm in radius is:

$$G_s = 740\,mJ \cdot m^{-2} \times 379.7\,m^2 \cdot kg^{-1} = 281\,J \cdot kg^{-1} \quad (4.144)$$

The potential energy (height h) necessary to create 281 J of surface energy is:

$$h = \frac{281\,J \cdot kg^{-1}}{1\,kg \cdot 9.81\,m \cdot s^{-2}} = 28.6\,m \quad (4.145)$$

That is to say, the minimum height to emulsify in the iron melt one kilogram of alumina generating inclusions of 2 μm of the radius is 28.6 m.

Commentaries: the mechanism of flotation (natural or forced with argon) is the method habitually used to eliminate the endogenous inclusions from the melt. This is one of the reasons to use the tundish before the casting of the melt in the continuous casting mold (Figs. 4.10, 4.11, and 4.12). The distance travelled by the inclusions of 4.0 μm in diameter by natural flotation is small and is no equivalent to the height of the melt occupied in the tundish (around 1 m). The elimination of these inclusions of small size to provide clean steel would be related both with the existence of mechanisms of forced flotation and with the coalescence.

Finally, it is possible to say that is very strange to find exogenous inclusions of slag trapped in the solidified product.

End of the exercise.

Exercise 4.10 One of the mechanisms to eliminate the alumina inclusions from the iron melt and to avoid their presence in the steel is to force their picking up by slags. One slag used to cover the tundish of the continuous casting machines has the following ternary composition: 20% Na_2O, 10% CaO, and 70% SiO_2. Estimate:

(a) What is the driving force of the catching process?
(b) The liquidus and solidus temperatures of the slag.
(c) Nature and proportion of the solid phases of the entrapment slag, at room temperature, compatible with the global composition of the slag.

Solution starts:

The driving force of the picking up of the alumina inclusion by the slag process would be related with the free energy of the mixture-dissolution of the alumina in the slag. The free energy, G_1, of the system before the entrapment process of a certain quantity of alumina moles, $n_{Al_2O_3}$, by a certain quantity of slag moles, n_{slag}, would be:

$$G_1 = n_{Al_2O_3} \cdot G^0(Al_2O_3) + n_{slag} \cdot G(slag) \quad (4.146)$$

The value of the free energy of the system, G_2, as a quantity of alumina moles, $n_{Al_2O_3}$, is dissolved-picked up by the slag, would be:

$$G_2 = n_{Al_2O_3} \cdot G^0(Al_2O_3) + n_{Al_2O_3} \cdot R \cdot T \cdot \ln(a_{Al_2O_3}) + n_{slag} \cdot G(slag) \quad (4.147)$$

Thus:

$$\Delta(G_2 - G_1) = \Delta G^M = n_{Al_2O_3} \cdot R \cdot T \cdot \ln(a_{Al_2O_3}) \quad (4.148)$$

that, expressed per mol of slag/alumina homogeneous mixture $(n_{Al_2O_3} + n_{slag})$ would be:

$$\Delta G_m^M = x_{Al_2O_3} \cdot R \cdot T \cdot \ln(a_{Al_2O_3}) \quad (4.149)$$

While the values of the molar fraction, $x_{Al_2O_3}$, and the activity of the alumina, $a_{Al_2O_3}$ are smaller than one, the process of dissolution-mixture-picking up, is favorable. It is possible that in some cases the activity might be close to the saturation, this means having a value for $a_{Al_2O_3}$ of 0.90, for instance, but always smaller than one, which would represent a free energy negative for the picking up process.

On the other hand, apart from the process of picking up, it is possible to consider the chemical reaction process, once trapped the alumina inclusion in the slag:

$$CaO(dis;\ slag) + Al_2O_3(inclusion) \leftrightarrow CaAl_4O_7(dis;\ slag) \quad (4.150)$$

reaction for which, the value of the standard free energy, Appendix 1, is always negative:

$$\Delta_r G(T) = -16.8 - 0.034 \cdot T(K)\left[kJ \cdot (mol\ CaO)^{-1}\right] \quad (4.151)$$

Using the Na_2O–CaO–SiO_2 ternary diagram, Fig. 4.14, it is possible to estimate the liquidus temperature of the picking up slag of the type I collected in Table 4.14 (16% Na_2O, 12% CaO, and 72% SiO_2). On the other hand, the solidus temperature, which indicates the appearance of the first liquid for this slag of type I, is the product of the following eutectic reaction at 790 °C:

$$NS_2(s) + SiO_2(s) + NC_3S_6 \leftrightarrow liquid \quad (4.152)$$

where N is the Na_2O, C is the CaO and S is the molecule of SiO_2. At the eutectic temperature of 790 °C, there is still 15% of solid to be dissolved.

Solid phases at room temperature compatible with the global composition indicated in the enunciation are those of the eutectic reaction: 47% $Na_2O \cdot 2SiO_2$, 35% $Na_2O \cdot 3CaO \cdot 6SiO_2$, and 18% SiO_2.

Commentaries: basic and reductant slags, due to the presence of carbon-graphite, are the materials used to lubricate and avoid the oxidation, nitriding, and hydrogenation of the melt, in the molds of the continuous casting machines, Figs. 4.8, 4.10, 4.11, and 4.12. The rest of the oxidized components of the casting powders are similar to those used in the exercise, Table 4.14.

End of the exercise.

Exercise 4.11 Most of the inclusions that can be found in steels and cast irons have endogenous nature, consequence of deoxidation and dephosphorizing processes. Determine the zones of stability of the oxide, sulfide, and metal, at 1823 K (1550 °C) for the following systems: Fe–O–S, Mn–O–S, Ca–O–S, and Mg–O–S, as a function of the oxygen and sulfur concentrations in the iron melt (solvent).

Solution starts:

In systems of three components, the number of degrees of freedom, L, is function of the number of phases, F, that are in equilibrium:

$$L = 5 - F \tag{4.153}$$

When we know the value of the temperature, the degrees of freedom of the system can be zero, when there are three condensed phases in equilibrium with the gas (in this case, the oxygen and sulfur partial pressures in the gas, of the so-called Kellogg diagrams in the roasting of sulfides, are replaced by the concentrations of these elements in other homogeneous phase, molten iron, where the atoms of oxygen and sulfur are dissolved).

In a first stage, the coordinates of the four invariant points are determined (zero degrees of freedom), for the systems indicated in the enunciation, that is to say, that point where the equilibrium of the three condensed phases is reached, Me–MeO–MeS, with the gas. The process of calculus that must be followed is the same for the four studied systems, and consequently, we are going to detail the steps for the Fe–O–S system, and we will show in a table and also graphically the position in a diagram where we represent the ordinates and abscissas corresponding to the concentrations of sulfur and oxygen in the iron melt, respectively.

For the determination of the coordinates of the invariant point of the Fe–O–S system, we used the data of the Appendices 1 and 3 (standard free energy) associated to the following reactions:

$$\mathrm{Fe(s)} + \frac{1}{2}\mathrm{O_2(g)} \leftrightarrow \mathrm{FeO(s)} \tag{4.154}$$

$$\frac{1}{2}\mathrm{O_2(g)} \leftrightarrow \mathrm{O}\big(\mathrm{dis;\ Fe},\ a_{\mathrm{H,O}} = 1\big) \tag{4.155}$$

The values of the standard free energy of the reactions (4.154) and (4.155) are, respectively:

$$\Delta_r G^0_{reaction(4.154)}(T) = -265000 + 65.0 \cdot T(K)\left[J \cdot (mol\ FeO)^{-1}\right] \quad (4.156)$$

$$\Delta_r G^0_{reaction(4.155)}(T) = -117000 - 3.14 \cdot T(K)\left[J \cdot (mol\ O)^{-1}\right] \quad (4.157)$$

that lead to the following reaction:

$$Fe(s) + O\left(dis; Fe, a_{H,O} = 1\right) \leftrightarrow FeO(s) \quad (4.158)$$

$$\Delta_r G^0_{reaction(4.158)}(T) = -148000 + 68.14 \cdot T(K)\left[J \cdot (mol\ FeO)^{-1}\right] \quad (4.159)$$

Considering that the activities of the iron and the ferrous oxide are equal to one (element and pure compound), the corresponding activity of Henry of the oxygen, $a_{H,O}$, in the melt in equilibrium is 0.2082.

In the same way, it is possible to calculate the standard free energy associated to the following reaction:

$$Fe(s) + S\left(dis; Fe, a_{H,S} = 1\right) \leftrightarrow FeS(s) \quad (4.160)$$

from the standard free energies associated to the following reactions

$$Fe(s) + \frac{1}{2}S_2(g) \leftrightarrow FeS(s) \quad (4.161)$$

$$\Delta_r G^0_{reaction(4.161)}(T) = -138000 + 38.0 \cdot T(K)\left[J \cdot (mol\ FeS)^{-1}\right] \quad (4.162)$$

$$\frac{1}{2}S_2(g) \leftrightarrow S\left(dis; Fe, a_{H,S} = 1\right) \quad (4.163)$$

$$\Delta_r G^0_{reaction(4.163)}(T) = -130000 + 21.0 \cdot T(K)\left[J \cdot (mol\ S)^{-1}\right] \quad (4.164)$$

Consequently, the standard free energy associated to the reaction (4.160):

$$\Delta_r G^0_{reaction(4.160)}(T) = -8000 + 17.0 \cdot T(K)\left[J \cdot (mol\ FeS)^{-1}\right] \quad (4.165)$$

When the activity of Henry of the iron and iron (II) sulfide is equal to one, the value of the activity of Henry, $a_{H,S}$, in the iron melt at 1823 K is 4.5581.

The reactions (4.158) and (4.160) lead to the equilibrium between the oxygen and the sulfur dissolved in the iron melt and the condensed phases FeO and FeS:

$$\mathrm{FeS(s) + O\left(dis; Fe, a_{H,O} = 1\right) \leftrightarrow FeO(s) + s\left(dis; Fe, a_{H,S} = 1\right)} \quad (4.166)$$

$$\Delta_r G^0_{reaction(4.166)}(T) = -140000 + 51.14 \cdot T(K)\left[J \cdot (mol\ FeO)^{-1}\right] \quad (4.167)$$

When the activity of the iron (II) oxide and sulfide is one, the relation is:

$$\frac{a_{H,S}}{a_{H,O}} = \frac{4.5581}{0.2082} = 21.89 \quad (4.168)$$

Consequently, the coordinates of the invariant point for the Fe–O–S system at 1823 K expressed as Henry's activities are: abscissa, $x, a_{H,O} = 0.2082$; ordinate, y, $a_{H,S} = 4.5581$.

When we talk about diluted solutions, the activities of Henry are identified with the concentration expressed in percentage. For that reason, as the most part of the diagram happens in the field of the diluted solutions, we represent log(%S) versus log(%O) instead of $\log\left(a_{H,S}\right)$ versus $\log\left(a_{H,O}\right)$.

The coordinates of the invariant point are indicated in Table 4.15 for the rest of the systems.

We represent in the enclosed Fig. 4.17 the zones of stability for the four studied systems: Fe–O–S, Mn–O–S, Ca–O–S and Mg–O–S at the temperature of 1823 K. Apart from the four invariant points (L = 0): letter A (Ca–O–S); letter B (Mg–O–S); letter C (Mn–O–S); and, letter D (Fe–O–S), we draw the lines that indicated the monovariant equilibriums (L = 1):

- Vertical lines of the Me–MeO equilibrium.
- Horizontal lines of the Me–MeS equilibrium.
- The slope equal to one of the MeO–MeS equilibrium.

Commentaries: the log(%S) versus log(%O) diagram at 1550 °C explains the modification-transformation that affects the MnS inclusions as it is indicated in the Fig. 4.17.

Finally, we want to indicate that from Exercise 4.6, we have tried to indicate some of the operating parameters that can have greater importance in the solidification process of iron melts by continuous casting. However, we have mentioned nothing about the microstructure and properties of the obtained steels. The attempt of relating and finding a greater continuity between these two fields of knowledge is the objective

Table 4.15 Coordinates of the invariant point for the rest of the systems

System	Ordinate, y, $a_{H,S}$	Abscissa, x, $a_{H,O}$	Temperature (K)
Mn–O–S	0.02387	3.10×10^{-4}	1823
Ca–O–S	1.04×10^{-9}	3.61×10^{-8}	1823
Mg–O–S	1.46×10^{-3}	178×10^{-7}	1823

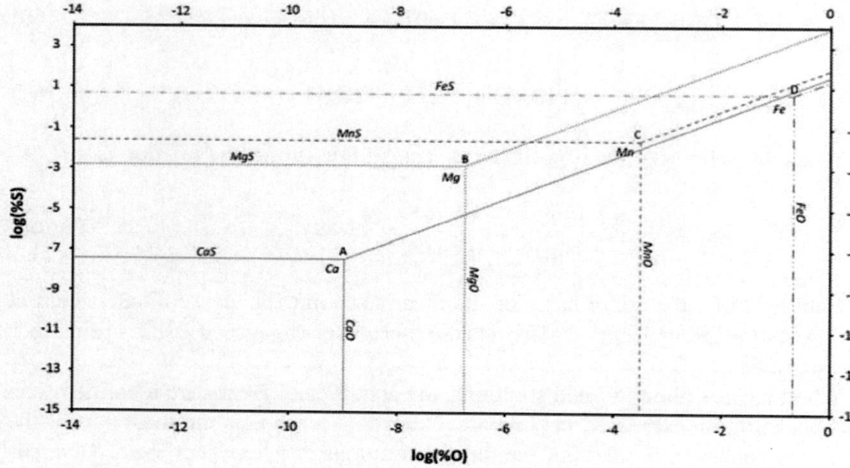

Fig. 4.17 Kellogg diagram for the different studied systems

of the research group in Ironmaking and Steelmaking, Metallurgy and Materials to which the authors of this manuscript belong by proposing solved exercises. This way, it is aimed at giving a first answer-approach to the relation that exists between processes, microstructure, properties, and applications of materials (Pero-Sanz et al. 2017, 2018, 2019; Verdeja et al. 2020).

End of the exercise.

4.3.4 Development of the Continuous Casting

The future challenges of the steel forming by continuous casting are mainly:

- Produce profiles, for long products, of similar characteristics to those of the final product.
- Reduce the thickness of the casting in the flat products.
- Link, of semi-continuous or continuous form, the casting operations with those of the hot plastic forming (compact steelworks).

Usual geometries of the continuous casting of long products (*bloom* or billet) have rectangular or squared sections. With these geometries, the objective is achieving the solidification formats similar to those of the finished product to reduce the cost of the hot plastic deformation. Nowadays, profiles with the shape of the dog bone (*beam blank*) to manufacture double T profiles are produced.

The solidification of thin slabs (<100 mm) was rapidly developed, and the installed capacity reached 26 million tons by the end of 1977. At the end of the XXth century

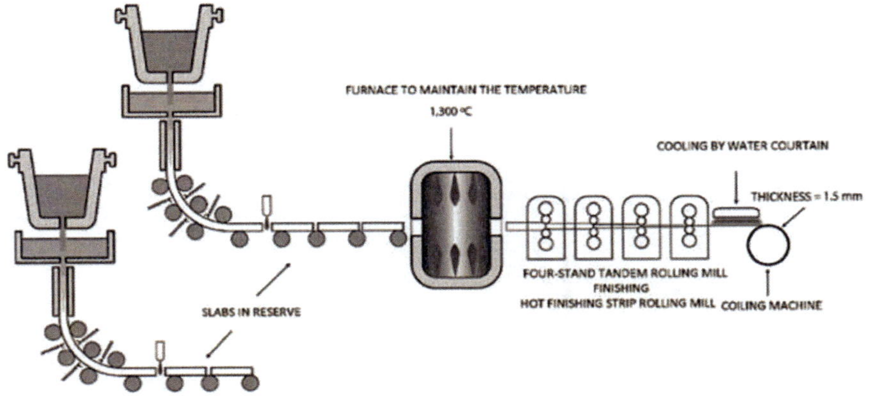

Fig. 4.18 Continuous casting of thin slabs. CSP process

and at the beginning of the current century the installed capacity reached the following values:

- 32Mt/year for flat products of 40–60 mm in thickness (thin slabs).
- 8 Mt/year for flat products of 80–100 mm in thickness (slabs that are classified as of intermediate thickness).

The technology of solidification in continuous of thin slabs is dominated by the CSP (*Continuous thin Slab casting Process*) process meaning 75% of the installed capacity (Fig. 4.18).

One of the operating advantages of the solidification of thin slabs is based on chaining up the process of the tertiary metallurgy with the hot rolling of flat products. The rough stand is disregarded in the hot finishing strip rolling mill semi-continuous (Fig. 4.18).

An example of the simplification of processes and iron and steelmaking operations is the compact steelworks (Fig. 4.19). Starting from the liquid iron refined in the electric furnace or in the converter, steel is fitted in composition during the secondary metallurgy and to straightaway perform the solidification and rolling in continuous until finishing with the obtaining of the hot-rolled coil using some of the following processes:

- *CSP*, developed by the SHS (Schlowmann Siemag AG, Düsseldorf, Germany).
- *ISP*, developed by the Mannesmann Demag-Germany in cooperation with the Arvedi-Italy.
- *FTSR*, developed by the Italian engineering company Danieli.

The limit to form thin flat or long products is the direct continuous casting of a product with thickness similar to that of the steel coil: 1.50–20.0 mm.

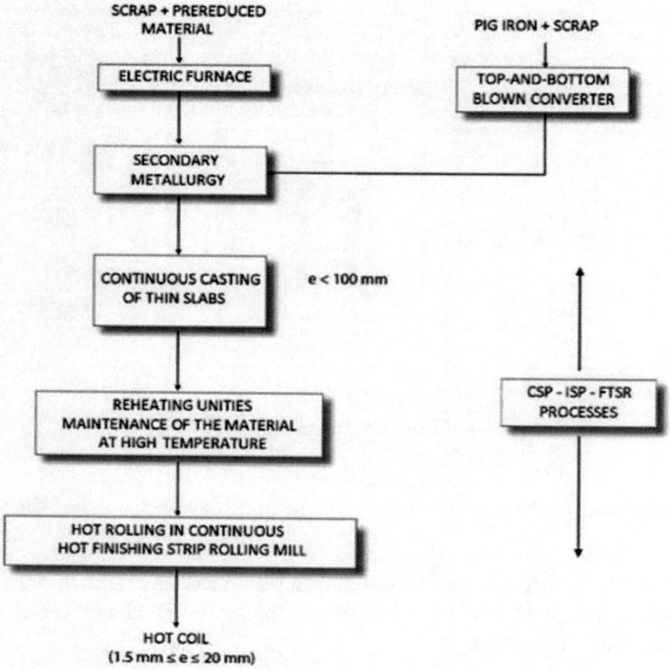

Fig. 4.19 Scheme of the compact steelworks for the manufacture of hot-rolled coil

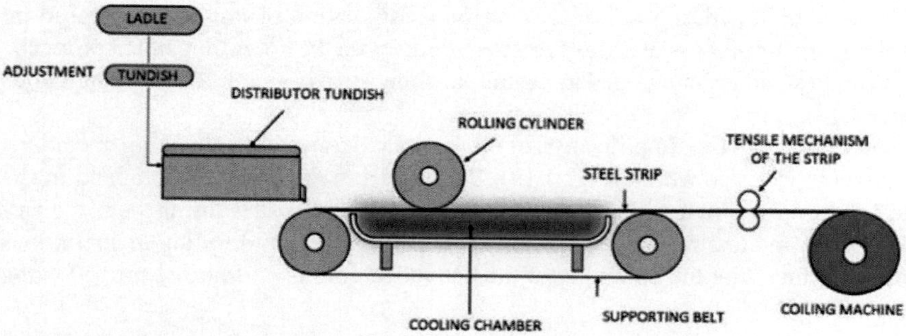

Fig. 4.20 Scheme of the DSC process for the obtaining of coil by means of direct solidification of steel strip

From 1986, Mannesmann Demag develops the *DSC* (*Direct Strip Casting*) process for the obtaining of steel strip or coil by direct casting, in continuous, of liquid iron (Fig. 4.20).

The most important feature of the DSC process is the manufacture of the steel band in continuous with thickness within 5 and 15 mm, with a very high productivity. The annual capacity of production of the process is around 2 million tons, which is very attractive for the future requirements of the integrated steel industry.

References

Ballester A, Verdeja LF, Sancho JP (2000) Metalurgia Extractiva, Fundamentos. Síntesis, Madrid, Spain

Davies GJ (1973) Solidification and casting, Ed. Wiley, New York, USA

Deo B, Boom R (1993) Fundamentals of steelmaking metallurgy. Prentice Hall International, Hertfordshire, UK

Fine HA, Geiger GH (1993) Handbook on material and energy balance calculations in Metallurgical Processes. TSM, Warrendale, USA

Flemings MC (1974) Solidification processing. McGraw-Hill, New York, USA

Flinn RA (1963) Fundamentals of metal casting. Addison-Wesley Pub. Co., London, UK

Kiessling R, Lange N (1978) Non-metallic inclusions in steel. The Metals Society, London, UK

Levin E, Robins CR, McMurdie HF (1969) Phase diagrams for ceramists. The American Ceramic Society, Columbus, Ohio, USA (and supplement of figures from the 2067 to the 4149 edited in 1969)

Pero-Sanz JA (2004) Aceros: Metalurgia Física, Selección y Diseño. Ed. Dossat 2000, 1st edn., Madrid, Spain

Pero-Sanz JA (2006) Ciencia e Ingeniería de Materiales: Estructura, Transformaciones, Propiedades y Selección. Ed. Dossat 2000, 5th edn., Madrid, Spain

Pero-Sanz JA, Quintana MJ, Verdeja LF (2017) Solidification and Solid-state Transformations of Metals and Alloys. Elsevier, 1st edn., Boston, USA

Pero-Sanz JA, Fernández-González D, Verdeja LF (2018) Physcal metallurgy of cast irons, 1st edn. Springer International Publishing, Cham, Switzerland

Pero-Sanz JA, Fernández-González D, Verdeja LF (2019) Structural materials: properties and selection, 1st edn. Springer International Publishing, Cham, Switzerland

Poirier DR, Poirier EJ (1992) Heat transfer fundamentals for metal casting. TMS, Pennsylvania, USA

Verdeja JI, Fernández-González D, Verdeja LF (2020) Operations and basic processes in ironmaking, 1st edn. Springer International Publishing, Cham, Switzerland

Winegard WC (1974) An introduction to the solidification of metals, Ed. Institute of Metals, London, UK

Chapter 5
Hot-Working Operations

Abstract Hot working is the initial step in the mechanical working of most metals and alloys. Not only does hot-working result in a decrease in the energy required to deform the metal and an increased ability to flow without cracking, but the rapid diffusion at hot-working temperatures aids in decreasing the chemical inhomogeneities of the cast-ingot structure. Blowholes and porosity are eliminated by the welding together of these cavities, and the coarse columnar grains of the casting are broken down and refined into smaller equiaxed recrystallized grains. These changes in structure from hot working improves ductility and toughness over the cast state.

However, there are certain disadvantages to hot-working. First, because high temperatures are usually involved, surface reactions between the metal and the furnace atmosphere become a problem. Ordinarily hot working is done in air, oxidation results, and a considerable amount of metal may be lost. Second, surface decarburization of hot-worked steel can be a serious problem, and frequently extensive surface finishing is required to remove the decarburized layer. Rolled-in oxide makes it difficult to produce good surface finishes on hot-rolled products, and because allowance must be made for expansion and contraction, the dimensional tolerances for hot-worked mill products are greater than for cold-worked products. Further, the structure and properties of hot-worked metals are generally not so uniform over the cross section as in metals which have been cold worked and annealed. Since the deformation is always greater in the surface layers, the metal will have a finer recrystallized grain size in this region.

This way, this chapter five is dedicated to the hot working of steel, and it is complemented with several exercises, problems, and case-studies related with the industrial practice.

5.1 Introduction

During the forming of liquid iron by continuous casting, a semi-finished product is obtained with geometrical characteristics very similar to those of the finished product. The objective is searching for the simplification of operations using the principles of

L. F. Verdeja González et al., *Operations and Basic Processes in Steelmaking*,
Topics in Mining, Metallurgy and Materials Engineering,
https://doi.org/10.1007/978-3-030-68000-8_5

the molding, which allows directly obtaining the finished product by solidification of the liquid (as-cast pieces).

Semi-finished products obtained by continuous casting of the liquid iron usually require from different processes or operations before their commercialization. According to their characteristics, they are classified in:

A. Hot forming processes:

- Rolling.
- Forging.

B. Finishing processes:

- Cold deformation: rolling, drawing, stamping, stretching, bending, deep drawing, or sizing.
- Welding.
- Coating.

C. Handling of the steel:

- Heat treatments.

Forging (or deformation under compression) is the most primary, simple, and also ancient process of forming. Normally, it is limited to the hot deformation although, in a restricted mode, the cold forge (stamping, stretching, deep drawing) is used in small parts, generally of low carbon content. The forge can be of two types: open forge or die forge (using blanking dies). The open forge is used to produce big volume parts by means of several steps of local compression using hydraulic press. In the die forge (closed forge), the forming of the part is performed within two blanking dies or stamps that when they close constitute the final mold of the part. To ensure the good occupancy of the empty space, an excess of metal is used and the excess flows through the empty space that separates both blanking dies to form scrap strip of small thickness. Rolling is the most important forming process to transform the formats of the liquid iron casting in semi-finished or finished products (Fig. 5.1).

We show in Fig. 5.1 an estimation of the distribution of the main iron and steelmaking products in 2015, for an overall of 1620 million tons of steel manufactured by casting, molding, and powder metallurgy.

The progressive introduction of continuous casting processes for the solidification of liquid iron (to manufacture mild steels, stainless steels, and highly alloyed steels) and the development of casting processes for either thin formats or geometries close to that of the finished product (near-net-shaped pieces), resulted in a remarkable reduction of the metallurgical losses during the forming of the metallic melt (Fig. 5.1). For steels with high demand (mild steels and certain stainless steels), the forming of the iron melt via continuous casting has imposed to the ingot casting due to the economic efficiency and mechanical properties-microstructural aspects.

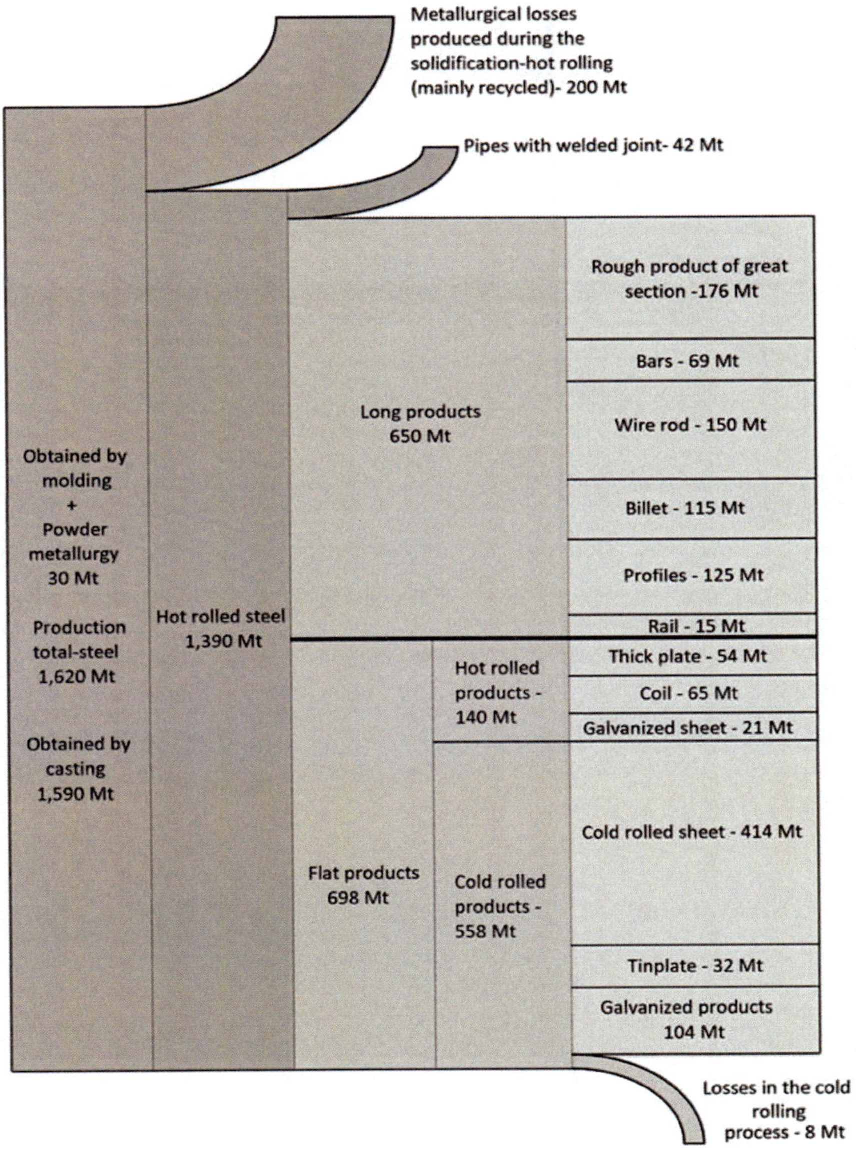

Fig. 5.1 Estimation of the distribution of the main iron and steelmaking products in 2015

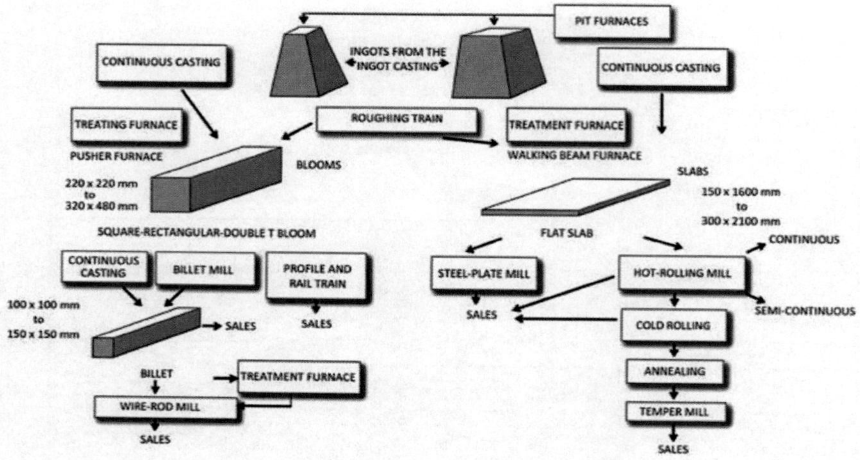

Fig. 5.2 Process steel forming by solidification in ingot mold, continuous casting, hot rolling, and finishing (by cold rolling)

Figure 5.2 indicates the products obtained in the solidification of liquid iron when composition and temperature are adjusted (secondary metallurgy): discontinuous casting (in ingots) and subsequent processes or continuous casting for flat (slabs) or long (blooms and billets) products.

Normally, the finishing of the parts by open forge is preceded by a process of hot rolling, although the open forge can be also directly performed on rough product of the ingot casting. We will describe in the next sections the physical-metallurgical principles of the hot-working.

5.2 Hot Working

When a metallic material, of recrystallized structure, is hot worked, at an adequate temperature T_H, a strain hardening, and a softening of the deformed material take place simultaneously. That is, two mechanisms compete at the same time, antagonist, to create and eliminate crystalline defects: the hot working conferred during this deformation and the dynamic recovery of the structure (or dynamic recrystallization) at the forming temperature.

Definitely, the material is less resistant, and more ductile, in hot state than in cold state. This allows—with less stress than that required for the cold forming—achieving greater deformations without cracks in the material.

Generally, there is a balance in the competition between the hot working and the dynamic recovery or recrystallization: it is possible to achieve an equilibrium regime—from a certain deformation—where when we apply a steady stress, the material deforms plastically without either softening or hardening.

Temperature T_H at which it is necessary to make the hot working to widely obtain these improvements of ductility should be $T_H > 0.75 \cdot T_E$ (in the case of the iron, whose melting point is 1538 °C, $T_H > 1811$ K·0.75 = 1358 K (1085 °C)), where T_E is the melting point of the material expressed in degrees Kelvin. Only above the temperature $0.5 \cdot T_E$ is possible the self-diffusion, and this is necessary to make possible the diffusion of the vacancies toward—or from—the dislocations. This way, it is possible to achieve a recovery/recrystallization, either dynamic or static, by annihilation and climbing of the dislocations produced during the forming.

The climbing of dislocations participates in an important way, together with other movements of dislocations, during any type of hot deformation. This participation is almost exclusive, with diffusion of vacancies, when the deformation is at high temperature, habitually lower than that used for the hot working, but always $T_H > 0.5 \cdot T_E$, by application of small tensions that produce slow deformation rates (as for instance in creep and superplastic behavior).

When at high temperatures very high stresses are used, dynamic recovery can be noticeable in some alloys, due to not only the softening produced by the climbing of dislocations but also because the great stress used, apart from producing dislocations, facilitates also their mobility at high temperature—by sliding and cross slip (apart from climbing)—to produce sub-grain structures, as happens with the ferrite. In the case of bcc metals and alloys and with high stacking-fault energy (stacking-fault energy >150 erg·cm^{-2}), such as aluminum, tend to lose the work hardening by polygonization, and thus, to widely recover by thermal activation, they dynamically recover during the hot working, but without dynamically recrystallizing. The dynamic recrystallization during the hot working takes place only in fcc metals and alloys with low stacking-fault energy (stacking-fault energy <150 erg·cm^{-2})—austenite, copper, nickel, alpha brass, alpha cupraluminum, cupronickel—for great deformation rates.

Figure 5.3a summarizes the hot extrusion behavior (which means great deformation rate) of metals or alloys with high stacking-fault energy: dynamic recovery during the deformation takes place and, subsequently, static recrystallization happens. This way, the structure of the extruded profile is formed by equiaxed grains.

In alloys with low stacking-fault energy, its behavior in extrusion processes differs from those of high stacking-fault energy (Fig. 5.4a). Apart from the dynamic recovery, dynamic recrystallization is produced during the deformation, followed by static recrystallization.

In Fig. 5.3b, the structural behavior of metals of high stacking-fault energy during a reduction of small quantity (lower deformation rate than in the extrusion) is mentioned, as happens in the hot rolling. In this process, dynamic recovery is produced during the deformation and static recovery is produced after the deformation. Thus, the slab, like this rolled, would have lengthened grains with morphology analogous to work-hardened grains (Fig. 5.5), although inside of the grains it would

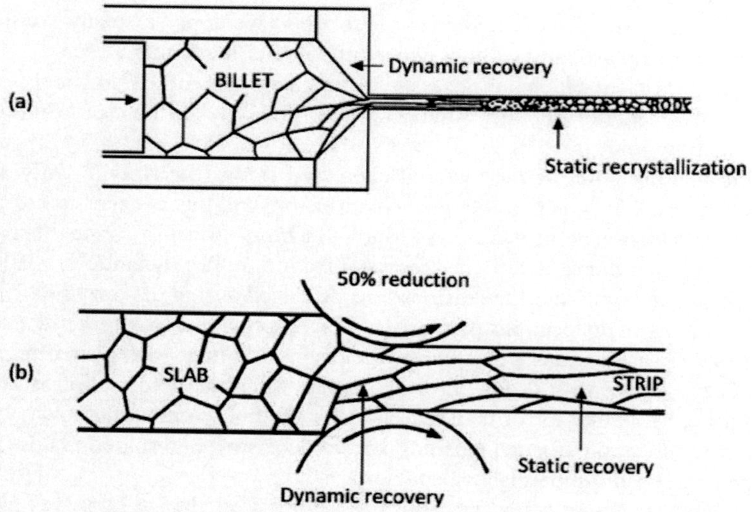

Fig. 5.3 Deformation at high temperature of metals with high stacking-fault energy **a** Extrusion, **b** Rolling

be possible to appreciate—by electron microscopy—the existence of equiaxed polygonized substructures (Fig. 5.6) (this question does not impede that, later, a static recrystallization between passes or at the end of the hot working could take place).

The average diameter (d) of these sub-grains decreases when the deformation temperature decreases, if we use the same deformation rate, $\dot{\varepsilon}$, (for instance, the same rolling rate to pass from a thickness h_0 to a thickness h).

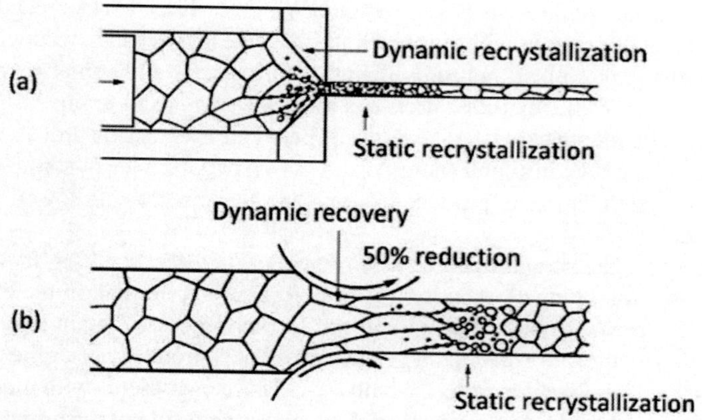

Fig. 5.4 Deformation at high temperature of metals with low stacking-fault energy **a** Extrusion, **b** Rolling

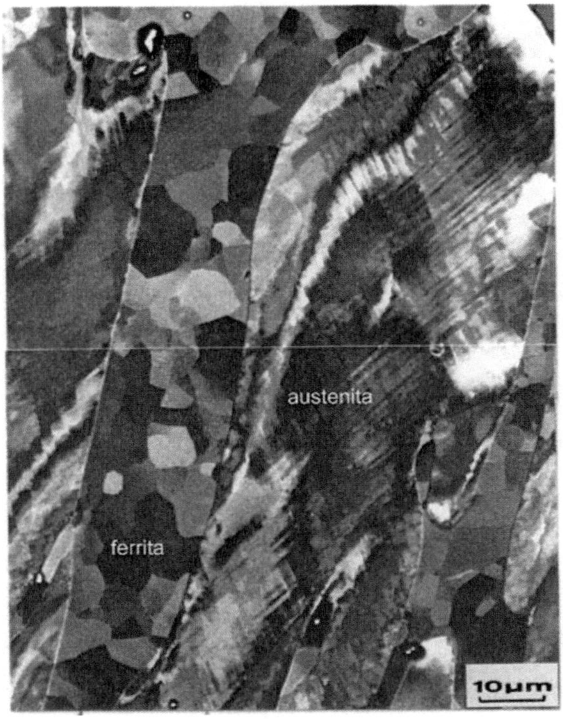

Fig. 5.5 Ferritic/austenitic stainless steel (T = 1000 °C, $\varepsilon = 1, \dot{\varepsilon} = 1\,s^{-1}$)

Fig. 5.6 Polygonization of ferritic steel (sub-grains) (T = 917 °C, $\varepsilon = 0.2$, $\dot{\varepsilon} = 5\,s^{-1}$)

We admit that the value of the diameter d (in μm) is calculated as follows:

$$d^{-1} = a + b \cdot \log Z \tag{5.1}$$

$$\sigma_s = a + b \cdot d^{-1} \tag{5.2}$$

where σ_s is the stationary stress, a and b are two empirical constants of the material, and Z (Zener-Hollomon parameter) is equal to:

$$Z = \dot{\varepsilon} \cdot \exp\left(\frac{Q}{R \cdot T}\right) \tag{5.3}$$

where T is the temperature, R is the gas constant, and Q is the activation energy for the diffusion processes.

In the hot rolling of low staking-fault energy alloys (Fig. 5.4b), dynamic recovery during the deformation takes place, while static recrystallization takes place after the deformation. Thus, the sheet would have a structure of regular equiaxed grains.

5.2.1 *Flow Stresses in the High Temperature Deformation. Constitutive Equation*

If we assume that, by hot extrusion (Fig. 5.3a), the cross section, S, of a steel billet is reduced to a cross section, s. The real deformation rate:

$$\dot{\varepsilon} = \frac{d\varepsilon}{dt} \tag{5.4}$$

will depend on both the stress used in extrusion (force/cross section) and temperature of the material. The greater the applied stress, the hot deformation could be performed either without increasing the work hardening (thus, keeping constant the applied stress), at lower temperatures (although greater than $0.5 \cdot T_E$), or with greater deformation rates.

This constant stress (stationary), σ, (that depends on the T and $\dot{\varepsilon}$), which makes possible the hot deformation without increasing the work hardening, due to an equilibrium between the creation of crystalline defects and the dynamic recovery, is called *stationary constant stress*. In the case of the hot working, the dependency between σ, T and $\dot{\varepsilon}$ (constitutive law) follows the next relationship:

$$A \cdot [\sinh(\alpha \cdot \sigma)]^k = \dot{\varepsilon} \cdot e^{\frac{Q}{R \cdot T}} = Z \tag{5.5}$$

where A and α are constants of the material, k is a constant that takes values close to 5 in most of the metals, Q is the activation energy for the dynamic recovery, R is the gas constant and T is the forming temperature expressed in Kelvin. The constitutive law depends on stress, temperature, strain rate, and no longer on deformation. The value Z, as it was previously indicated, is the Zener–Hollomon parameter.

As opposed to the forming at high temperature, the cold forming of metallic materials depends chiefly only on the work hardening (on the deformation, ε, and the strain hardening coefficient, n). The cold behavior law is of the type: $\sigma = f(\varepsilon)$. It does not currently depend on $\dot{\varepsilon}$ (except for superplastic behavior), that is to say, it does not depend on the time, and is usually fitted to the following relationship proposed by Ludwik-Hollomon:

$$\sigma = k \cdot \varepsilon^{n} \tag{5.6}$$

We should take into account that for any metallic material, the applied stress, σ, depends on the temperature, T, and on the deformation rate, $\dot{\varepsilon}$—for a given temperature, the value of σ is greater the required $\dot{\varepsilon}$—but it does not depend on the reduction degree, ε (always and when a certain critical limit of deformation is not exceeded).

We should also notice that for each applied stress, the deformation rate exponentially increases with the temperature:

$$\dot{\varepsilon} = \text{constant} \cdot e^{-\frac{Q}{R \cdot T}} \tag{5.7}$$

It is possible to see in Fig. 5.7 the deformation rate as a function of the temperature.

The deformation rate follows an Arrhenius type equation as happens in other situations, for instance in the diffusion of a liquid in other, oxidation of metals, corrosion, grain growth in the solidification, etc. In the hot law of behavior, deformation rate is determined by the mass diffusion.

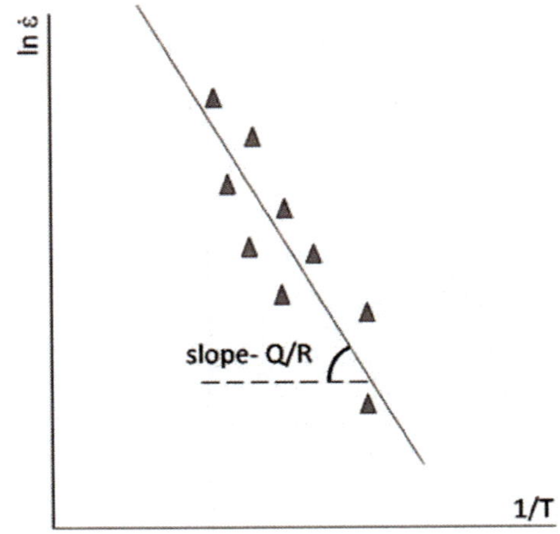

Fig. 5.7 Deformation rate, $\dot{\varepsilon}$, as a function of the temperature (for applied stress constant, σ_1)

It is important to mention that the law of behavior for the deformation at high temperature and grain size greater than 1 μm is universal: it is valid either for the hot working or the deformation as a result of the creep, even when both processes differ on the applied stresses and deformation mechanisms.

For very fine-grained materials and superplastic materials, m approaches to 1 ($0.3 < m < 1$), the value of σ also depends on the grain diameter, d. The law of behavior is, in this case, as follows:

$$\dot{\varepsilon} = \frac{A \cdot \sigma \cdot \exp\left(-\frac{Q}{R \cdot T}\right)}{d^2} \tag{5.8}$$

We propose an example of superplastic behavior in a structural ship building steel at the end of this chapter.

5.2.2 The Hot Forming ($T_H > 0.75 \cdot T_E$)

We call *hot working or forming* to that process carried out at temperature higher than $0.75 \cdot T_E$, which produces great deformations (ε between 0.5 and 5) at deformation rates $\dot{\varepsilon}$ within 10^{-2} and 10^3 s^{-1}. This requires applying stresses according to the constitutive law. This happens, for instance, in the following processes:

Hot rolling: at rates, $\dot{\varepsilon}$, between 1 s^{-1} and 10^2 s^{-1}. The process is performed in successive passes whose deformation, ε, is comprised within 0.2 (which is equivalent—we should remember that $\varepsilon = \ln(L_t/L_0)$—to $L_t = 1.22 \cdot L_0$, that is, is equivalent to a reduction of the cross section of the 22%) and 0.5 (which is equivalent to a reduction of the cross section of the 65%). Cumulated deformation, in successive passes, can reach values of ε of around 5.

Hot extrusion, in a single pass: at rates $\dot{\varepsilon}$ between 10 s^{-1} and $5 \cdot 10^2$ s^{-1}, this allows deformations ε between 2 and 5; or which is the same, final sections between 7 and 15 times smaller than the initial.

Hot forge: at rates $\dot{\varepsilon}$ between 10^{-2} s^{-1} and 5 s^{-1}, with deformations ε, each time, between 0.2 and 0.5 (sections after each reduction between 1.22 and 1.65 times smaller than the initial). With accumulated deformations, sum of the ε experienced each pass, between 0.5 and 2; or which is the same, with a final section equal to the initial section divided by a value comprised within 1.65 and 7.4.

Hammer forge: $10\ s^{-1} < \dot{\varepsilon} < 10^3\ s^{-1}$ and $0.2 < \sum \varepsilon < 2$.

For hot deformations, with values of $\alpha \cdot \sigma$ greater than 1.2, the constitutive law can be simplified and expressed as follows:

$$C \cdot e^{k \cdot \alpha \cdot \sigma} = \dot{\varepsilon} \cdot e^{\frac{Q}{R \cdot T}} = Z \tag{5.9}$$

or:

$$\dot{\varepsilon} = C \cdot e^{k \cdot \alpha \cdot \sigma} \cdot e^{-\frac{Q}{R \cdot T}} \tag{5.10}$$

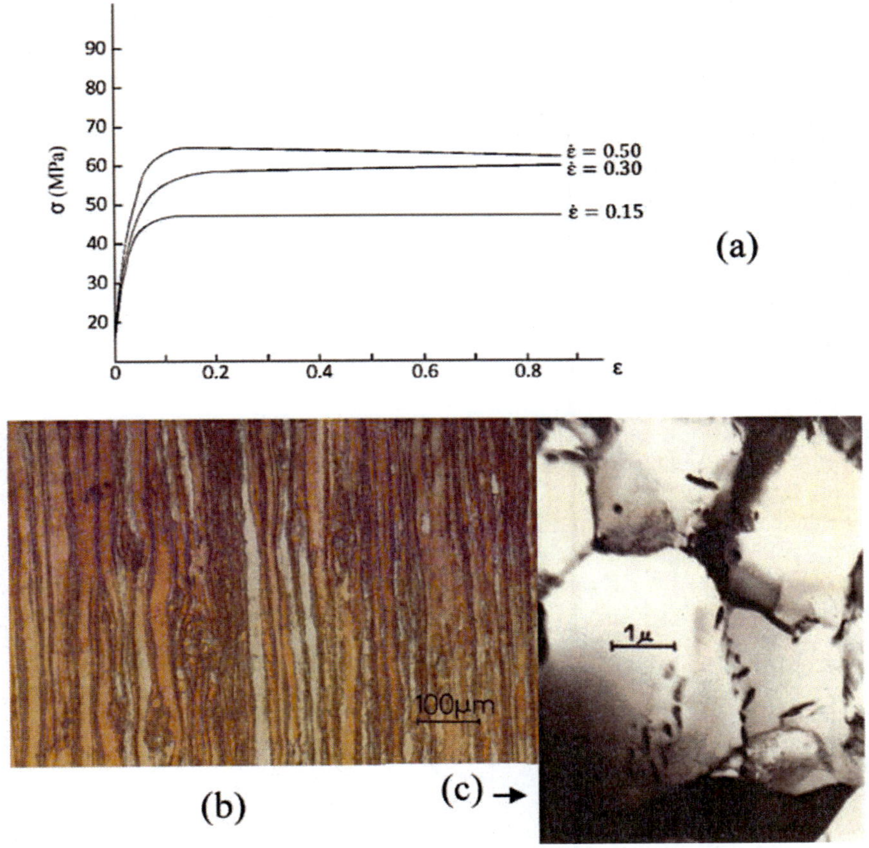

Fig. 5.8 **a** Dynamic recovery at 917 °C; **b** Transverse section parallel to the hot-rolled direction of the sheet of stainless ferritic steel (0.061% C–16.67% Cr) deformed at 900 °C; **c** TEM of this sheet showing a recovery substructure

We should remember that:

$$\text{sinhx} = \frac{\left(e^{x} - e^{-x}\right)}{2} \tag{5.11}$$

Thus, for great values of x, e^{-x} is close to 0 and, consequently:

$$\sinh x = \frac{e^{x}}{2} \text{ (for medium and low values of x, } \sinh x = \frac{(1 + x) - (1 - x)}{2} \simeq x, \text{ creep and superplastic behavior)} \tag{5.12}$$

It is possible to see in Fig. 5.8 the stress-deformation curves (Urcola Galarza and Fuentes Pérez 1980a, b, 1981), corresponding to the deformation at 917 °C of a ferritic stainless steel for different deformation rates. It is possible to appreciate in each one of the curves that a stationary regime is reached for σ—due to equilibrium between hardening and softening (with values of the work hardening coefficient, n, equal to zero)—when the plastic deformation ε exceeds around 0.2. Before reaching the stationary regime, it is observed a transient regime during which the stress grows very fast due to the work hardening.

The values of σ, during the transient regime, are usually comprised between $\sigma_y/50$ (for high temperatures and low $\dot{\varepsilon}$) and $\sigma_y/5$ (for low temperatures and great $\dot{\varepsilon}$). Both these values of σ as those corresponding to the transient regime grow when increasing the deformation rate $\dot{\varepsilon}$. Stress level, in one and the other regime, grows when the deformation temperature decreases. σ_y is the yield strength on the cold deformed material.

In previous lines, we have considered that during the deformation at high temperature only dynamic recovery is produced. But, when there is dynamic recrystallization, the aspect of the stress–strain curves is different from those previously indicated. Figure 5.9 shows the stress–strain curves for steel with 0.25% C deformed in austenitic state at 1100 °C ($0.76{\cdot}T_E$). For high values of $\dot{\varepsilon}$, it is possible to see the typical shape of the dynamic recrystallization: after a peak corresponding to the transient regime, it is observed that stress falls, a softening is produced as a result

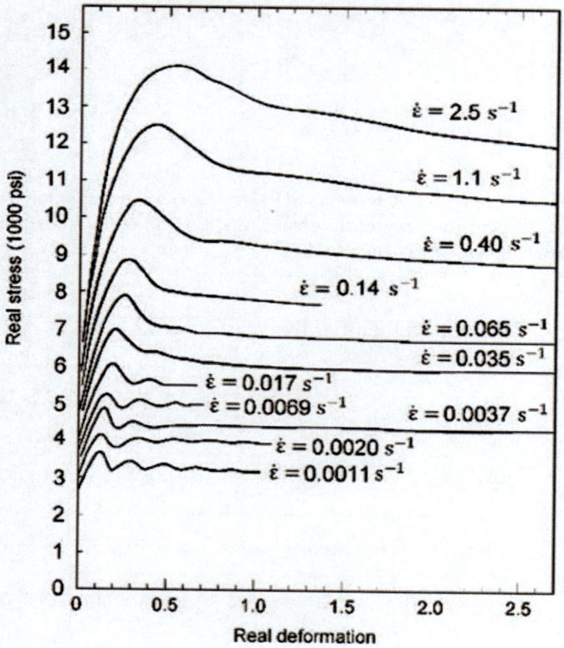

Fig. 5.9 Dynamic recrystallization (True Stress-True Strain curves at different strain rates)

of the dynamic recrystallization, followed by a stationary stress as a consequence of dynamic recrystallization.

With respect to the materials that dynamically recrystallize, everything seems to indicate that the law of behavior is the same as that indicated in Eq. (5.11) if the following in equation is verified:

$$\frac{d \ln Z}{d\varepsilon} < 2 \tag{5.13}$$

As hot ductility is concerned, that is to say, the maximum admissible deformation without producing the breaking of the material, everything seems to indicate that is almost independent of the deformation rate, $\dot{\varepsilon}$, and depends on the temperature following Arrhenius's law:

$$\varepsilon_r = \text{constant} \cdot e^{-\frac{Q}{R \cdot T}} \tag{5.14}$$

However, some of the factors that have limited influence on the hot mechanical resistance have, on the contrary, noticeably influence on the ductility. This happens, for instance, with few deformable inclusions—alumina, aluminates, some types of sulphides, etc.—that noticeably reduce the ductility due to the formation of cavities during the deformation—damage—in the interface of inclusion and matrix.

Biphasic structures ($\alpha + \beta$) have also influence on the ductility. This way, two phases whose individual ductility was high, have a noticeably lower ductility when they are part of a biphasic material. This is probably due to the weakness of α/β interphases. Fractographic study of austenitic-ferritic steels shows a special tendency for the clustering of cracks in the interface that separates both phases, Fig. 5.5.

It is obvious that chemical segregations also noticeably affect the hot ductility. It is sufficient to remember the possibility of intergranular melting or burning effects and the interest of homogenization heat treatments before the forge in the case of as-cast structures, as well.

5.2.3 *Improvements by Hot Forming of Solidification Structures*

As-cast structures usually have physical—columnar structure, macro-shrinkage, micro-shrinkage, blow holes, porosity, internal cracks, etc.—and chemical—micro-segregation, dendritic segregation—heterogeneities.

Hot forming, whose primary aim is to configurate the final shape of the part, allows, complementarily, correcting some of the physical and chemical heterogeneities of the as-cast structure.

This way, for instance, squeezing of dendritic grains at high temperature approaches the nucleus of the crystals to the periphery. With this, the homogenization of the dendritic structure is favored as the distance that the atoms should cover during the diffusion is smaller. Obviously, the major segregation of an ingot cannot be eliminated by hot forging (it cannot be eliminated by homogenization annealing).

However, work hardening conferred by the deformation, if the temperature of conformation is adequate, induces the recrystallization of the structure. Internal discontinuities, when free surfaces—squeezed and compressed between them as a result of the conformation—are completely clean, almost disappear by solid welding. Definitely, the part is mechanically and metallurgically free of defects.

Recrystallization nucleus, located at one and other side of the joints/grain boundaries, can cross, during their growth, that boundary, and the discontinuity is welded. On the contrary, when the free joint surface is oxidized by contact with the air, the cavity cannot be welded by recrystallization during the hot mechanical forming.

As a result of an adequate combination of deformation rate and temperature, it is possible to correct the physical heterogeneities and it is possible to achieve a grain refinement with respect to that of the as-cast solidification.

Recrystallization makes also possible the obtaining of metallic materials from powders. This requires the previous obtaining of powders, subsequent compaction, and final sintering. Interconnected porosity of the compacted part, but not sintered, is reduced during the sintering. This way, the density of the part is increased, and the properties improve if compared with those of the simply compacted product due to diffusion of vacancies and recrystallization of deformed powders. After the sintering process, remaining porosity is not connected. This porosity appears in the form of isolated pores, whose distribution, shape, and size depend on the degree of cold deformation, temperature, sintering time, and atmosphere used in the process.

Forming of ingots and semi-finished products obtained by solidification orientates the intermetallic precipitates, if there are, and the possible inclusions. If the inclusions are plastic, they deform in the direction of forming; if they are brittle, they can be broken, apart from aligned, in that direction.

All this confers a forge fiber structure and an anisotropy of properties in the lengthwise and crosswise directions. As a consequence of the fiber structure, lengthwise characteristics, and concretely, toughness, are greater than the crosswise characteristics. It is convenient to mention that the forge fiber structure, of microscopic nature, is different, although it can be complementary, than the texture (whose nature is crystalline).

5.2.4 Banded Structure (Mechanical Fibering)

It is convenient to mention here a particular type of fibered structure (composite), obtained by hot forming, called banded structure. It is the heritage of the as-cast state, particularly if peritectic reactions take place in the progress of the solidification. It usually appears after hot forming of solid solutions of low partition coefficient, when the solvent metal has different allotropic states, at the forming temperature and at

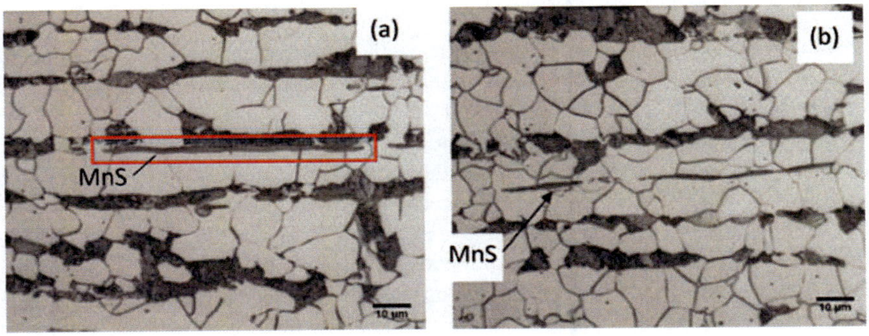

Fig. 5.10 **a** Banded structure in ferritic-pearlitic steel. Transverse section parallel to the rolling direction (MnS in the pearlite). **b** Banded structure in ferritic-pearlitic steel. Transverse section parallel to the rolling direction (MnS inclusions in the ferrite)

room temperature, and if the solute elements favor one or another allotropic state from forming temperature to room temperature (see Pero-Sanz et al. 2017).

Figure 5.10a and b show this banded structure—alternated bands of ferrite and pearlite—which is habitual in some low wrought alloy steels, as-forged (hot rolled).

The genesis of the banded structure can be summarized as follows. During the forming process of steel, the contours of the dendrites are enriched in elements in solid solution (we should remember that the partition coefficient, K, of the elements in the steel is always <1), either alphagenous or gammagenous.

During the forming of the as-cast steel, in gamma state, carbon easily diffuses. Thus, carbon content in the austenite is similar everywhere, but it is not the case of other elements (additions or impurities) that are segregated in grain boundaries, and where the solidification takes place in the first moment of the process.

Manganese sulphides usually appear as precipitates or as constituents of a eutectic phase in the periphery of the dendrites. These manganese sulphides align later in the forming direction. The zones that accommodate the sulphides (that corresponds to the zones that were the contour of the dendrites) have always a greater content in elements that are in solid solution. This is because most of the solutes, as opposed to carbon, hardly diffuse during the forming (at least if—before the forming—steel was not subjected to a long duration and noneconomic homogenization annealing heat treatment).

The alphagenous or gammagenous character of the elements in solid solution is, precisely, responsible for the banded structure, as the values of the A_3 (and also A_1) are different in the segregated zones and in the non-segregated zones.

We should assume, for instance, that elements in solid solution were only alphagenous, v. gr. small silicon and phosphorus contents, as happens in the carbon steels of Fig. 5.10b. When the temperature decreases, once the steel was hot formed, the first zones that are transformed from gamma into alpha are, these ones, as a consequence of the higher A_3 temperature, enriched in the alphagenous elements. As a result of this, carbon atoms expelled from these ferrite grains pass later to the contiguous austenite. This austenite is enriched in carbon and, for that reason, the A_3 is reduced. Therefore, these zones are later transformed than the first ones during the continuous cooling, and pearlite bands are produced.

If solute elements were all of them gammagenous or counterbalance alphagenous elements, instead of alphagenous, the banded structure would have been also produced. It would result—by analogous reasoning—that the segregated zones would transform in the last place and would give pearlite. These pearlite bands, at the same time, would be flanked by the bands of ferrite grains previously formed, and sulphides would appear located in the pearlite bands instead of being in the ferrite ones (this happens, as said above, when the alloying elements have gammagenous nature), Figs. 5.10a and 5.11.

Sometimes, there is not banded structure for slow cooling from the austenitic state if the chemical composition of the steel is such one that, for an adequate dosage of alphagenous and gammagenous elements (antagonist), it is possible to transform at the same time, from gamma to alpha, both the segregated and nonsegregated zones (as a result of achieving the same value of A_3 in both zones).

However, there are chemical heterogeneities in this case, although bands of ferrite and pearlite would not appear. Bands are only hided/disguised (ghost microstructure). Segregated zones—rich in elements in solid solution—have more hardenability than the nonsegregated zones. For that reason, if the structure is austenitized again, and then is continuously and rapidly cooled, it is possible to appreciate these differences in the hardenability: there will appear—depending on the TTT curves corresponding to the segregated and non-segregated zones—alternated bands of either martensite and bainite, or bainite and pearlite, or martensite and pearlite, etc.

The structure in bands is non-desired for the lengthwise machining. For milling, for instance, the surface finishing is worse if the structure is banded (Fig. 5.11a) than if the structure is not banded (Fig. 5.11b). Ferrite bands have less hardness than pearlite bands and, for that reason, chip removal is different depending on the zone. In the turning, on the contrary, as a consequence of not being a lengthwise operation, the degree of surface finishing has less dependency on the banded structure.

Banded structure is also non-desirable when the vibration wants to be reduced. This way, for instance, gears with banded structure are noisier than those without banded structure.

Factors that contribute to achieve a homogeneous chemical composition, at microscopic scale, before the forming, are favorable to avoid the banded structure. The contrary also occurs. This is extremely difficult to obtain as a consequence of nonequilibrium solidification derived from peritectic reaction $\delta + L \leftrightarrow \gamma$.

Thus, a remedy to make disappearing the banded structure is the homogenization annealing at very high temperature (which results more effective whether it is applied after a forge applied before the definitive hot forming). The counterpart is coarse grain microstructure that impairs mechanical properties, chiefly strength, and toughness, simultaneously.

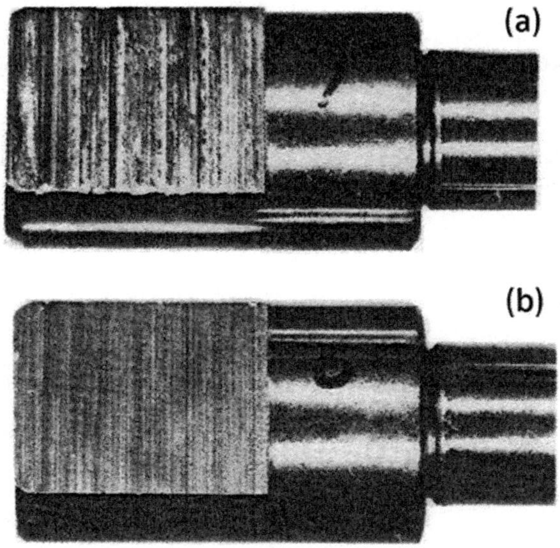

Fig. 5.11 Influence of the banded structure in the milling

Logically, the chemical heterogeneity responsible for the bands is less accused when the solidification takes place with crystallization in dendrites of small size. Steels with fine solidification grains, achieved by an increase of the nucleation rate—by solidification in ingots of small size, or by means of nucleation agents, grain refinement agents, etc.—do not usually have structure in bands because they homogenize easier during the hot forming. And, in any case, the chemical heterogeneity can be faster removed by a homogenization annealing.

Ultra-low carbon steels, with carbon content <0.02%, do not show banded structure because it is not possible the formation of pearlite. Neither the banded structure is showed in steels with carbon content greater than 0.6%, which do not experiment peritectic reaction in solidification. It is possible to appreciate the banded structure in steels whose carbon content is comprised within 0.1 and 0.5%. These steels, whose solidification takes place by peritectic reaction, are more prone to the banded structure than the high-carbon steels that directly solidify in gamma without peritectic reaction.

On another note, as example, in steel with 0.15% C, proportions of delta solid (approximately 0.1% C) and gamma solid (approximately 0.2% C) are almost the same at the end of the peritectic solidification. Later, once the delta→gamma allotropic transformation took place, the ancient regions of gamma still contain almost twice the carbon content than the others. Thus, apart from the heterogeneity referred to the solute's composition, it is added a heterogeneity referred to the carbon content.

On another note, always when there is peritectic reaction, both in ferrous and nonferrous alloys, equilibrium is not complete. Consequently, also in those steels, whose carbon content is comprised within 0.17 and 0.53% and whose primary structure of solidification should be only gamma, exhibit a discontinuity in the chemical composition as a consequence of the peritectic reaction (liquid + delta → gamma) that takes place before the end of the solidification.

Even steels whose equilibrium solidification structure is only delta, v. gr. steels with carbon content <0.09%, can also present peritectic reaction: by nonequilibrium solidification (Pero-Sanz et al. 2017).

Thus, the tendency to give banded structure depends, markedly, on the delta and gamma proportions, and, consequently, on the carbon content in the steel.

When it was not possible to eliminate the chemical heterogeneity responsible of the bands, it is possible to disguise the banded structure as it is indicated in the following paragraphs. With this, it is possible to facilitate the regularity in machining operations, although the bands would survive, and they can appear in posterior heat treatments.

For instance, the banded structure will not appear if the cooling of the steel, instead of being performed in the air from the forming temperature, is isothermally carried out in hot salts: at temperatures—below A_1—corresponding to the upper pearlite zone of the TTT curve. This way, it is possible to simultaneously transform both the segregated and non-segregated zones to give only pearlitic structure (more favorable for machining operations). Though, proceeding in that form, and as the heterogeneity regarding the chemical composition still exists, if this structure is austenitized again, and later, an air cooling was performed from this temperature—for example, a normalizing treatment after the machining—the banded structure would appear again. Thus, it would be possible to check that the above-indicated isothermal transformation, or the utilization of an isothermal annealing treatment, would have led to only disguise/hide the banded structure.

The bands can be also disguised by means of a full austenization annealing if the permanence at temperatures higher than A_3 is extended to coarse the gamma grain. The objective is to achieve that the diameter of the austenite grain was greater than the distance that separates the segregated and non-segregated bands, adjacent and parallel. This way, if the width of the chemical heterogeneity is located inside of the austenitic grain, bands will not appear during the gamma-alpha transformation that takes place during the continuous cooling. Anyway, the coarsening homogenization annealing—which has associated low toughness inherited from the coarse austenite grain—only has justification to facilitate the machining, and is, for that reason, an intermediate treatment, not a final treatment.

As said, steel with alternated bands of ferrite and pearlite can be considered as a composite material because pearlite fibers (which, at the same time, comprise ceramic material, cementite, embedded in a metallic material, ferrite) which in turn are also surrounded by a metallic matrix of ferrite.

The banded structure accentuates the anisotropy of properties between the forming direction and the perpendicular direction, which, as it was previously indicated, can be unfavorable. However, as happens with textures, this anisotropy is advantageous for some directional behaviors. For example, in a banded structure, toughness is enhanced when specimens are machined parallel to the rolling direction than in the transverse one. The same accounts for toughness behavior (V-notch in the thickness direction).

Knowledge about the formation of the bands provides the fundamentals to obtain, by means of the conventional processes, products with directional properties.

5.2.5 *Thermal–Mechanical Treatments with Phase Transformation*

We have previously referred to the structural consequences of forming at a temperature that allowed either the dynamic recovery, the static recrystallization, or the dynamic recrystallization of one steel. However, we have not paid much attention to that, additionally, the base metal can experience allotropic transformations, or phase transformations resulted from solubility changes, whose effects complement the recovery and/or recrystallization during the hot forming.

This, on one side, provides the basis to improve the properties of the alloy by thermal–mechanical treatment (thermal–mechanical controlled rolled processes, TMCRP). On the other side, this justifies some of the structural particularities that produce anisotropy, such as the forge fiber structures, or the above-mentioned banded structure (mechanical fibering).

5.2.5.1 Thermal–Mechanical Treatments in Steels

When a metal has several allotropic states, for instance, iron and titanium, its dynamic recovery or its static or dynamic recrystallization can be previous, simultaneous, or posterior to the phase transformation.

These varied possibilities of thermal–mechanical forming substantiate the research and development of different thermal–mechanical processes of the austenite to—for the same chemical composition—improve the mechanical characteristics of the steel.

The thermal–mechanical processes of the austenite can take place either before the transformation of the austenite (*controlled rolling (TMCRP) and ausforming*) or during the transformation of the austenite (*isoforming, Transformation Induced Plasticity steels (TRIP steels)*) (Pero-Sanz 2004).

Other thermal–mechanical treatments are carried out after the transformation of the austenite. This is the case of either the hot rolling of mild steel, at temperatures lower than A_1, to give non-recrystallized grains of alpha at the end of the rolling, or the pearlite-forming and the marforming.

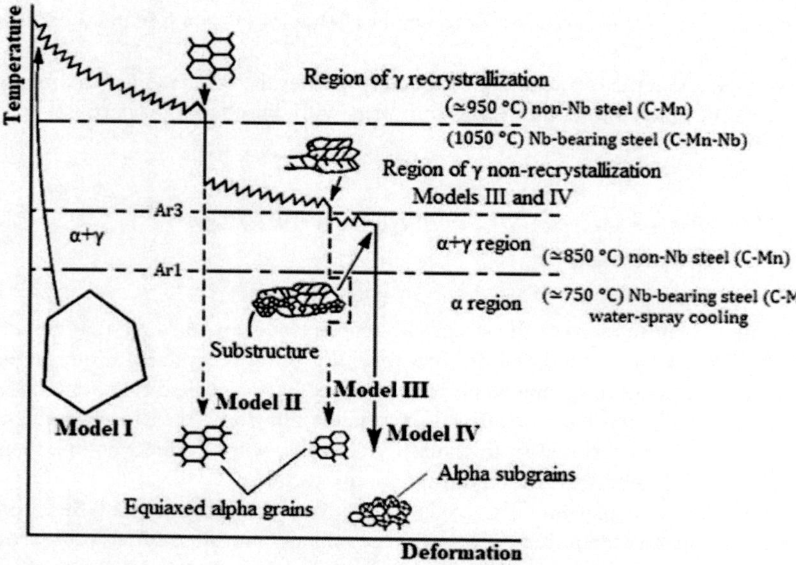

Fig. 5.12 Scheme of three processes of controlled rolling and structures

Figure 5.12 schematically shows the changes of structure from gamma to alpha in four rolling processes, of the same steel, in gamma phase. By means of the suitable schedule of rolling it is possible to delay the recrystallization of the austenitic structure and to obtain at temperature higher than A_3, a non-recrystallized (deformed) structure at the end of the rolling. They are the following (Tanaka 1981):

Model I: Very high-temperature deformation. High coarse austenite grain size. Coarse ferrite size after $\gamma \rightarrow \alpha$ phase transformation. Low mechanical properties.

Model II: High-temperature deformation. Coarse austenite grain size. Less coarse ferrite grain size after $\gamma \rightarrow \alpha$ phase transformation. Commercial properties.

Model III: Roughing and finishing schedule of rolling (TMCRP). Finishing operation is conducted in the range of non-recrystallized austenite, accumulating deformation up to approximately 0.7. After $\gamma \rightarrow \alpha$ phase transformation, ferrite microstructure is very fine (about 5 μm or less). Nb is essential.

Model IV: Idem that *Model III* but steel experiences some $\gamma \rightarrow \alpha$ phase transformation at final finishing. Irregular but fine ferrite (deformed and non-deformed) microstructure. High strength but low toughness than *Model III*.

Obtained by this controlled rolling, the fine gamma microstructure of non-recrystallized austenite, when is later allotropically transformed during the cooling, provides regular grains of alpha phase. The size of the alpha grains is very fine (Fig. 5.13), noticeably smaller than that obtained if the austenite grains, at the end of the rolling, would be recrystallized. The refinement of the like this achieved structure, provides significant improvements of both yield strength and toughness of the steel

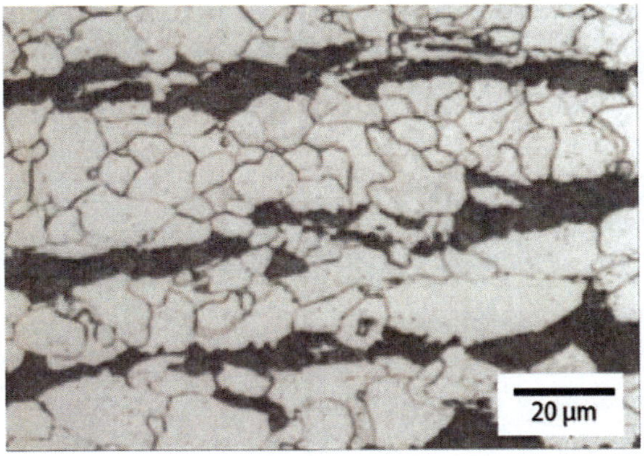

Fig. 5.13 Micro-alloyed steel (0.17% C, 1.36% Mn, 0.026% Ti, 0.033% Nb, 0.028% Al, 0.45% Si, 0.022% P, 0.009% S). Obtained by hot-controlled rolling (ship building plate 1-inch thickness)

(micro-alloyed HSLA steels). The presence of micro-alloyed Nb is fundamental; also, of Ti and V, all in contents <0.1% (Ti $\simeq$ 0.02%; V $\simeq$ 0.8%).

5.3 Forces and Geometrical Relationships in Rolling

Rolling is a metal forming process by which the section of a slab or billet is reduced by compressive plastic conformation. When a slab is rolled (see Fig. 5.14) the section is reduced from h_0 to h_1 over a length l as it passes through the rolls. At first sight, it might appear that there would be no sliding (and thus no friction) between the slab and the rolls since the rolls move with the slab. But the metal is elongated in the rolling direction, so it speeds up as it passes through the rolls and some slipping is inevitable. If the rolls are polished and lubricated (as they are for precision and cold rolling), the frictional losses are small. We shall ignore them here (though all detailed

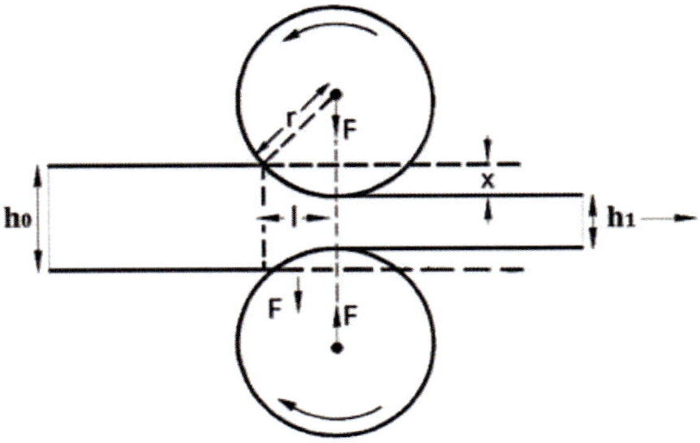

Fig. 5.14 The rolling of metal sheet

treatments of rolling include them) (see Dieter 1976) and calculate the rolling force and the rolling torque for perfectly lubricated rolls.

From the geometry of Fig. 5.14:

$$l^2 + (r - x)^2 = r^2 \rightarrow l^2 \simeq 2 \cdot r \cdot x = r \cdot \Delta h \rightarrow l = \sqrt{r \cdot \Delta h} \tag{5.15}$$

if:

$$x = \frac{1}{2} \cdot (h_0 - h_1) \tag{5.16}$$

is small (as it almost always is).

The rolling force F must cause the metal to yield over the length l and width w (normal to the figure). Thus:

$$F = \sigma_y \cdot w \cdot l \tag{5.17}$$

If the reaction on the rolls appears halfway along the length marked l, as shown on the lower roll, the torque (T) is:

$$T = \frac{F \cdot l}{2} = \frac{\sigma_y \cdot w \cdot l^2}{2} = \frac{\sigma_y \cdot w \cdot r \cdot \Delta h}{2} \tag{5.18}$$

The torque required to drive the rolls increases with yield strength, σ_y, so hot rolling (when σ_y is small) takes less power than cold rolling. It obviously increases with the reduction in section $(h_0 - h_1)$ and it increases with roll diameter 2r; this is one of the reasons why small-diameter rolls often backed-by two or more rolls of larger diameter (simply to stop them bending) are used.

5.3.1 Solved Exercises

Exercise 5.1 Constitutive law and flow rules in rolling.

Solution starts:

From the mechanical point of view, the process of pure rolling takes place under conditions of plain strain (Fig. 5.15). Considering a volume element: 1, rolling direction; 2, width direction; 3, thickness direction, this deformation conditions involve:

$$\varepsilon_1 = -\varepsilon_3 \tag{5.19}$$

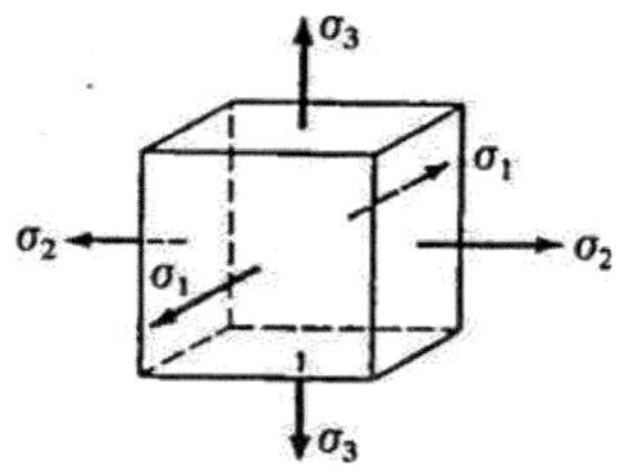

Fig. 5.15 Principal stresses, positives

$$\varepsilon_2 = 0 \tag{5.20}$$

Plastic deformation takes place at constant volume:

$$\varepsilon_1 + \varepsilon_2 + \varepsilon_3 = 0 \tag{5.21}$$

Assuming 1, 2, and 3 as principal directions of stresses and deformations (parallel between them), we apply the criterion of Von Mises:

$$\bar{\sigma}^2 = {\sigma_y}^2 = \frac{1}{2}\cdot\left[(\sigma_1-\sigma_2)^2+(\sigma_1-\sigma_3)^2+(\sigma_2-\sigma_3)^2\right] = \mathrm{F}(\text{plastic function}) \tag{5.22}$$

where σ_y is the yield strength calculated in tension.

Considering also the law of plastic flow (Levy-Von Mises):

$$\frac{d\varepsilon_1}{\frac{\partial F}{\partial \sigma_1}} = \frac{d\varepsilon_2}{\frac{\partial F}{\partial \sigma_2}} = \frac{d\varepsilon_3}{\frac{\partial F}{\partial \sigma_3}} = \lambda \tag{5.23}$$

We can deduce as:

$$\varepsilon_2 = 0 \tag{5.24}$$

due to there is no deformation in the width, that:

$$\frac{\partial F}{\partial \sigma_2} = 0 \tag{5.25}$$

and we can establish the following state of stresses:

$$\frac{\partial F}{\partial \sigma_2} = \frac{\partial\left\{\frac{1}{2}\cdot\left[(\sigma_1-\sigma_2)^2+(\sigma_1-\sigma_3)^2+(\sigma_2-\sigma_3)^2\right]\right\}}{\partial \sigma_2} = 0 \tag{5.26}$$

$$\frac{\partial\{\frac{1}{2}\cdot[(\sigma_1{}^2-2\cdot\sigma_1\cdot\sigma_2+\sigma_2{}^2)+(\sigma_1{}^2-2\cdot\sigma_1\cdot\sigma_3+\sigma_3{}^2)+(\sigma_2{}^2-2\cdot\sigma_2\cdot\sigma_3+\sigma_3{}^2)]\}}{\partial\sigma_2}=0 \quad (5.27)$$

$$\frac{1}{2}\cdot[(0-2\cdot\sigma_1+2\cdot\sigma_2)+0+(2\cdot\sigma_2-2\cdot\sigma_3+0)]=0 \quad (5.28)$$

as

$$\sigma_1=0 \quad (5.29)$$

$$\frac{1}{2}\cdot[(0-2\cdot 0+2\cdot\sigma_2)+0+(2\cdot\sigma_2-2\cdot\sigma_3+0)]$$
$$=0\rightarrow 2\cdot\sigma_2+2\cdot\sigma_2=2\cdot\sigma_3\rightarrow\sigma_2=\frac{\sigma_3}{2}\ \text{(compresive as } \sigma_3) \quad (5.30)$$

$$\sigma_3<0 \quad (5.31)$$

Moreover, the equivalent deformation expressed in the principal axis of the deformation:

$$\bar{\varepsilon}=\sqrt{\frac{2}{3}\cdot\left(d\varepsilon_1{}^2+d\varepsilon_2{}^2+d\varepsilon_3{}^2\right)} \quad (5.32)$$

that results in $\bar{\sigma}=\sigma_y$ (equivalent stress) and $\bar{\varepsilon}$, the general law that expresses the general relation between equivalent stresses and deformations in any forming process, referred to the principal directions of forming. Now, we are going to demonstrate several questions related to the forming process:

(a) Demonstrate that $\sigma_3=\sigma_{PD}(\text{plain strain}-\text{condition})=\left(2/\sqrt{3}\right)\cdot\sigma_y=1.15\cdot\sigma_y$ and $\varepsilon_{PD}=\left(\sqrt{3}/2\right)\cdot\bar{\varepsilon}$.

Considering the above-indicated criteria of plasticity:

$$\sigma_y{}^2=\frac{1}{2}\cdot\left[(\sigma_1-\sigma_2)^2+(\sigma_1-\sigma_3)^2+(\sigma_2-\sigma_3)^2\right]\rightarrow\sigma_1=0;\ \sigma_2=\frac{\sigma_3}{2} \quad (5.33)$$

$$\sigma_y{}^2=\frac{1}{2}\cdot\left[\left(0-\frac{\sigma_3}{2}\right)^2+(0-\sigma_3)^2+\left(\frac{\sigma_3}{2}-\sigma_3\right)^2\right] \quad (5.34)$$

$$\sigma_y{}^2=\frac{1}{2}\cdot\left[\left(-\frac{\sigma_3}{2}\right)^2+(-\sigma_3)^2+\left(-\frac{\sigma_3}{2}\right)^2\right] \quad (5.35)$$

$$\sigma_y{}^2=\frac{1}{2}\cdot\left(\frac{\sigma_3{}^2}{4}+\sigma_3{}^2+\frac{\sigma_3{}^2}{4}\right) \quad (5.36)$$

$$\sigma_y{}^2 = \frac{1}{2} \cdot \left(\frac{\sigma_3{}^2 + 4 \cdot \sigma_3{}^2 + \sigma_3{}^2}{4} \right) \quad (5.37)$$

$$\sigma_y = \sqrt{\frac{1}{2} \cdot \left(\frac{6 \cdot \sigma_3{}^2}{4} \right)} \rightarrow \sigma_y = \sqrt{\frac{3}{4} \cdot \sigma_3{}^2} \rightarrow \sigma_y = \frac{\sqrt{3}}{2} \cdot \sigma_3 \rightarrow \sigma_3 = \frac{2}{\sqrt{3}} \cdot \sigma_y \quad (5.38)$$

which is the stress that it is necessary to apply to deform the sheet.

In the case of the deformation, we proceed as follows:

$$\bar{\varepsilon} = \sqrt{\frac{2}{3} \cdot \left(d\varepsilon_1{}^2 + d\varepsilon_2{}^2 + d\varepsilon_3{}^2 \right)}; \; d\varepsilon_2 = 0; \; d\varepsilon_1 = -d\varepsilon_3 \quad (5.39)$$

$$\bar{\varepsilon} = \sqrt{\frac{2}{3} \cdot \left(d\varepsilon_1{}^2 + 0^2 + (-d\varepsilon_1)^2 \right)} \rightarrow \bar{\varepsilon} = \sqrt{\frac{2}{3} \cdot \left(d\varepsilon_1{}^2 + d\varepsilon_1{}^2 \right)} \rightarrow \bar{\varepsilon} = \sqrt{\frac{4 \cdot d\varepsilon_1{}^2}{3}} \quad (5.40)$$

$$\bar{\varepsilon} = \frac{2 \cdot d\varepsilon_1}{\sqrt{3}} \quad (5.41)$$

that is the equivalent deformation in rolling.

(b) Deformation rate and angles involved in the rolling process.

During a rolling pass, the deformation rate, $\dot{\varepsilon}$, varies within a value zero at the entrance of the stand until a value $\dot{\varepsilon}_{max}$ at the exit of the box. If the conditions of plain deformation are verified in differential elements in the arc of contact between solid and cylinder (deformation zone), it is possible to assume that the instantaneous deformation rate is given by the value of the deformation rate of a system where plain strain conditions are verified. That is to say:

$$\varepsilon = \ln\left(\frac{h}{h_0} \right) \quad (5.42)$$

and if we derivate with respect to t:

$$\dot{\varepsilon} = \frac{dh}{dt} \cdot \frac{1}{h} \quad (5.43)$$

where h is the thickness of the band in a point of the deformation zone.

From the geometry of the deformation in the rolling (Fig. 5.16):

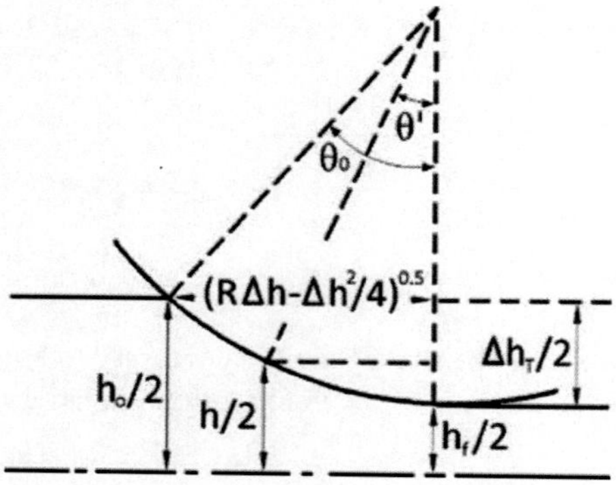

Fig. 5.16 Geometry of the rolling

$$\frac{dh}{dt} = 2 \cdot v \cdot \sin\theta \tag{5.44}$$

where v is the peripheric speed of the rolling cylinders. Additionally, we assume that during the hot rolling conditions of adherent friction are verified.

Moreover, the following equation is verified due to geometrical considerations:

$$h = h_f + 2 \cdot R \cdot (1 - \cos\theta) \tag{5.45}$$

where R is the radius of the rolling cylinder.

Thus:

$$\dot{\varepsilon} = \frac{2 \cdot v \cdot \sin\theta}{h_f + 2 \cdot R \cdot (1 - \cos\theta)} \tag{5.46}$$

If we subdivide the arc of contact in an appropriate number of sectors in such a way that the angle from the center of the cylinder is $d\theta$, it is possible to draw the variation of $\dot{\varepsilon}$ along the zone of deformation.

It is habitual to calculate the average deformation rate for each pass to determine the flow stress in the zone of deformation. The equation is calculated as a function of the time as follows:

$$\begin{aligned}
\dot{\varepsilon} &= \frac{1}{\theta_0} \cdot \int_0^{\theta_0} \frac{2 \cdot v \cdot \sin\theta}{h_f + 2 \cdot R \cdot (1 - \cos\theta)} \cdot d\theta \\
&= \frac{1}{\theta_0} \cdot \int_0^{\theta_0} \frac{2 \cdot v \cdot \sin\theta}{h_f + 2 \cdot R \cdot (1 - \cos\theta)} \cdot d\theta \cdot \frac{2 \cdot R}{2 \cdot R} \\
&= \frac{1}{\theta_0} \cdot \frac{2 \cdot v}{2 \cdot R} \cdot \int_0^{\theta_0} \frac{2 \cdot R \cdot \sin\theta}{h_f + 2 \cdot R - 2 \cdot R \cdot \cos\theta} \cdot d\theta = \left(\int \frac{u'}{u} = \ln u\right) \\
&= \frac{v}{R \cdot \theta_0} \cdot [\ln(h_f + 2 \cdot R - 2 \cdot R \cdot \cos\theta)]_0^{\theta_0} = \frac{v}{R \cdot \theta_0} \cdot [\ln h]_{h_f}^{h_0} \\
&= \frac{v}{R \cdot \theta_0} \cdot \ln\left(\frac{h_0}{h_f}\right) \simeq \frac{v}{\sqrt{R \cdot h}} \cdot \ln\left(\frac{h_0}{h_f}\right)
\end{aligned} \tag{5.47}$$

as (see Fig. 5.16):

$$\frac{h_0}{2} = \frac{h_f}{2} + R \cdot (1 - \cos\theta) \rightarrow h$$
$$= 2 \cdot R \cdot (1 - \cos\theta) \simeq 4 \cdot R \cdot \sin^2\left(\frac{\theta}{2}\right) \simeq 4 \cdot R \cdot \frac{\theta^2}{4}$$
$$= \theta^2 \cdot R \rightarrow \theta_0 = \sqrt{\frac{h}{2 \cdot R}} \tag{5.48}$$

when $1 - \cos\ \theta = 2 \cdot \sin^2(\theta/2)$ and as $\theta \rightarrow 0$ thus $\sin^2(\theta/2) = (\theta/2)^2$

The angle of contact between the rolling cylinders and sheet will be equal to:

$$\theta_0 = \sqrt{\frac{\Delta h}{2 \cdot R}} \simeq 10^\circ \text{(see next exercise)} \tag{5.49}$$

and, this way, the real arc $R \cdot \theta$ might be coincident with the projected arc $\sqrt{R \cdot \Delta h}$ in the rolling direction.

Finally, the conditions of adherent friction imposed (not exactly) require that the peripheral speed of the cylinder to be equal to the flow of mater rate:

$$v = 2 \cdot \pi \cdot R \cdot \frac{rpm}{60} \tag{5.50}$$

Commentaries: The above-calculated formulas will be used in the following exercises. We collect them now:

$$\bar{\varepsilon} = 1.155 \cdot \ln\left(\frac{h_0}{h_f}\right) = \frac{2 \cdot \varepsilon_{PD}}{\sqrt{3}} \tag{5.51}$$

$$\bar{\sigma} = \frac{\sqrt{3}}{2} \cdot \sigma_{PD} \tag{5.52}$$

$$\dot{\bar{\varepsilon}} = \frac{1.155 \cdot v}{\sqrt{R \cdot \Delta h}} \cdot \ln\left(\frac{h_0}{h_f}\right) \tag{5.53}$$

$$v_L = \omega \cdot R = 2 \cdot \pi \cdot R \cdot \frac{rpm}{60} \tag{5.54}$$

End of the exercise.

Exercise 5.2 A slab of structural steel has 1250 mm in width, and it experiences a rolling pass to reduce the thickness from 40 to 24 mm at 1000 °C. The working cylinders have 865 mm in diameter and rotate at 66 rpm. The average flow stress of deformation is 183 MPa. We want to calculate the load and torque that must hold up the box. The value of the constant Q in the equation of Sims is 1.395 approximately.

Solution starts:

The force is:

$$\begin{aligned} F =& Q \cdot \text{width of the band} \cdot \text{lenght of contact} \cdot \text{average flow stress} \\ &= 1.395 \cdot w \cdot l \cdot \sigma_y \\ &= 1395 \cdot w \cdot \sqrt{r \cdot \Delta h} \cdot \sigma_y \\ &= 1.395 \cdot 1250\ \text{mm} \cdot \sqrt{\left(\frac{865\ \text{mm}}{2}\right) \cdot (40-24)\ \text{mm}} \cdot \frac{183\ \text{N}}{\text{mm}^2} \\ &= 26.55 \cdot 10^6\ \text{N} \cdot \frac{1\ \text{kg}}{9.81\ \text{N}} \cdot \frac{1\ \text{t}}{1000\ \text{kg}} = 2706\ \text{t} \end{aligned} \tag{5.55}$$

The torque is calculated as follows:

$$\begin{aligned} T =& F \cdot \frac{l}{2} = \frac{\sigma_y \cdot w \cdot l^2}{2} = \frac{\sigma_y \cdot w \cdot r \cdot h}{2} \\ &= \frac{1.395 \cdot \frac{183\ \text{N}}{\text{mm}^2} \cdot 1250\ \text{mm} \cdot \left(\frac{865\ \text{mm}}{2}\right) \cdot (40-24)\text{mm}}{2} \cdot \frac{1\ \text{m}}{1000\ \text{mm}} \\ &= 1.10 \cdot 10^6 \text{N} \cdot \text{m} \cdot \frac{1\ \text{kg}}{9.81\ \text{N}} \cdot \frac{1\ \text{t}}{1000\ \text{kg}} = 112.55\ \text{t} \cdot \text{m} \end{aligned} \tag{5.56}$$

Finally, the rolling power is:

$$\begin{aligned} W =& T \cdot v = 1.10 \cdot 10^6 \text{N} \cdot \text{m} \cdot 66 \frac{\text{rev}}{\text{min}} \cdot \frac{2 \cdot \pi}{\text{rev}} \cdot \frac{1\ \text{min}}{60\ \text{s}} \\ &= 7.63 \cdot 10^7 \frac{\text{J}}{\text{s}} \cdot \frac{\text{w} \cdot \text{s}}{\text{J}} \cdot \frac{1\ \text{kW}}{1000\ \text{w}} = 7631\ \text{kW} \end{aligned} \tag{5.57}$$

End of the exercise.

5.4 Rolling Mills

The basic element of the rolling is known with the name "stand." A stand comprises, essentially, rolls and a structure known as "housing" that is the frame for the rolls.

When the stand comprises two rolls, it is called two-high rolling stand. When big reductions are required, the called "four-high rolling stands" are used, and they are equipped with four rolls, two rolls of small diameters that are responsible for the reduction, that is to say, which are in contact with the steel (work rolls), and two rolls that support the other two (back up rolls) (Fig. 5.17). If a good dimensional finishing is required, the "universal stands" are used, which are equipped with horizontal and vertical rolls.

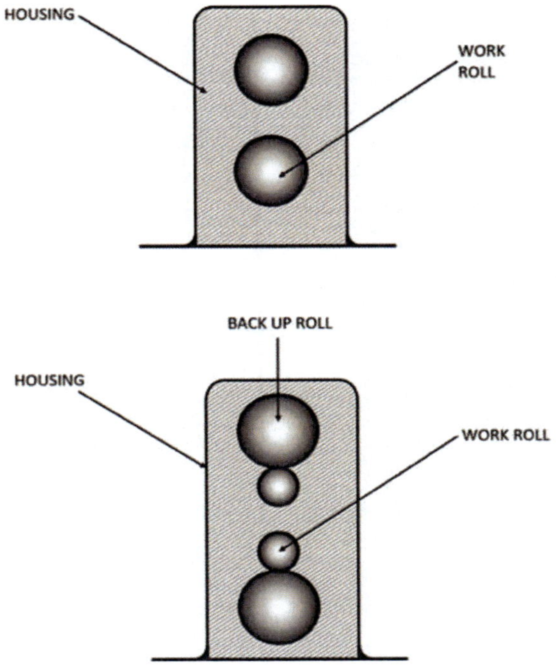

Fig. 5.17 Detail of a two-high rolling stand (above) and detail of a four-high rolling stand

A "rolling mill" is the ensemble formed by one or several stands with some other additional elements such are: the "roller table", shears, cooling tables, engines, and handling (instrumentation) and control equipment. A rolling mill formed by only one stand that can operate in both directions is reversible (Fig. 5.18). If we want to obtain a product of very small section (wire or sheet) (if compared with the initial dimensions of the product), it is not possible to use a reversible mill with only one stand because of the large length of the rolled product. The problem is solved using a "continuous rolling mill," that is to say, several stands located one after the other (Fig. 5.19).

A "roughing mill" is a two-high rolling mill reversible where the ingots of the conventional casting are transformed either in squared rough product that is fed to the "blooming train" or in flat products that are processed by the "slabbing train."

The "hot rolling mill" is the installation used to produce hot coil (flat products) from the slabs. For an annual production equal or greater than 3.5 million of tons, a disposition of four-stand mill in continuous: hot rolling mill in continuous is used. On the contrary, when the capacity of production is smaller than 3.5 million of tons per year, a configuration formed by a reversible universal stand for roughing and a continuous six-stand or seven-stand mill (with the configuration of four-high rolling stand), that is to say, "semi-continuous" hot rolling mill for finishing (Fig. 5.20).

The production of stainless-steel sheet can be also carried out in hot rolling mills which imply great investments and a volume of production greater than 2 million of

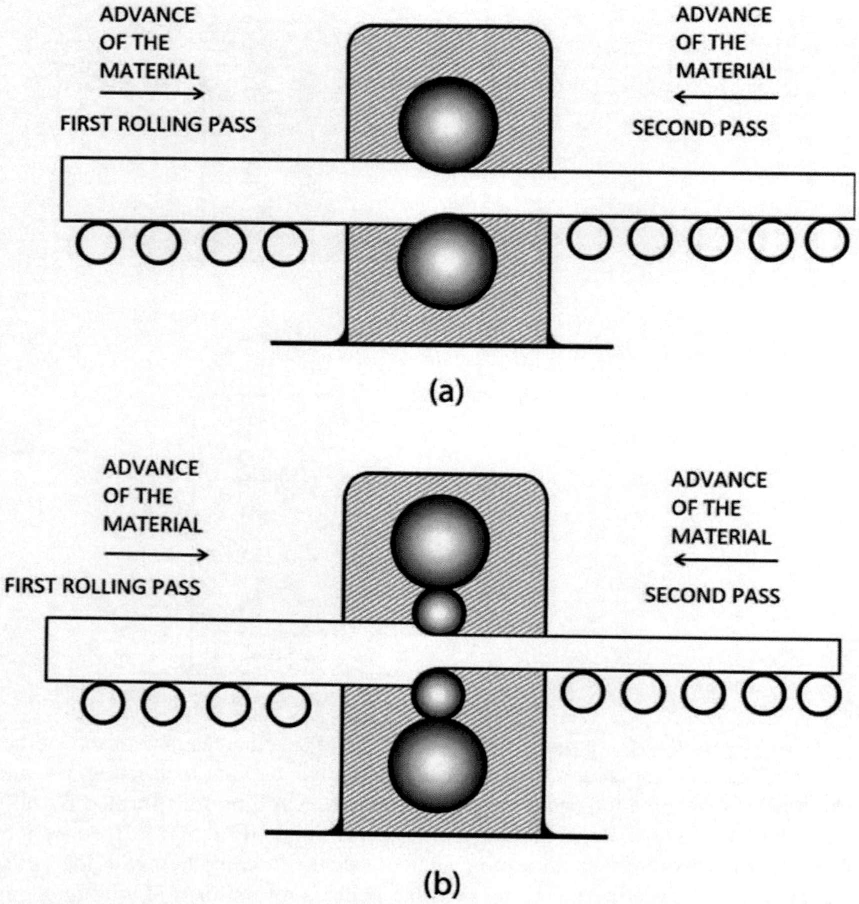

Fig. 5.18 Scheme of a two-high reversing mill (**a**) and a four-high reversing mill (**b**)

tons per year. For capacities of production close to 1 million of tons, it is possible to use the following configuration:

- A four-high reversing roughing mill that reduces the thickness of the slabs of the continuous casting down to a value of 25 mm.
- A Steckel finishing rolling mill of one six-high reversing stand that allows reaching thicknesses in the final product of around 2 mm. The strip is collected in the coiling machines situated inside of furnaces used to hold the temperature, when considerable increases the length of the strip/band due to the successive rolling passes.

The qualities of steel currently producible using a hot rolling mill are:

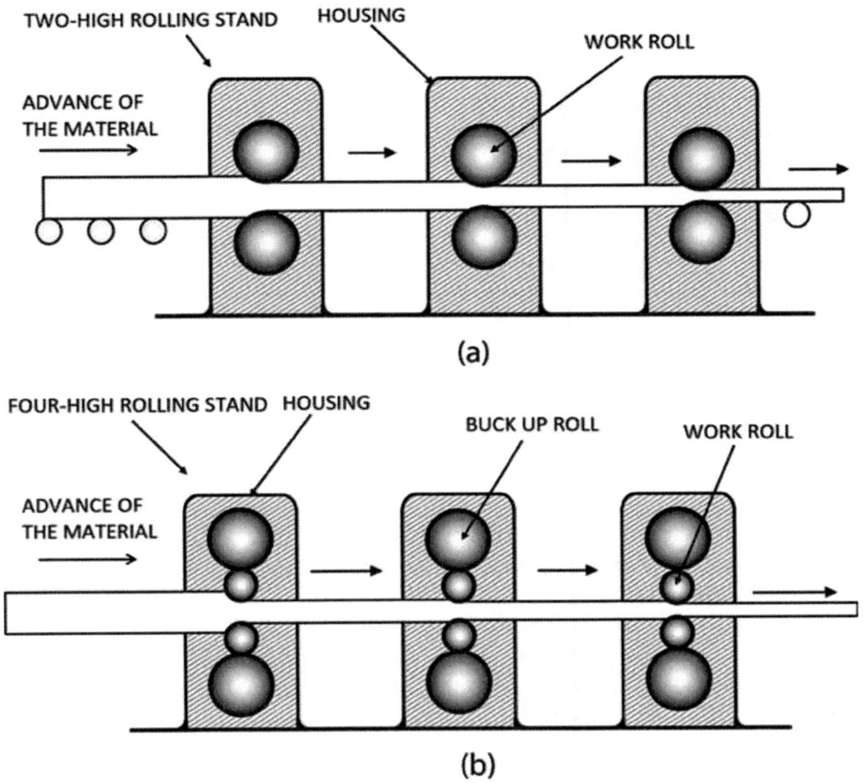

Fig. 5.19 Scheme of a continuous two-high rolling stand mill (**a**) and a continuous four-high rolling stand mill (**b**) (increasing speed as the material advances to the right in the figure)

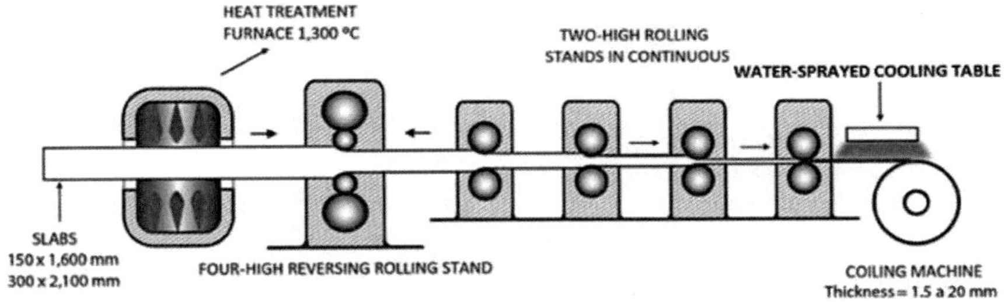

Fig. 5.20 Scheme of a semi-continuous hot rolling mill

- Extra low carbon steel, ELC.
- Ultra-low carbon steel, ULC.
- Low-alloy steel, LAS.
- High-strength low-alloy steel, HSLA.
- Dual phase steel, DP.
- Interstitial free steel, IF.

The production of hot-rolled coil using a hot rolling mill in an integrated steel plant, compared with a CSP process (continuous thin slab casting process) (Fig. 5.21) integrated in an electric steelwork (compact steelworks), implies:

- Lower energy consumption as 7.5 GJ/t of coil are required in the compact process (electric steelworks + CSP) versus the 21.7 GJ/t of the conventional process (integrated steel plant + hot rolling mill).
- Improvement of the productivity estimated in around 3000 tons per worker and year, for the sequence electric steelworks + CSP.
- Improvement of the transformation costs in around 20% against the sequence integrated steel plant + hot rolling mill.

On the other hand, it is not possible to obtain all the qualities demanded by the market (manufactured by the route steel plant + hot rolling mill) using the sequence electric steelworks + CSP.

Other possibility of treatment is the heavy plate mill that uses a two-high rolling stand in continuous to descale the steel and a four-high reversing stand to manufacture the steel sheet, suitable for the construction of ships, pressure vessels, oil pipelines, and oil platforms (Fig. 5.22).

A fine grain size in steels is a parameter of great importance, the same as in the design of those structural materials that must not work under creep conditions, as not only the ductile–brittle transition temperature is decreased but also hardness, yield strength, and ultimate tensile strength are improved. Table 5.1 collects the grain sizes that can be obtained as a function of the processes that are used in the manufacture of the finished product.

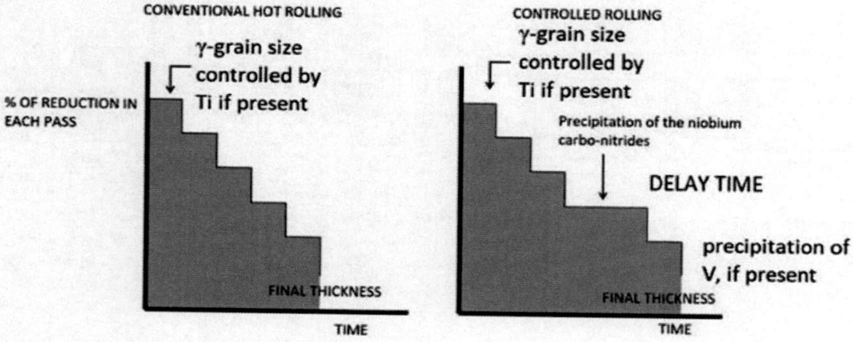

Fig. 5.21 Diagram reduction-time for a conventional rolling versus a controlled rolling

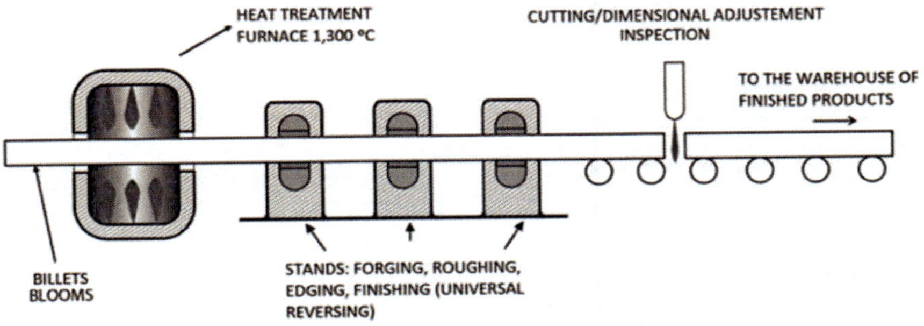

Fig. 5.22 Scheme of a heavy plate mill (flat products)

Table 5.1 Grain size of the steels as a function of the mill-techniques used for their production

Techniques of production	Grain size
Molded steel	2 ASTM (150 μm)
As-cast steel	7 ASTM (25 μm)
Hot rolled and micro-alloyed steel	10 ASTM (10 μm)
Hot rolled and micro-alloyed steel (controlled rolling)	12 ASTM (5 μm)
Rapid solidification techniques	16 ASTM (1 μm)

During the hot rolling process of flat rough product (slabs) to produce thick plate (Fig. 5.22), the grain refinement is considerable, around 500 μm for the average grain size in the slab and around 40 μm (6 ASTM) in the final product.

However, these results can be improved by adding small quantities of niobium that when precipitates in austenite phase in the form of carbo-nitrides restrain the grain growth of the austenite. During the process of controlled rolling (Fig. 5.21), once reached a certain reduction of thickness with respect to the initial thickness, the rolling process is interrupted, and the product is directed to a stand in parallel with the train, and thus the progress is not altered, and the rolling of another slab can start. During the delay-time, the precipitation of niobium carbo-nitrides avoids the austenitic grain growth consequence of the static recrystallization. Finally, the rough product returns to the heavy plate reversing rolling mill and the process of thickness reduction continues until reaching the desired thickness (Fig. 5.21).

In the "profile and rail mill" the pass through the different stands of the square rough products of the continuous casting or those coming from the roughing mill produces their transformation in profiles and in rails (long products). The profile mill has the following stands (Fig. 5.23):

- Two-high reversing forging stand.
- Universal roughing reversing stand.
- Two-high reversing ending stand.
- Universal reversing finishing stand.

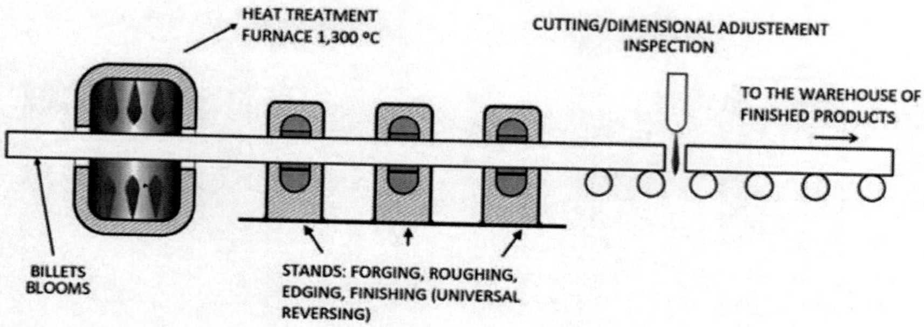

Fig. 5.23 Scheme of a train of rolling of profiles and rails (long products)

A mill to roll rails comprises the following stands:

- Two-high reversing forging stand.
- Two-high reversing roughing stand.
- Two-high reversing finishing stand.

The "wire-rod mill" is the mill where the billet is transformed into steel wire inside of an ensemble of more than twenty stands that operate continuously (Fig. 5.24).

From the group of stands that form the wire-rod mill, it is necessary to differentiate the following groups of stands:

- Roughing stands
- Wire-rod stands.
- Finishing stands.

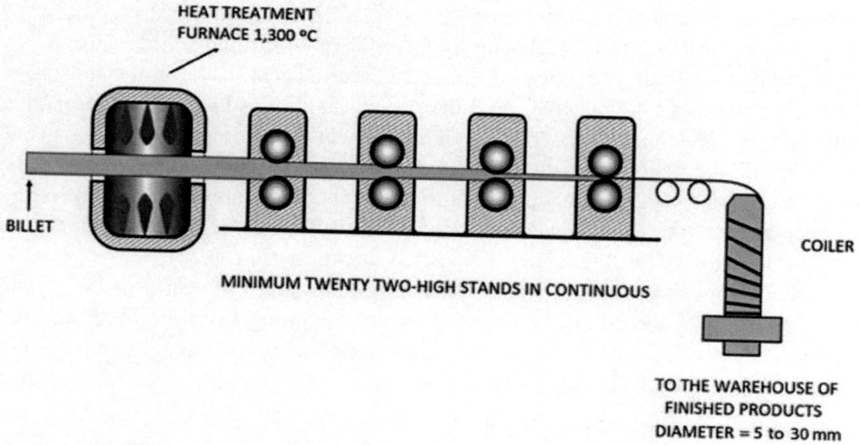

Fig. 5.24 Scheme of a wire-rod mill

Rolled steel bundle of wires must be cooled in special conditions to achieve the desired microstructure and, thus, wire-rod mills have sophisticated cooling systems for cooling (recrystallization intermediate annealing heat treatments, patenting) the coils that form the bundle of wires using air or water to guarantee the proper structural transformation of the material.

All the processes previously considered of hot rolling of steel take place in austenitic state. There are initiatives to develop mixed rolling processes or operate exclusively in ferritic state. For instance, roughing is performed in austenitic phase for the obtaining of hot coil while finishing is in ferritic state. Even though rolling in ferritic phase requires more power in the mills, it has the following advantages:

- Lower energy consumption.
- Less formation of mill scale in the heating furnaces.
- Reduction of the wear of rolls.
- Reduction of the oxide layer in the cooling stage before the coiling.

5.4.1 *Solved Exercises*

Exercise 5.3 It is assumed that the mechanical properties of the final product, univocal function of its microstructure, are characterized in the current case by a single parameter: grain size. This is the reason for the importance of calculating the evolution of grain size along the rolling process with the aim of obtaining the suitable mechanical properties (yield strength, ultimate tensile strength, ductility, ductile–brittle transition temperature).

We consider a slab of 240 mm in thickness, homogenized at 1160 °C, with 300 μm of grain size, it is subjected to 9 passes with 15% of reduction in each pass in a two-high reversing mill (Fig. 5.20) until a final thickness of 55 mm. The peripheric rate, v, of the cylinders and the radius, R, of the cylinders are 1200 mm/s and 500 mm, respectively. The time between passes is 20 s. We use the empirical relations determined by C. M. Sellars and J. A. Whitemann for the calculation in each pass of the Zener-Hollomon (Z) parameter, the time of recrystallization ($t_{x=0.95}$) and the time of grain growth (Sellars and Whitemann 1979; Dutta and Sellars 1986, 1987):

$$Z = \dot{\varepsilon} \cdot \exp\left(\frac{37440}{T}\right)\left[s^{-1}\right] \tag{5.58}$$

$$t_{x=0.95} = K \cdot Z^{-0.375} \cdot \varepsilon^{-4} \cdot {D_0}^2 \cdot \exp\left(\frac{Q_{rex}}{R \cdot T}\right) \tag{5.59}$$

where Q_{rex} represents a constant of the material, which, in this case, is equal to 57,400 J/mol (activation energy for the recrystallization); and K depends on the fraction of recrystallized material, X, in this case, X = 95% of mild steel, and the value of K is:

$$K = 3.54 \cdot 10^{-21}\left[s^{0.625} \cdot \mu m^{-2}\right] \quad (5.60)$$

The time of growth is calculated as follows:

$$t_{growth} = t_{between\ passes} - t_{x=0.95} \quad (5.61)$$

Additionally, we take as temperature between passes the average temperature between them.

The mean grain size at the end of the recrystallization process depends only on two factors: the energy stored during the deformation, and the density of possible sites for the nucleation. This way, the size of the recrystallized grain is:

$$D_{rex} = K' \cdot \left[\frac{1}{n \cdot \alpha} \cdot \ln\left(\frac{Z}{A}\right)\right]^{-0.67} \cdot \varepsilon^{-1} \cdot {D_0}^{0.5} \quad (5.62)$$

where:

$$K' = 25\left[\left(m^2/MN\right)^{0.67} \cdot \mu m^{1/2}\right] \quad (5.63)$$

$$n \cdot \alpha = 6.7 \cdot 10^{-2}\left[m^2/MN\right] \quad (5.64)$$

$$A = 8.5 \cdot 10^{9}\left[s^{-1}\right] \quad (5.65)$$

for mild steel.

Regarding grain growth, with short times of permanence at high temperature, the variation with the time of the grain diameter is adjusted by the following equation:

$$D^{10} - {D_{rex}}^{10} = M \cdot t_{growth} \quad (5.66)$$

where M is given by an equation similar to that of thermally activated processes:

$$M = K'' \cdot \exp\left(-\frac{Q}{R \cdot T}\right) \quad (5.67)$$

and in the case of mild steel:

$$M = 3.87 \cdot 10^{32} \cdot \exp\left(-\frac{400000\,J/mol}{R \cdot T}\right) \quad (5.68)$$

We also use the following equations (see at the end of the Exercise 5.1):

– Linear speed of the rolls:

$$v = 2 \cdot \pi \cdot R \cdot \frac{rpm}{60} \tag{5.69}$$

– Equivalent deformation:

$$\bar{\varepsilon} = 1.155 \cdot \ln\left(\frac{h_0}{h_f}\right) \tag{5.70}$$

– Mean equivalent deformation rate:

$$\dot{\bar{\varepsilon}} = \frac{1.155 \cdot v}{\sqrt{R \cdot \Delta h}} \cdot \ln\left(\frac{h_0}{h_f}\right) \tag{5.71}$$

Start of the exercise:

We collect in the following tables (Tables 5.2, 5.3, 5.4, 5.5, 5.6, 5.7, 5.8, 5.9 and 5.10) the results for each pass using the equations indicated in the enunciation of the exercise.

We appreciate in Table 5.11 a stagnation of the grain size in around 60 μm due to the noticeable growth of the specific surface area of the grain, that stimulates

Table 5.2 Results for the pass 1

Pass 1	
h_0 (mm)	240
h_f (mm)	204
$\bar{\varepsilon}$	0.188
$\dot{\bar{\varepsilon}}$ (s^{-1})	1.679
T (K)	1423
Z	$3.73 \cdot 10^{11}$
$t_{x=0.95}$ (s)	2.9
t_{growth} (s)	17.08
D_0 (μm)	300
D_{rex} (μm)	154.7
D_{final} (μm)	154.7
M	$8.02 \cdot 10^{17}$

Table 5.3 Results for the pass 2

Pass 2	
h_0 (mm)	204
h_f (mm)	173
$\bar{\varepsilon}$	0.188
$\dot{\bar{\varepsilon}}$ (s^{-1})	1.821
T (K)	1403
Z	$5.86 \cdot 10^{11}$
$t_{x=0.95}$ (s)	1.2
t_{growth} (s)	18.8
D_0 (μm)	154.7
D_{rex} (μm)	103.0
D_{final} (μm)	103.7
M	$4.95 \cdot 10^{17}$

Table 5.4 Results for the pass 3

Pass 3	
h_0 (mm)	173
h_f (mm)	147
$\bar{\varepsilon}$	0.188
$\dot{\bar{\varepsilon}}$ (s^{-1})	1.975
T (K)	1383
Z	$9.30 \cdot 10^{11}$
$t_{x=0.95}$ (s)	0.8
t_{growth} (s)	19.2
D_0 (μm)	103.7
D_{rex} (μm)	78.7
D_{final} (μm)	82.7
M	$3.02 \cdot 10^{17}$

the growth between passes. After the allotropic transformation $\gamma \rightarrow \alpha$ (at around 890 °C), each grain of austenite generates between 2 and 3 grains of ferrite (Tanaka 1981; Tamura et al. 1988). That is to say, the grain size in the ferritic-pearlitic steel, at room temperature, will be comprised within 30 (7 ASTM) and 20 (8 ASTM) μm. It is not possible to obtain finer grains, except in the case of the controlled rolling (as we will see in the next exercises).

End of the exercise.

Table 5.5 Results for the pass 4

Pass 4	
h_0 (mm)	147
h_f (mm)	125
$\bar{\varepsilon}$	0.188
$\dot{\bar{\varepsilon}}$ (s^{-1})	2.142
T (K)	1363
Z	$1.49 \cdot 10^{12}$
$t_{x=0.95}$ (s)	0.8
t_{growth} (s)	19.2
D_0 (μm)	82.7
D_{rex} (μm)	65.9
D_{final} (μm)	74.1
M	$1.81 \cdot 10^{17}$

Table 5.6 Results for the pass 5

Pass 5	
h_0 (mm)	125
h_f (mm)	106
$\bar{\varepsilon}$	0.188
$\dot{\bar{\varepsilon}}$ (s^{-1})	2.324
T (K)	1343
Z	$2.42 \cdot 10^{12}$
$t_{x=0.95}$ (s)	0.9
t_{growth} (s)	19.1
D_0 (μm)	74.1
D_{rex} (μm)	58.7
D_{final} (μm)	69.2
M	$1.07 \cdot 10^{17}$

Exercise 5.4 Now we analyze the microstructural evolution of a carbon-manganese steel (class Q0, type St42 44) obtained by controlled rolling, whose chemical composition is: $0.13 < \%C < 0.15$, $0.60 < \%Mn < 0.80$, max. 0.3% Si, max. 0.020% S, max. 0.025% P and max. 0.050% Al.

Roughing is performed in 9 passes (four-high reversing mill) and finishing in 7, semicontinuous hot rolling mill, whose geometry and treatment process is:

- Dimensions of the rough product (slab): 230 mm × 1579 mm × 7123 mm.
- Intermediate plate (slabbing mill): 40 mm × 41,797 mm (plate obtained at the end of the roughing).

Table 5.7 Results for the pass 6

Pass 6	
h_0 (mm)	106
h_f (mm)	91
$\bar{\varepsilon}$	0.188
$\dot{\bar{\varepsilon}}$ (s^{-1})	2.520
T (K)	1323
Z	$3.98 \cdot 10^{12}$
$t_{x=0.95}$ (s)	1.3
t_{growth} (s)	18.7
D_0 (μm)	69.2
D_{rex} (μm)	53.7
D_{final} (μm)	65.1
M	$6.23 \cdot 10^{16}$

Table 5.8 Results for the pass 7

Pass 7	
h_0 (mm)	91
h_f (mm)	77
$\bar{\varepsilon}$	0.188
$\dot{\bar{\varepsilon}}$ (s^{-1})	2.734
T (K)	1303
Z	$6.62 \cdot 10^{12}$
$t_{x=0.95}$ (s)	1.8
t_{growth} (s)	18.2
D_0 (μm)	65.1
D_{rex} (μm)	49.3
D_{final} (μm)	61.2
M	$3.56 \cdot 10^{16}$

– Weight: 20,306 kg.

Homogenization treatment: 326 min (time of permanence in the furnace), temperature at the exit of the furnace: 1256 °C. Initial grain size: 500 μm.

Roughing mill: diameter of the working cylinders, 1118.2 mm; diameter of the back (supporting) cylinders, 1304.5 mm.

Values of t_{pause} (s)are known for the roughing.

In the case of the finishing, the distance between cages being about 5500 mm, and the t_{pause} (s) is calculated as follows:

Table 5.9 Results for the pass 8

Pass 8	
h_0(mm)	77
h_f (mm)	65
$\bar{\varepsilon}$	0.188
$\dot{\bar{\varepsilon}}$ (s^{-1})	2.965
T (K)	1283
Z	$1.12 \cdot 10^{13}$
$t_{x=0.95}$ (s)	2.6
t_{growth} (s)	17.4
D_0 (μm)	61.2
D_{rex} (μm)	45.5
D_{final} (μm)	57.4
M	$2.00 \cdot 10^{16}$

Table 5.10 Results for the pass 9

Pass 9	
h_0(mm)	65
h_f (mm)	56
$\bar{\varepsilon}$	–
$\dot{\bar{\varepsilon}}$ (s^{-1})	–
T (K)	1273
Z	–
$t_{x=0.95}$ (s)	–
t_{growth} (s)	–
D_0 (μm)	–
D_{rex} (μm)	–
D_{final} (μm)	–
M	–

$$t_{pause}(s) = \frac{5500\,\text{mm}}{v_{linear}(m/min) \cdot \frac{1000\,\text{mm}}{1\,\text{m}} \cdot \frac{1\,\text{min}}{60\,\text{s}}} \tag{5.72}$$

Solution starts:

We use the same equations that in the Exercise 5.3 with $D_0 = 300$ mm. We collect in the following tables (Tables 5.12, 5.13, 5.14, 5.16, 5.17, 5.18, 5.19 and 5.20) the results for each pass of the roughing process.

We collect in Table 5.21 a summary of the roughing process.

Table 5.11 Rolling process of mild steel

Pass	h_0 (mm)	h_f (mm)	$\bar{\varepsilon}$	$\dot{\bar{\varepsilon}}$ (s^{-1})	T (K)	$t_{x=0.95}$ (s)	t_{growth} (s)	D_0 (μm)	D_{rex} (μm)	D_{final} (μm)
1	240	204	0.188	1.679	1423	2.9	17.1	300.0	154.7	154.7
2	204	173	0.188	1.821	1403	1.2	18.8	157.4	103.0	103.7
3	173	147	0.188	1.975	1383	0.8	19.2	103.7	78.7	82.7
4	147	125	0.188	2.142	1363	0.8	19.2	82.7	65.9	74.1
5	125	106	0.188	2.324	1343	0.9	19.1	74.1	58.7	69.2
6	106	91	0.188	2.520	1323	1.3	18.7	69.2	53.7	65.1
7	91	77	0.188	2.734	1303	1.8	18.2	65.1	49.3	61.2
8	77	65	0.188	2.965	1283	2.6	17.4	61.2	45.5	57.4
9	65	56	–	–	1273	–	–	–	–	–

Table 5.12 Results for the pass 1 (roughing)

Pass 1	
h_0 (mm)	230
h_f (mm)	205
$\bar{\varepsilon}$	0.133
$v_{cylinder}$ (rpm)	38.9
v_{linear} (mm/s)	2278
$\dot{\bar{\varepsilon}}$ (s^{-1})	2.560
T (K)	1511
Z	$1.5 \cdot 10^{11}$
$t_{x=0.95}$ (s)	5.8
t_{growth} (s)	6.0
t_{pause} (s)	11.8
D_0 (μm)	500
D_{rex} (μm)	340.6
D_{final} (μm)	340.6
M	$3.40 \cdot 10^{19}$

Table 5.13 Results for the pass 2 (roughing)

Pass 2	
h_0 (mm)	205
h_f (mm)	180
$\bar{\varepsilon}$	0.150
$v_{cylinder}$ (rpm)	44.4
v_{linear} (mm/s)	2600
$\dot{\bar{\varepsilon}}$ (s^{-1})	3.302
T (K)	1503
Z	$2.2 \cdot 10^{11}$
$t_{x=0.95}$ (s)	1.7
t_{growth} (s)	7.9
t_{pause} (s)	9.6
D_0(μm)	340.6
D_{rex} (μm)	228.4
D_{final} (μm)	228.4
M	$3.75 \cdot 10^{19}$

Table 5.14 Results for the pass 3 (roughing)

Pass 3	
h_0 (mm)	180
h_f (mm)	156
$\bar{\varepsilon}$	0.165
$v_{cylinder}$ (rpm)	47.0
v_{linear} (mm/s)	2752
$\dot{\bar{\varepsilon}}$ (s^{-1})	3.925
T (K)	1497
Z	$2.9 \cdot 10^{11}$
$t_{x=0.95}$ (s)	0.6
t_{growth} (s)	9.4
t_{pause} (s)	10.0
D_0 (μm)	228.4
D_{rex} (μm)	161.1
D_{final} (μm)	161.1
M	$3.96 \cdot 10^{19}$

Table 5.15 Results for the pass 4 (roughing)

Pass 4	
h_0 (mm)	156
h_f (mm)	132
$\bar{\varepsilon}$	0.193
$v_{cylinder}$ (rpm)	51.3
v_{linear} (mm/s)	3004
$\dot{\bar{\varepsilon}}$ (s^{-1})	5.002
T (K)	1488
Z	$4.2 \cdot 10^{11}$
$t_{x=0.95}$ (s)	0.2
t_{growth} (s)	9.0
t_{pause} (s)	9.2
D_0 (μm)	161.1
D_{rex} (μm)	107.9
D_{final} (μm)	109.4
M	$3.12 \cdot 10^{19}$

Table 5.16 Results for the pass 5 (roughing)

Pass 5	
h_0 (mm)	132
h_f (mm)	108
$\bar{\varepsilon}$	0.232
$v_{cylinder}$ (rpm)	55.8
v_{linear} (mm/s)	3267
$\dot{\bar{\varepsilon}}$ (s^{-1})	6.535
T (K)	1480
Z	$6.3 \cdot 10^{11}$
$t_{x=0.95}$ (s)	0.04
t_{growth} (s)	9.2
t_{pause} (s)	9.2
D_0 (μm)	109.4
D_{rex} (μm)	69.3
D_{final} (μm)	88.4
M	$2.66 \cdot 10^{19}$

Table 5.17 Results for the pass 6 (roughing)

Pass 6	
h_0 (mm)	108
h_f (mm)	83
$\bar{\varepsilon}$	0.304
$v_{cylinder}$ (rpm)	55.9
v_{linear} (mm/s)	3273
$\dot{\bar{\varepsilon}}$ (s^{-1})	8.416
T (K)	1469
Z	$9.9 \cdot 10^{11}$
$t_{x=0.95}$ (s)	0.01
t_{growth} (s)	6.6
t_{pause} (s)	6.6
D_0 (μm)	88.4
D_{rex} (μm)	44.5
D_{final} (μm)	82.7
M	$1.50 \cdot 10^{19}$

Table 5.18 Results for the pass 7 (roughing)

Pass 7	
h_0 (mm)	83
h_f (mm)	63
$\bar{\varepsilon}$	0.318
$v_{cylinder}$ (rpm)	60.8
v_{linear} (mm/s)	3560
$\dot{\bar{\varepsilon}}$ (s^{-1})	10.717
T (K)	1458
Z	$1.5 \cdot 10^{12}$
$t_{x=0.95}$ (s)	0.01
t_{growth} (s)	10.2
t_{pause} (s)	10.2
D_0 (μm)	82.7
D_{rex}(μm)	38.8
D_{final} (μm)	84.3
M	$1.81 \cdot 10^{19}$

Table 5.19 Results for the pass 8 (roughing)

Pass 8	
h_0 (mm)	63
h_f (mm)	49
$\bar{\varepsilon}$	0.290
$v_{cylinder}$ (rpm)	63.7
v_{linear} (mm/s)	3730
$\dot{\bar{\varepsilon}}$ (s^{-1})	12.233
T (K)	1436
Z	$2.6 \cdot 10^{12}$
$t_{x=0.95}$ (s)	0.02
t_{growth} (s)	7.6
t_{pause} (s)	7.6
D_0 (μm)	84.3
D_{rex} (μm)	40.2
D_{final} (μm)	77.8
M	$8.13 \cdot 10^{18}$

Table 5.20 Results for the pass 9 (roughing)

Pass 9	
h_0 (mm)	49
h_f (mm)	40
$\bar{\varepsilon}$	–
$v_{cylinder}$ (rpm)	–
v_{linear} (mm/s)	–
$\dot{\bar{\varepsilon}}$ (s^{-1})	–
T (K)	1414
Z	–
$t_{x=0.95}$ (s)	–
t_{growth} (s)	–
t_{pause} (s)	–
D_0 (μm)	–
D_{rex} (μm)	–
D_{final} (μm)	–
M	–

After the roughing process in 9 passes, there is a time delay of 44 s between the roughing and finishing processes. In this period, the grain grows up to 87.1 μm (from 77.8 μm at the end of the roughing process), while the temperature decreases from 1141 to 1091 °C. We present in the following tables (Tables 5.22, 5.23, 5.24, 5.25, 5.26, 5.27, 5.28 and 5.29) the results corresponding to the finishing stage. We also use the data indicated in the enunciation, as in the Exercise 5.3.

We appreciate in Table 5.29 a stagnation of the austenitic grain size in around 44.4 μm due to the noticeable growth of the specific surface area of the grain, that stimulates the growth between passes. After the allotropic transformation $\gamma \rightarrow \alpha$ (at around 890 °C), each grain of austenite generates between 2 and 3 grains of ferrite (Tanaka 1981; Tamura et al. 1988). That is to say, the grain size in the ferritic-pearlitic steel, at room temperature, will be comprised within 18 (8.5 ASTM) and 21 (7.5 ASTM) μm. It is not possible to obtain finer grains, except in the case of the controlled rolling (as we will see in the next exercise).

The final microstructure of the material is around 20 μm (ferritic grain size), being very difficult to obtain finer grains if the steel is not micro-alloyed with niobium. It is possible to estimate the non-recrystallization temperature of the austenite (Barbosa et al. 1987; Belzunce et al. 1995):

$$\begin{aligned} T_{nr}(^{\circ}C) = {} & 887 + 464 \cdot \%C + \left[6445 \cdot \%Nb - 644 \cdot (\%Nb)^{1/2}\right] \\ & + \left[732 \cdot \%V - 230 \cdot (\%V)^{1/2}\right] \\ & + 890 \cdot \%Ti + 363 \cdot \%Al - 357 \cdot \%Si \end{aligned} \tag{5.73}$$

Table 5.21 Rolling process (roughing) of a carbon-manganese steel (class Q0, type St4244)

Pass	h_0 (mm)	h_f (mm)	$\bar{\varepsilon}$	$\dot{\bar{\varepsilon}}$ (s^{-1})	T (K)	$t_{x=0.95}$ (s)	t_{pause} (s)	t_{growth} (s)	D_0 (μm)	D_{rex} (μm)	D_{final} (μm)
1	230	205	0.133	2.560	1511	5.8	11.8	6.0	500.0	340.6	340.6
2	205	180	0.150	3.302	1503	1.7	9.6	7.9	340.6	228.4	228.4
3	180	156	0.165	3.925	1497	0.6	10.0	9.4	228.4	161.1	161.1
4	156	132	0.193	5.002	1488	0.2	9.2	9.0	161.1	107.9	109.4
5	132	108	0.232	6.535	1480	0.04	9.2	9.2	109.4	69.3	88.4
6	108	83	0.304	8.416	1469	0.01	6.6	6.6	88.4	44.5	82.7
7	83	63	0.318	10.717	1458	0.01	10.2	10.2	82.7	38.8	84.3
8	63	49	0.290	12.233	1436	0.02	7.6	7.6	84.3	40.2	77.8
9	49	40	–	–	1414	–	–	–	–	–	–

Table 5.22 Results for the pass 1 (finishing)

Pass 1	
h_0 (mm)	40
h_f (mm)	24.21
$\bar{\varepsilon}$	0.580
v_{linear} (m/min)	47.2
$\dot{\bar{\varepsilon}}$ (s^{-1})	5.417
T (K)	1189
Z	$2.6 \cdot 10^{14}$
$t_{x=0.95}$ (s)	0.9
t_{growth} (s)	6.1
t_{pause} (s)	6.1
D_0 (μm)	87.1
D_{rex} (μm)	13.8
D_{final} (μm)	38.0
M	$1.01 \cdot 10^{15}$

Table 5.23 Results for the pass 2 (finishing)

Pass 2	
h_0 (mm)	24.21
h_f (mm)	14.38
$\bar{\varepsilon}$	0.602
v_{linear} (m/min)	79.4
$\dot{\bar{\varepsilon}}$ (s^{-1})	14.239
T (K)	1201
Z	$4.9 \cdot 10^{14}$
$t_{x=0.95}$ (s)	0.1
t_{growth} (s)	4.1
t_{pause} (s)	4.2
D_0 (μm)	38.0
D_{rex} (μm)	8.4
D_{final} (μm)	38.0
M	$1.52 \cdot 10^{15}$

The composition of our steel is: %C, 0.14; %Mn, 0.7; %Si, 0.15; %Al, 0.025. The non-recrystallization temperature of the austenite is:

$$T_{nr}(°C) = 887 + 464 \cdot 0.14 + 363 \cdot 0.025 - 357 \cdot 0.15 = 907.5\,°C \qquad (5.74)$$

Table 5.24 Results for the pass 3 (finishing)

Pass 3	
h_0(mm)	14.38
h_f(mm)	9.01
$\bar{\varepsilon}$	0.540
v_{linear} (m/min)	126.8
$\dot{\bar{\varepsilon}}$ (s^{-1})	27.542
T (K)	1188
Z	$1.3 \cdot 10^{15}$
$t_{x=0.95}$ (s)	0.1
t_{growth} (s)	2.5
t_{pause} (s)	2.6
D_0 (μm)	38.0
D_{rex} (μm)	8.8
D_{final} (μm)	34.6
M	$9.80 \cdot 10^{14}$

Table 5.25 Results for the pass 4 (finishing)

Pass 4	
h_0 (mm)	9.01
h_f (mm)	6.08
$\bar{\varepsilon}$	0.454
v_{linear} (m/min)	187.8
$\dot{\bar{\varepsilon}}$ (s^{-1})	45.443
T (K)	1165
Z	$4.1 \cdot 10^{15}$
$t_{x=0.95}$ (s)	0.4
t_{growth} (s)	1.4
t_{pause} (s)	1.8
D_0 (μm)	34.6
D_{rex} (μm)	9.4
D_{final} (μm)	30.2
M	$4.40 \cdot 10^{14}$

Austenite recrystallizes above 907.5 ºC in our steel, and as the temperature of entrance to the last stand is greater (1025 ºC), thus all austenite is recrystallized, and all the process of controlled rolling takes place in recrystallized austenite phase.

Finally, the temperature Ar_3 (temperature of the transformation of the austenite into ferrite during the cooling) of the steel is:

Table 5.26 Results for the pass 5 (finishing)

Pass 5	
h_0 (mm)	6.08
h_f (mm)	4.55
$\bar{\varepsilon}$	0.335
v_{linear} (m/min)	250.8
$\dot{\bar{\varepsilon}}$ (s^{-1})	62.145
T (K)	1251
Z	$6.2 \cdot 10^{14}$
$t_{x=0.95}$ (s)	0.1
t_{growth} (s)	1.2
t_{pause} (s)	1.3
D_0 (μm)	30.2
D_{rex} (μm)	13.3
D_{final} (μm)	39.6
M	$7.54 \cdot 10^{15}$

Table 5.27 Results for the pass 6 (finishing)

Pass 6	
h_0 (mm)	4.55
h_f (mm)	3.63
$\bar{\varepsilon}$	0.261
v_{linear} (m/min)	314.7
$\dot{\bar{\varepsilon}}$ (s^{-1})	77.889
T (K)	1298
Z	$2.6 \cdot 10^{14}$
$t_{x=0.95}$ (s)	0.1
t_{growth} (s)	1.0
t_{pause} (s)	1.1
D_0 (μm)	39.6
D_{rex} (μm)	20.6
D_{final} (μm)	44.4
M	$3.04 \cdot 10^{16}$

Table 5.28 Results for the pass 7 (finishing)

Pass 7	
h_0 (mm)	3.63
h_f (mm)	3.23
$\bar{\varepsilon}$	–
v_{linear} (m/min)	–
$\dot{\bar{\varepsilon}}$ (s^{-1})	–
T (K)	1239
Z	–
$t_{x=0.95}$ (s)	-
t_{growth} (s)	–
t_{pause} (s)	–
D_0 (μm)	–
D_{rex} (μm)	–
D_{final} (μm)	–
M	–

$$Ar_3(°C) = 910 - 310 \cdot \%C - 80 \cdot \%Mn + 0.35 \cdot (t - 8) \quad (5.75)$$

where t is the thickness plate (mm):

$$Ar_3(°C) = 910 - 310 \cdot 0.14 - 80 \cdot 0.7 + 0.35 \cdot (3.23 - 8) = 808.9\,°C \quad (5.76)$$

Both the Ar_3 and the T_{nr} are lower than the final finishing temperature of the steel and, for that reason, the process takes place, as it was mentioned, in austenitic phase. There is always recrystallization of the austenite in each pass. At the end, we will have a ferritic-pearlitic structure at room temperature with grain size of around 20 μm.

Now we are going to calculate the yield strength and the ductile–brittle transition temperature. We use the following equations (Quintana 2013; Fernández et al. 2013; Pickering 1978; Petite et al. 1998):

$$\sigma_y(MPa) = 15.4 \cdot \left(3.5 + 2.1 \cdot \%Mn + 5.4 \cdot \%Si + 23 \cdot \%N_{free}{}^{1/2} + 1.13 \cdot d^{-1/2}\right) \quad (5.77)$$

and:

$$DBTT(°C, 27J) = -19 + 44 \cdot \%Si + 700 \cdot \%N_{free}{}^{1/2} - 11.5 \cdot d^{-1/2} + 2.2 \cdot f_P \quad (5.78)$$

Table 5.29 Rolling process (finishing) of a carbon-manganese steel (class Q0, type St42 44)

Pass	h_0 (mm)	h_f (mm)	$\bar{\varepsilon}$	$\dot{\bar{\varepsilon}}$ (s^{-1})	T (K)	$t_{x=0.95}$ (s)	t_{pause} (s)	t_{growth} (s)	D_0 (μm)	D_{rex} (μm)	D_{final} (μm)
1	40	24.2	0.580	5.417	1189	0.9	6.7	6.1	87.1	13.8	38.0
2	24.21	14.38	0.602	14.239	1201	0.1	4.2	4.1	38.0	8.4	38.0
3	14.38	9.01	0.540	27.542	1188	0.1	2.6	2.5	38.0	8.8	34.6
4	9.01	6.08	0.454	45.443	1165	0.4	1.8	1.4	34.6	9.4	30.2
5	6.08	4.55	0.335	62.145	1251	0.1	1.3	1.3	30.2	13.3	39.6
6	4.55	3.63	0.261	77.889	1298	0.1	1.1	0.9	39.6	20.6	44.4
7	3.63	3.23	–	–	1239	–	–	–	–	–	-

where d is the grain diameter expressed in millimeters, %Si and $\%N_{free}$ are the compositions in mass percent, f_P is the pearlite fraction. The percentage of pearlite in our steel is approximately 20%. This way (ferrite grain size is 20 μm):

$$\sigma_y(\text{MPa}) = 53.9 + 17.4 \cdot \left(20\ \mu\text{m} \cdot \frac{1\ \text{mm}}{10^3\ \mu\text{m}}\right)^{-1/2} = 176.9\ \text{MPa} \tag{5.79}$$

$$\text{DBTT}\ (^\circ\text{C}, 27\ \text{J}) = -19 - 11.5 \cdot \left(20\ \mu\text{m} \cdot \frac{1\ \text{mm}}{10^3\ \mu\text{m}}\right)^{-1/2} + 2.2 \cdot 20 = -56.3\ ^\circ\text{C} \tag{5.80}$$

End of the exercise.

Exercise 5.5 It is assumed that mechanical properties of the final product, univocal function of its microstructure, are characterized in the current case by a single parameter: the grain size. This is the reason for the importance of calculating the evolution of grain size along the rolling process with the aim of obtaining the suitable mechanical properties (yield strength, ultimate tensile strength, ductility, ductile–brittle transition temperature).

We analyze in this case the microstructural evolution of a carbon-manganese steel micro-alloyed with niobium (ND class, API X-70 type) obtained by controlled rolling. The chemical composition of the steel is: max. 0.15%C, max. 1.60%Mn, max. 0.3% Si, max. 0.01% S, max. 0.02% P, max. 0.04% Al, max. 0.05% Nb and max. 0.07% V. Roughing is carried out in 9 passes while finishing requires 7 passes, and the process is carried out in a semicontinuous mill whose geometry and treatment process are described as follows:

- Dimensions of the rough product (slab): 230 mm × 1559 mm × 8988 mm.
- Intermediate plate thickness (roughing mill): 45 mm × 45,128 mm (plate obtained at the end of the roughing).
- Weight: 25,299 kg.

Homogenization treatment: 240 min (time of permanence in the furnace), temperature at the exit of the furnace: 1263 °C. Initial grain size: 500 μm.

Roughing mill: diameter of the rolling cylinders, 1115.8 mm; diameter of the backup cylinders, 1304.5 mm.

We will use the equations proposed by Sellars and Whitemann (1979) to calculate the different parameters of the rolling process:

- For T > 1004 °C:

$$t_{x=0.50} = 2.52 \cdot 10^{-19} \cdot \varepsilon^{-4} \cdot {D_0}^2 \cdot \exp\left(\frac{325000\ \text{J/mol}}{\text{R} \cdot \text{T}}\right) \tag{5.81}$$

- For 891 °C < T < 1004 °C:

$$t_{x=0.50} = 5.94 \cdot 10^{-38} \cdot \varepsilon^{-4} \cdot D_0{}^2 \cdot \exp\left(\frac{780000\,\text{J/mol}}{R \cdot T}\right) \tag{5.82}$$

- For T< 891 °C:

$$t_{x=0.50} = 9.24 \cdot 10^{-9} \cdot \varepsilon^{-4} \cdot D_0{}^2 \cdot \exp\left(\frac{130000\,\text{J/mol}}{R \cdot T}\right) \tag{5.83}$$

For a different recrystallized fraction:

$$X = 1 - \exp\left[(\ln 0.5) \cdot \left(\frac{t}{t_{0.5}}\right)^2\right] \rightarrow t_{0.95} = t_{0.5} \cdot \left(\frac{\ln 0.05}{\ln 0.5}\right)^{1/2} \tag{5.84}$$

The time of growth is calculated as follows (the value of the t_{pause} is known in the roughing process):

$$t_{growth} = t_{pause} - t_{x=0.95} \tag{5.85}$$

In the case of the finishing, the distance between cages being about 5500 mm, and the $t_{pause}(s)$is calculated as follows:

$$t_{pause}(s) = \frac{5500\,\text{mm}}{v_{linear}(\text{m/min}) \cdot \frac{1000\,\text{mm}}{1\text{m}} \cdot \frac{1\,\text{min}}{60\text{s}}} \tag{5.86}$$

Additionally, we take as temperature between passes the mean temperature between them.

The mean grain size at the end of the recrystallization process depends only on two factors: the energy stored during the deformation, and the density of possible sites for the nucleation. This way, the size of the recrystallized grain is:

$$D_{rex} = 0.9 \cdot \varepsilon^{-0.67} \cdot D_0{}^{0.67}\,[\mu\text{m}] \tag{5.87}$$

Regarding the grain growth, with short times of permanence at high temperature, the variation with the time of the grain diameter is adjusted by the following equation:

$$D^{10} - D_{rex}{}^{10} = 2.15 \cdot 10^{65} \cdot \exp\left(-\frac{1.33 \cdot 10^6\,\text{J/mol}}{R \cdot T}\right) \cdot t_{growth}\,[\mu\text{m}] \tag{5.88}$$

We also use the following equations, indicated at the end of the Exercise 5.1:

- Linear speed of the rolls:

$$v = 2 \cdot \pi \cdot R \cdot \frac{\text{rpm}}{60} \tag{5.89}$$

– Equivalent deformation:

$$\bar{\varepsilon} = 1.155 \cdot \ln\left(\frac{h_0}{h_f}\right) \tag{5.90}$$

– Average equivalent deformation rate:

$$\dot{\bar{\varepsilon}} = \frac{1.155 \cdot v}{\sqrt{R \cdot \Delta h}} \cdot \ln\left(\frac{h_0}{h_f}\right) \tag{5.91}$$

Solution starts:

We use in this case the equations indicated in the enunciation for the case of the steel micro-alloyed with niobium. We collect in the following tables (Tables 5.30, 5.31, 5.32, 5.33, 5.34, 5.35, 5.36, 5.37 and 5.38) the results for each pass of the roughing process.

We see in Table 5.30 that the recrystallized fraction is 37.20%, this way, most of the deformation accumulates to the next pass, and in the second pass $t_{x=0.50}$ and D_{rex} are calculated for the sake of simplicity with the addition of the deformation of the first and the second passes (0.156 + 0.173).

We see in Table 5.37 that the recrystallized fraction is 30.50%, this way, the deformation should accumulate to the next passes.

We collect in Table 5.39 a summary of the roughing process.

Table 5.30 Results for the pass 1 (roughing)

Pass 1	
h_0 (mm)	230
h_f (mm)	201
$\bar{\varepsilon}$	0.156
$v_{cylinder}$ (rpm)	36
v_{linear} (mm/s)	2103
$\dot{\bar{\varepsilon}}$ (s^{-1})	2.573
T (K)	1518
Z	$1.3 \cdot 10^{11}$
$t_{x=0.50}$ (s)	16.6
$t_{x=0.95}$ (s)	34.5
t_{growth} (s)	-20.9
t_{pause} (s)	13.6
X	0.3720
D_0 (μm)	500.0
D_{rex} (μm)	-
D_{final} (μm)	500.0

Table 5.31 Results for the pass 2 (roughing)

Pass 2	
h_0 (mm)	201
h_f (mm)	173
$\bar{\varepsilon}$	0.156 + 0.173
$v_{cylinder}$ (rpm)	40.1
v_{linear} (mm/s)	2103
$\dot{\bar{\varepsilon}}$ (s^{-1})	3.247
T (K)	1509
Z	$1.9 \cdot 10^{11}$
$t_{x=0.50}$ (s)	1.0
$t_{x=0.95}$ (s)	2.0
t_{growth} (s)	7.4
t_{pause} (s)	9.4
X	1
D_0 (μm)	500.0
D_{rex} (μm)	121.9
D_{final} (μm)	124.1

Table 5.32 Results for the pass 3 (roughing)

Pass 3	
h_0 (mm)	173
h_f (mm)	145
$\bar{\varepsilon}$	0.204
$v_{cylinder}$ (rpm)	44.3
v_{linear} (mm/s)	2588
$\dot{\bar{\varepsilon}}$ (s^{-1})	4.222
T (K)	1501
Z	$2.9 \cdot 10^{11}$
$t_{x=0.50}$ (s)	0.5
$t_{x=0.95}$ (s)	0.1
t_{growth} (s)	9.6
t_{pause} (s)	10.6
X	1
D_0 (μm)	124.1
D_{rex} (μm)	66.0
D_{final} (μm)	100.3

Table 5.33 Results for the pass 4 (roughing)

Pass 4	
h_0 (mm)	145
h_f (mm)	116
$\bar{\varepsilon}$	0.258
$v_{cylinder}$ (rpm)	46.9
v_{linear} (mm/s)	2740
$\dot{\bar{\varepsilon}}$ (s^{-1})	5.551
T (K)	1491
Z	$4.5 \cdot 10^{11}$
$t_{x=0.50}$ (s)	0.1
$t_{x=0.95}$ (s)	0.3
t_{growth} (s)	7.9
t_{pause} (s)	8.2
X	1
D_0 (μm)	100.3
D_{rex} (μm)	49.0
D_{final} (μm)	91.5

Table 5.34 Results for the pass 5 (roughing)

Pass 5	
h_0 (mm)	116
h_f (mm)	88
$\bar{\varepsilon}$	0.319
$v_{cylinder}$ (rpm)	50.4
v_{linear} (mm/s)	2945
$\dot{\bar{\varepsilon}}$ (s^{-1})	7.515
T (K)	1479
Z	$7.4 \cdot 10^{11}$
$t_{x=0.50}$ (s)	0.1
$t_{x=0.95}$ (s)	0.1
t_{growth} (s)	11.5
t_{pause} (s)	11.6
X	1
D_0 (μm)	91.5
D_{rex} (μm)	39.9
D_{final} (μm)	87.0

Table 5.35 Results for the pass 6 (roughing)

Pass 6	
h_0 (mm)	88
h_f (mm)	65
$\bar{\varepsilon}$	0.350
$v_{cylinder}$ (rpm)	42.4
v_{linear} (mm/s)	2477
$\dot{\bar{\varepsilon}}$ (s^{-1})	7.650
T (K)	1463
Z	$9.9 \cdot 10^{11}$
$t_{x=0.50}$ (s)	0.1
$t_{x=0.95}$ (s)	0.1
t_{growth} (s)	6.1
t_{pause} (s)	6.2
X	1
D_0 (μm)	87.0
D_{rex} (μm)	36.3
D_{final} (μm)	72.6

Table 5.36 Results for the pass 7 (roughing)

Pass 7	
h_0 (mm)	65
h_f (mm)	55
$\bar{\varepsilon}$	0.193
$v_{cylinder}$ (rpm)	44.2
v_{linear} (mm/s)	2582
$\dot{\bar{\varepsilon}}$ (s^{-1})	6.669
T (K)	1439
Z	$1.3 \cdot 10^{12}$
$t_{x=0.50}$ (s)	0.6
$t_{x=0.95}$ (s)	1.3
t_{growth} (s)	13.1
t_{pause} (s)	14.4
X	1
D_0 (μm)	72.6
D_{rex} (μm)	47.8
D_{final} (μm)	65.6

Table 5.37 Results for the pass 8 (roughing)

Pass 8	
h_0 (mm)	55
h_f (mm)	50
$\bar{\varepsilon}$	0.110
$v_{cylinder}$ (rpm)	44.5
v_{linear} (mm/s)	2600
$\dot{\bar{\varepsilon}}$ (s^{-1})	5.417
T (K)	1399
Z	$2.3 \cdot 10^{12}$
$t_{x=0.50}$ (s)	10.2
$t_{x=0.95}$ (s)	21.2
t_{growth} (s)	−13.8
t_{pause} (s)	7.4
X	0.3050
D_0 (μm)	65.6
D_{rex} (μm)	–
D_{final} (μm)	65.6

Table 5.38 Results for the pass 9 (roughing)

Pass 9	
h_0 (mm)	50
h_f (mm)	45
$\bar{\varepsilon}$	0.110 + 0.122
$v_{cylinder}$ (rpm)	–
v_{linear} (mm/s)	–
$\dot{\bar{\varepsilon}}$ (s^{-1})	–
T (K)	1364
Z	–
$t_{x=0.50}$ (s)	–
$t_{x=0.95}$ (s)	–
t_{growth} (s)	–
t_{pause} (s)	–
X	–
D_0 (μm)	–
D_{rex} (μm)	–
D_{final} (μm)	–

Table 5.39 Rolling process (roughing) of a carbon-manganese steel (class ND, type API X-70)(*second pass accumulates the deformation of the first pass)

Pass	h_0 (mm)	h_f (mm)	$\bar{\varepsilon}$	$\dot{\bar{\varepsilon}}$ (s^{-1})	T (K)	$t_{x=0.50}$ (s)	$t_{x=0.95}$ (s)	t_{pause} (s)	t_{growth} (s)	X	D_0 (μm)	D_{rex} (μm)	D_{final} (μm)
1	230	201	0.156	2.573	1518	16.6	34.5	13.6	-20.9	0.372	500.0	-	500.0
2	201	173	0.173*	3.247	1509	1.0	2.0	9.4	7.4	1	500.0	121.9	124.1
3	173	145	0.204	4.222	1501	0.5	1.0	10.6	9.6	1	124.1	66.0	100.3
4	145	116	0.258	5.551	1491	0.1	0.3	8.2	7.9	1	100.3	49.0	91.5
5	116	88	0.319	7.515	1479	0.1	0.1	11.6	11.5	1	91.5	39.9	87.0
6	88	65	0.350	7.650	1463	0.1	0.1	6.2	6.1	1	87.0	36.3	72.6
7	65	55	0.193	6.669	1439	06	1.3	14.4	13.1	1	72.6	47.8	65.6
8	55	50	0.110	5.417	1399	10.2	21.23	7.4	−13.8	0.304	65.6	–	65.6
9	50	45	0.122		1364								

Table 5.40 Results for the pass 1 (finishing)

Pass 1	
h_0 (mm)	45
h_f (mm)	33.79
$\bar{\varepsilon}$	0.331
v_{linear} (m/min)	51.1
$\dot{\bar{\varepsilon}}$ (s^{-1})	3.96
T (K)	1243
Z	$4.8 \cdot 10^{13}$
$t_{x=0.50}$ (s)	13.3
$t_{x=0.95}$ (s)	27.6
t_{growth} (s)	−21.2
t_{pause} (s)	6.5
X	0.1508
D_0 (μm)	65.6
D_{rex} (μm)	–
D_{final} (μm)	65.6

After the roughing process in 9 passes, there is a time delay of 56 s between the roughing and the finishing process. Niobium at all inhibits grain growth during the time between the roughing and the finishing. We present in the following tables (Tables 5.40, 5.41, 5.42, 5.43, 5.44, 5.45 and 5.46) the results corresponding to the finishing stage. We use the data indicated in the enunciation.

We see in Table 5.40 that the recrystallized fraction is 15.08%, this way, the deformation accumulates to the next pass, and in the second pass $t_{x=0.50}$ and D_{rex} are calculated with the addition of the deformation of the first and the second passes (0.331 + 0.332).

We see in Table 5.42 that the recrystallized fraction is 19.73%, this way, the deformation accumulates to the next pass, and in the second pass $t_{x=0.50}$ and D_{rex} are calculated with the addition of the deformation of the first and the second passes (0.307 + 0.207).

We see in Table 5.44 that the recrystallized fraction is 0.02%, this way, the deformation accumulates to the next pass, and in the fifth pass $t_{x=0.50}$ and D_{rex} are calculated with the addition of the deformation of the fourth and the fifth passes (0.146 + 0.075).

We see in Table 5.45 that the recrystallized fraction is 0.02%,

We appreciate in Table 5.46 (Table 5.45) a stagnation of the grain size in around 10.3 μm. After the allotropic transformation $\gamma \rightarrow \alpha$ (at around 890 °C), each grain of austenite generates between 2 and 3 grains of ferrite (Tanaka 1981; Tamura et al.1988). That is to say, the grain size in the ferritic-pearlitic steel, at room temperature, will be around 4–5 μm (12 ASTM) μm.

The final microstructure of the material is around 4–5 μm (ferritic grain size), being possible this fine microstructure due to the presence of niobium. It is possible to

Table 5.41 Results for the pass 2 (finishing)

Pass 2	
h_0 (mm)	33.79
h_f (mm)	25.34
$\bar{\varepsilon}$	0.331 + 0.332
v_{linear} (m/min)	68.2
$\dot{\bar{\varepsilon}}$ (s^{-1})	7.28
T (K)	1230
Z	$1.2 \cdot 10^{14}$
$t_{x=0.50}$ (s)	1.8
$t_{x=0.95}$ (s)	3.8
t_{growth} (s)	1.0
t_{pause} (s)	4.8
X	1
D_0 (μm)	65.6
D_{rex} (μm)	19.5
D_{final} (μm)	19.5

Table 5.42 Results for the pass 3 (finishing)

Pass 3	
h_0 (mm)	25.34
h_f (mm)	19.42
$\bar{\varepsilon}$	0.307
v_{linear} (m/min)	88.9
$\dot{\bar{\varepsilon}}$ (s^{-1})	10.45
T (K)	1220
Z	$2.2 \cdot 10^{14}$
$t_{x=0.50}$ (s)	6.6
$t_{x=0.95}$(s)	13.7
t_{growth} (s)	−10.0
t_{pause} (s)	3.7
X	0.1973
D_0 (μm)	19.5
D_{rex} (μm)	–
D_{final} (μm)	19.5

Table 5.43 Results for the pass 4 (finishing). *We assume 95% of recrystallization

Pass 4	
h_0 (mm)	19.42
h_f (mm)	16.23
$\bar{\varepsilon}$	0.307 + 0.207
v_{linear} (m/min)	106.4
$\dot{\bar{\varepsilon}}$ (s^{-1})	1.29
T (K)	1210
Z	$3.1 \cdot 10^{14}$
$t_{x=0.95}$ (s)	1.6
$t_{x=0.95}$ (s)	3.3
t_{growth} (s)	-0.2*
t_{pause} (s)	3.1
X	0.93 ($\simeq$0.95)
D_0 (μm)	19.5
D_{rex} (μm)	10.3
D_{final} (μm)	10.3

Table 5.44 Results for the pass 5 (finishing)

Pass 5	
h_0 (mm)	16.23
h_f (mm)	14.3
$\bar{\varepsilon}$	0.146
v_{linear} (m/min)	12.8
$\dot{\bar{\varepsilon}}$ (s^{-1})	11.64
T (K)	1197
Z	$4.5 \cdot 10^{14}$
$t_{x=0.50}$ (s)	156.5
$t_{x=0.95}$ (s)	325.4
t_{growth} (s)	-322.7
t_{pause} (s)	2.7
X	0.0002
D_0 (μm)	10.3
D_{rex} (μm)	-
D_{final} (μm)	10.3

Table 5.45 Results for the pass 6 (finishing)

Pass 6	
h_0 (mm)	14.3
h_f (mm)	13.4
$\bar{\varepsilon}$	0.146 + 0.075
v_{linear} (m/min)	128.8
$\bar{\dot{\varepsilon}}$ (s^{-1})	926
T (K)	1186
Z	$4.7 \cdot 10^{14}$
$t_{x=0.50}$ (s)	61.7
$t_{x=0.95}$ (s)	128.4
t_{growth} (s)	−125.8
t_{pause} (s)	2.6
X	0.0012
D_0 (μm)	10.3
D_{rex} (μm)	–
D_{final} (μm)	10.3

Table 5.46 Results for the pass 7 (finishing)

Pass 7	
h_0 (mm)	13.4
h_f (mm)	13.29
$\bar{\varepsilon}$	–
v_{linear} (m/min)	-
$\bar{\dot{\varepsilon}}$ (s^{-1})	–
T (K)	1173
Z	–
$t_{x=0.95}$ (s)	–
t_{growth} (s)	–
t_{pause} (s)	–
D_0 (μm)	–
D_{rex} (μm)	–
D_{final} (μm)	–
M	–

Table 5.47 Rolling process (finishing) of a carbon-manganese steel (class ND, type API X-70) (*second, fourth and sixth passes accumulate the deformation of the previous pass,)

Pass	h_0 (mm)	h_f (mm)	$\bar{\varepsilon}$	$\dot{\bar{\varepsilon}}(s^{-1})$	T (K)	$t_{x=0.50}$ (s)	$t_{x=0.95}$ (s)	t_{pause} (s)	t_{growth} (s)	X	D_0 (μm)	D_{rex} (μm)	D_{final} (μm)
1	45	33.79	0.331	3.96	1243	13.3	27.6	6.5	-21.2	0.1508	65.6	–	65.6
2	33.79	25.34	0.332	7.28	1230	1.8	3.8	4.8	1.0	1	65.6	19.5	19.5
3	25.34	19.42	0.307	10.45	1220	6.6	13.7	3.7	-10.0	0.1973	19.5	–	19.5
4	19.42	16.23	0.207	11.29	1210	1.6	3.3	3.1	-0.2	0.9299	19.5	10.3	10.3
5	16.23	14.3	0.146	11.64	1197	156.5	325.4	2.7	-322.7	0.0002	10.3	–	10.3
6	14.3	13.4	0.075	9.26	1186	61.7	128.4	2.6	-125.8	0.0012	10.3	–	10.3
7	13.4	13.29	–										

estimate the non-recrystallization temperature of the austenite (Barbosa et al. 1987; Belzunce et al. 1995):

$$T_{nr}(°C) = 887 + 464 \cdot \%C + \left[6445 \cdot \%Nb - 644 \cdot (\%Nb)^{1/2}\right] + \left[732 \cdot \%V - 230 \cdot (\%V)^{1/2}\right] + 890 \cdot \%Ti + 363 \cdot \%Al - 357 \cdot \%Si \tag{5.92}$$

Our steel has the following composition: %C, 0.15; %Mn, 1.6; %Si, 0.3; %Al, 0.040; %Nb, 0.050; %V, 0.070%. This way, the austenite non-recrystallization temperature is:

$$T_{nr}(°C) = 887 + 464 \cdot 0.15 + \left[6445 \cdot 0.050 - 644 \cdot (0.050)^{1/2}\right] + \left[732 \cdot 0.070 - 230 \cdot (0.070)^{1/2}\right] + 363 \cdot 0.040 - 357 \cdot 0.3 = 1131\ °C \tag{5.93}$$

The austenite recrystallizes above 1131 °C in our steel. We can see in previous tables that some of the passes of the roughing and finishing processes take place in presence of non-recrystallized austenite.

Finally, the Ar_3 temperature (temperature of the transformation of the austenite into ferrite during the cooling) is:

$$Ar_3(°C) = 910 - 310 \cdot \%C - 80 \cdot \%Mn + 0.35 \cdot (t - 8) \tag{5.94}$$

where t is the thickness plate (mm). In our case:

$$Ar_3(°C) = 910 - 310 \cdot 0.15 - 80 \cdot 1.6 + 0.35 \cdot (13.29mm - 8) = 737.4\ °C \tag{5.95}$$

All the deformation took place in austenitic phase, although there were several passes where the rolling was carried out in presence of non-recrystallized austenite. There would be at the end of the process certain heterogeneity regarding the grain size (see Fig. 5.25).

Now we are going to calculate the yield strength and the ductile–brittle transition temperature. We use the following equations (Quintana 2013; Fernández et al. 2013; Pickering 1978; Petite et al. 1998):

$$\sigma_y(MPa) = 15.4 \cdot \left(3.5 + 2.1 \cdot \%Mn + 5.4 \cdot \%Si + 23 \cdot \%N_{free}{}^{1/2} + 1.13 \cdot d^{-1/2}\right) \tag{5.96}$$

and:

$$DBTT(°C, 27J) = -19 + 44 \cdot \%Si + 700 \cdot \%N_{free}{}^{1/2} - 11.5 \cdot d^{-1/2} + 2.2 \cdot f_P \tag{5.97}$$

Fig. 5.25 Micrograph of the steel at the end of the rolling process (roughing and finishing)

where d is the grain diameter expressed in millimeters, %Si and $\%N_{free}$ are the compositions in mass percent, f_P is the pearlite fraction. The percentage of pearlite in our steel is approximately 20%. This way (ferrite grain size is 5 μm):

$$\sigma_y(\text{MPa}) = 53.9 + 17.4 \cdot \left(5\ \mu\text{m} \cdot \frac{1\ \text{mm}}{10^3\ \mu\text{m}}\right)^{-1/2} = 299.97\ \text{MPa} \tag{5.98}$$

$$\text{DBTT}\ (°\text{C}, 27\ \text{J}) = -19 - 11.5 \cdot \left(5\ \mu\text{m} \cdot \frac{1\ \text{mm}}{10^3\ \mu\text{m}}\right)^{-1/2} + 2.2 \cdot 20 = -137.6\ °\text{C} \tag{5.99}$$

To achieve the 500 MPa of yield strength of the API-X70, it is necessary to take into account the nanometric precipitates of Nb carbonitrides (austenitic phase) and V carbides (ferritic phase) that appear during the rolling and fast cooling of the steel down to the room temperature (Leslie 1981, pp. 193–199).

End of the exercise.

5.5 Homogenization Heat Treatment (Soaking): Oxidation and Decarburizing of the Steel

A previous stage to all processes of hot rolling of the steel consists of heating ingots, slabs, billets, or blooms in treatment furnaces to achieve that the material could reach the temperature of the forming process: in austenitic phase (commonly), in ferritic phase (ferritic stainless steels, Fe-3% Si) or in austenitic-ferritic phase. As it was previously indicated (Pero-Sanz et al. 2017, 2019), the homogenization heat treatment (soaking) has the following objectives:

- conditioning the steel to facilitate the subsequent treatment of hot forge.
- eliminating or palliating the effects of the minor segregation that are consequence of nonequilibrium solidification: alloying elements or impurities tend to concentrate in zones that solidify in the last moment (coring).
- avoiding, if possible, the effects of the oxidation-decarburizing that are unavoidable at high temperatures, that reduce the yield of the operation and oblige to remove the altered surface (oxidized and decarburized).

In integrated steel plants, the combustible used in the process is a mixture of blast furnace and coke oven gases, while in electric steelworks they usually employ natural gas or fuel oil. The types of furnaces are:

- Pit furnaces, for ingots and slabs.
- Car-hearth furnaces, for ingots and rough products of large section.
- Walking beam furnaces, for the heating of the formats of continuous casting.
- Pusher furnaces, for ingots and continuous casting formats.

In any case, it is necessary that the material has a suitable temperature both in the surface and inside. We show in Fig. 5.26 the cross section of one slab during the heating process in a walking beam furnace at 1350 °C (rather high for a homogenization annealing), the solution to the problem of heat transfer in nonsteady state is found using the software based on the finite element method (ANSYS), under the boundary conditions indicated in the figure. We show in Fig. 5.27 the distribution of the isotherms in the steel after a treatment of three hours, when the mean temperature of the slab that is charged in the furnace is 25 °C.

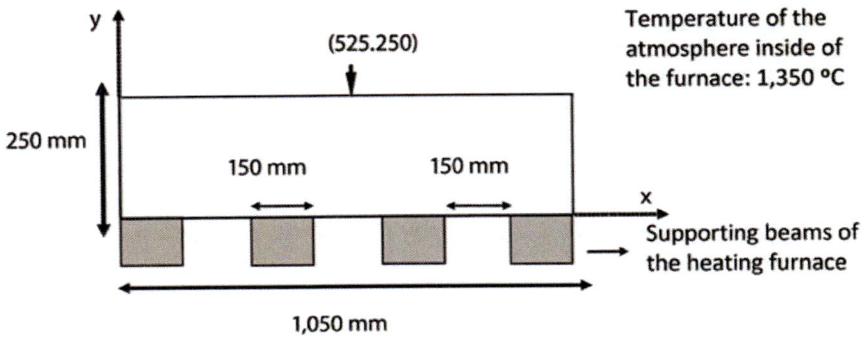

Fig. 5.26 Cross section of a slab inside of a walking beam furnace (four supports). Boundary conditions of the problem of heat transfer: temperature of the furnace, 1350 °C; thermal conductivity of the steel (0 °C $\le$ T $\le$ 1450 °C), $\lambda = 29.8 + 1.45 \cdot 10^{-3} \cdot T(°C)$ [$W \cdot m^{-1} \cdot K^{-1}$]; enthalpy of the steel (0 °C $\le$ T $\le$ 1450 °C), $H = 5.69 \cdot 10^{-6} \cdot T(°C)$ [$J \cdot m^{-3}$]; upper horizontal heat transfer coefficient, $h_{hs} = 2.4 \cdot \Delta T^{0.25}$ [$W \cdot m^{-2} \cdot K^{-1}$]; vertical heat transfer coefficient, $hv = 1.7 \cdot \Delta T^{0.25}$ [$W \cdot m^{-2} \cdot K^{-1}$]; lower horizontal heat transfer coefficient, $h_{hi} = 1.2 \cdot \Delta T^{0.25}$ [$W \cdot m^{-2} \cdot K^{-1}$]

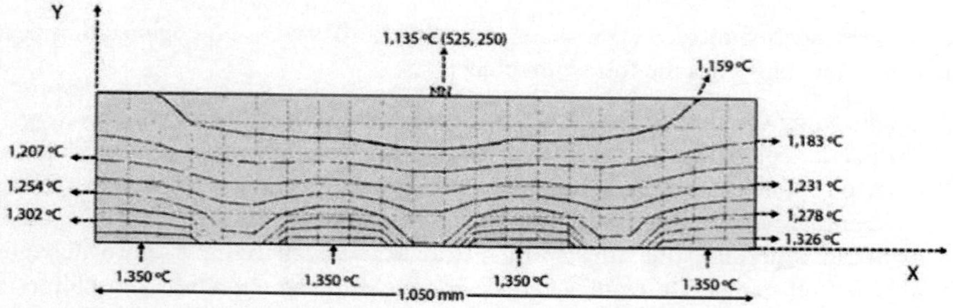

Fig. 5.27 Distribution of temperatures in the slab after three hours. The temperature of the material at the beginning of the process is 25 °C

As it is possible to check, even after three hours, there is not thermal homogeneity inside of the material, there are hot zones in contact with the supports (beams) of the furnace, while around the central upper zone of the rough product the coldest points can be found.

For equality of stresses on the surface of the rough product (rolling pressure), the periphery of the slab (zone of maximum temperature) suffers more severe deformation than the central zone (nucleus). This phenomenon gives origin to the known "fishtail" effect when flat (coil and thick plates) or long (profiles and rail) products are hot rolled.

Figure 5.28 shows the variations of temperature, in the node of coordinates (525, 250, 0), with the time as a function of the average temperature of entrance of the rough product into the heat treatment furnace.

Logically, the possibility of chaining operations of continuous casting of slabs with their subsequent hot plastic deformation (compact steelworks) involves, apart from an improvement in the productivity, a reduction of the operation costs.

As a consequence of the annealing homogenization heat treatment previous to the hot deformation of the steel, it is produced:

– the decarburizing of the steel, especially in ferritic-pearlitic and pearlitic qualities (medium and high-carbon).
– the oxidation of the material (formation of mill scale).

In the heating furnaces, previous to the forge, apart from the oxidation, there is decarburization of the steel as:

$$a_C\ (\text{gas}) < a_C\ (\text{steel}) \tag{5.100}$$

Thus, the transportation of the carbon from the steel to the gaseous phase is thermodynamically feasible. We show in Fig. 5.29 a scheme of the situation that would be established for steel, in the heart of the treatment furnace, after a time t:

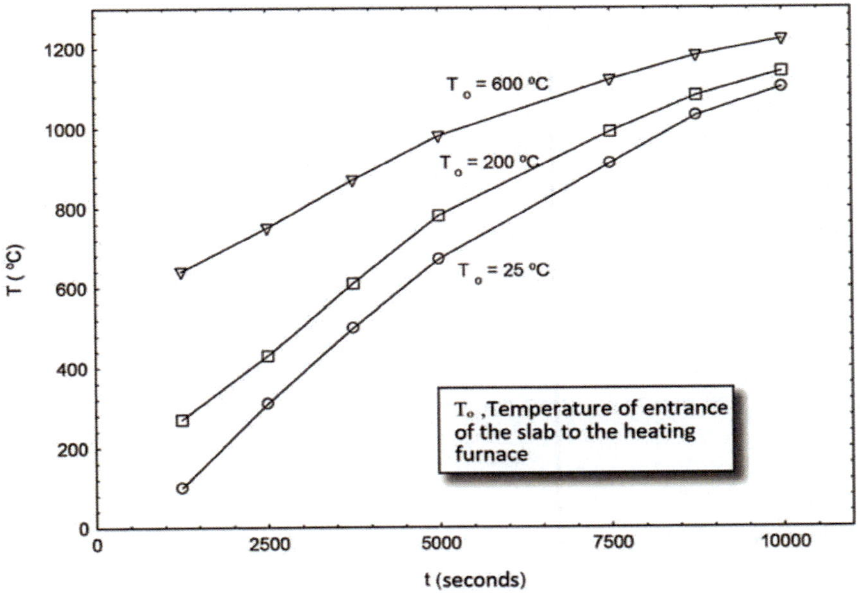

Fig. 5.28 Variation of the temperature, in the node of minimum temperature (Fig. 5.27), of a rectangular rough product of steel, with the time as a function of the initial mean temperature of the slab. Working temperature of the treatment furnace: 1350 °C

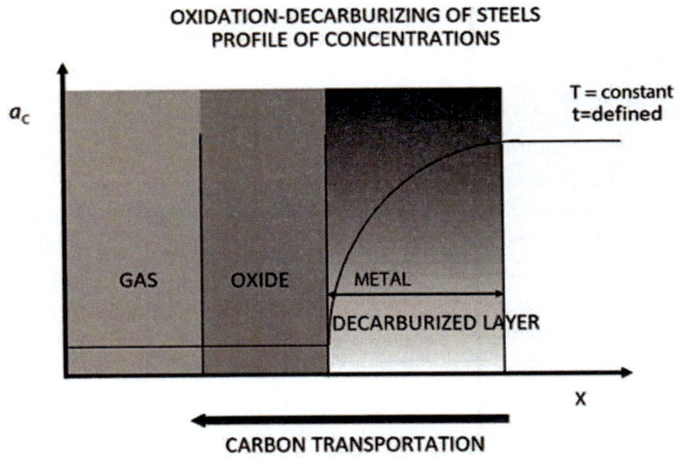

Fig. 5.29 Profile of activity-concentration-penetration of the carbon in the steel during the oxidizing/decarburization process

– The formation of an oxide layer whose thickness is estimated as follows:

$$x^2 = k_{o/p} \cdot t \tag{5.101}$$

where $k_{o/p}$ is the reaction rate constant of the oxidation process (parabolic model).

– The presence of a decarburized layer, whose thickness is defined as the distance for which the carbon content reaches 75% of the global concentration in the steel.

If the potential of the atmosphere in contact with the steel is not oxidizing, carbon concentration, as a function of the position and time, will be expressed using the second Law of Fick:

$$\frac{dC_C}{dt} = D_C{}^{\gamma} \cdot \frac{d^2C_C}{dx^2} \tag{5.102}$$

The solution to the differential equation (Eq. 5.102) includes the error function (Appendix 7) and is equal to:

$$\frac{C_C^0 - C_C}{C_C^0 - C_C^g} = 1 - \operatorname{erf}\left(\frac{x}{2 \cdot \sqrt{D_C{}^{\gamma} \cdot t}}\right) \tag{5.103}$$

where C_C^0 is the global carbon concentration in the steel, C_C^g is the carbon concentration (activity) in the gas, consequence of the relation of partial pressures of the carburizing gases (CO/CO_2 or CH_4/H_2), C_C is the carbon concentration in the steel as a function of the time and depth x, and $D_C{}^{\gamma}$ is the diffusion coefficient of the carbon in the steel.

Normally, the conditions of the atmosphere of the heating furnaces for the formats of the continuous casting are oxidizing. The gasification of the carbon is not performed in the steel-gas interface, it is carried out in the oxide-gas interface (Fig. 5.30), and the carbon not only diffuses through the decarburized layer but also through the FeO layer. The variation of the concentration with the time and position does not correspond to Eq. (5.103) as it does not include the diffusional transportation of the carbon through the oxide layer. The decarburization of the steel under oxidizing conditions would be equal to:

$$\frac{C_C^0 - C_C}{C_C^0 - C_C^s} = \frac{1 - \operatorname{erf}\left(\frac{x}{2 \cdot \sqrt{D_C{}^{\gamma} \cdot t}}\right)}{1 - \operatorname{erf}\left(\frac{1}{2} \cdot \sqrt{\frac{k_O}{D_C{}^{\gamma}}}\right)} \tag{5.104}$$

where C_C^s is the concentration (activity) of the carbon in the FeO-steel interface and k_O is the oxidation constant of the metal.

5.5.1 Solved Exercises

Exercise 5.6 Homogenization treatment: dendritic segregation.

Solution starts:

As discussed earlier, even single-phase alloys possess microsegregation after dendritic solidification. This microsegregation (coring) can be reduced by a high-temperature treatment termed homogenization. The index of residual microsegregation, δ_i, is a useful parameter to evaluate it:

$$\delta_i = \frac{C_M - C_m}{C_M^0 - C_m^0} \quad (5.105)$$

where C_M is the maximum solute concentration of the element i (in interdendritic spaces) at time t, C_m is the minimum solute concentration of the element i (in center of dendrite arms) at time t, C_M^0 is the maximum initial solute concentration of the element i, C_m^0 is the minimum initial solute concentration of the element i, and δ_i is the index of residual microsegregation of the element i at the instant t.

According to a simple model of homogenization (sinusoidal variation of elements between center and periphery of dendrites (Flemings 1974; Porter and Easterling 1992), where the minimum segregation is observed in the nucleus and the maximum in the periphery) the value of δ_i after a homogenization treatment depends on the simple dimensionless group $D_S \cdot t/l^2$, where D_S is the diffusion coefficient in the solid at the temperature of homogenization of the element (assumed here a constant), t is the homogenization time and l is a characteristic diffusion distance of the order of the dendrite arm spacing. It results:

$$\delta_i = e^{-\pi^2 \cdot \left(\frac{D_S \cdot t}{l_0^2}\right)} \quad (5.106)$$

This equation is useful for approximate prediction of times and temperatures required to homogenize a given as-cast structure.

Application to an ingot 0.4% C–2% Ni with interdendritic spacing of 500 μm. How long will be the time required to reduce the coring of carbon by 90%? And to reduce the coring of nickel by the same amount?

Data (at 1200 °C): $D_C^\gamma = 2.23 \cdot 10^{-6} \mathrm{cm^2/s}$; $D_{Ni}^\gamma = 7.03 \cdot 10^{-11} \mathrm{cm^2/s}$

Using Eq. (5.106):

$$\delta_i = e^{-\pi^2 \cdot \left(\frac{D_S \cdot t}{l_0^2}\right)} \rightarrow \ln\delta_i = -\pi^2 \cdot \left(\frac{D_S \cdot t}{l_0^2}\right) \rightarrow t = -\frac{\ln\delta_i \cdot l_0^2}{\pi^2 \cdot D_S} \quad (5.107)$$

– For carbon:

$$t = -\frac{\ln\delta_i \cdot l_0{}^2}{\pi^2 \cdot D_S} = -\frac{\ln(1-0.9)\cdot(0.05\text{cm})^2}{\pi^2 \cdot 2.23 \cdot 10^{-6}\text{cm}^2/\text{s}} = 262\text{s}(< 5\text{min}) \quad (5.108)$$

that is viable.

- For nickel:

$$t = -\frac{\ln\delta_i \cdot l_0{}^2}{\pi^2 \cdot D_S} = -\frac{\ln(1-0.9)\cdot(0.05\text{cm})^2}{\pi^2 \cdot 7.03 \cdot 10^{-11}\text{cm}^2/\text{s}} = 8296609\text{s}(2305\text{h}) \quad (5.109)$$

2305 h (>>1 day), which is non-viable.

End of the exercise.

Exercise 5.7 Homogenization treatment. Steel 0.75% C (billet) experiences a homogenization treatment at 1250 °C for 2 h and 30 min (150 min). We make the following assumptions:

(a) thermal and chemical homogeneity between billets and treatment atmosphere.
(b) there is no formation of mill scale in the surface (suitable atmosphere).
(c) From t = 0 carbon concentration in the solid/gas interface is zero.

We want to calculate the thickness of the decarburized layer when the carbon concentration is reduced to 75% of the nominal composition.

Data: the diffusion coefficient of the carbon in the austenite is equal to: $D_C^\gamma = 0.246 \cdot \exp(-17546.2/T)\text{cm}^2/\text{s}$.

Solution starts:

In this case, the solution to the second Fourier's equation (deoxidizing and decarburizing atmosphere, Fig. 5.30):

$$\frac{\partial C}{\partial t} = D \cdot \frac{\partial^2 C}{\partial x^2} \quad (5.110)$$

is the following one:

$$\frac{C_S - C_x}{C_S - C_0} = \text{erf}\left(\frac{x}{2 \cdot \sqrt{D_C \cdot t}}\right) \quad (5.111)$$

where C_S is the carbon concentration in the surface (which takes the value of zero), C_0 is the carbon nominal concentration (which takes the value of 0.75) and C_x is the concentration where the carbon content in the steel is 0.75·0.75 = 0.563% C.

We calculate the diffusion coefficient at 1250 °C:

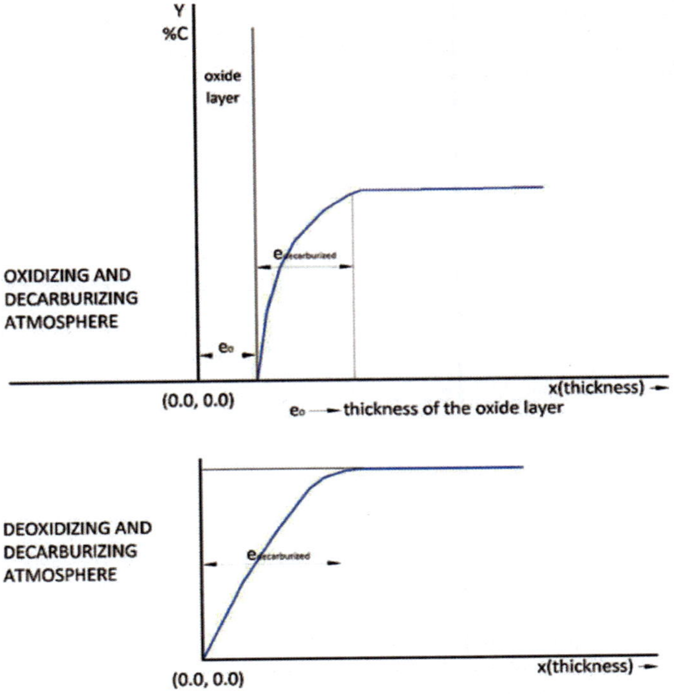

Fig. 5.30 Profile of carbon concentration

$$D_C^{\gamma} = 0.246 \cdot \exp\left(\frac{-17546.2}{1250 + 273}\right) = 2.44 \cdot 10^{-6}\mathrm{cm}^2/\mathrm{s} \tag{5.112}$$

We solve Eq. (5.111):

$$\begin{aligned} \frac{0-0.563}{0-0.75} &= \mathrm{erf}\left(\frac{x}{2\cdot\sqrt{2.44\cdot 10^{-6}\frac{\mathrm{cm}^2}{\mathrm{s}}\cdot 150\ \mathrm{min}\cdot\frac{60\ \mathrm{s}}{1\ \mathrm{min}}}}\right) \to 0.7507 \\ &= \mathrm{erf}\left(\frac{x}{2\cdot\sqrt{2.44\cdot 10^{-6}\frac{\mathrm{cm}^2}{\mathrm{s}}\cdot 150\ \mathrm{min}\cdot\frac{60\ \mathrm{s}}{1\ \mathrm{min}}}}\right) \to 0.8146 \\ &= \frac{x}{2\cdot\sqrt{2.44\cdot 10^{-6}\frac{\mathrm{cm}^2}{\mathrm{s}}\cdot 150\ \mathrm{min}\cdot\frac{60\ \mathrm{s}}{1\ \mathrm{min}}}} \to x = 0.2414\ \mathrm{cm}(2.4\ \mathrm{mm}) \end{aligned} \tag{5.113}$$

that is the thickness of the decarburized layer.

Solution ends.

Exercise 5.8 In the same conditions that in the previous exercise, estimate the appearance of an oxide layer in the steel surface, through which the carbon diffuses toward the heat treatment atmosphere. The value of the constant of parabolic oxidation $k_{o/p}$ is $3.56 \cdot 10^{-6}\text{cm}^2/\text{s}$ and D_C^γ is $2.44 \cdot 10^{-6}\text{cm}^2/\text{s}$ at 1250 °C.

Solution starts:

The solution to the second equation of Fick is in this case (mixed decarburizing -deoxidizing), see Fig. 5.30:

$$\frac{C_C^0 - C_x}{C_C^0 - C_C^s} = \frac{1 - \text{erf}\left(\frac{x}{2\cdot\sqrt{D_C\cdot t}}\right)}{1 - \text{erf}\left(\frac{1}{2}\sqrt{\frac{k_{o/p}}{D_C{}^\gamma}}\right)} \tag{5.114}$$

where C_C^0 is the nominal carbon content in the steel, C_x is the carbon content after the treatment at a distance x from the metal/oxide interface and C_C^s is the carbon content in the metal/oxide interface (zero).

Thus:

$$\frac{0.75 - 0.56}{0.75 - 0} = \frac{1 - \text{erf}\left(\frac{x}{2\cdot\sqrt{D_C{}^\gamma\cdot t}}\right)}{1 - \text{erf}\left(\frac{1}{2}\cdot\sqrt{\frac{k_{o/p}}{D_C{}^\gamma}}\right)} \to 0.2533 = \frac{1 - \text{erf}\left(\frac{x}{2\cdot\sqrt{D_C{}^\gamma\cdot t}}\right)}{1 - \text{erf}\left(\frac{1}{2}\cdot\sqrt{\frac{k_{o/p}}{D_C{}^\gamma}}\right)} \tag{5.115}$$

First, we solve:

$$1 - \text{erf}\left(\frac{1}{2}\cdot\sqrt{\frac{k_{o/p}}{D_C{}^\gamma}}\right) = 1 - \text{erf}\left(\frac{1}{2}\cdot\sqrt{\frac{3.56\cdot 10^{-6}\frac{\text{cm}^2}{\text{s}}}{2.44\cdot 10^{-6}\frac{\text{cm}^2}{\text{s}}}}\right)$$
$$= 1 - \text{erf}(0.6039) = 1 - 0.6069 = 0.3931 \tag{5.116}$$

$$0.2533 = \frac{1 - \text{erf}\left(\frac{x}{2\cdot\sqrt{D_C^\gamma\cdot t}}\right)}{0.3931} \to 0.9004 = \text{erf}\left(\frac{x}{2\cdot\sqrt{D_C^\gamma\cdot t}}\right) \to 1.1645$$
$$= \frac{x}{2\cdot\sqrt{D_C^\gamma\cdot t}} \to x \tag{5.117}$$
$$= 1.1645\cdot 2\cdot\sqrt{2.44\cdot 10^{-6}\frac{\text{cm}^2}{\text{s}}\cdot 150\text{ min}\cdot\frac{60\text{ s}}{1\text{ min}}} \to x$$
$$= 0.3451\text{ cm } (\simeq 3.4\text{ mm})$$

Thus, the total thickness (addition of the decarburized and deoxidized layer) is:

$$x = x_o + x_d \tag{5.118}$$

The thickness of the oxidized layer is:

$$x_o = \sqrt{k_{o/p} \cdot t} = \sqrt{3.56 \cdot 10^{-6} \frac{cm^2}{s} \cdot 150\,min \cdot \frac{60s}{1\,min}} = 0.1790\,cm\ (\simeq 1.8\,mm) \quad (5.119)$$

The thickness of the decarburized layer is:

$$x_d = x - x_o = 3.4 - 1.8 = 1.9\,mm \quad (5.120)$$

Commentaries: In the previous exercise, non-oxidizing atmosphere: $x_o = 0$ and $x_d = 2.4\,mm$. In this exercise, oxidizing atmosphere: $x_o = 1.8\,mm$ and $x_d = 1.6\,mm$. If the atmosphere is oxidizing (usual situation), the thickness of the decarburized layer diminishes.

We are going to demonstrate Eq. (5.114) considering Fig. 5.31.

First, we calculate the thickness of the oxidized layer (x_o) from:

$$x_o = \sqrt{k_{o/p} \cdot t} \quad (5.121)$$

and:

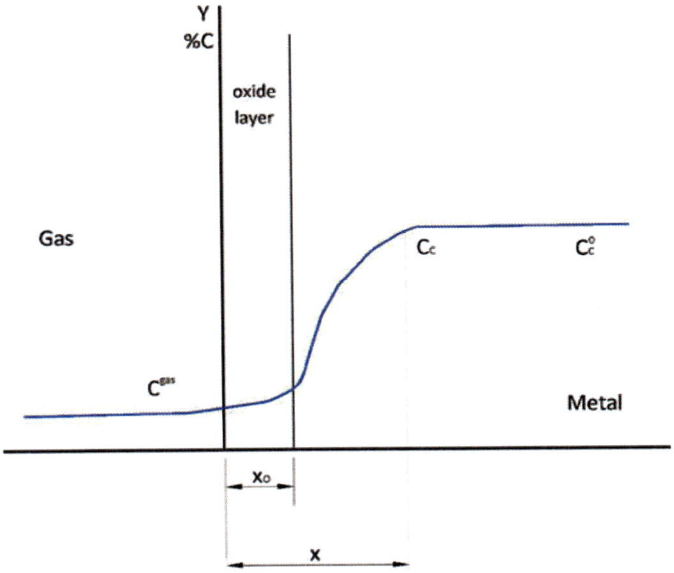

Fig. 5.31 Decarburizing with oxidation

$$C(x,t) = A + B \cdot \mathrm{erf}\left(\frac{x}{\sqrt{4 \cdot D_C{}^{\gamma} \cdot t}}\right) \tag{5.122}$$

We have as boundary conditions:

$$\forall t \text{ and } x = x_o \rightarrow C_C^{gas} = A + B \cdot \mathrm{erf}\left(\frac{x_o}{\sqrt{4 \cdot D_C{}^{\gamma} \cdot t}}\right) \tag{5.123}$$

$$\forall t \text{ and } x = \infty \rightarrow C_C^0 = A + B \tag{5.124}$$

If we subtract to Eq. (5.123) the Eq. (5.124):

$$C_C^{gas} - C_C^0 = B \cdot \left[\mathrm{erf}\left(\frac{x_o}{\sqrt{4 \cdot D_C{}^{\gamma} \cdot t}}\right) - 1\right] \tag{5.125}$$

$$B = \frac{C_C^{gas} - C_C^0}{\mathrm{erf}\left(\frac{x_o}{\sqrt{4 \cdot D_C{}^{\gamma} \cdot t}}\right) - 1} \tag{5.126}$$

$$A = C_C^0 - B = C_C^0 - \frac{C_C^{gas} - C_C^0}{\mathrm{erf}\left(\frac{x_o}{\sqrt{4 \cdot D_C{}^{\gamma} \cdot t}}\right) - 1} \tag{5.127}$$

$$C_C = C_C^0 + \frac{C_C^{gas} - C_C^0}{\mathrm{erf}\left(\frac{x_o}{\sqrt{4 \cdot D_C{}^{\gamma} \cdot t}}\right) - 1} \cdot [\mathrm{erf}(z) - 1] \tag{5.128}$$

$$\frac{C_C - C_C^0}{C_C^{gas} - C_C^0} = \frac{\mathrm{erf}(z) - 1}{\mathrm{erf}\left(\frac{x_o}{\sqrt{4 \cdot D_C{}^{\gamma} \cdot t}}\right) - 1} \tag{5.129}$$

$$\frac{C_C^0 - C_C}{C_C^0 - C_C^{gas}} = \frac{1 - \mathrm{erf}(z)}{1 - \mathrm{erf}\left(\frac{x_o}{\sqrt{4 \cdot D_C{}^{\gamma} \cdot t}}\right)} \tag{5.130}$$

with $x_o = \sqrt{k_{o/p} \cdot t}$:

$$\frac{C_C^0 - C_C}{C_C^0 - C_C^{gas}} = \frac{1 - \mathrm{erf}(z)}{1 - \mathrm{erf}\left(\frac{1}{2} \cdot \sqrt{\frac{k_O}{D_C{}^{\gamma}}}\right)} \tag{5.131}$$

End of the exercise.

5.6 Superplasticity

In the research of micro-alloyed ferritic-pearlitic weldable steels with grain sizes of around 12 ASTM ($\simeq$5 μm), hot formed up to temperatures equivalents to those of the $\gamma \rightarrow \alpha$ transformation (or slightly lower), it is possible to identify an intermediate behavior between the typical of the hot deformation (great $\dot{\varepsilon}$, $A(\sinh\alpha\sigma)^k \simeq \dot{\varepsilon} \cdot \exp(Q/R \cdot T) = Z$) and the deformation due to creep (very small $\dot{\varepsilon}$, $\dot{\varepsilon} = A' \cdot \sigma^5 \cdot \exp(-Q/R \cdot T)$), which is called superplastic behavior, characterized by a great sensitivity to the applied stress at the deformation rate:

$$\sigma \simeq k \cdot \dot{\varepsilon}^m \tag{5.132}$$

with values of the coefficient m of around 0.5. The conditions that a material should have to exhibit superplastic behavior (in our case steel for ship building sheet of 1" of thickness) are:

- fine grain size, stable during the hot deformation (there must not be any tendency to the grain growth). Stability of the microstructure.
- deformation by intergranular sliding and rotation of the ensemble of grains, accommodated by the dislocations to avoid the cavitation or appearance of cracks and intergranular porosity.
- resistance to the appearance of localized necking in tensile test. It is possible the appearance of one or several necks, but they move along the specimen without appearing, prematurely, localized necking in a certain section: there is diffuse necking in superplasticity.
- with a constitutive law of the type:

$$\sigma = k \cdot \varepsilon^n \cdot \dot{\varepsilon}^m \tag{5.133}$$

the deformation of the specimen will be stable (superplastic) if:

$$\sigma = S \cdot \dot{\varepsilon}^m \tag{5.134}$$

where S is the instantaneous cross section of the specimen. In other words, if:

$$\frac{dS}{dt} \tag{5.135}$$

is constant along the specimen, it is verified that:

$$\varepsilon_u = \frac{n}{1-m} \tag{5.136}$$

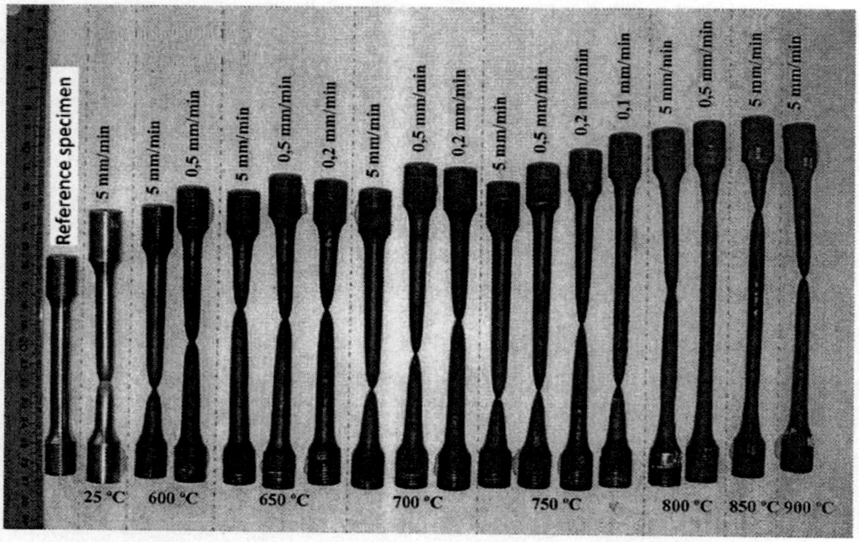

Fig. 5.32 Batch of specimens tested in tension at different temperatures and cross-head speed (L_0 = 50 mm)

and for great values of m (0.3 < m < 1), the uniform deformation (ε_u) can be as great as 1000%.

Figure 5.32 shows a batch of specimens tested in tension at different temperatures and different cross-head speed (v_T) and Fig. 5.33 shows the representation of $\log\sigma/\log\varepsilon$ measured at 800 °C: deformations reached 300%. Tests were performed at ambient atmosphere (without protecting atmosphere; using protecting atmosphere would have allowed obtaining greater elongations, avoiding the phenomenon of oxidation-corrosion observed in the experimentation) (Quintana 2013).

5.6.1 Solved Exercises

Exercise 5.9 If the constitutive law of a material follows a law like that showed below:

$$\sigma = k \cdot \varepsilon^n \cdot \dot{\varepsilon}^m \tag{5.137}$$

demonstrate that:

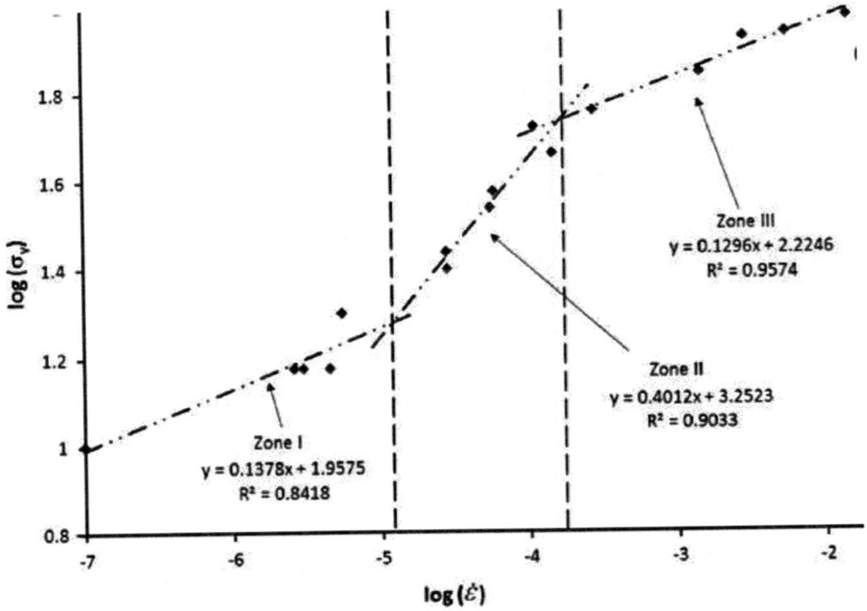

Fig. 5.33 Influence of strain rate on yield stress in superplastic behavior at 800 °C

– if the material has superplastic behavior (n ↓, m ↑), that is to say very sensitive to the deformation rate, the maximum uniform deformation is:

$$\varepsilon_u = \frac{n}{(1 - m)} \tag{5.138}$$

being the condition for a superplastic behavior:

$$\frac{dS}{dt} = \text{constant} \tag{5.139}$$

(the material deforms uniformly from one side to the other of the specimen).

ε_u is the maximum true uniform deformation in tension, n is the strain hardening coefficient and m is the strain rate sensitivity coefficient.

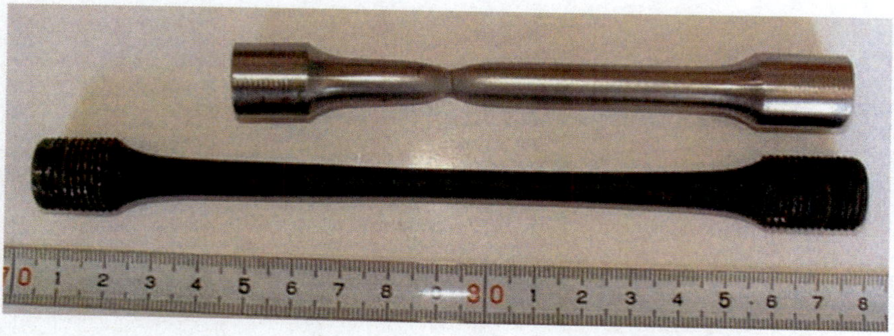

Fig. 5.34 Specimens with localized necking (upper specimen) and diffuse necking (lower specimen) ($L_0 = 50$ mm)

Solution starts:

The conditions for the superplastic behavior of a material (steel) are (Daviess et al. 1970):

- stress must be very sensitive to the deformation rate.
- necking should be diffused; necking must not be in a weak section as it is the typical situation (Fig. 5.34).

The generalized equation of the law of Ludwik–Hollomon is:

$$\sigma = k \cdot \varepsilon^n \cdot \dot{\varepsilon}^m \tag{5.140}$$

and relates stresses, strains, and true deformation rates. Habitually, $m \ll 0.1$ and the simplified equation in tension is:

$$\sigma = k \cdot \varepsilon^n \tag{5.141}$$

It indicates that the maximum force applied in the moment of the appearance of necking passes through a maximum and ε_u (maximum uniform deformation is equal to n). In any moment of the test:

$$F = \sigma \cdot S \tag{5.142}$$

where S is the instantaneous section of the specimen at the beginning of the necking. Necking begins when we reach the maximum of applied force ($dF = 0$):

$$dF = 0 \rightarrow \sigma \cdot dS + S \cdot d\sigma = 0 \tag{5.143}$$

balancing both the geometrical softening ($dS < 0$) and the strain hardening ($d\sigma > 0$). This way, we have:

$$\frac{d\sigma}{\sigma} = -\frac{dS}{S} = d\varepsilon \rightarrow \frac{d\sigma}{d\varepsilon} = \sigma \rightarrow \frac{d\sigma}{d\varepsilon} = n \cdot k \cdot \varepsilon^{n-1} \tag{5.144}$$

If we replace in the simplified equation of Ludwik-Hollomon:

$$n \cdot k \cdot \varepsilon^{n-1} = k \cdot \varepsilon^{n} \tag{5.145}$$

$$\varepsilon_u = n \tag{5.146}$$

However, if the steel exceptionally admits superplastic behavior, a complementary condition is verified:

$$\frac{dS}{dt} = k \tag{5.147}$$

which implies constancy in volume during the plastic deformation. Then:

$$S \cdot l = \text{constant} \rightarrow \frac{dS}{dt} \cdot l + \frac{dl}{dt} \cdot S = 0 \tag{5.148}$$

By definition:

$$\frac{dl}{l \cdot dt} = \dot{\varepsilon} \tag{5.149}$$

Then, if we multiply and divide by l:

$$\frac{dS}{dt} \cdot \frac{l}{l} + \frac{dl}{dt} \cdot \frac{S}{l} = 0 \rightarrow k = -\dot{\varepsilon} \cdot S \tag{5.150}$$

As:

$$\varepsilon = \ln\left(\frac{S_0}{S}\right) \rightarrow e^{\varepsilon} = \frac{S_0}{S} \rightarrow S = \frac{S_0}{e^{\varepsilon}} \rightarrow S = S_0 \cdot e^{-\varepsilon} \tag{5.151}$$

This way:

$$k = -\dot{\varepsilon} \cdot S \rightarrow k = -\dot{\varepsilon} \cdot S_0 \cdot e^{-\varepsilon} \rightarrow k_1 = \dot{\varepsilon} \cdot e^{-\varepsilon} \rightarrow \dot{\varepsilon} = k_1 \cdot e^{\varepsilon} \tag{5.152}$$

We also know:

$$\frac{d\dot{\varepsilon}}{d\varepsilon} = k_1 \cdot e^{\varepsilon} = \dot{\varepsilon} \tag{5.153}$$

We know from Eq. (5.142) that:

$$F = \sigma \cdot S \tag{5.154}$$

We obtain, considering Eq. (5.151) and the generalized equation of Ludwik–Hollomon:

$$F = \sigma \cdot S = k \cdot \varepsilon^n \cdot \dot{\varepsilon}^m \cdot S_0 \cdot e^{-\varepsilon} = k_2 \cdot \varepsilon^n \cdot \dot{\varepsilon}^m \cdot e^{-\varepsilon} \tag{5.155}$$

The condition of necking establishes $dF = 0$ (maximum of F), this way:

$$\begin{aligned} \frac{dF}{d\varepsilon} = 0 &\rightarrow \frac{d\left(k_2 \cdot \varepsilon^n \cdot \dot{\varepsilon}^m \cdot e^{-\varepsilon}\right)}{d\varepsilon} = 0 \rightarrow \frac{d\left[k_2 \cdot \varepsilon^n \cdot (k_1 \cdot e^{\varepsilon})^m \cdot e^{-\varepsilon}\right]}{d\varepsilon} = 0 \\ &\rightarrow \frac{d\left[k_2 \cdot \varepsilon^n \cdot k_1^m \cdot e^{\varepsilon \cdot m} \cdot e^{-\varepsilon}\right]}{d\varepsilon} = 0 \\ &\rightarrow n \cdot \varepsilon^{n-1} \cdot k_2 \cdot k_1^m \cdot e^{\varepsilon \cdot m} \cdot e^{-\varepsilon} + m \cdot e^{\varepsilon \cdot m} \cdot k_2 \cdot \varepsilon^n \cdot k_1^m \cdot e^{-\varepsilon} \\ &- k_2 \cdot \varepsilon^n \cdot k_1^m \cdot e^{\varepsilon \cdot m} \cdot e^{-\varepsilon} = 0 \end{aligned} \tag{5.156}$$

considering that:

$$\dot{\varepsilon} = k_1 \cdot e^{\varepsilon} \tag{5.157}$$

$$\begin{aligned} &n \cdot \varepsilon^{n-1} \cdot k_2 \cdot k_1^m \cdot e^{\varepsilon \cdot m} \cdot e^{-\varepsilon} + m \cdot e^{\varepsilon \cdot m} \cdot k_2 \cdot \varepsilon^n \cdot k_1^m \cdot e^{-\varepsilon} - k_2 \cdot \varepsilon^n \cdot k_1^m \cdot e^{\varepsilon \cdot m} \cdot e^{-\varepsilon} \\ &= 0 \rightarrow n \cdot \varepsilon^{n-1} \cdot k_2 \cdot \dot{\varepsilon}^m \cdot e^{-\varepsilon} + m \cdot k_2 \cdot \varepsilon^n \cdot \dot{\varepsilon}^m \cdot e^{-\varepsilon} - k_2 \cdot \varepsilon^n \cdot \dot{\varepsilon}^m \cdot e^{-\varepsilon} = 0 \end{aligned} \tag{5.158}$$

we divide by Eq. (5.155):

$$\frac{n \cdot \varepsilon^{n-1} \cdot k_2 \cdot \dot{\varepsilon}^m \cdot e^{-\varepsilon} + m \cdot k_2 \cdot \varepsilon^n \cdot \dot{\varepsilon}^m \cdot e^{-\varepsilon} - k_2 \cdot \varepsilon^n \cdot \dot{\varepsilon}^m \cdot e^{-\varepsilon}}{k_2 \cdot \varepsilon^n \cdot \dot{\varepsilon}^m \cdot e^{-\varepsilon}} = 0 \tag{5.159}$$

$$\frac{n \cdot \varepsilon^{n-1}}{\varepsilon^n} + m - 1 = 0 \tag{5.160}$$

$$n \cdot \varepsilon^{n-1} \cdot \varepsilon^{-n} + m - 1 = 0 \tag{5.161}$$

$$\varepsilon_u = \frac{n}{1 - m} \tag{5.162}$$

Fig. 5.35 Micrograph of steel tested at 800 °C, 0.1 mm/min crosshead speed, $L_0 = 50$ mm, and cooled in air (superplastic behavior)

If m is great ($0 < m < 1$), the uniform deformations are elevated, $\varepsilon_u \gg n$ and the behavior is superplastic. It is based on grain-boundary sliding with diffusional accommodation. In so doing, grains suffer accommodation strains and translate past each other by sliding at their boundaries.

We show in Fig. 5.35 the micrograph of steel tested under superplastic behavior. It is possible to compare with the same steel after roughing and finishing rolling Fig. 5.13: after superplasticity condition, the former banded microstructure is nearly removed.

End of the exercise.

References

Barbosa R, Boratto F, Yue S, Jonas JJ (1987) The influence of chemical composition on the recrystallisation behaviour of microalloyed steels. In: Proceedings symposium on processing, microstructure and properties of HSLA steel, Pittsburgh, pp 51–56

Belzunce FJ, Rodríguez C, Fernández B (1995) Rolling procedure and thickness effects on strength and fracture behavior of microalloyed steels. Mater Sci Technol 11:252–257

Davies GJ, Edington JW, Cutler CP, Padmanabhan KA (1970) Superplasticity: a review. J Mater Sci 5:1091–1102

Dieter GE (1976) Mechanical metallurgy, 2nd edn. McGraw Hill, New York, USA

Dutta B, Sellars CM (1986) Strengthening of austenite by niobium during hot rolling of microalloyed steel. Mater Sci Technol 2:146–153

Dutta B, Sellars CM (1987) Effect of composition and process variables on Nb (C, N) precipitation in niobium microalloyed austenite. Mater Sci Technol 3:197–205

Fernández S, Quintana MJ, García JO, Verdeja LF, González R, Verdeja JI (2013) Superplastic HSLA steels: microstructure and failure. J Failure Anal 13:368–376

Fernández-Fernández S (2012) La Superplasticidad en Aceros de Bajo Contenido en Carbono, Master Thesis, June 2012. University of Oviedo, Oviedo, Asturias, Spain, Escuela Técnica Superior de Ingenieros de Minas de Oviedo

Flemings MC (1974) Solidification processing. Ed. McGraw Hill, New York, United States of America

Leslie WC (1981) The physical metallurgy of steels,1st edn. McGraw Hill International Book Company, London, United Kingdom

Quintana Hernández MJ (2013) Room temperature and high temperature behavior of low carbon steels manufactured by controlled rolling process, PhD Thesis, September 2013, Escuela Técnica Superior de Ingenieros de Minas de Oviedo, University of Oviedo, Oviedo, Asturias, Spain

Pero-Sanz JA (2004) Aceros: Metalurgia Física, Selección y Diseño. Ed. CIE Dossat 2000, Madrid, Spain

Pero-Sanz JA, Quintana-Hernández MJ, Verdeja LF (2017) Solidification and Solid-State Transformations of Metals and, Alloys. Elsevier, Boston, United States of America

Pero-Sanz JA, Fernández-González D, Verdeja LF (2019) Structural materials: properties and, selection. Springer International Publishing, Cham

Petite MM, Monsalve A, Gutiérrez I, Zaitegui J, Larburu JI (1998) Modelización de la evolución microestructural durante el recocido continuo de chapas de acero bajo en carbono. Predicción de las propiedades mecánicas. Revista De Metalurgia De Madrid 34:333–337

Pickering FB (1978) Physical metallurgy and design of steels, Ed. Applied Science Publishers, London, United Kingdom

Porter DA, Easterling KE (1992) Phase transformations in metals and alloys, 2nd edn. Nelson Thornes Ltd., Cheltenham, United Kingdom

Sellars CM, Whiteman JA (1979) Recrystallization and grain growth in hot rolling. Metal Sci 187–194

Tamura I, Ouchi C, Tanaka T, Sekine H (1988) Thermomechanical processing of high strength low alloy steels. Ed. Butterworths, London, United Kingdom

Tanaka T (1981) Controlled rolling of steel plate and strip. Int Metals Rev 4:185–212

Urcola Galarza JJ, Fuentes Pérez M (1980a) Deformación en caliente. I parte: Principios fundamentales. Revista De Metalurgia Del CENIM 16:263–267

Urcola Galarza JJ, Fuentes Pérez M (1980b) Deformación en caliente. II parte: Resistencia mecánica y ductilidad. Revista De Metalurgia Del CENIM 16:337–342

Urcola Galarza JJ, Fuentes Pérez M (1981) Deformación en caliente. III parte: Cambios microstructurales durante y después de la deformación. Revista De Metalurgia Del CENIM 17:9–15

Chapter 6
Cold Rolling. Deep Drawing. Wire Drawing. Coatings

Abstract Not all the hot-rolled products are used as finished products and most of them are subjected to additional treatments aimed at both increasing their mechanical properties, adjusting their dimensions, or increasing their resistance to corrosion. This way, this chapter is focused on different processes of great importance in steelmaking with the purpose of accomplishing the above-mentioned requirements. These processes are the cold rolling of long products, the deep drawing or cold forming the wire drawing, and the coatings. After these processes, a product of very small thickness with good surface finishing and narrow margin of dimensional tolerance is obtained in the case of the cold rolling or deep drawing (in this case with the suitable shape conferred by the action of a punch), a product with suitable mechanical properties for structural applications in the case of the wire drawing, and a protection against corrosion and superficial wear in the case of the coating.

6.1 Introduction

Not all the hot-rolled products are used as finished products and a great part of them is processed again to improve their metallurgical properties, to adjust their dimensions, or to increase their resistance to corrosion. For that reason, hot-worked products are subjected again to a series of finishing processes, which on many occasions are interrelated, and that are usually classified in:

- Cold rolling of long products. Cold forming (deep drawing and stretching).
- Wire drawing.
- Coatings.

Certainly, the requirement of reaching a dimensional specification can make necessary to perform a cold rolling treatment on a hot-rolled coil, but if the dimensional specification required for automotive steel is 1 mm in thickness, the alternatives are:

I. Using the sequence blast furnace, converter, conventional continuous casting, hot rolling mill, and cold rolling (minimum installed capacity of 3.0 Mt/year).

L. F. Verdeja González et al., *Operations and Basic Processes in Steelmaking*, Topics in Mining, Metallurgy and Materials Engineering,
https://doi.org/10.1007/978-3-030-68000-8_6

II. Using the sequence electric furnace (80% scrap + 20% Direct Reduction Iron, DRI), continuous casting of thick plate and CSP process (Compact Strip Production) (minimum installed capacity 1.0 Mt/year).
III. Using the sequence electric furnace (80% scrap + 20% DRI), continuous casting of thin plate, CSP process, and cold rolling to improve the surface finishing (minimum installed capacity 1.0 Mt/year).

From all of them, the most economical is the second option, then the third alternative and, finally, the first option, with independence of the galvanizing and pre-painting processes to which the coil is subjected in the finishing process. That is to say, if the structural and dimensional requirements defined by the automotive company are accomplished, the most economical method to produce the steel sheet is the second alternative because it does not require finishing by cold rolling.

This example of competitiveness that electric steelworks offer in comparison with the integrated steel plants has led the traditional steel installations to reduce costs, even when they are aware of their undeniable leadership about quality.

6.2 Cold Rolling of Long Products

When a very small thickness (tenths of millimeter) product with good surface finishing and narrow margin of dimensional tolerance is required, thus it is necessary to turn to the cold rolling without previous heating of the steel. The cold plastic deformation of a polycrystalline aggregate, by a rolling process, is translated into a permanent deformation of it and, especially (in the case of steel), apart from the variation of the external shape of the material, some of the intrinsic characteristics are modified as the material acquires work hardening, which is a characteristic property of the metallic state that have neither the polymers nor the ceramic materials. In a simple way, it is possible to say that the cold-working (work-hardening) is the increase of the hardness that a metal (steel in particular) acquires because of the cold deformation.

The increase in the steel hardness after the cold rolling can lead to a wrong conclusion: assuming that the cold deformation produces either a certain consolidation or a kind of increase in the density of the material that is responsible for the increase in the hardness. The metallurgical reasons for the work hardening are the line defects of structure (dislocations) that exist inside of the crystals and their multiplication during the cold-working and reactions between them during the deformation (Pero-Sanz 2004). The density of dislocations can grow six orders of magnitude ($10^5 \rightarrow 10^{11}$ dis/cm^2).

The same as during the hot rolling, the work performed in the steel, ΔW, is consumed in increasing the internal energy of the steel, ΔU, and in the dissipation of the heat, ΔQ. However, as the process is performed at room temperature, all the internal energy accumulated by the material, as punctual, linear or modification of the grain boundaries, cannot be released during the process (dynamic recrystallization), and produces a growing degree of cold-working (work-hardening) in the material. This happens when $T_D < T_M/3$.

Before the cold rolling of the material in a tandem mill (Figs. 6.1 and 6.2), the hot coil must be pickled in a sulfuric or chlorhydric medium to eliminate the oxide layer that is adhered to the surface of the steel, and that is produced as a consequence of the high temperatures that were reached during the coiling of the steel strip. Once pickled and cleaned, the steel is coiled again applying a film of oil to either protect it from the oxidation during the storage (hot-rolled and pickled coil) or to act as a lubricant if the coil passes to a cold rolling process.

The cold rolling reduces the thickness of the band controlling very closely the flatness and dimensional tolerances of the sheet in two types of installations:

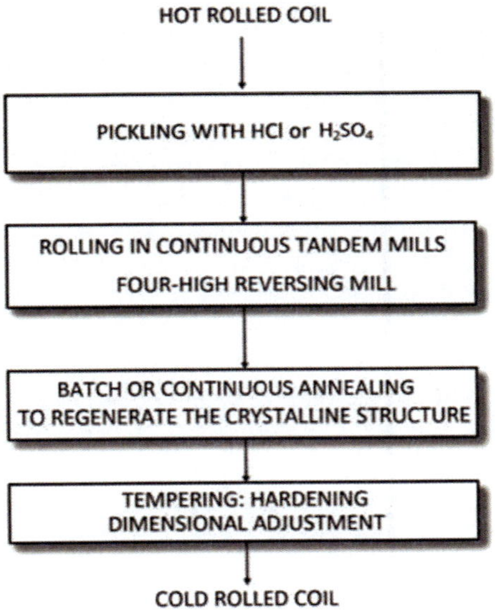

Fig. 6.1 Stages of the cold rolling process for flat products

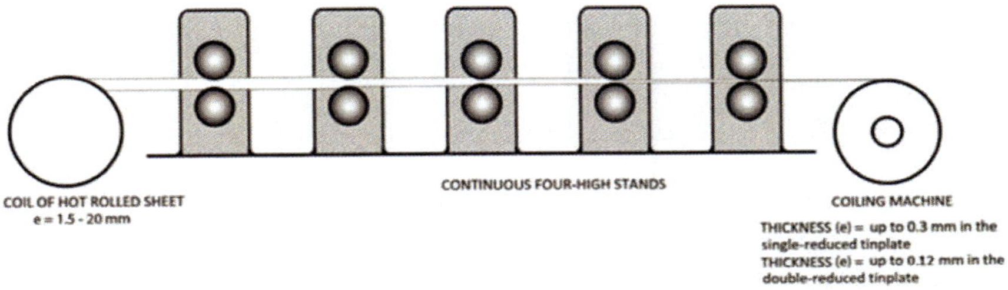

Fig. 6.2 Scheme of a tandem mill. Backup rolls are not represented in the figure

- Reversing mills with only one four-high stand, where several passes are performed until obtaining the final product.
- Continuous tandem mills formed by four or five four-high stands (Fig. 6.2).

The regeneration of the equiaxed crystalline structure that the steel had before the cold rolling is achieved using the annealing furnaces to perform the following annealing treatments:

- Batch annealing, discontinuous heat treatment in bell furnaces.
- Continuous annealing or flash annealing.

Before the annealing treatment of the steel strip, it is necessary to remove the rests of oil from the surface that are consequences of the cold rolling process. The cleaning of the strip is achieved by means of several continuous passes through a series of baths with detergents that perform cleaning of mechanical type. Equally, and if it is required, it is possible to use an electrolytic cleaning line after the mechanical degreasing. Finally, the clean and dry coil is ready to be treated via continuous annealing or batch annealing.

However, if the objective is to eliminate the main organic impurities (the oil layer of the cold-rolled coils), it is possible to achieve it using a N_2/H_2 atmosphere (650 °C, coming from the dissociation of the ammonia), a thermal pickling of the layer (that apart from decomposing the organic macromolecules, reduces the iron oxides that are still unaltered after the pickling). Many galvanizing lines combine chemical pickling with the thermal pickling.

The batch annealing is an operation that is usually performed stacking up several coils under the protecting cover of an indirect heating furnace (Fig. 6.3). If the heating was direct, the atmosphere of the furnace would be reductant. An atmosphere that

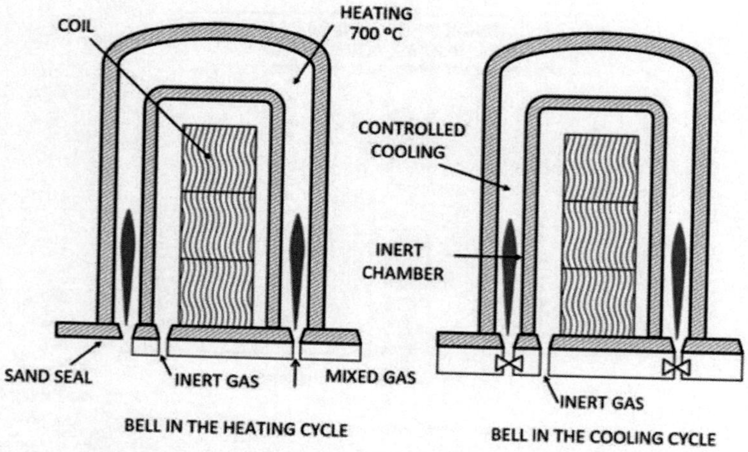

Fig. 6.3 Scheme of a batch annealing furnace during the cycles of heating, holding, and cooling

comprises 90% N_2-10% H_2, apart from improving the heat transfer, is very suitable to achieve excellent surface characteristics as the automotive industry requires.

The cycle of the batch annealing (heating, holding, and cooling) can take at least several days.

The continuous annealing (Fig. 6.4) is the fastest option (minutes), and the heating, holding, and cooling are optimized. The main differences with respect to the products that are recrystallized using the batch annealing are:

- Improvement of the surface characteristics by both removing the surface carbon and improving the flatness of the product.
- The faster cooling of the continuous annealing provides an equiaxial crystalline structure instead of a structure of lengthened grains (pancake structure), typical from steels annealed using the batch annealing treatment.
- The recrystallized structure of lengthened grain of steels annealed using the batch annealing is not as favorable as the equiaxial structure to avoid micro-segregations.
- Increase of the productivity, reduction of the workforce, and less consumption of energy in the continuous annealing installations.

The cold-rolled and annealed sheet can have in the stress-strain curve a yield plateau that indicates the practical yield strength of the material as it is possible to estimate an upper yield strength, σ_y (upper), and a lower yield strength, σ_y (lower), which are not desirable in the final product, especially if the final destination of the material is the cold-working operation of deep drawing.

The tensile behavior of a cold-rolled, and annealed steel strip (Pero-Sanz 2004) is a consequence of the aging of materials by the formation of Cottrell atmospheres (strain aging). When a metallic material is a solid solution (it does not matter if it is substitutional or interstitial), there are several interactions between the solute atoms and the dislocations of the solvent metal. The concurrence of several atoms of solute in the same point, close to the dislocation slip plane, constitutes the Cottrell atmospheres, which contribute to affix the edge dislocation increasing the stiffness of the material (yield strength): this causes the appearance of the yield plateau and

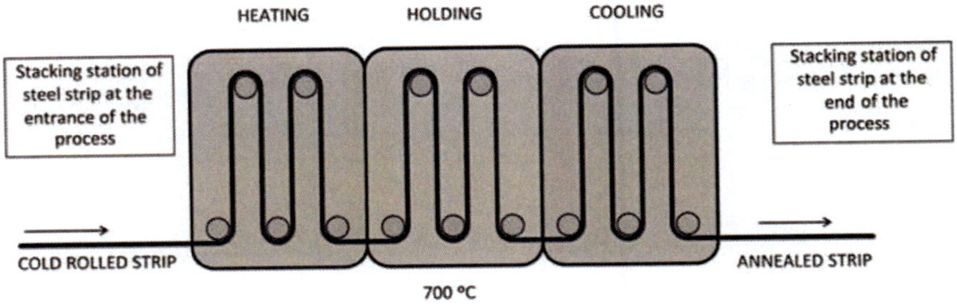

Fig. 6.4 Scheme of a continuous annealing processing line

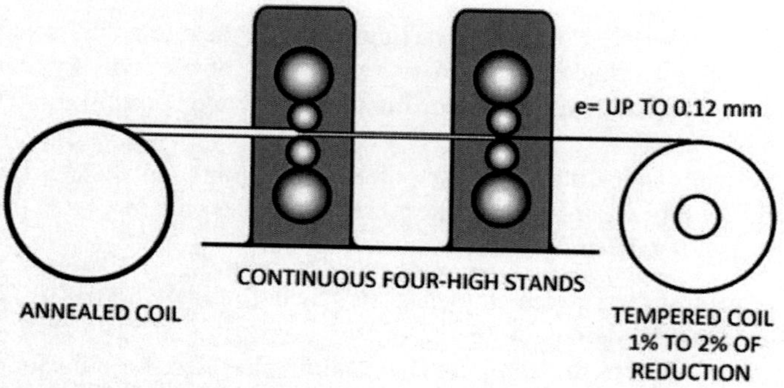

Fig. 6.5 Scheme of a temper mill

the Lüders bands (Piobert-Lüders). Rimmed steels are very prone to this effect, it is not the case of the killed steels except for prolonged stages of storage.

The upper yield strength, σ_y (upper), is the stress that is necessary to break the pinning of the dislocations by nitrogen, carbon, or oxygen atoms dissolved in the steel. Once broken the affixes, the plastic deformation starts with a smaller stress, σ_y (lower). The yield plateau is produced by the liberated dislocations but also by the new dislocations generated via Frank-Read mechanism. During the plastic deformation of the yield plateau, bands where this deformation is localized (Lüders bands) appear macroscopically on the surface and heterogeneously distributed along the specimen.

The finishing of the cold worked, and annealed coils takes place in the Temper mill. The tempering or skin-pass is performed in a continuous train with one or two four-high stands (Fig. 6.5). It is a soft rolling, carried out in dry conditions, and that produces an elongation of 1–2%. With Temper mills with greater power and adequate rolling oils, it is possible to achieve reductions of 30–50%, which can make feasible the manufacture of double-reduced tinplate (final thickness of 0.12–0.14 mm, although it is estimated that in the close future the final thickness might be in the range 0.08–0.06 mm).

The cold-rolled, annealed, and tempered sheet is used as raw material for the coating processes that do not necessarily have annealing furnaces, as the lines of electroplating: of tin (tinplating), of zinc (zinc plating), or chromium (chrome plating). Equally, it is also possible to commercialize packages or coil of cold-rolled, annealed, and tempered sheet.

6.3 Metallurgical Fundamentals of the Cold-Working

Previous chapters underline the scientific and technical fundamentals of the raw materials, obtaining of pig iron, steels, ferroalloys, and hot rolling of ironmaking products: this last one constitutes the last step in the elaboration of structural steel (ship building sheet, pipes, profiles, steel for concrete, etc.). The final microstructure of any metallic material (in this case, steel), and, for that reason, the mechanical properties in service are the result of both the chemical composition, the solidification structures, and the alterations by heat treatment, forming, or other modifications that the steel can experience in solid-state.

The industrial history of steel almost always begins with a solidification process. In some cases, it can start in other way: for example, the direct reduction of iron ore (DRI alternatives) or the method of iron carbide (cementite), the subsequent compaction of the solid sponge or that of metallic powder (Verdeja et al. 2020).

There are many parts directly obtained by molding (used in service without posterior mechanical forming, except machining). This is the case, for instance, of the grey cast iron parts (they approximately represent more than 5% (in weight) of the metallic materials produced every year worldwide). Many other parts are also directly manufactured by molding, both of steel and non-ferrous alloys. These products are used in service without subsequent mechanical forming. They sometimes require heat treatments—their fundamentals have been previously indicated to modify the as-molded structure (Pero-Sanz et al. 2017, 2018, 2019). Definitely, the Physical Metallurgy of the solidification and the phase transformations lies upon the elaboration of steels, cast irons, and powder metallurgy products of base iron.

Mechanical forming of steels can be also carried out in cold, at room temperature. In other cases (see Chap. 5), the deformation is carried out at high temperatures, from ingots, blooms or rough products previously obtained by solidification (conventional casting or continuous casting and hot forming).

This last chapter is aimed at studying the microstructural consequences of the cold forming or plastic deformation of the steel (which are not significantly different from those of the rest of metallic materials).

The cold plastic deformation ($T < T_M/4$, where T_M is the melting point in Kelvin, $\simeq$250 °C for the steel) of a steel (iron) polycrystalline aggregate by any forming process (rolling, stretching, drawing, deep drawing, compaction of powders, bending, straightening, etc.) can be translated into the appearance of a permanent deformation (work hardening), either inside of the grains or in the grain boundaries, which must assimilate the deformation. However, the cold deformation together with the progressive reduction of the sheet's thickness (i.e., rolling processes) or the external variation of the geometry of the product modifies, in an important manner, some intrinsic properties of the material (steel): the product acquires work hardening by cold forming.

The work hardening is a property that is characteristic of the metallic state, which is observed neither in polymers nor in ceramic materials. To illustrate this property, we assume that low carbon steel is cold-rolled, and different degrees of reduction in thickness are measured (the volume is constant in all the process). The steel will be harder the smaller the final thickness is, that is to say, the greater the cold reduction ($\varepsilon = \ln(h_0/h_f)$, where h_0 is the initial thickness and h_f is the final thickness), the harder the material will be. In a first approximation it is possible to say that the work hardening is the increase in hardness that the steel acquires because of the cold forming of the steel. On the contrary, we have seen when we studied the hot forming

(Chap. 5) that the steel, when it exceeds a certain critical deformation ($\varepsilon \approx 0.2$), it deforms at constant stress, σ_s, and acquires a stationary regime of plasticity, either by recovery or recrystallization (dynamic or static).

The degree of work hardening that the steel acquires depends both on extrinsic factors (characteristics of the stress (load), deformation rate) and on intrinsic factors (chemical composition, microstructure, crystalline defects (vacancies and dislocations, packing defects energy, reaction between dislocations, etc.)).

The hardness of the steel progressively grows after the cold rolling, and this might lead to a wrong conclusion: assuming that the cold rolling produces a certain consolidation, a kind of increase of the density of the material, which would be responsible of the increase in the hardness of the material (this question is observed in the case of the powder metallurgy, by cold isostatic pressing). The metallurgical fundamentals of the work hardening are, mainly, the crystalline linear defects (dislocations) that exist inside of the metals, their multiplication during the cold deformation, and the reactions between them during the deformation.

We have previously indicated (Pero-Sanz et al. 2019) that some of the properties of metallic materials (steels) are insensible to the microstructure (grain size) and the possible crystalline defects of them (melting point, specific heat, elasticity constants, density, etc.); while other properties as the cold deformation, hardness, mechanical resistance, electric resistivity, etc., are related with the crystalline geometry (we should remember that there are three allotropic varieties in the case of the iron), the chemical composition and the microstructural defects (vacancies, dislocations, grain boundaries, inclusions).

6.3.1 *Constitutive Laws*

The theory of plasticity establishes that, it does not matter the complexity of the system of loads to which the material is subjected, its state of principal stresses in any element of the volume is defined by three vectors: σ_1, σ_2, and σ_3, which are perpendicular between them in the dimensions x (x(1)), y (y(2)), and z (z(3)) (principal directions of the stress (deformation)), Fig. 6.6. Due to convention, we

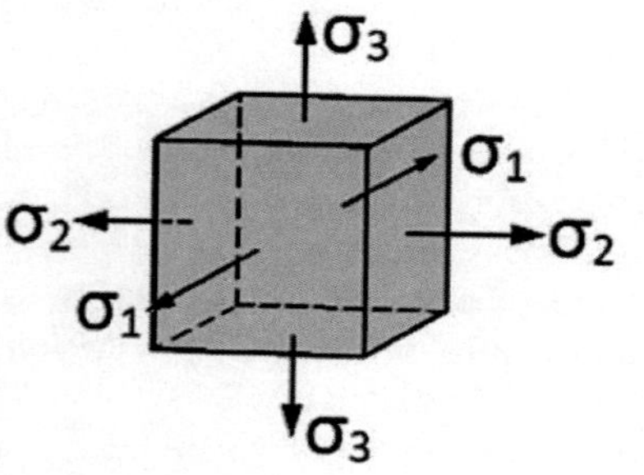

Fig. 6.6 Principal directions, positives

use the positive sign to represent the tension stresses and the negative sign for the compression stresses. We admit that the volume of material before and after the plastic deformation remains constant. That is to say, $\varepsilon_1 + \varepsilon_2 + \varepsilon_3 = 0$, being ε the true elongations in the above indicated principal directions.

According to the Von Mises criterion, it is possible to deduce that the stresses state or required efforts to achieve that a metallic material (steel) could start and continue its cold plastic deformation is related with the yield strength of the material determined in the tensile test, σ_y, by means of the following equation:

$$\sigma_y^2 = \frac{1}{2} \cdot \left[(\sigma_1 - \sigma_2)^2 + (\sigma_1 - \sigma_3)^2 + (\sigma_2 - \sigma_3)^2\right] = F(\sigma_y) \qquad (6.1)$$

where it is also verified that the plastic function $F(\sigma_y)$ accomplishes the following relation of plastic flow:

$$d\lambda \cdot \frac{\partial F}{\partial \sigma_i} = d\varepsilon_i \qquad (6.2)$$

In other words:

$$\frac{d\varepsilon_1}{(\partial F/\sigma_1)} = \frac{d\varepsilon_2}{(\partial F/\sigma_2)} = \frac{d\varepsilon_3}{(\partial F/\sigma_3)} = d\lambda \qquad (6.3)$$

or which is the same:

$$\frac{d\varepsilon_1}{[2 \cdot \sigma_1 - (\sigma_2 + \sigma_3)]} = \frac{d\varepsilon_2}{[2 \cdot \sigma_2 - (\sigma_1 + \sigma_3)]} = \frac{d\varepsilon_3}{[2 \cdot \sigma_3 - (\sigma_1 + \sigma_2)]} = \text{cte} \qquad (6.4)$$

where the equivalent plastic deformation is:

$$\overline{d\varepsilon} = \sqrt{\frac{2}{3} \cdot \left(d\varepsilon_1^2 + d\varepsilon_2^2 + d\varepsilon_3^2\right)} \qquad (6.5)$$

applicable when the principal directions of stresses and deformations (σ_i, ε_i) coincide. Other criterion of plasticity, formulated by Tresca, establishes that the plastic deformation occurs in those planes where the shearing stress is maximum:

$$\frac{\sigma_{max} - \sigma_{min}}{2} = \frac{\sigma_y}{2} \qquad (6.6)$$

where σ_{max} and σ_{min} are the maximum and minimum principal stresses. Otherwise, the difference between the criterions of plasticity of Von Mises and Tresca does not differ in the obtained values in more than 15%: we are going to apply one or other criterion depending on the case.

Application to different processes of cold forming (Fig. 6.7):

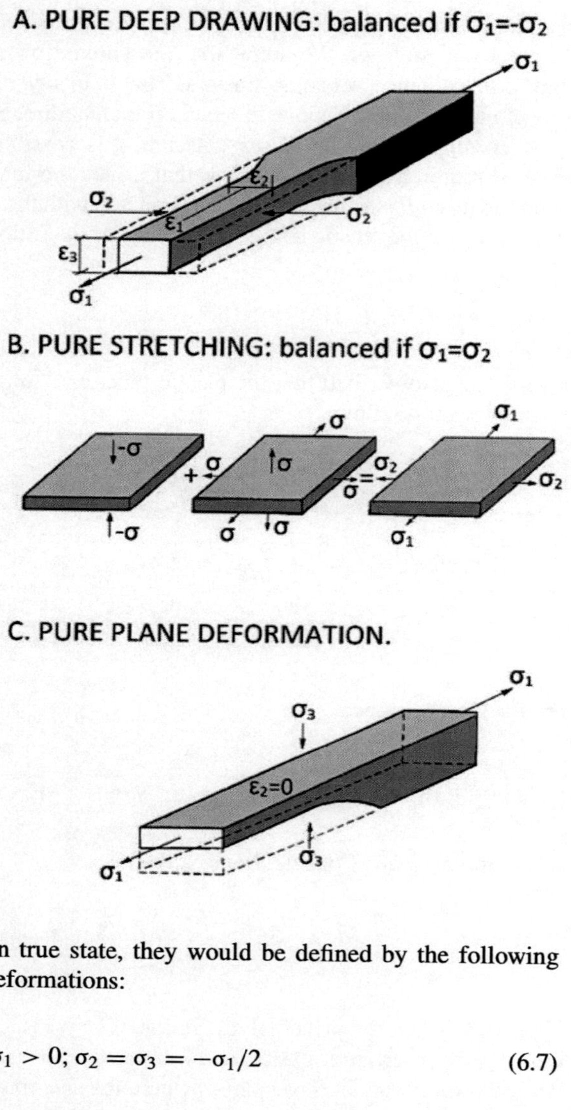

Fig. 6.7 Deep drawing, Stretching, Rolling

1. Extrusion and wiring: In true state, they would be defined by the following system of stresses and deformations:

$$\sigma_1 > 0; \sigma_2 = \sigma_3 = -\sigma_1/2 \tag{6.7}$$

$$\varepsilon_1 > 0; \varepsilon_2 = \varepsilon_3 = -\varepsilon_1/2 \tag{6.8}$$

Consequently, the uniaxial effort required for the beginning of the extrusion or drawing is smaller than that required for the deformation in tension. If we apply the previous equations (criterion of plasticity and relation of plastic flow/increase of deformation):

$$\sigma_y = \sqrt{\frac{1}{2} \cdot \left[\left(\sigma_1 + \frac{\sigma_1}{2} \right)^2 + \left(\sigma_1 + \frac{\sigma_1}{2} \right)^2 + 0 \right]} = \frac{3}{2} \cdot \sigma_1 \rightarrow \sigma_1 = \frac{2}{3} \cdot \sigma_y \tag{6.9}$$

$$\frac{d\varepsilon_1}{\sigma_1} = \frac{d\varepsilon_2}{\sigma_1/2} = \frac{d\varepsilon_3}{\sigma_1/2} \rightarrow \varepsilon_2 = \varepsilon_3 = -\frac{\varepsilon_1}{2}; \varepsilon_1 > 0 \tag{6.10}$$

$$\bar{\varepsilon} = \sqrt{\frac{2}{3} \cdot \left(\varepsilon_1^2 + \varepsilon_2^2 + \varepsilon_3^2 \right)} = \varepsilon_1 \tag{6.11}$$

2. Pure stretching: (state of plane stress) it is defined by the following state of stresses/deformations:

$$\sigma_1 = \sigma_2 > 0; \sigma_3 = 0 \tag{6.12}$$

$$\varepsilon_1 = \varepsilon_2 > 0; \varepsilon_3 = -2 \cdot \varepsilon_1 \tag{6.13}$$

The effort that is necessary to make for the pure stretching, or the pure compression, is equivalent to that obtained in the tensile test:

$$\sigma_1 = \sigma_2 = \sigma_y \tag{6.14}$$

We indicate the equivalence between the states of pure compression and the state of plane stress (as the stresses by hydrostatic pressures do not contribute to the plastic deformation) in Fig. 6.8.

3. Pure rolling: also called or characterized by a state of plane deformation:

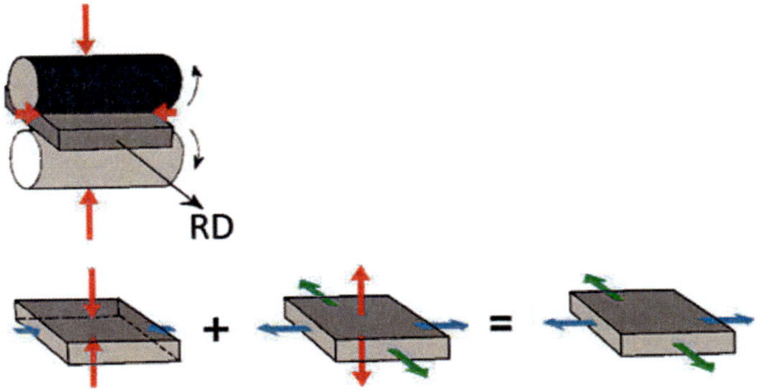

Fig. 6.8 Equivalence of stresses in rolling and biaxial tension. These stress states differ only by a hydrostatic tension which does not affect yielding

$$\sigma_1 = 0; \sigma_2 = -\sigma_3/2; \sigma_3 < 0 \tag{6.15}$$

$$\varepsilon_1 = -\varepsilon_3; \varepsilon_2 = 0 \tag{6.16}$$

deformations according to the rolling direction and according to the thickness are equal and of opposite sign; deformation according to the width is equal to zero. That is to say, to start and continue the plastic deformation by rolling, the stress or load that is necessary, considering the criterion of Von Mises, is smaller than the yield strength in tension of the material:

$$\sigma_y = \sqrt{\frac{1}{2} \cdot \left[\left(\frac{\sigma_3}{2}\right)^2 + \sigma_3^2 + \left(\frac{\sigma_3}{2}\right)^2\right]} = \frac{\sqrt{3}}{2} \cdot \sigma_3 \tag{6.17}$$

$$\bar{\varepsilon} = \sqrt{\frac{2}{3} \cdot \left(2 \cdot \varepsilon_3^2\right)} = \frac{2}{\sqrt{3}} \cdot \varepsilon_3 \tag{6.18}$$

4. Deep drawing: the pure drawing, equilibrated, is characterized by the following state of stresses and deformations, as it takes place in the flange of a cup in the drawing process:

$$\sigma_1 = -\sigma_2; \sigma_3 = 0 \tag{6.19}$$

$$\varepsilon_1 = -\varepsilon_2; \varepsilon_3 = 0 \tag{6.20}$$

That is to say, it is a state of pure shearing: the sheet is radially elongated and contracts circumferentially, without a change in thickness. The drawing requires smaller stresses than the deformation in tension. If we apply the equation of Von Mises:

$$\sigma_y = \sqrt{\frac{1}{2} \cdot \left[(2 \cdot \sigma_1)^2 + 2 \cdot \sigma_1^2\right]} \rightarrow \sigma_1 = 0.58 \cdot \sigma_y \tag{6.21}$$

From the previously mentioned, it is possible to deduce an important consequence: if we know the behavior in tension of a metallic material (steel):

$$\sigma = k \cdot \varepsilon^n \tag{6.22}$$

where σ is the true stress, ε is the true deformation, n is the strain hardening exponent or the work hardening coefficient, it is possible to calculate the state of stresses and deformations corresponding to any system of cold forming (rolling, wire drawing, drawing). It is sufficient to use the constitutive law in tension of generalized manner:

$$\overline{\sigma} = k \cdot \overline{\varepsilon}^{n} \tag{6.23}$$

$$\overline{\sigma} = \sqrt{\frac{1}{2} \cdot \left[(\sigma_1 - \sigma_2)^2 + (\sigma_1 - \sigma_3)^2 + (\sigma_2 - \sigma_3)^2\right]} \tag{6.24}$$

$$\overline{\varepsilon} = \sqrt{\frac{2}{3} \cdot \left(\varepsilon_1^2 + \varepsilon_2^2 + \varepsilon_3^2\right)} \tag{6.25}$$

where $\overline{\sigma}$ and $\overline{\varepsilon}$ are the equivalent stresses and deformations associated to any process of forming (σ_i, ε_i).

Other important consequence is that the cold forming process causes, especially in the flat and cylindrical products (sheet, wire rod), a strong texture of deformation: grains stop to be randomly oriented with respect to the principal directions of rolling. Some crystallographic planes tend to be situated parallel to the rolling plane; and some crystallographic directions tend to do the same with respect to the rolling direction (or wire drawing). This way, the sheet losses the isotropic behavior (the yield strength depends on the orientation that is chosen). The texture is defined by the crystallographic indexes $\{h\ k\ l\} < u\ v\ w >$: $\{h\ k\ l\}$ represents the family of crystallographic planes oriented preferentially in the rolling plane; $<u\ v\ w>$ is the family of crystallographic directions oriented preferentially in the rolling direction. For instance, in rolled steel sheets to be drawn, there appear the components of textures: $\{1\ 0\ 0\} < 0\ 0\ 1 > -\{1\ 0\ 0\} < 1\ 1\ 1 >$ and $\{1\ 1\ 1\} < 1\ 0\ 0 > -\{1\ 1\ 1\} < 1\ 1\ 2 >$, the so-called α and γ fibre, respectively, see Figs. 6.9 and 6.10. This way, in sheets destined to final operations (after recrystallization annealing) of stretching and drawing (deep drawing quality), it is defined as a normal anisotropy coefficient or Lankford coefficient:

$$R = \frac{\varepsilon_2}{\varepsilon_3} \tag{6.26}$$

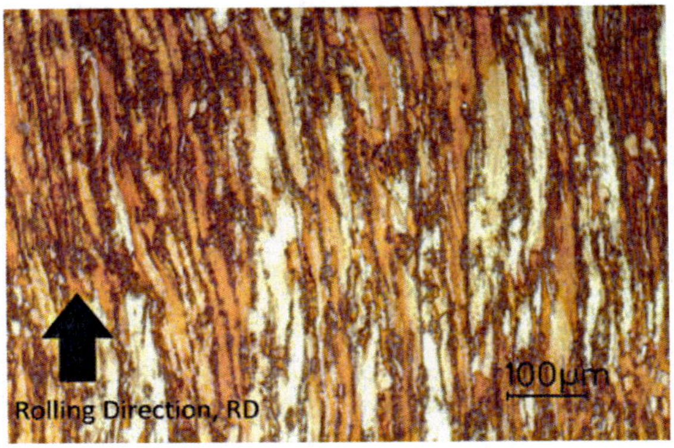

Fig. 6.9 Ferritic stainless steel 0.061% C–16.67% Cr. Cold-rolled (70%). Micrograph of the cross section parallel to the rolling direction, taken with polarized light. Grains are very elongated in the rolling direction

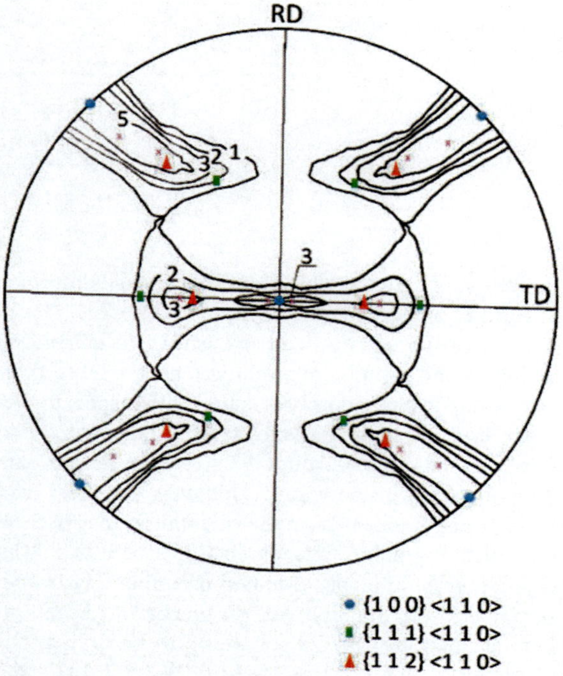

Fig. 6.10 Pole Fig. {2 0 0} of the sheet showed in Fig. 6.9 at medium thickness. α-fiber is a stronger component of texture in cold-rolled steels heavily deformed (70–90%). γ-fiber is weaker (x-random)

relation of deformations in width and in thickness of sheets deformed in tension. Sheets with R > 1 are interesting, which deform more according to the width than according to the thickness (this differs from the necking and breaking). From this point of view, sheets with texture {1 1 1} ($R \simeq 2$) behave much better in drawing than those with texture mainly of the type {1 0 0} (R < 1).

When the sheet is not isotropic in the rolling plane, it is possible to define an average coefficient of Lankford:

$$\overline{R} = \frac{R_{0^\circ} + R_{90^\circ} + 2 \cdot R_{45^\circ}}{4} \tag{6.27}$$

where R_{0°, R_{90°, and R_{45° are the values of the Lankford coefficient (measured in tensile tests performed at 0°, 45°, and 90° to the rolling direction).

It is also interesting to measure the plane anisotropy (variation of the Lankford index) in the rolling plane; it is used as a parameter:

$$\overline{\Delta R} = \frac{(R_{0^\circ} + R_{90^\circ} - 2 \cdot R_{45^\circ})}{2} \tag{6.28}$$

where $R_{0°}$, $R_{90°}$, and $R_{45°}$ are the values of the Lankford index, measured, as in the previous case, in the plane of the sheet at 0°, 90°, and 45°, respectively, to the rolling direction. A great coefficient of plane anisotropy increases the height of the ears in the formed sheets, decreasing the yield of the operations (losses because of trimmings or crops) after the drawing.

The sheet will exhibit "drawing ears" at 0° and 90° if $R_{0°}$ and $R_{90°} > R_{45°}$. In the opposite case, it will exhibit "drawing ears" at 45°. The ideal case will be when $\overline{\Delta R} = 0$ and $\overline{R} > 1$.

Exercise 6.1 9 sheets of very low carbon (≃0.003) and stabilized interstitials free steel are plastically deformed by rolling in such a way that the samples are lengthened and thinned down, without modification of the width (plane deformation). The increases in length are 1, 10, 20, 30, 40, 50, 60, 70, and 100%, respectively. Additionally, Vickers hardness indentations were made in the nine sheets after rolling, and the following results were obtained (Table 6.1).

In the assumption that the hardness test introduces a supplementary deformation of 8% in the samples, and that the value of the hardness is three times the real yield strength of the sample with an 8% of additional deformation, plot the apparent stress-strain curve (engineering). Besides, calculate:

1. Ultimate tensile strength of the steel (engineering and true).
2. Deformation at which the necking starts (engineering and true).
3. Reduction of area at the beginning of the breaking (necking).
4. Work that is necessary to break in tension 1 m^3 of steel.
5. If the rolling process introduces a density of dislocations equal to 10^{12} cm ⊥/cm^3, calculate the energy stored in the metal as a result of these defects. Compare this value with that calculated in the previous point. What consequences can be deduced?

Data: $G = 6.4 \cdot 10^4$ MPa (300 K), $b = 2.48 \cdot 10^{-10}$ m (Burgers vector, minimum distance between closest atoms), $a = 2.86 \cdot 10^{-10}$ m (lattice parameter).

Solution starts:

First, we plot the engineering stress-strain curve (Fig. 6.11) with the data collected in Table 6.2.

Now we plot the true stress-strain curve (σ, ε), Fig. 6.12, using the data of Table 6.3, which is obtained using the following equations:

$$\sigma = s \cdot (1 + e_T) \tag{6.29}$$

Table 6.1 Vickers hardness for each sheet after a certain deformation

Deformation (expressed as per unit)	0.01	0.1	0.2	0.3	0.4	0.5	0.6	0.7	1.0
HV (MPa)	690	812	897	960	1007	1049	1080	1190	1178

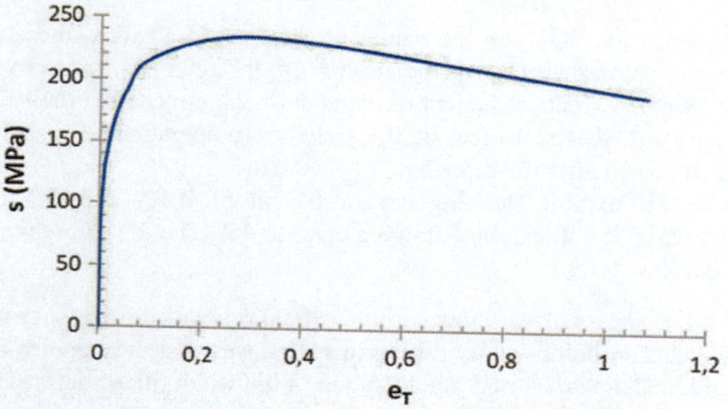

Fig. 6.11 Engineering stress-strain curve (s, e_T)

Table 6.2 Data to plot the engineering stress-strain curve (s, e)

e	e_T (e + 0.08)	$s = HV/[3 \cdot (1 + e_T)]$ (MPa)
0.01	0.09	211
0.1	0.18	229
0.2	0.28	234
0.3	0.38	232
0.4	0.48	227
0.5	0.58	221
0.6	0.68	214
0.7	0.78	208
1.0	1.08	189

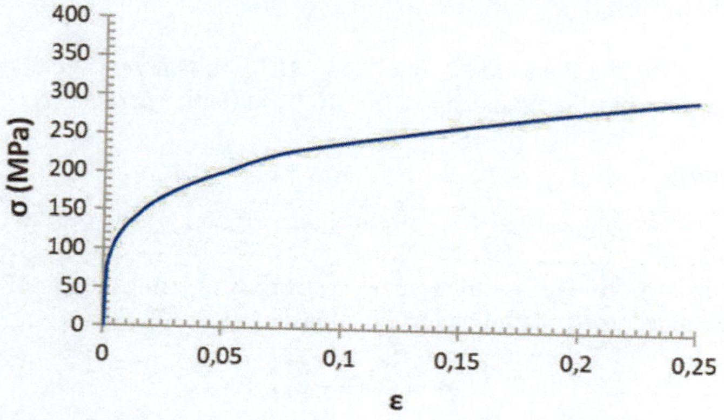

Fig. 6.12 True stress-strain curve (σ, ε)

Table 6.3 Data to plot the true stress-strain curve (σ, ε)

$\varepsilon = \ln(1+e)$	$\sigma = s\cdot(1+e)$
0.0861	230
0.1655	270
0.2468	300
0.3221	320
0.3920	336
0.4574	349
0.5188	360
0.5766	370
0.7324	393

$$\varepsilon = \ln(1 + e_T) \tag{6.30}$$

The equation that relates the stress and strain is:

$$\sigma = k \cdot \varepsilon^n \rightarrow \sigma = 424.3 \cdot \varepsilon^{0.25} \tag{6.31}$$

where n and k are calculated using the graph of Fig. 6.13, where the linear trend line is obtained using Excel. This equation is:

$$\ln \sigma = n \cdot \ln \varepsilon + \ln k \rightarrow \ln \sigma = 0.2504 \cdot \ln \varepsilon + 6.0514 \rightarrow n \simeq 0.25 \tag{6.32}$$

k is calculated when $\varepsilon = 1$:

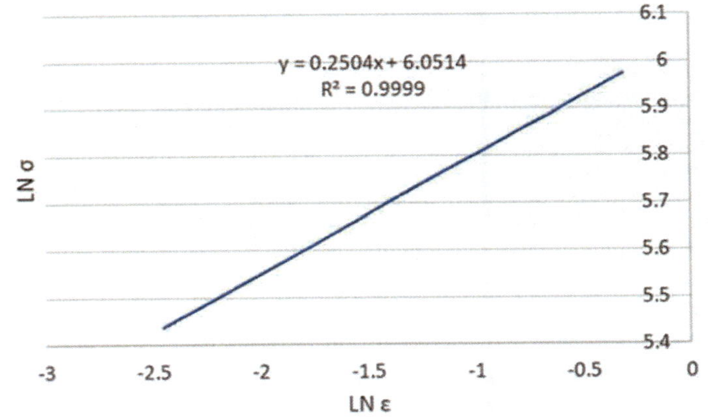

Fig. 6.13 Representation of ln σ-ln ε

$$\ln \sigma_0 = n \cdot \ln 1 + \ln k \rightarrow k = \sigma_0 \quad (6.32)$$

thus:

$$\ln \sigma_0 = 6.0514 \rightarrow k = \sigma_0 = \exp 6.0514 = 424.71 \quad (6.33)$$

Finally:

$$\sigma = 424.3 \cdot \varepsilon^{0.25} \quad (6.34)$$

1. We calculate the ultimate tensile strength (true):

$$\sigma = 424.3 \cdot 0.25^{0.25}(\varepsilon_u = n) \rightarrow \sigma_u \simeq 300\ \text{MPa} \quad (6.35)$$

and the yield strength ($\varepsilon = 0.002 \simeq e$):

$$\sigma_y = 424.3 \cdot 0.002^{0.25} \rightarrow \sigma_{y,0.2\%} = 90\ \text{MPa} \quad (6.36)$$

It is a stabilized interstitial free steel nearly without free C (0.003%) and N (< 0.002%, picked up by titanium (0.04%), with grain size of around 7 ASTM or greater).

2. Deformation at which the necking starts:

$$\varepsilon_u = 0.25 = \ln(1 + e_u) \rightarrow e_u \simeq 28\% \quad (6.37)$$

very close to $(e_u)_M \simeq 0.30$ that can be achieved in ultralow carbon interstitial free sheet.

3. The reduction of area at the beginning of the breaking:

The volume is constant:

$$S_0 \cdot l_0 = S \cdot l \rightarrow \frac{S}{S_0} = \frac{l_0}{l} = -\ln(1 + e_u) \simeq 0.25 \quad (6.38)$$

$$\sum \simeq \frac{S_0 - S}{S_0} \simeq 75\%\text{(great necking)} \quad (6.39)$$

4. Work necessary for the deformation per unit of volume (until the necking):

$$W = \int_0^{0.25} \sigma \cdot d\varepsilon = 424.3 \int_0^{0.25} \varepsilon^{0.25} \cdot d\varepsilon = 424.3 \cdot \left[\frac{\varepsilon^{1.25}}{1.25}\right]_0^{0.25} \simeq \frac{424.3 \cdot 0.25^{1.25}}{1.25}$$

$$\simeq 60\ \mathrm{MJ} \cdot \mathrm{m}^{-3} \tag{6.40}$$

5. The energy stored in the metal by the dislocations:

$$\rho_{\perp} = 10^{12}\mathrm{cm}^{\perp} \cdot \mathrm{cm}^{-3} = 10^{16}\mathrm{m}^{\perp} \cdot \mathrm{m}^{-3} \tag{6.41}$$

$$E_{\perp} = \frac{1}{2} \cdot G \cdot b^2 = \frac{1}{2} \cdot 6.4 \cdot 10^4 \frac{\mathrm{MN}}{\mathrm{m}^2} \cdot \left(2.48 \cdot 10^{-10}\mathrm{m}\right)^2 = 2.0 \cdot 10^{-16}\mathrm{MN} \tag{6.42}$$

where:

$$\begin{aligned} b(\text{minimum distance between atoms}) &= a \cdot \tfrac{\sqrt{3}}{2} = 2.86 \cdot \tfrac{\sqrt{3}}{2} \\ &= 2.48\ \text{Å}(0.248\ \mathrm{nm}) \end{aligned} \tag{6.43}$$

$$\begin{aligned} W_{deform} &= \text{Driving force for the recrystallization} \\ &= 2.0 \cdot 10^{-16}\mathrm{MN} \cdot 10^{16}\mathrm{m}^{\perp} \cdot \mathrm{m}^{-3} \simeq 2\ \mathrm{MJ} \cdot \mathrm{m}^{-3} \end{aligned} \tag{6.44}$$

that is to say, only 2/60 $\simeq$ 4% is used to deform the metal: the rest is released as heat (in a tensile test performed in cold condition, it is possible to check that the specimen heats).

The sheet is subjected to a skin-pass to increase its yield strength to around 160 MPa before forming. What percentage of deformation is necessary?

Considering Eq. 6.34:

$$160\ \mathrm{MPa} = 424.3 \cdot \varepsilon^{0.25} \rightarrow \varepsilon \simeq 0.02\ (2\%) \tag{6.45}$$

In conditions of plane deformation (rolling):

$$\varepsilon_1 = -\varepsilon_3;\ \varepsilon_2 = 0 \tag{6.46}$$

And thus:

$$\begin{aligned} \bar{\varepsilon} = \sqrt{\tfrac{2}{3} \cdot \left(\varepsilon_1^2 + \varepsilon_2^2 + \varepsilon_3^2\right)} &= \tfrac{2}{\sqrt{3}} \cdot \varepsilon_{PD} = 0.02 \rightarrow \varepsilon_{PD} \simeq 0.0175(\simeq 0.02) \\ &= \ln\left(\tfrac{h_i}{h_f}\right) \rightarrow \tfrac{h_i}{h_f} = 1.018 \rightarrow \tfrac{h_i - h_f}{h_i} \simeq 2\% \end{aligned} \tag{6.47}$$

End of the exercise.

Exercise 6.2 Crystallographic textures of metallic materials. Pole figures obtained by X-ray diffraction.

Solution starts:

Preferred orientation derived from cold deformation and recrystallization processes is well described by means of a pole figure. This is a stereographic projection which shows the variation in pole density with pole orientation for a selected set of crystal planes. This method of describing textures was first used by German metallurgist, Schulz arrangement (reflection method) and Decker et al. arrangement (transmission method). And its meaning can best be illustrated by the following simple example. Suppose we have a very coarse-grained sheet of a cubic metal, and that we determined the orientation of each of 10 grains by one punctual method (i.e., Laue arrangement). We decide to represent the orientations of all these grains together by plotting the positions of their {1 0 0} on a single stereographic projection, with the projection plane parallel to the sheet surface. Since each grain has three {1 0 0} poles, there will be a total of $3 \cdot 10 = 30$ poles plotted in the projection. If the grains have completely random orientations, these poles will be distributed nearly uniformly over the projection as indicated in Fig. 6.14a (Cullity 1959). But if preferred orientation is present, the poles will tend to cluster together into certain areas of the projection, leaving other areas virtually unoccupied. For example, might take the particular shown in Fig. 6.14b (Cullity 1959). This is called the "cube texture," because each grain is oriented with its (1 0 0) planes nearly parallel to the sheet surface and the [1 0 0] direction in these planes nearly parallel to the rolling direction. This simple texture may be described by the shorthand notation {1 0 0} < 1 0 0 > , which form the recrystallization texture of many FCC metals and alloys under suitable conditions. If we have chosen to construct a {1 1 1} pole figure, by plotting only (1 1 1) poles, the resulting pole figure would look entirely different from Fig. 6.14b (Cullity 1959) for the same preferred orientation {1 0 0} < 1 0 0 > cube texture; in fact, it would consist of four "high-intensity" areas, one near the center of each quadrant, Fig. 6.15 (Verdeja and Fernández 1992). This illustrates the fact that the appearance of a pole figure depends on the indices of the poles plotted, and the choice of indices depends

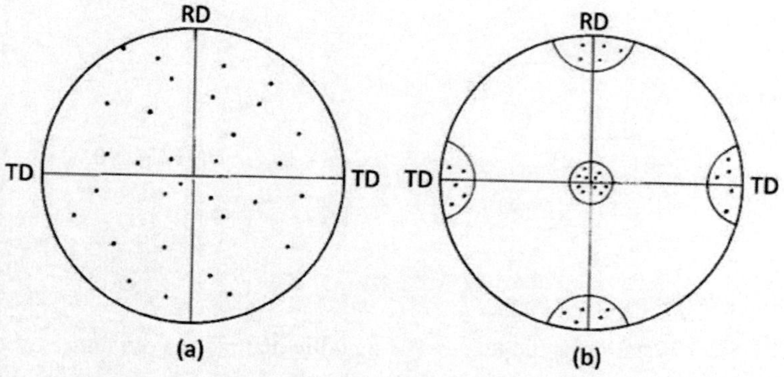

Fig. 6.14 (1 0 0) pole figures for sheet material, illustrating **a** random orientation and **b** preferred orientation. R. D. (rolling direction) and T.D. (transverse direction) are the reference directions in the plane of the sheet

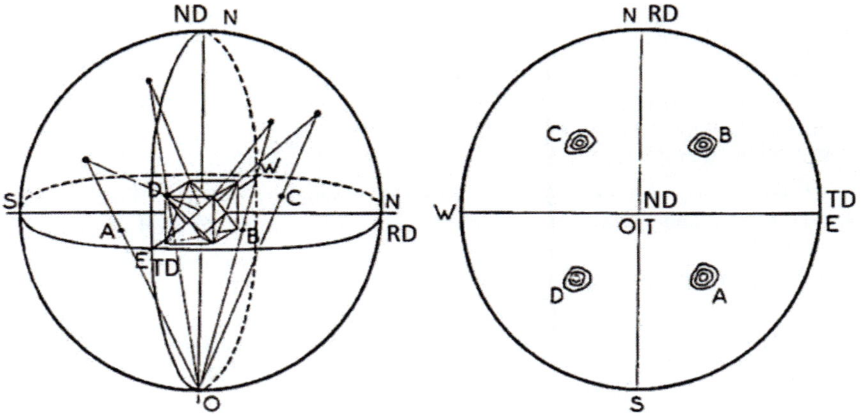

Fig. 6.15 Representation of the {1 0 0} <1 0 0> in the (1 1 1) pole figure same texture, Fig. 6.14b

on which aspect of the texture one wishes to show most clearly. For example in describing the texture of BCC metals (low carbon steel for deep drawing operation) it is particularly worth to represent the {2 0 0} pole figure which shows the particular distribution of crystallographic planes {2 0 0} over the stereographic projection.

Concerning inverse pole figures and the calculation of the oriented distribution function of crystallites obtained by X-ray diffraction analyses (experimental), the reader is addressed to Bocos-González (2002).

For further information about orientation imaging microscopy based on the electro backscattered diffraction technique the reader is referred to Gourgues-Lorenzon (2007).

End of the exercise.

6.3.2 Recovery and Recrystallization. Deep Drawing Steels

We have previously indicated that the cold forming ($T < T_M/4$, where T_M is the melting point in Kelvin) confers work hardening to metallic materials. Cold-worked state implies an increase of the energy with respect to that of the polycrystalline aggregate in equilibrium, which was not cold deformed. Metallic crystals are energetically more stables when they do not have dislocations. For that reason, the cold-worked structures, with the support of the thermal agitation, tend to transform into others with smaller number of non-oriented dislocations.

Cold-worked crystals do not spontaneously cede their excess of energy to pass to the recrystallized state. With that purpose, it is necessary to reach the activation energy, Q, through external heating, which made possible the mass transfer (diffusion of atoms) through the vacancy's mechanism at the heating temperature. For steels Q $\simeq 18 \cdot R \cdot T_M \simeq 270$ kJ/mol.

The heat treatment to eliminate the work hardening of a metal or alloy is usually called annealing against the work hardening. It is also known by the name of full annealing, and in certain cases, it is known as annealing recovery.

When a metal (steel), sufficiently cold deformed, is subjected to suitable heating, it gradually recovers the properties that had before the deformation: metal softens, mechanical resistance decreases, ductility increases, texture of cold deformation changes, definitely, material recrystallizes, and a microstructure of more or less equiaxed grains appears, with low density of dislocations (~10^5 cm ⊥/cm^3).

The new structure receives the name of recrystallized structure. Its formation takes place when the heating is performed above a certain temperature T_r, called temperature of static primary recrystallization. This, generally greater than $T_M/3$, mainly depends on the cold deformation previously conferred, and, for that reason, it is not a constant temperature. Other criterion consists in considering T_r of around $T_M/2$ (in the case of the steel is $T_r > 600$ °C) (Pero-Sanz et al. 2017). See micrographs in Fig. 6.16.

6.3.2.1 Recovery

During this process ($T \simeq T_M/3$, slow heating), the microstructure and texture of the corresponding cold-worked state are not substantially modified, but the following modifications can be appreciated:

(a) Partial release, as heat, of the internal energy stored in the steel.
(b) Disappearance of internal stresses (stress relieving) (Pero-Sanz et al. 2018).
(c) Recovery of electrical properties: the cold deformation increases the electrical resistivity.
(d) Partial recovery of mechanical properties: for instance, loss of hardness and increase of the ductility (appreciable elongation in the tensile test).

In body-centered cubic metals (ferrite in steels), the polygonization of the deformed structure can take place. It is possible to see, by transmission electronic microscopy (in some occasions by simple optical previous to the adequate etching) (Fig. 6.17), the presence of sub-grains, very perfect, few disoriented one sub-grain with respect to others, separated by walls of dislocations (Fig. 6.17), which allows thinking in both the elimination and reorganization of these defects during the recovery process. It comes accompanied by the previously mentioned problems (losses of hardness, reduction of the tensile strength, increase of the ductility, etc.), but we insist that this happens without appreciable modification of the microstructure (see Fig. 6.18, here we show an example for an aluminum single crystal, similar to the alpha-iron, Lacombe 1963).

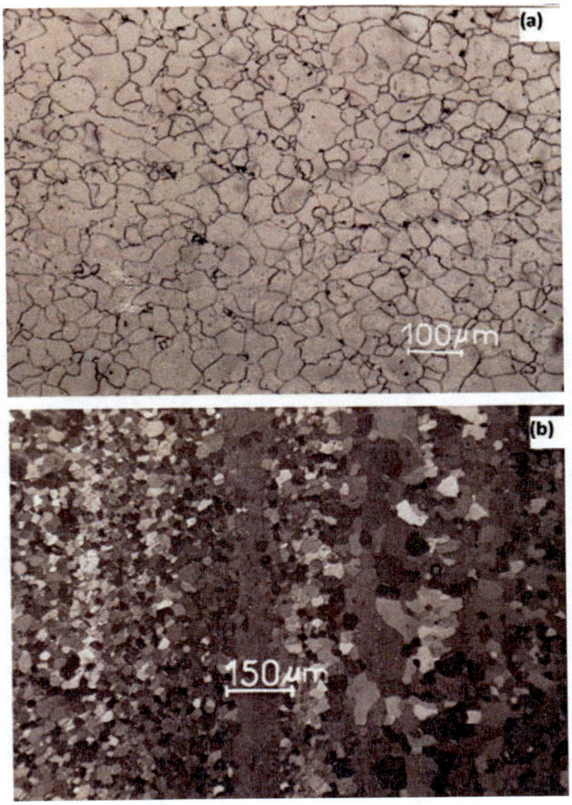

Fig. 6.16 **a** Ferritic alloy 0.027%C–3.14% Si. As hot-rolled, recrystallized. Micrograph at surface; **b** Ferritic alloy 0.027%C–3.14% Si. Cold-rolled (70%) and annealed (4 h, 820 °C), recrystallized. Micrograph at mid-thickness. Polarized light

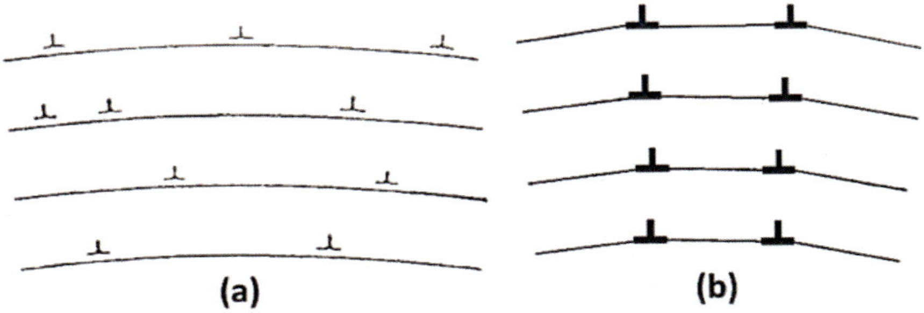

Fig. 6.17 Polygonization by thermal activation

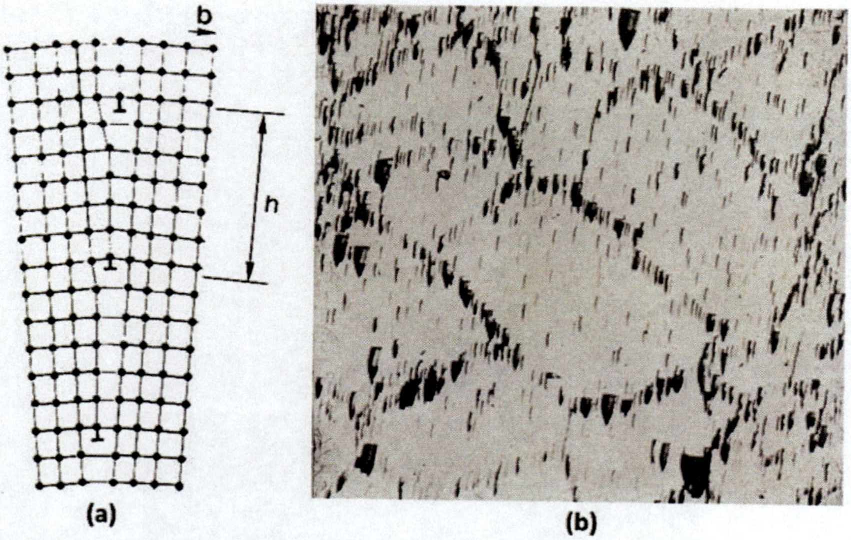

Fig. 6.18 **a** Polygonization. **b** Polyhedral alignments of etch pits showing the polygonization sub-boundaries of an aluminum single crystal

6.3.2.2 Recrystallization. Drawing Steels

The degree of deformation of the steel, the temperature that is necessary to reach to start the recrystallization, and the time of permanence at this temperature until obtaining a regular structure (recrystallized grain size, d) are related between them (Fig. 6.19).

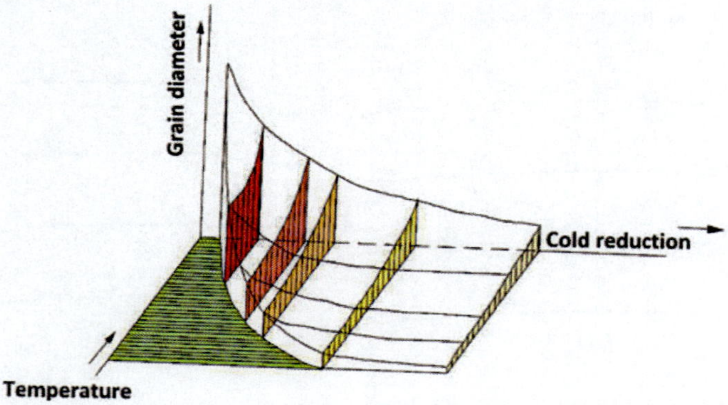

Fig. 6.19 Correlation between work hardening, temperature of recrystallization, and grain size diameter

It is usual that the recrystallization annealing of a work-hardened structure to be performed, in the practice, by continuous heating. Even if the annealing furnace is at room temperature, the sample acquires more or less rapidly the temperature chosen for the recrystallization.

In what follows, we will make reference to the recrystallization at constant temperature with the objective of obviating the possible previous recovery (in the heating), which was indicated in the previous section, that has a significant influence on the properties of ferritic steels to be used in drawing.

When the heating conditions (temperature-time) are adequate to begin what is called primary recrystallization, the loss of work hardening of the metal by heating is accompanied by the appearance of small regular crystalline grains (more or less equiaxed) inside of the work-hardened matrix.

During the primary recrystallization, there are generally two consecutive stages. First, a time t is required for the formation of the "strain free" clusters (nuclei) of recrystallization. Later, the clusters grow by migration of the atoms (diffusion) from the formerly strained matrix, crossing the borders that the nucleus/cold-worked matrix interfaces imply. The new boundary advances by removing dislocations (sweeping out) and, for that reason, the cluster grows. Finally, the cold-worked matrix disappears; it is completely replaced by a new structure of regular grains more or less equiaxed when the grain boundaries of the different crystals in growth impinge one another.

Recrystallization does not represent a phase change: there is not allotropic change, but there exists a simple reordering of atoms (without change of crystalline structure), by diffusion, with a density of dislocations (10^4–10^5 cm ⊥/cm^3) very much smaller than that of the work hardened state of reference (10^{10}–10^{11} cm ⊥/cm^3).

The energy of activation required to mobilize the atoms and form "recrystallized" groups, with low density of dislocations, is smaller the more deformed in cold (work hardening) the steel is (metal) before the heating. The isothermal temperature that is necessary to reach for the beginning (clustering) of the recrystallization is inversely proportional to the work hardening of the metal. Thus, the smaller the degree of deformation in cold, the greater the temperature should be to start the recrystallization.

It is necessary a minimum previous work hardening (it is around 5% in steel) to achieve the isothermal recrystallization of a steel at the temperature T_1. This strain is called critical strain for the temperature T_1 (Fig. 6.20).

In recrystallization, as it is involved the bulk diffusion, temperature and time are complementaries. This way, when we increase the annealing time, for a certain degree of work hardening, the temperature required for the recrystallization decreases. The minimum critical work hardening for an isothermal recrystallization at the temperature T decreases when the holding at this temperature is extended, because the probability of formation of new grains from the nucleus increases. In the limit situation, there should be at least one nucleus that, by subsequent growth, gives a monocrystalline sample or with a grain size very big.

Consequently, the recrystallization is an irreversible process as opposed to the situation in other solid-state transformations. Thus, in the solidification of a pure

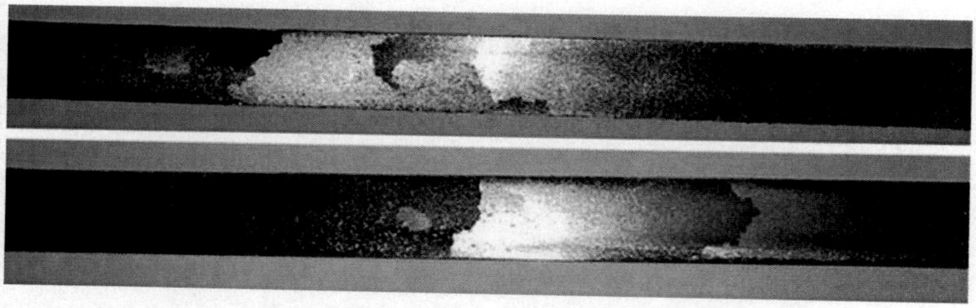

Fig. 6.20 High purity iron, critical strained. Large grains are observed. Polished and macro-etched with Nital-5% (specimens 125 mm × 10 mm)

metal, for example, the solid melts at $T > T_M$ after a certain time, while below this temperature, the liquid solidifies again. Something similar happens in the allotropic transformations: the pure Fe passes from the allotropic variety α to the γ when the temperature of 912 °C is exceeded; but goes back to the variety α below this temperature. It analogously happens in the precipitation of a new phase by solubility change: the precipitates of the alloy (steel) dissolve, to form solid solution, when the temperature is greater than the critical one (solvus line), but they reprecipitate when the temperature descends below T_s. On the contrary, once achieved the recrystallization of a work hardened structure, a cooling below the reached temperature T_1 does not give again the work-hardened state. Anyway, there are other similitudes between the solidification processes, allotropic changes, and precipitation (Pero-Sanz et al. 2017), which follow the model of nucleation and growth, with those of the recrystallization; all of them are controlled by the diffusion (thermal activation of the atoms) and by the force or driving energy of the process: in the recrystallization by the difference of existing energy in the form of dislocations between the work-hardened state and the recrystallized state. The nuclei of recrystallization are already present in the form of sub-grains in the microstructure (texture) of the deformed matrix, no matter the critical clusters grow by either an oriented nucleation or by an oriented-dependent grain growth mechanism. Those nuclei are preferentially formed in regions where the degree of deformation is greater (heterogeneous nucleation) as in the solidification. For example, ancient grain boundaries, deformation bands, inclusions, twin boundaries of thermal origin, and peripheric zones close to the free surfaces.

On the other hand, for an isothermal temperature T_1 of recrystallization, the number of nucleus formed after a time t_1 (nucleation rate) increases with the degree of cold deformation of steel. For that reason, the greater the degree of work hardening and, for that reason, greater the number of nucleus of recrystallization that will be formed, smaller the size of the nucleus of recrystallization at the end of the primary recrystallization at temperature T_1, Fig. 6.19. For that reason, with small work hardening, the recrystallized grain size is very big (especially when the work hardening is of around the above-mentioned critical).

The average recrystallized grain size does not depend exclusively on the nucleation rate motivated by the cold work hardening, but also by the impurities: atoms in solid solution and precipitated phases, especially when they are nanometric ($\ll$1 μm).

Atoms in solid solution usually delay the advance of the limit surface of the nucleus, curved, due to the elastic interactions between atoms and dislocations.

Precipitates or particles of the second phases also obstacle the advance of the border nucleus/work hardening matrix. This effect is called "pinning." We have mentioned, due to the important industrial relevance, the inhibition caused by the AlN on the sub-boundaries of grain {1 0 0}, which favor the selective growth of sub-grains with texture {1 1 1} very favorable for drawing.

With respect to the nucleation, it is possible to say that the impurities (inclusions) or precipitated second phases in the work hardened matrix, accelerate the formation of nucleus if the size of the second phases is great (of around microns): this is due to the local concentration of stresses and deformations that the cold deformation produces in the work-hardened matrix that surrounds the particle.

On the contrary, precipitates retard the nucleation if they are nanometric because they block the mobility of the sub-grain boundaries in alloys that can be recovered by polygonization (this is the case of the work-hardened ferrite) and impede that the sub-grains could become in an effective nucleus (with a size greater than the critical one) for the recrystallization.

In some cases, as it was seen in the Chap. 5, nanometric precipitates cannot only delay the nucleation but also impede or increase the recrystallization temperature.

This is the case of small quantities of microalloying element Nb ($\simeq$0.04%, 400 ppm), which can precipitate in hot conditions in the form of Nb(C, N) in structural ferritic–pearlitic weldable steels, called micro-alloyed steels. When rolling these steels at high temperature, in austenitic phase, this last one is deformed and acquires work hardening, but it does not recrystallize during the last passes of the hot rolling (controlled rolling of micro-alloyed steels, HSLA).

In these steels, the inhibition of the recrystallization of the austenite is very advantageous because the work-hardened austenite transforms later in ferrite during the cooling, and very fine grains of ferrite are obtained (more than 30000 grains/mm^2), which cannot be achieved if we start from recrystallized austenite (Chap. 5).

These micro-alloyed steels low in carbon (<0.25%), weldable, have, for that reason, high yield strength ($300 < \sigma_y < 500$ MPa) and excellent toughness (DBTT $\ll -20$ °C) at low temperature, and make possible their utilization in pipelines for petroleum and gas in artic regions, off-shore platforms, icebreaking ships, etc.

On the other hand, the growth of the grains, once formed, is delayed as a consequence of the pinning effect that obstacles the migration of the grain boundaries, the more numerous and smaller the precipitates are (nanometric). The size of the recrystallized grain, d, is according to Zener directly proportional to the average size of the precipitates, $\bar{x}$, and inversely proportional to the volumetric fraction of the precipitates, f_v:

$$d \simeq \frac{\bar{x}}{f_v} \tag{6.48}$$

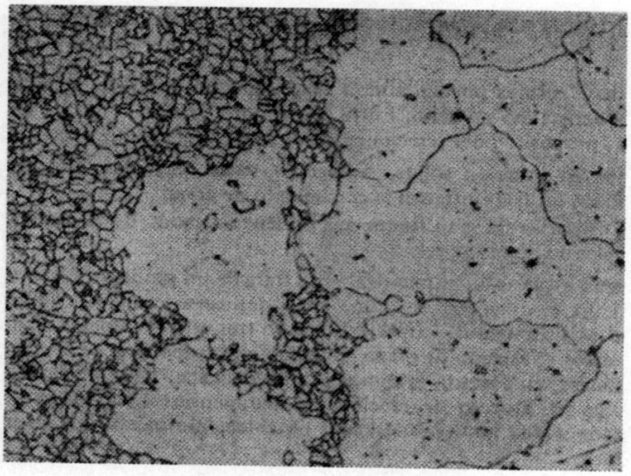

Fig. 6.21 Secondary recrystallization

Thus, a grain size of 30 μm can be stable against the presence of precipitates with an average grain size of 30 nm and $f_v \simeq 1‰$. If not, in presence of precipitates partially or totally redissolved, some grains might abnormally grow (exaggeratedly); the microstructure will be very heterogeneous and, for that reason, this situation is undesirable for the mechanical properties of the steel (abnormal grain growth, see Fig. 6.21).

To conclude with this section, we will talk about the kinetics of the static and isothermal recrystallization. The percentage of recrystallized steel, X_v, static and isothermally, at the temperature T_1 of recrystallization increases with the time of permanence at that temperature. As in all the transformations that take place as a consequence of nucleation and growth (Chap. 3, Pero-Sanz et al. 2017), this process follows a sigmoidal law type Johnson-Mehl-Avrami:

$$X_v = 1 - \exp\left(-B \cdot t^k\right) \tag{6.49}$$

where B and k are constants that depend on the transformation type, X_v is the recrystallized fraction and t is the time.

Thus, the Johnson-Mehl-Avrami law adjusts to a straight line of ordinate in the origin equal to log B and slope k (it takes values of approximately 2, 3, or 4 depending on the anisotropy of growth), when we represent in ordinates the values of $\log[\ln(1/(1 - X_v))]$ and in abscissas log t. Two examples of application of the Johnson-Mehl-Avrami were indicated in the Chap. 5 (modelling of the hot deformation of two structural steels: C-Mn and C-Mn-Nb).

Exercise 6.3 TTT curves of the Fe-α (ferritic steels) isothermal recrystallization. Following the same process that in the solidification of a pure metal, calculate the nucleation rate, I, the growing rate, v, and the solidified fraction (X = 1%, X = 99%), Johnson–Mehl–Avrami, as a function of the temperature.

Data: G = 67 GPa; density of dislocations in the work-hardened iron: $\rho_\perp = 10^{11}$ cm $\perp$/cm^3; Burgers vector in the Fe-α, $2.5 \cdot 10^{-10}$ m; coefficient of self-diffusion in the α-Fe, $D_\alpha = 2 \cdot \exp(240000\text{J} \cdot R^{-1} \cdot T^{-1})$; interfacial energy α-work hardened/α-recrystallized, 250 erg/cm^2; θ = 0°.

Solution starts:

The following equation is used to calculate the nucleation rate:

$$I = n \cdot \exp\left(-\frac{\Delta G^*}{k \cdot T}\right) \cdot \frac{D_\alpha}{b^2} \cdot \omega^* \tag{6.50}$$

where n is the number of atoms per volume unit, D_α is the diffusion coefficient at a temperature T and b is the Burgers vector, k is the Stefan–Boltzmann constant and ω^* is the number of atoms in the reacted-unreacted interface.

We also know that the critical free energy is equal to:

$$\Delta G^* = \frac{16}{3} \cdot \frac{\pi \cdot \gamma^3}{\Delta G_v^2} \tag{6.51}$$

where ΔG_v is the driving force for the nucleation, which is equal to the energy associated to a dislocation multiplied by the density of dislocations:

$$\Delta G_v = \frac{1}{2} \cdot G \cdot b^2 \cdot \rho_\perp = \frac{1}{2} \cdot 67 \cdot 10^9 \frac{N}{m^2} \cdot \left(2.5 \cdot 10^{-10}\ m\right)^2 \cdot \frac{10^5\ dyn}{1N} \cdot 10^{11} \frac{cm^\perp}{cm^3}$$
$$\simeq 2 \cdot 10^7 \frac{erg}{cm^3} \tag{6.52}$$

This way, we calculate ΔG^*:

$$\Delta G^* = -\frac{\frac{16}{3} \cdot \left(250\ erg \cdot cm^{-2}\right)^3}{\left(2 \cdot 10^7\ erg \cdot cm^{-3}\right)^2 \cdot 1.38 \cdot 10^{-16}\left(erg \cdot K^{-1}\right) \cdot T(K)} = -\frac{4.7 \cdot 10^9}{T} \tag{6.53}$$

Additionally:

$$n = \frac{6.023 \cdot 10^{23} atoms \cdot mol^{-1}}{7.01\ cm^3 \cdot mol^{-1}} = 8.59 \cdot 10^{22}\ atoms \cdot cm^{-3} \tag{6.54}$$

because:

$$V_s = \frac{1\ cm^3}{7.97\ g} \cdot \frac{55.85\ g}{1\ mol} = 7.01\ cm^3 \cdot mol^{-1} \tag{6.55}$$

Thus:

$$I = 8.59 \cdot 10^{22} \frac{\text{atoms}}{\text{cm}^3} \cdot \frac{2 \cdot \exp\left(-\frac{240000\ \text{J}}{8.31\ \text{J} \cdot \text{mol}^{-1} \cdot \text{K}^{-1} \cdot \text{T}}\right) \text{cm}^2 \cdot \text{s}^{-1}}{\left(2.5 \cdot 10^{-8}\ \text{cm}\right)^2} \cdot \exp\left[-\frac{4.7 \cdot 10^9}{\text{T}} \cdot f(\theta)\right] \tag{6.56}$$

this last exponential tends to 1 ($\omega^* = 1$) because there is perfect wetting $\theta = 0$ (heterogeneous nucleation). Thus:

$$I \simeq 2.74 \cdot 10^{38} \cdot \exp\left(-\frac{240000\ \text{J}}{8.31\ \text{J} \cdot \text{mol}^{-1} \cdot \text{K}^{-1} \cdot \text{T}}\right) \text{nuclei} \cdot \text{cm}^{-3} \cdot \text{s}^{-1} \tag{6.57}$$

The growth rate is calculated as follows:

$$v = \Omega \cdot \rho_{\text{planar}} \cdot \frac{D_\alpha}{b^2} \cdot \frac{\Delta G_v}{k \cdot T} \tag{6.58}$$

where Ω is the atomic volume and ρ_{planar} is the number of atoms per surface unit of a dense plane. Ω is calculated as follows:

$$b = 2.5 \cdot 10^{-10}\ \text{m} = \frac{a \cdot \sqrt{3}}{2} \rightarrow a = \frac{5}{\sqrt{3}} \text{Å} \rightarrow \Omega = \frac{\left(\frac{5}{\sqrt{3}} \text{Å}\right)^3}{2\ \text{atoms}} \simeq 12.07 \frac{\text{Å}^3}{\text{atoms}} \tag{6.59}$$

and:

$$\rho_{\text{planar}} = \frac{2\ \text{atoms}}{\left(\frac{5}{\sqrt{3}} \text{Å}\right) \cdot \sqrt{2} \cdot \left(\frac{5}{\sqrt{3}} \text{Å}\right)} = 0.1697 \frac{\text{atoms}}{\text{Å}^2} \tag{6.60}$$

this way:

$$\begin{aligned} v &= \Omega \cdot \rho_{\text{planar}} \cdot \frac{D_\alpha}{b^2} \cdot \frac{\Delta G_v}{k \cdot T} \\ &= 12.07\ \text{Å}^3 \cdot 0.1697 \frac{\text{atoms}}{\text{Å}^2} \\ &\cdot \frac{2 \cdot \exp\left(-\frac{240000\ \text{J}}{8.31\ \text{J} \cdot \text{mol}^{-1} \cdot \text{K}^{-1} \cdot \text{T}}\right) \text{cm}^2 \cdot \text{s}^{-1}}{\left(2.5\ \text{Å}\right)^2} \cdot \frac{1\ \text{Å}}{10^{-8}\ \text{cm}} \\ &\cdot \frac{2 \cdot 10^7 \frac{\text{erg}}{\text{cm}^3} \cdot \frac{1\ \text{cm}^3}{7.97\text{g}} \cdot \frac{55.85\ \text{g}}{1\ \text{mol}} \cdot \frac{1\ \text{mol}}{6.023 \cdot 10^{23}\ \text{atoms}}}{1.38 \cdot 10^{-16} \left(\text{erg} \cdot \text{K}^{-1}\right) \cdot \text{T(K)}} \end{aligned}$$

$$= 1.11 \cdot 10^7 \cdot \frac{1}{\mathrm{T(K)}} \cdot \exp\left(-\frac{240000\ \mathrm{J}}{8.31\mathrm{J} \cdot \mathrm{mol}^{-1} \cdot \mathrm{K}^{-1} \cdot \mathrm{T}}\right) \mathrm{cm} \cdot \mathrm{s}^{-1} \tag{6.61}$$

The recrystallized fraction, X, after a certain time, t, at the temperature T is:

$$t_X = \sqrt[4]{-\frac{4 \cdot \ln(1 - X)}{I \cdot v^3}} \tag{6.62}$$

We know I and v at each temperature T, and we calculate t_X when X = 0.01 and X = 0.99. Finally, we represent T-t_X, which is the TTT curve (Temperature, Time, Transformation) (Fig. 6.22).

Note: The temperature that would be reached for a recrystallization time of about 1 h will be around 500 °C. At around 750 °C only a time of few seconds (flash annealing) will be required.

Finally, we plot the recrystallized fraction as a function of the time, see Fig. 6.23.

End of the exercise.

Exercise 6.4 A mild steel sheet is cold rolled down to half the initial thickness (50%) while the other sheet is cold rolled down to 1/4 the initial thickness (75%). They are subjected to isochrone recrystallization annealing. Determine:

- What of the two sheets would need a greater recrystallization temperature? Why?
- If the recrystallization temperature is the same, T_r, what of both sheets would have a smaller grain size for the same annealing time?

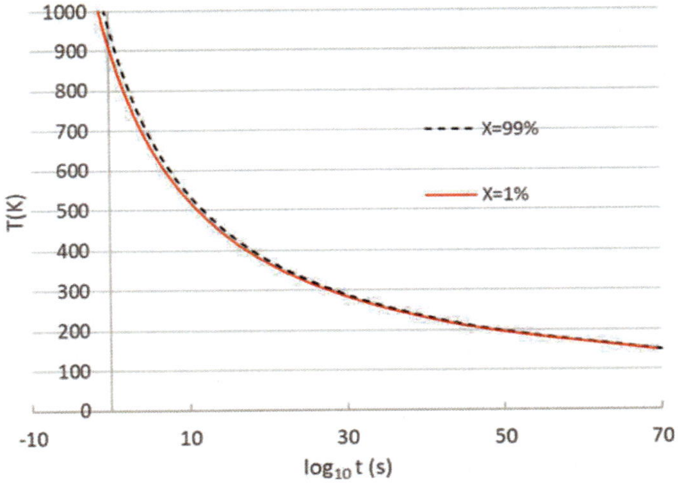

Fig. 6.22 TTT curve (Temperature-Transformation-Time)

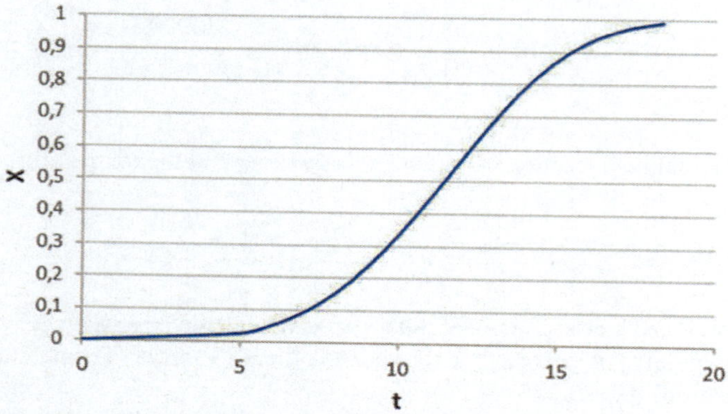

Fig. 6.23 Recrystallized fraction (X) versus time (s) (T = 600 °C (873 K))

- Would happen, in any case, that the chosen temperature for the recrystallization, allowed the recrystallization of any of them, but not the other? If it were like that, what would be the reason? What would be the non-recrystallized sheet?

Solution starts:

- If we consider Fig. 6.19 (Pero-Sanz 2004), that connects, T_r, d, and cold reduction, the less cold deformed sheet (50%) would require a greater temperature of recrystallization, because of having less stored energy or driving force for the recrystallization (associated to the density of dislocations).
- The most deformed sheet (75%) would possess less grain size because of having greater nucleation rate (more energy of deformation stored).
- Yes, less cold deformed sheet (50%) would not recrystallize, or it does partially if $T_r \simeq T_M/3$. The most deformed sheet (75%) would possess at least a more recrystallized fraction.

End of the exercise.

6.3.2.3 Deep Drawing Steels

Besides, in low carbon steels, aluminum-killed, destined to deep-drawing operations (DDQ, Deep Drawing Quality, EDDQ, Extra Deep Drawing Quality) with $\overline{R} \geq 1.5$, a favorable texture can be obtained: {1 1 1} (diagonal planes to the cube preferentially situated in the rolling plane, Figs. 6.24 and 6.25, Verdeja and Fernández 1992; Bocos-Fernández 2002; Blickwede 1984; Romano-Sobrino 2000), when the following parameters are rigorously controlled in the process of hot rolling of the sheets: temperature of coiling, percentage of cold reduction and recrystallization annealing (temperature and time).

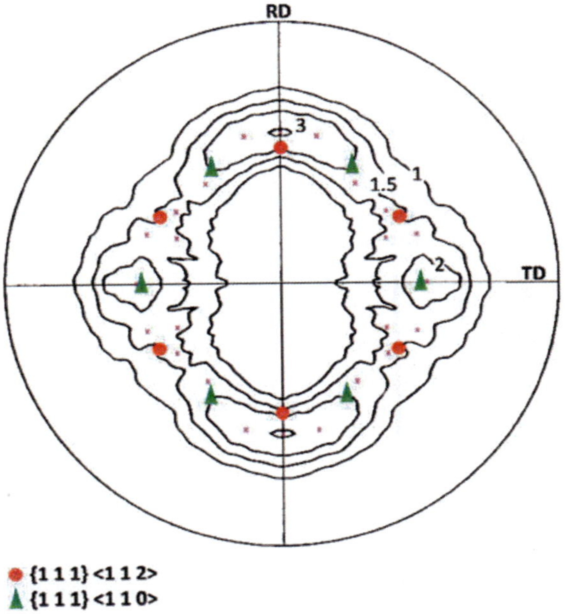

Fig. 6.24 {2 0 0} poles figure direct of aluminum-killed steel and batch annealed (recrystallized). $\overline{R} = 1.5$ (x-random)

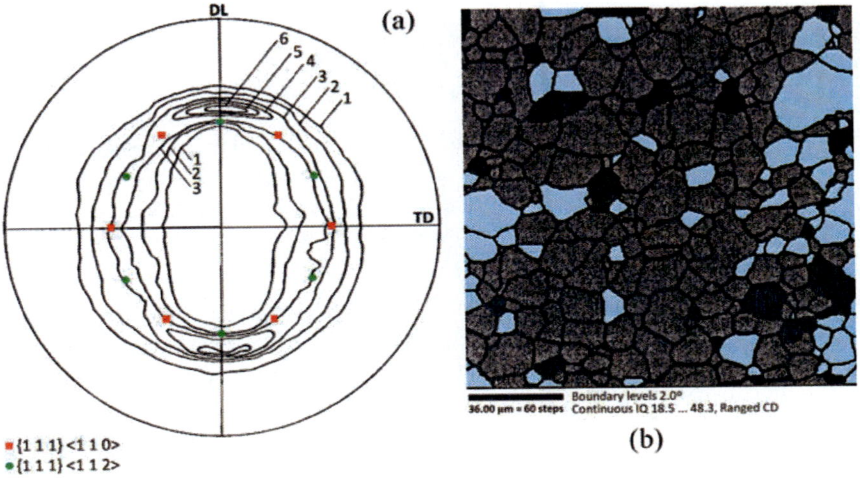

Fig. 6.25 **a** {2 0 0} poles figure direct of an interstitial free steel treated with titanium (recrystallized). $\overline{R} = 2.1$ (x-random). **b** Interstitial free. The OIM (Orientation Imaging Microscopy) based on the EBSD (Electron Backscattered Diffraction) image reveals the orientations of the grains: black grains {1 0 0}, gray-dotted grains {1 1 1} and blue grains {others}

The structure, at the end of the hot rolling process, is recrystallized and is austenitic: the atoms of nitrogen and aluminum are in solid solution in the austenite (T > 900 °C). After the hot rolling, sheets must be rapidly cooled (water shower) with the aim of avoiding the precipitation of the AlN during the allotropic transformation $\gamma \rightarrow \alpha$ and subsequent cooling. The coiling is carried out at 550–570 °C. Nitrogen and aluminum remain in solid solution in the ferrite as they were in the austenite. Later, the coils are unrolled, descaled and they are cold rolled (70% of reduction in thickness). Sheets maintain their surface quality; nitrogen and aluminum atoms are still in solid solution in the ferrite. On the contrary, it is intended that the AlN precipitates during the slow heating rate (20–40 °C/h) that precedes the isothermal annealing (batch annealing) performed at around 650 °C, under protecting atmosphere N_2–H_2. During the heating, it seems that the AlN nanometric precipitates lies preferentially on the sub-grain boundaries {1 0 0} from the texture of the cold-rolled sheet and impede their growth; these precipitates favor, on the contrary, the clustering and growth of the grains with texture {1 1 1} at the expense of the grains with orientation {1 0 0}. It is a combined process of oriented nucleation and growth with the aim of obtaining the most favorable texture of recrystallization, {1 1 1} < u v w >, for the final forming of the sheets by drawing.

In the opposite case, if the heating rates were very fast (>100 °C/min), as it is the case of the continuous annealing processing line (CAPL), the oriented crystallization and growth of the recrystallized α grains are not promoted: AlN precipitates are submicroscopic but not nanometric, and they do not interfere with the cold rolling texture. The accelerated subcritical or supercritical annealing treatments are very appreciated in the manufacture of tinplate for packaging. They are quasi-isotropic products ($\overline{R} \simeq 1$), with few tendence to earing (increase of the yielding), the available texture {1 0 0} < 1 1 0 > with tendency to form ears at 45° RD (rolling direction) balances the tendency of the texture {1 1 1} < u v w > to form ears at 0 and 90° RD (rolling direction).

Obviously, low carbon rimmed steels are not deoxidized with aluminum and they do not contain AlN precipitates, and, for that reason, there is not the possibility of suitable inducing the precipitation of AlN to favor the {1 1 1} textures and great $\overline{R}$ values (the value of $\overline{R}$ is around 1). These steels are called CQ (commercial quality) when we refer to drawing processes.

Exercise 6.5 Influence of textures {1 1 1} and {1 0 0}, parallel to the rolling plane of the sheets, in the coefficient R of normal anisotropy of Lankford. Justify the microstructural differences and their influence on the mechanical properties (drawability), in two very low carbon steels (<0.02%), cold-rolled ($\simeq$70% reduction) and subsequently recrystallized, one in the batch furnace and the other annealed in continuous.

Note: Enclosed Fig. 6.26 (Hosford and Backofen 1964) represents M versus r and M curves for textures which are rotationally symmetric about the sheet normal. Curves are shown for sheets in which {1 1 1}, {1 1 0}, and {1 0 0} are parallel to the rolling plane, and for a randomly oriented sheet. M is the so-called Taylor factor that is equal to the quotient of the deformation by shearing (due to the dislocations), $d\gamma$, and the

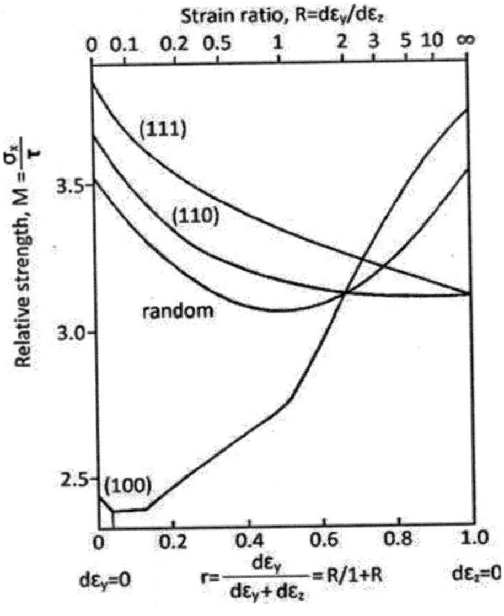

Fig. 6.26 M versus r curves for textures which are rotationally symmetric about the sheet normal. Curves are shown for sheets in which (1 1 1), (1 1 0), and (1 0 0) are parallel to the rolling plane, and for a randomly oriented sheet

deformation experienced in tension, $d\varepsilon_x$, or which is the same, serves to relate the tensile flow strength, σ_x, of grain to the basic shear strength τ for slip. M would be minimum according to the principle of virtual work:

$$dW = \tau \cdot d\gamma = \sigma_x \cdot d\varepsilon_x \tag{6.63}$$

If the tensile axis x is taken parallel to one of the principal axes of anisotropy, yet one of the strain components, $d\varepsilon_y$ **or** $d\varepsilon_z$ remains unspecified and this may be incorporated in a useful parameter:

$$r = \frac{d\varepsilon_y}{d\varepsilon_y + d\varepsilon_z} = \frac{R}{1 + R} \tag{6.64}$$

by assuming different values of r and calculating the corresponding value of:

$$M = \frac{d\gamma}{d\varepsilon_x} \tag{6.65}$$

thus, M versus r plot may be generated.

The minimum of M of such a plot represents the least slip or least work with which the strain increment $d\varepsilon_x$ could be produced and therefore corresponds to the expected behavior. The values of M and r at the minimum identify the relative strength σ_x/τ and the Lankford coefficient:

$$R = \frac{r}{1 - r} \tag{6.66}$$

M versus r curve for a material consisting of several texture components a, b, c may be found from the weighted average:

$$M = f_a \cdot M_a + f_b \cdot M_b + f_c \cdot M_c + \cdots \tag{6.67}$$

where $f_a, f_b, f_c, \ldots$ are the volume fraction of components a, b, c.... The minimum will occur when:

$$\frac{\partial M}{\partial r} = f_a \times \frac{\partial M_a}{\partial r} + f_b \times \frac{\partial M_b}{\partial r} + f_c \times \frac{\partial M_c}{\partial r} + \ldots = 0 \tag{6.68}$$

and the R value of the sheet could be calculated (approximately).

Solution starts:

Considering the Fig. 6.26, it may be related to the observations of Whitely and Wise on steel sheets. When these authors evaluated the relative amounts of {1 1 1}, {1 1 0}, and {1 0 0} they found that {1 1 1} contributed to high R, {1 1 0} was neutral, while a small amount of {1 0 0} strongly depressed R.

Such findings are consistent with the slopes of the three curves in the range 1/2 < r < 2/3 (R = 1 to 2) that characterized their own steels. If only {1 1 1} and {1 0 0} texture components were present the minimum M would occur when:

$$f_{100} \cdot \frac{\partial M(100)}{\partial r} + f_{111} \cdot \frac{\partial M(111)}{\partial r} = 0 \tag{6.69}$$

At R = 1.5:

$$\frac{\partial M(100)}{\partial r} \simeq +0.9 \tag{6.70}$$

and:

$$\frac{\partial M(111)}{\partial r} \simeq -0.2 \tag{6.71}$$

so that the effect of a {1 0 0} component is depressing R would balance the elevating effect of 4–5 times as much {1 1 1} component.

Reaching a $\overline{R} = 2$(limit industrially obtained) would be equivalent to obtaining a quotient between the fraction of planes:

$$\frac{f_{111}}{f_{100} + f_{111}} > 0.9(90\%) \tag{6.72}$$

In recrystallization annealing of extra-mild steels performed in the CAPL (Continuous Annealing Processing Line, CAPL), high heating rates and short times of holding favor the nucleation and growth of grains with texture {1 0 0} or randomly, and shift the value:

$$\frac{f_{111}}{f_{111} + f_{100} + f_{random}} \tag{6.73}$$

from value $\overline{R} \simeq 2$ ($r = 0.66$), toward values closer to 1:

$$\begin{aligned}(\overline{R} \simeq 1 - 1.2) &\rightarrow f_{111}(-0.4) + f_{100}(0.4) + f_{random}(0.1) = 0 \rightarrow f_{111} \\ &= f_{100} + 0.25 \cdot f_{random}(0.5 < r < 0.55)\end{aligned} \tag{6.74}$$

It is possible to obtain quasi-isotropic (without ears) with all the added advantages of the CAPL with respect to those of the batch annealing (lower energy costs, greater productivity, savings of workforce, etc.).

End of the exercise.

Exercise 6.6 Calculate the temperature of starting of the aluminum nitride, AlN, precipitation when the quantities of aluminum and nitrogen are respectively 250 and 100 ppm. If the final temperature of the precipitation process is estimated in 800 °C, calculate the volume fraction of AlN in the steel.

Data: the density of the AlN is 3200 kg·m^{-3} and that of the steel is 7950 kg·m^{-3}, while the solubility of the nitrogen in the iron (solvent) at 800 °C is 30 ppm.

Solution starts:

We used data from Table 6.4 (Turkdogan 1996) to calculate the temperature at which the process of AlN precipitation starts in steel (austenite):

$$\log[(\%Al) \cdot (\%N)] = -\frac{6770}{T} + 1.03 \tag{6.75}$$

$$\log[(0.025) \cdot (0.010)] = -\frac{6770}{T} + 1.03 \rightarrow T = 1462\ \text{K}\ (1189\ ^\circ\text{C}) \tag{6.76}$$

Table 6.4 Solubility product of nitrides, carbides, and carbonitrides in gamma iron and alpha iron

Solubility product	$\log K_{Fe-\alpha}$	$\log K_{Fe-\gamma}$
$C_{Al}C_N$	2.05–8790/T	1.03–6770/T
C_BC_N	4.61–14250/T	5.24–13970/T
$C_{Nb}C_N$	4.91–12170/T	3.79–10150/T
$C_{Nb}C_C^{0.87}$	4.33–9830/T	2.81–7020/T
$C_{Nb}C_C^{0.7}C_N^{0.2}$	5.57–12120/T	4.12–9450/T
$C_{Ti}C_N$	6.40–18420/T	5.40–15790/T
$C_{Ti}C_C$	4.45–10230/T	2.75–7000/T
C_VC_N	3.90–9720/T	2.86–7700/T

If we take into account that the stoichiometry of the precipitation reaction is:

$$Al(dis;\, s) + N(dis;\, s) \leftrightarrow AlN(s) \tag{6.77}$$

the quantity of free nitrogen in the metal, $(\%N)_f$, to react will be:

$$(\%N)_f = (\%Al)_f \cdot \frac{1\ \text{mol Al}}{26.98\ \text{g Al}} \cdot \frac{1\ \text{mol N}}{1\ \text{mol Al}} \cdot \frac{14.01\ \text{g N}}{1\ \text{mol N}} - 0.0030 \tag{6.78}$$

where 0.0030% represents the nitrogen solubility (in atomic form) that the iron (ferrite) can dissolve at 800 °C.

On the other hand, at 800 °C, the solubility product of the AlN, product of the aluminum and nitrogen concentrations in equilibrium with the precipitated AlN (Table 6.4) is:

$$\begin{aligned} \log[(\%Al)\cdot(\%N)] &= -\tfrac{6770}{T} + 1.03 = -\tfrac{6770}{1073} + 1.03 = -5.2794 \\ &\rightarrow (\%Al)\cdot(\%N) = 5.255\cdot 10^{-6} \end{aligned} \tag{6.79}$$

Consequently, the quantity of aluminum dissolved in the iron that is not combined with the nitrogen, $(\%Al)_f$, is obtained by means of the following equation (replacing Eq. (6.78) in the solubility product, Eq. (6.79)):

$$(\%Al)_f \cdot \left[(\%Al)_f \cdot \frac{14.01}{26.98} - 0.0030\right] - 5.255\cdot 10^{-6} = 0 \tag{6.80}$$

$$\left[(\%Al)_f\right]^2 \cdot 0.5204 - 0.003\cdot(\%Al)_f - 5.255\cdot 10^{-6} = 0 \tag{6.81}$$

The result for the free aluminum and nitrogen, once reached the 800 °C (end of the precipitation) is:

$$(\%Al)_f = 0.0072\ (72\ \text{ppm}) \tag{6.82}$$

$$(\%N)_f = 0.00074\ (7.4\ \text{ppm}) \tag{6.83}$$

Consequently, the quantity of AlN precipitates will be:

$$(0.025 - 0.0072) + (0.010 - 0.00074) = 0.027\% \tag{6.84}$$

Finally, the volume fraction of AlN will be:

$$f_v = 0.027\% \cdot \frac{7950\ \text{kg}\cdot\text{m}^{-3}}{3200\ \text{kg}\cdot\text{m}^{-3}} = 0.067\%\ (0.7‰) \tag{6.85}$$

Commentaries: the volume fraction of inclusions that is required for a "clean steel" is lower than 1% and lower than 0.5% in the case of an "extra-clean steel." Steel is habitually considered not clean when the volume fraction of inclusions is greater than 5%.

End of the exercise.

6.3.3 Von Mises Criterion for Anisotropic Materials. Applications

Following Von Mises, the anisotropic plasticity theory of Hill is useful for yielding under other conditions of combined stress. In applying the theory to sheet materials for deep drawing and other operations (under other conditions of combined stress). In focusing the theory on sheet materials, it is generally convenient assuming a texture that is rotationally symmetric about the sheet normal and an absence of the Bauschinger effect. Hill's yield criterion and associated flow rules reduce to:

$$\left(\sigma_z - \sigma_y\right)^2 + (\sigma_x - \sigma_z)^2 + R \cdot \left(\sigma_x - \sigma_y\right)^2 = (1 + R) \cdot X^2 \tag{6.86}$$

and:

$$d\varepsilon_x : d\varepsilon_y : d\varepsilon_z = (R + 1) \cdot \sigma_x - R \cdot \sigma_y : (R + 1) \cdot \sigma_y - R \cdot \sigma_x - \sigma_z : 2 \cdot \sigma_z - \sigma_x - \sigma_y \tag{6.87}$$

where x(1) and y(2) are the two directions of principal stress in the plane of the sheet and z(3) is the sheet normal: x, y, and z are the uniaxial yield stresses in the x (rolling), y (transverse), and z (normal) directions.

For the special case of loading in the plane of the sheet, $\sigma_z = 0$, the above equations become in (yield stresses on the sheet plane):

$$\sigma_x^2 + \sigma_y^2 - \sigma_x \cdot \sigma_y \cdot \left(\frac{2 \cdot R}{R + 1}\right) = X^2 = Y^2 \tag{6.88}$$

$$d\varepsilon_x : d\varepsilon_y : d\varepsilon_z = (R + 1) \cdot \sigma_x - R \cdot \sigma_y : (R + 1) \cdot \sigma_y - R \cdot \sigma_x : -\sigma_x - \sigma_y \tag{6.89}$$

where R is the normal coefficient of anisotropy of Lankford if the texture is rotationally symmetric in the plane of the sheet $\Delta R = 0$. As it is in tension:

$$\sigma_x \neq 0, \quad \sigma_y = \sigma_z = 0 \rightarrow d\varepsilon_{yy}/d\varepsilon_{zz} = R \tag{6.90}$$

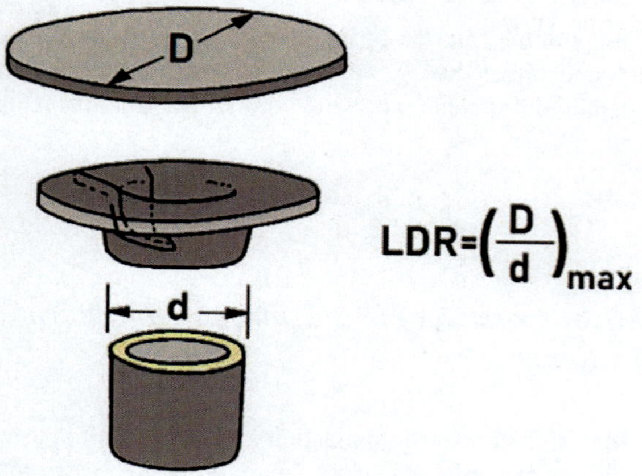

Fig. 6.27 Deep drawing

A test usually employed in the characterization of the drawability of materials is the Swift test: it consists of drawing circular discs of growing diameter D with a punch die of diameter d. We call limit drawing ratio (LDR) to the quotient of the diameter of the disc with the greatest diameter, D_{max}, that can be drawn with a punch of diameter d:

$$LDR = \frac{D_{max}}{d} \tag{6.91}$$

See Fig. 6.27 (Stuwe and Faustmann 1969).

Exercise 6.7 Correlation of drawing limits with strain ratios. Demonstrate that:

$$\ln\left(\frac{D_{max}}{d}\right) = \sqrt{\frac{R+1}{2}} \tag{6.92}$$

The reason for such a correlation with R is most apparent in the drawing of flat bottom cups (see Fig. 6.28, Martínez-Pérez 2003).

Solution starts:

In this operation, there are two regions of special importance: the flange where deformation occurs and force requirements originate, and the cup wall which must support the necessary forces without tearing. Often, there is little thickening in the flange, yielding being governed by the plain strain with $d\varepsilon_z = 0$. Plain strain is also imposed on the wall, but now $d\varepsilon_y = 0$ because a solid punch prevents thinning. For maximum drawability in ductile metals (steels), the wall strength should be as high as possible relative to that of the flange. In this way, the largest reduction can be accomplished before wall failure by necking. Drawing limit must be governed by an increase with the ratio β of the two plane strain strengths:

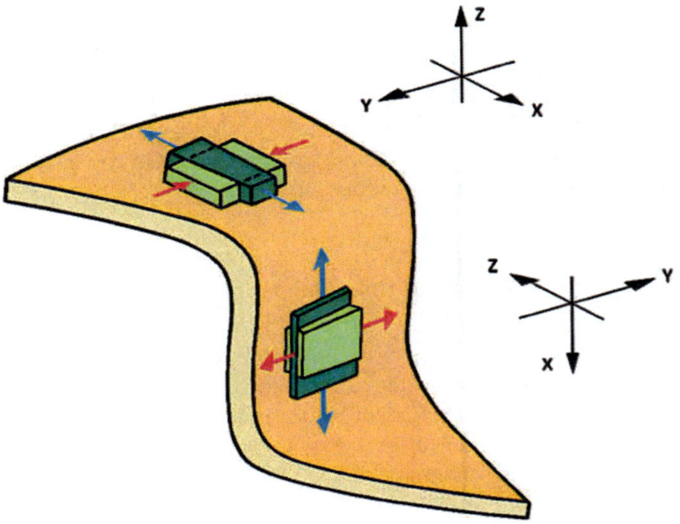

Fig. 6.28 Stresses state in the flange and walls of a cylindrical cup during the process of deep drawing by contraction

$$\beta = \frac{\sigma_x(\text{wall},\quad d\varepsilon_y = 0)}{\sigma_x(\text{flange},\quad d\varepsilon_z = 0)} \tag{6.93}$$

Reference to the Hill criteria for anisotropic sheets ($\sigma_z = 0$) indicates that the wall strength increases with R while the flange strength decreases with R. Hence β and drawability increase with R.

- Cup flange in drawing (pure shear), $d\varepsilon_z = 0 \rightarrow \sigma_x/X = \sqrt{2} \cdot (R+1)/(2 \cdot R+1)$. See Table 6.5.
- Cup wall in drawing (plane strain), $d\varepsilon_y = 0 \rightarrow \sigma_x/X = (R+1)/\sqrt{(2 \cdot R+1)}$. See Table 6.6.

So:

Table 6.5 Cup flange in drawing (pure shear)

R	0	1	2	3	5
σ_x/X	0.707	0.572	0.548	0.535	0.522

Table 6.6 Cup wall in drawing (plane strain)

R	0	1	2	3	5
σ_x/X	1.000	1.154	1.348	1.512	1.809

$$\beta = \frac{\frac{(R+1)}{\sqrt{(2\cdot R+1)}}}{\sqrt{\frac{2\cdot(R+1)}{(2\cdot R+1)}}} = \sqrt{\frac{R+1}{2}} \tag{6.94}$$

This correlation had been studied in detail by Whiteley (1960). The following expression for drawability was derived with an analysis by Goodman and Hu (1970) based on Hill's anisotropic plasticity criterion:

$$\ln\left(\frac{D}{d}\right) = \frac{1}{1+\eta}\cdot\sqrt{\frac{R+1}{2}} \tag{6.95}$$

where D is the largest blank diameter which can be drawn into a cup of diameter d without being torn out. R is the average strain ratio from the tensile test at different angles of the sheet (Lankford coefficient) and η is a friction parameter with a value comprised between 0.2 and 0.3. Rigorously, the analysis applies to what be best termed pure radial drawing with no bending and unbending around tool radii.

The dependence of D/d on $\overline{R}$ was experimentally demonstrated for a series of materials with a range of $\overline{R}$ values from 0.5 to 4. The LDR (Limit Drawing Ratio) can vary approximately between 2 and 3.7.

Note: Demonstration of the value of β.

With $\sigma_z = 0$, the criterion of Hill is:

$$\sigma_x^2 + \sigma_y^2 + R\cdot(\sigma_x - \sigma_y)^2 = X^2\cdot(1+R) = Y^2\cdot(1+R) \tag{6.96}$$

$$\sigma_x^2 + \sigma_y^2 - \sigma_x\cdot\sigma_y\cdot\left(\frac{2\cdot R}{R+1}\right) = X^2 = Y^2 \tag{6.97}$$

- Wall ($d\varepsilon_y = 0$):

$$0 = \frac{\partial f}{\partial \sigma_y} \to \sigma_y - R\cdot(\sigma_x - \sigma_y) = 0 \tag{6.98}$$

$$\sigma_y\cdot(1+R) = R\cdot\sigma_x \tag{6.99}$$

$$\frac{\sigma_y}{\sigma_x} = \frac{R}{R+1} = \alpha \tag{6.100}$$

Substituting in the Hill's criterion:

$$\sigma_x^2 + \alpha^2\cdot\sigma_x^2 - \sigma_x^2\cdot\alpha\cdot\left(\frac{2\cdot R}{R+1}\right) = X^2 \tag{6.101}$$

$$\left(\frac{\sigma_x}{X}\right)^2\cdot\left[1+\frac{R^2}{(R+1)^2} - \frac{2\cdot R^2}{(R+1)^2}\right] = 1 \to \left(\frac{X}{\sigma_x}\right)^2$$

$$= \frac{(R+1)^2 + R^2 - 2 \cdot R^2}{(R+1)^2} \rightarrow \frac{\sigma_x}{X} = \frac{R+1}{\sqrt{2 \cdot R + 1}} \quad (6.102)$$

– Flange ($d\varepsilon_z = 0$):

$$\sigma_z \neq 0; \quad \sigma_y = 0; \quad d\varepsilon_z = 0 \rightarrow \partial f / \partial \sigma_z = 0 \quad (6.103)$$

$$2 \cdot \sigma_z = \sigma_x + \sigma_y \rightarrow \sigma_z = \frac{\sigma_x}{2} \quad (6.104)$$

We apply the criterion of Hill:

$$\left(\frac{\sigma_x}{2}\right)^2 + \left(\frac{\sigma_x}{2}\right)^2 + R \cdot \sigma_x^2 = X^2 \cdot (1+R) \rightarrow \frac{2 \cdot R + 1}{2} \cdot \sigma_x^2 = X^2 \cdot (1+R) \rightarrow \frac{\sigma_x}{X} = \sqrt{\frac{2 \cdot (R+1)}{2 \cdot R + 1}} \quad (6.105)$$

Thus:

$$\left(\frac{\sigma_x}{X}\right)_{wall} = \frac{R+1}{\sqrt{2 \cdot R + 1}} \quad (6.106)$$

$$\left(\frac{\sigma_x}{X}\right)_{flange} = \sqrt{\frac{2 \cdot (R+1)}{2 \cdot R + 1}} \quad (6.107)$$

Finally:

$$\beta = \frac{\left(\frac{\sigma_x}{X}\right)_{wall}}{\left(\frac{\sigma_x}{X}\right)_{flange}} = \frac{\frac{R+1}{\sqrt{2 \cdot R+1}}}{\sqrt{\frac{2 \cdot (R+1)}{2 \cdot R+1}}} = \sqrt{\frac{R+1}{2}} \quad (6.108)$$

End of the exercise.

6.3.4 Forming Limit Diagrams

The experimental concept of the LDR (Limit Drawing Ratio) calculation using the Swift test has been enriched in the last decades. Since 1965, a group called American Deep Drawing Research Group (this Society was created in Europe some years before) has been developing and testing different methods to help people from a variety of areas related to sheet metal forming. They developed a technique based upon the use of circle (or square) grids engraved on the sheets to be stamped out and calculated their forming limit diagrams. A regular pattern of circles (preferentially

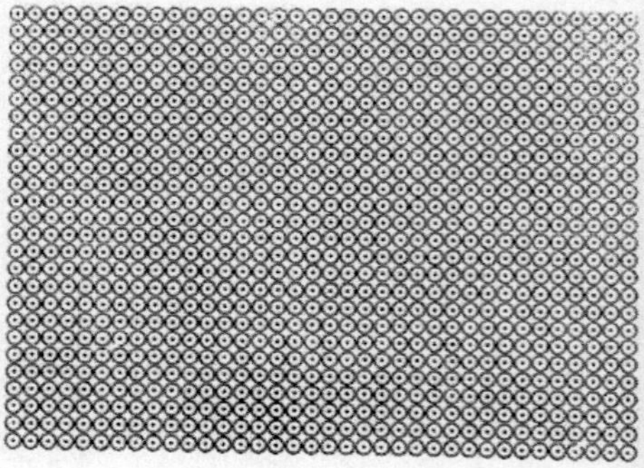
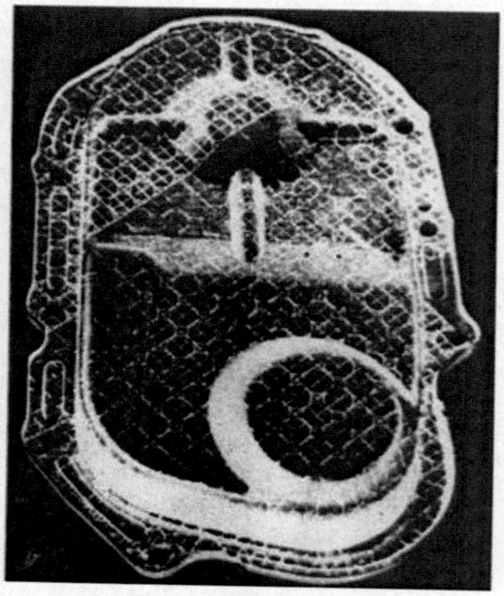

Fig. 6.29 Circle grid pattern (left); Pattern grid in a drawn part (right)

2–3 mm in diameter) is marked by electro-etching or manual methods (Fig. 6.29). Then we determine the largest amount that a given circle is stretched and the minor stretch that a given circle is stretched (circles began ellipsoids). A FLD (Forming Limit Diagram) describes the limits that sheet metal can be stretched under different conditions (path trajectories, according to the deformation type (tension, plane strain, biaxial tension, etc.)). The FLD underlies the necking failures of a formed part by thinning abruptly in a narrow-localized area. An extreme case of necking is splitting. We represent $(e_1)_M$ versus $(e_2)_m$ in a FLD diagram where %stretch is the percentage of final length minus original length/original length. The subscripts M and m call for the maximal (M) and minimal (m) stretch a sheet can withstand before necking or split becomes. Note that the minor stretch ($e_2 \gtrless 0$) is always at a perpendicular or 90º direction from the major stretch.

There can be the paths trajectories for special cases of stretch:

(a) The first is in tension e_2 (transversal direction of tensile axis)/ e_1 (parallel direction to tensile axis):

$$\rho = \frac{e_2}{e_1} = \frac{e_2}{-e_2 - e_3} = \frac{1}{-1 - \frac{1}{R}} = -\frac{R}{1+R} \tag{6.109}$$

(b) Pure shear: $e_1 = -e_2$. Path trajectory:

$$\rho = \frac{-e_2}{e_1} = -1 \tag{6.110}$$

(c) Plane strain: $e_2 = 0 \rightarrow \rho = 0$. When the minor stretch is zero (or the ellipse width equals to the original circle diameter) this is called also "plane stretch". This is the lower point of the FLD, so the more dangerous to be arrived at.

(d) Plain stretch. Very often we find a circle becomes larger instead of an ellipse. The major stretch and minor stretch are both positives and approximately equal. We call this "balanced stretch" because the major and minor stretches are balanced or nearly equal.

Enclosed Fig. 6.30 represents a complete FLD diagram (Dinda et al. 1981).

The FLD map can be divided into different zones. The zone above the top curve is the area where some percentages of necking, tearing, or localized thinning occur. The area between the two curves is called the marginal zone. If any circle grids fall in this zone, failure may not occur today but could be tomorrow. In other words, the part is too close to failure. Without a visual failure, you as a die tryout person would not know this. Your FLD map warns you, however, of this danger. The marginal zone means that today may have no trouble, but tomorrow you may be in trouble,

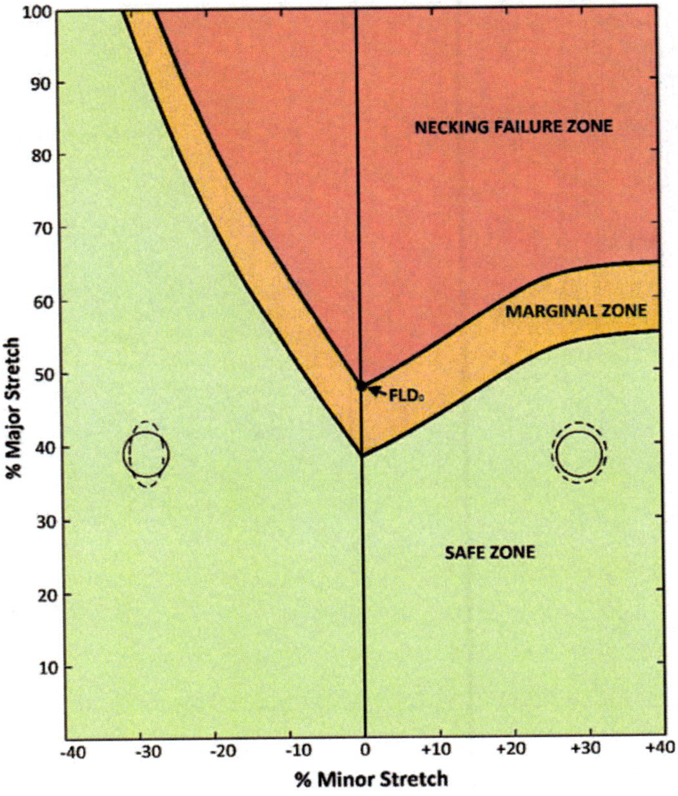

Fig. 6.30 FLD diagram

as forming conditions could be changed (different steel, different friction, lubricant change, different metal surface, change in drawbeads and dies, etc.).

There are different FLD maps for different thickness steels, different strength steels and different lubricant conditions.

As example of FLD charts, we attach the FLD's (Figs. 6.31 and 6.32, Dinda et al. 1981) of low strength steel of different thickness and for 0.04" (1 mm) thick steel of various yield strength. What consequences can be deduced?

- Note here that we are showing only the top line of the FLD which separates the necking failure zone from the safe marginal zone. If we included both lines, figures would be too confusing.
- The greater the thickness, the higher are the curve on the FLD. It may be a consequence of the Unwin equation in tension: the total elongation $e_T = e_u + (\sqrt{S_0}/l_0) \cdot$ constant, where e_u is the uniform elongation, $\sqrt{S_0}/l_0$ is the constant of the test or similarity geometric index that increases with the section $S_0 = w_0 \cdot t_0$, that is to say, with the thickness t_0.

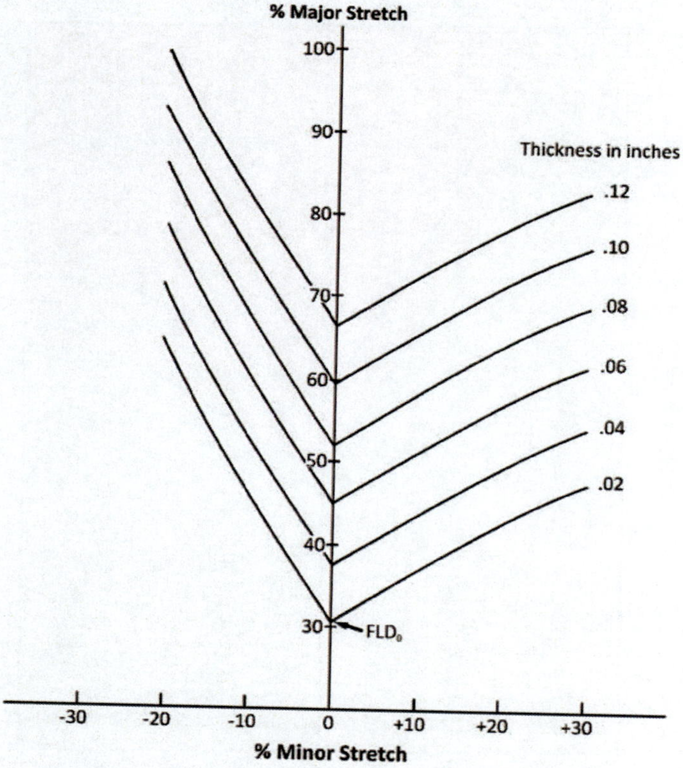

Fig. 6.31 FLD's for low strength steels of different thickness

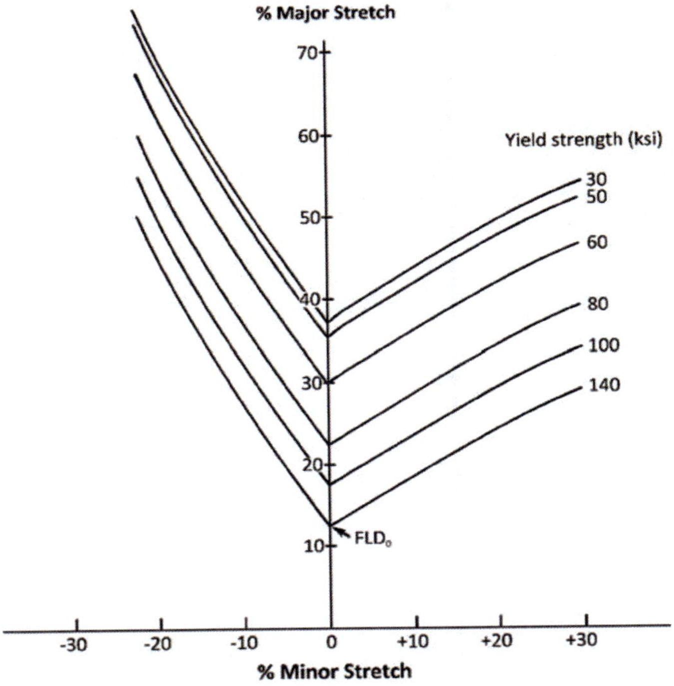

Fig. 6.32 FLD's for 0.04 inches thick steel of various yield strengths

– Higher strength steels have lower FLD's than the low strength steel of equal thickness. For instance, comparing the FLD for the same thickness of sheet (0.04" $\simeq$ 1 mm in thickness), % minor stretch FLD_0 of about 38% correspond to 30 ksi steel ($\simeq$210 MPa). For a steel with high yield strength (80 ksi $\simeq$560 MPa), the FLD_0 (plane strain conditions: $\varepsilon_M \simeq 23\%$, $\varepsilon_m \simeq 0$). This variation corresponds with the decrease of the ductility ($\simeq$ n index) and with the increase of the value of the yield strength (Fig. 6.33).

Exercise 6.8 Utilization of circle grids (Fig. 6.29) and FLD diagrams (Fig. 6.30).

Solution starts:

The technique that uses circle grids and FLD diagrams is known as Circle Grid Analysis. It is a very useful technique for die tryout people. The network of circles allows to have a grid properly oriented for whatever direction the sheet is stretched. This way, no matter how the sheet is stretched, the circle will have a diameter lined up with the direction of maximum and minimum stretch. Thus, the circle diameter can be used for the original length measurement and the stretched diameter (positive or negative) can be used as the final length measurement (the circle will change into an ellipse, Fig. 6.34, Dinda et al. 1981).

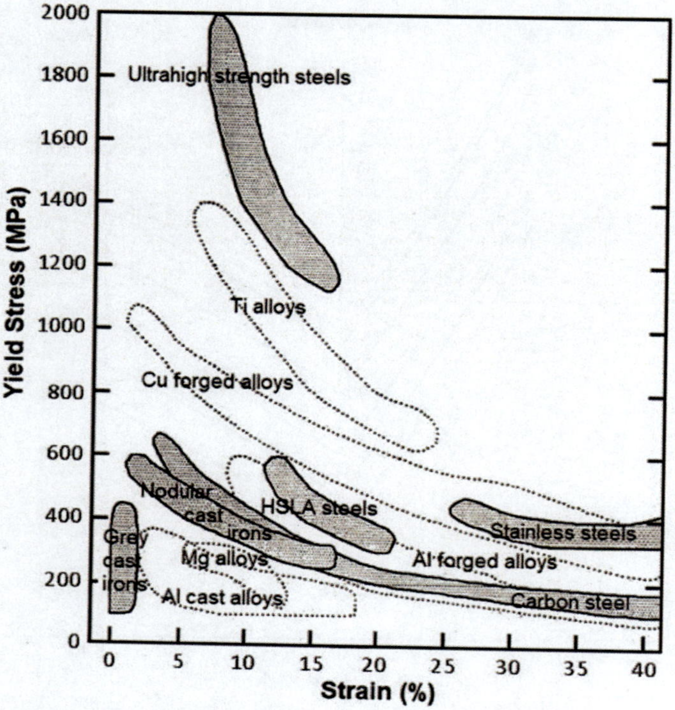

Fig. 6.33 Yield strength-elongation for different materials

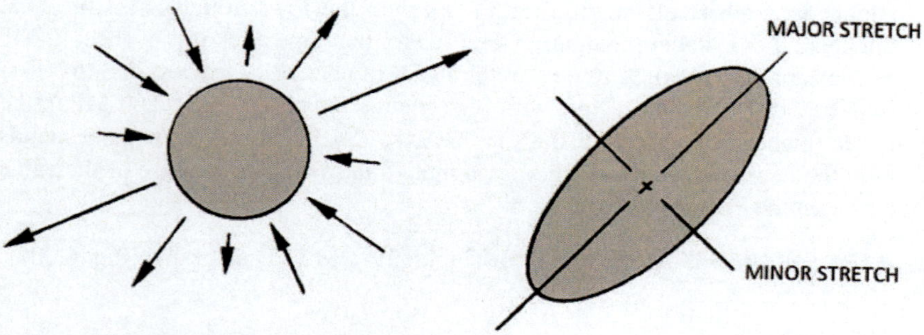

Fig. 6.34 Transformation of the circle into an ellipse as a result of the drawing

The ellipse will have two important lines. One is the long axis of the ellipse (length) and is called the major stretch. This is always the largest amount of stretch the circle undergoes and is always positive. The shortest axis of the ellipse (width) is the minor stretch. In some cases, this short axis may be longer than the original circle diameter; this is known as the positive minor stretch (Fig. 6.35, case A, Dinda et al.

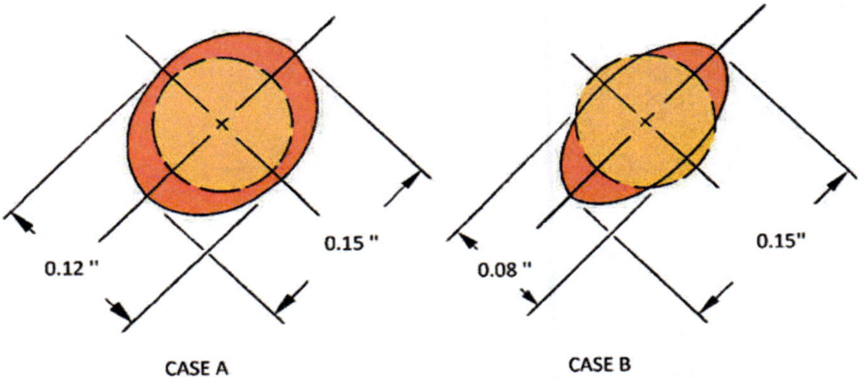

Fig. 6.35 Minor and major stretches in common cases (example)

1981). In other cases, this short axis may be shorter than the original circle diameter; this is known as negative minor stretch (Fig. 6.35, case B, Dinda et al. 1981). If the original circle has 0.10 inches in diameter and we consider Fig. 6.35 (Dinda et al. 1981), we can determine the minor and major stretches in both cases:

– Case A:

$$\text{Major Stretch} = \frac{0.15'' - 0.10''}{0.10''} \cdot 100 = +50\% \text{ (positive)} \tag{6.111}$$

$$\text{Minor Stretch} = \frac{0.12'' - 0.10''}{0.10''} \cdot 100 = +20\% \text{ (positive)} \tag{6.112}$$

– Case B:

$$\text{Major Stretch} = \frac{0.15'' - 0.10''}{0.10''} \cdot 100 = +50\% \text{ (positive)} \tag{6.113}$$

$$\text{Minor Stretch} = \frac{0.08'' - 0.10''}{0.10''} \cdot 100 = -20\% \text{ (negative)} \tag{6.114}$$

The minor stretch is always perpendicular to the major stretch direction.
There are other special cases. Those are shown in Fig. 6.36 (Dinda et al. 1981).

– Case A:

$$\text{Major Stretch} = \frac{0.15'' - 0.10''}{0.10''} \cdot 100 = +50\% \text{ (positive)} \tag{6.115}$$

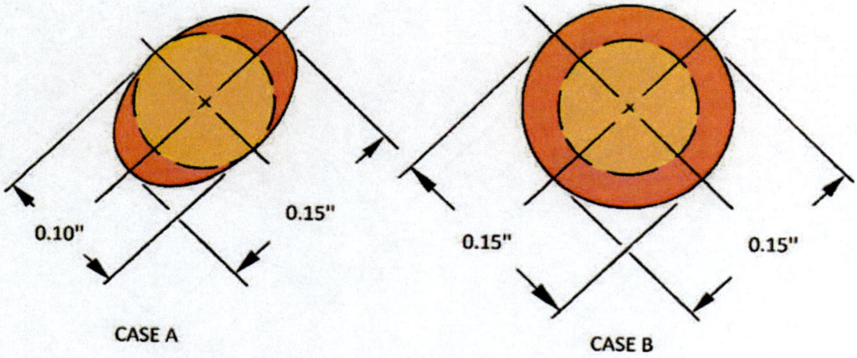

Fig. 6.36 Minor and major stretches in special cases (example)

$$\text{Minor Stretch} = \frac{0.10'' - 0.10''}{0.10''} \cdot 100 = 0\% \tag{6.116}$$

when the minor stretch is zero (or the ellipse width equal to the original circle diameter), this is called plane stretch.

- Case B:

$$\text{Major Stretch} = \frac{0.15'' - 0.10''}{0.10''} \cdot 100 = +50\% \text{ (positive)} \tag{6.117}$$

$$\text{Minor Stretch} = \frac{0.15'' - 0.10''}{0.10''} \cdot 100 = +50\% \text{ (negative)} \tag{6.118}$$

it is also possible that the circle becomes larger than the original, instead of an ellipse. The major and minor stretches are positives and equal. This case is called balanced stretch because both stretches are equal.

Circles are uniform in diameter. Therefore, we have only to select several of the ellipses of interest on the deformed sheet. The preferred method of measurement tool is the Mylar tape with diverging lines (this tape is placed over the ellipse and then shifted until the diverging lines line up with major axis of the ellipse, Fig. 6.37 (Dinda et al. 1981), and the stretch can be read directly from the tape; the tape is next turned 90° to read the minor stretch).

Finally, we locate the coordinates of the ellipses (minor stretch, major stretch) on an FLD diagram as that of Fig. 6.30 (Dinda et al. 1981), and we check whether they fall on the safe, marginal, or necking failure zones.

End of the exercise.

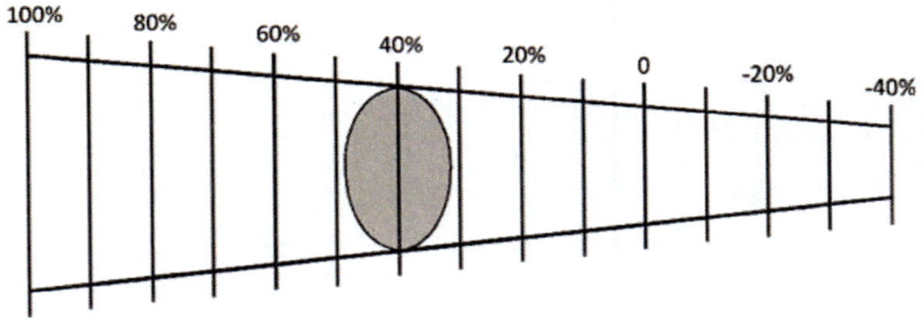

Fig. 6.37 Mylar tape with diverging lines

6.4 Wire Drawing

Wire drawing is a cold finishing process by which the wire rod is passed through wire drawing dies manufactured in abrasion-resistant materials (metallic carbides, oxides, or nitrides) with a diameter smaller than that of the starting wire rod. The advantages of this operation include:

- Reduction of the wire rod section until the defined diameter.
- Remarkable improvement in the mechanical properties of the steel. It is possible to obtain wires by combination of the wire drawing and heat treatments that exceed the 3000 MPa of ultimate tensile strength.
- The possibility of obtaining wires of small dimensions: with diameter of around 0.15 mm, such as the steel used to manufacture radial tires, or with diameters greater than 10 mm, such as rod for pre-stressed concrete (rebars).
- Wires with excellent surface finishing and strict dimensional tolerances.
- The possibility of wire drawing a great variety of steels, from the ferritic steels to those of high carbon content (including stainless steels).

In the manufacture of wires of small sections, it is possible to reach so high work hardening values that any subsequent deformation is not possible. It is necessary a regeneration of the crystalline structure of the ferrite in continuous annealing furnaces, under reductant atmosphere, and, in some cases, non-decarburizing atmosphere, or using a patenting treatment (Verdeja et al. 2020).

A subcritical annealing is performed in ferritic wires by heating the product up to a temperature slightly lower than that of the eutectoid point. As there are no allotropic transformations, the cooling should give optimal productivity and reliability of the process. The resulting structure comprises globular cementite dispersed in a matrix of ferrite (Pero-Sanz et al. 2019).

In the ferritic-pearlitic and pearlitic wires, the heat treatment that is given to the material before its introduction in the finishing wire drawing dies is called patenting, where the austenitized wire is submerged in molten lead bath or in a salts bath at the temperature corresponding to the lower zone of the pearlitic zone of the TTT curve of the steel, to transform the austenite into very fine pearlite (separation between lamellas of cementite of 0.1–0.2 μm), which allows ultimate tensile strengths up to 1600 MPa.

Exercise 6.9 A pearlitic steel of high resistance for wires has the following chemical composition: 0.85% C, 0.85% Mn, 0.50% Si, 0.40% Cr, and 0.20% Mo. Demonstrate that the structure of the steel, at room temperature, is that of a hypoeutectoid steel.

Data: the maximum solubility of the carbon in the austenite (Fe-C metastable diagram) at the eutectoid temperature of 727 °C is 0.77% C. Assume that the data of the metastable equilibrium can be used in the equilibrium of the Fe-C stable system. The coefficient of activity of the carbon at infinite dilution in the solid solution of the carbon of the Fe-γ at 727 °C (1000 K) is 14.00.

Solution starts:

Regarding the microstructure of the steel, and whether it is that with pearlite dispersed in a matrix of ferrite, it is necessary that (at 727 °C) the carbon solubility in the austenite to be greater than 0.85% C. This way, for an equilibrium cooling below 727 °C, ferrite grains can nucleate and grow in the grain boundaries of the austenite with carbon content greater than 0.85%.

To estimate the value of the maximum solubility of the carbon in the austenite in equilibrium with the carbon-cementite, the non-linear equation to be solved is the following:

$$\ln x_C + \ln \gamma_C^* + \varepsilon_C^C \cdot x_C + \Omega = 0 \tag{6.119}$$

where:

$$\Omega = \varepsilon_C^{Mn} \cdot x_{Mn} + \varepsilon_C^{Si} \cdot x_{Si} + \varepsilon_C^{Cr} \cdot x_{Cr} + \varepsilon_C^{Mo} \cdot x_{Mo} \tag{6.120}$$

being the concentration of the different solutes in solid solution expressed in atomic fraction, x_i. We use the data of Table 6.7 referred to the interaction parameters of the first order of the solutes in the austenite.

We calculate the molar fractions. The denominator of the molar fraction is:

$$D = 0.85\ \text{g C} \cdot \frac{1\ \text{mol C}}{12.01\ \text{g C}} + 0.85\ \text{g Mn} \cdot \frac{1\ \text{mol Mn}}{54.94\ \text{g Mn}} + 0.50\ \text{g Si} \cdot \frac{1\ \text{mol Si}}{28.09\ \text{g Si}}$$
$$+ 0.40\ \text{g Cr} \cdot \frac{1\ \text{mol Cr}}{52\ \text{g Cr}} + 0.20\ \text{g Mo} \cdot \frac{1\ \text{mol Mo}}{95.95\ \text{g Mo}} + 97.2\ \text{g Fe}$$

Table 6.7 Interaction parameters of the first order of the solutes in the austenite

Element	C	Mn	Si	Cr	Mn
e_C^i(727 °C)	0.2045	−0.0223	0.1014	−0.0900	−0.0480
ε_C^i(727 °C)	10.91	−5.04	12.24	−18.93	−19.71

$$\cdot \frac{1\ \text{mol Fe}}{55.85\ \text{g Fe}} = 1.8542 \quad (6.121)$$

The molar fractions are:

$$x_C = \frac{0.85\ \text{g C} \cdot \frac{1\ \text{mol C}}{12.01\ \text{g C}}}{D} = 0.0382 \quad (6.122)$$

$$x_{Mn} = \frac{0.85\ \text{g Mn} \cdot \frac{1\ \text{mol Mn}}{54.94\ \text{g Mn}}}{D} = 0.0083 \quad (6.123)$$

$$x_{Si} = \frac{0.50\ \text{g Si} \cdot \frac{1\ \text{mol Si}}{28.09\ \text{g Si}}}{D} = 0.0096 \quad (6.124)$$

$$x_{Cr} = \frac{0.40\ \text{g Cr} \cdot \frac{1\ \text{mol Cr}}{52.00\ \text{g Cr}}}{D} = 0.0041 \quad (6.125)$$

$$x_{Mo} = \frac{0.20\ \text{g Mo} \cdot \frac{1\ \text{mol Mo}}{95.95\ \text{g Mo}}}{D} = 0.0011 \quad (6.126)$$

Using the values of Table 6.7 and the values of the molar fractions, and considering that under these conditions of temperature and composition, the coefficient of activity at infinite dilution of the carbon in the austenite is 14.00, the molar fraction of the carbon, x_C, that solves the Eq. (6.119) is:

$$x_C = 0.0449 \quad (6.127)$$

which corresponds to a value of the carbon concentration:

$$x_C = \frac{\%C \cdot \frac{1\ \text{mol C}}{12.01\ \text{g C}}}{D} = 0.0449 \rightarrow \%C = \frac{0.0449 \cdot D}{\frac{1\ \text{mol C}}{12.01\ \text{g C}}} = \frac{0.0449 \cdot 1.8542}{\frac{1\ \text{mol C}}{12.01\ \text{g C}}}$$
$$= 1.00\% \left(> 0.85\%, \text{the steel is a hypoeutectoid's one}\right) \quad (6.128)$$

keeping constant the number of Mn, Si, Cr, and Mn moles and adjusting the corresponding number of iron moles (solvent) until reaching the same number of moles existing in the composition of the high resistance steel.

Commentaries: it is possible that, apart from the alloying elements, it may also have an influence on the eutectoid carbon quantity, the steel cooling rate, value that displaces the eutectoid point toward the left (lower both carbon content and temperature). However, the greatest difficulty, consequence of the low temperatures and reduced values of the diffusion coefficients, is to reach reliable values for the thermodynamic variables involved in the eutectoid reaction at temperatures of around 700 °C.

End of the exercise.

6.4.1 Wire Drawing of Carbon Steels

Wire drawing of wires of circular section, aximetric reduction of the section by passing through a draw die, generally conic, pulling the wire (Fig. 6.38) is one of the simplest operations of forming by plastic deformation. However, the control still depends on many empirical or semi-empirical factors. As any other process of forming (Fig. 6.39), it is a very complex function of the interaction of several mechanical and metallurgical components despite the formal easiness of the nucleus of the operation, the pass through the draw die. These components or areas of the process are:

- The nature and characteristics of the material to be processed, defined by all its previous history: composition, method of obtaining, operations before the forming, heat treatments, etc. In our case, we are talking about the carbon steel wire.
- The analysis of the deformation zone, the field of study of the continuum mechanics: calculation of stresses, deformations and deformation rates, stresses on the tools, variations of temperature by quasi-adiabatic dissipation of the deformation work, etc., as well as the calculation of the operation limits.

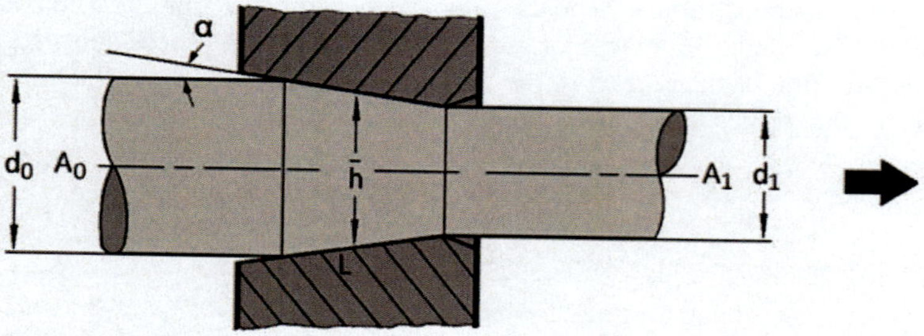

Fig. 6.38 Pulling of the wire

Fig. 6.39 Deformation processing system

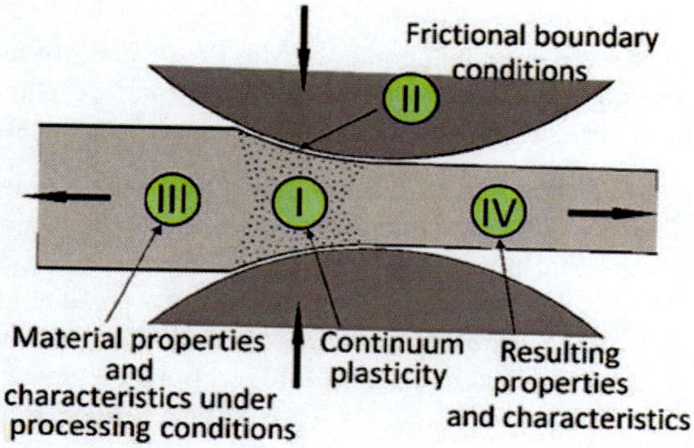

- The conditions of the material-tool interface: lubrication, friction, wear, heat resulted from the dissipation of the friction work, heat transfer through the interface, etc.
- The properties of the formed material, steel wire in this case, resulted from the pass of the initial material through the deformation zones: resistance and ductility, rugosity, resistance to fatigue, resistance to corrosion, electrical resistivity, etc., conditioned by the process of plastic deformation: deformation of the grain and grain boundaries, fibering of the inclusions, structural damage, substructure of dislocations, crystallographic texture, etc.
- The process itself as a mechanical system: from the simplest monoblock wire drawing machine to the analysis of the wire drawing in a series of draw dies disposed in tandem.

Unavoidably, the information presented here of the areas and components previously mentioned will be brief and partial. It is possible to find a wide bibliographic review about the wire drawing in Dieter (1981).

6.4.2 General Scheme of the Wire Drawing Process

This section is aimed at studying the composition, dimensions, and treatment of the wire rod. The diameter of the hot-rolled wire rod (Stelmor process) varies within 5 and 20 millimetere, approximately. Carbon content comprises all the range of steels, from extra-mild steel to eutectoid composition steel or even to slightly "hypereutectoid" steel (see Exercise 6.9). It is not specified and it is not required a minimum content of other elements, although limits are defined for the certain elements as manganese (1.65%), silicon (0.60%), and copper (0.40%) to maintain the specification of carbon steel wire rod. Figure 6.40 shows a general scheme of the wire drawing of steels.

Wire rod from mild steel is wire drawn with the structure resulted from the hot rolling or after the subcritical annealing of recrystallization intermediate between passes (Pero-Sanz et al. 2019).

Wire rods with medium and high carbon contents, especially those used to manufacture high resistance wires, are wire drawn from a pearlitic lamellar structure as

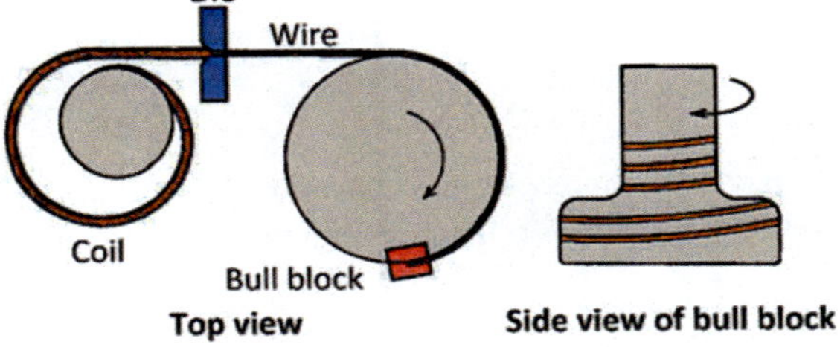

Fig. 6.40 Wire drawing equipment (schematic)

finest, homogeneous and perfect as possible, obtained by isothermal transformation, at around 550 °C, in a lead bath (patenting) or by controlled cooling from the finishing temperature of the rolling process (Stelmor process and similar). It is possible to demonstrate that the interlamellar space of the pearlite, S_0, is inversely proportional to the temperature of the eutectoid transformation of the steel (550 °C, "nose" of the pearlitic transformation: S_0 (μm) $\simeq$ 17/ΔT. If ΔT $\simeq$ 170 = (720–550) °C → $S_0 \simeq$ 0.1 μm) (Pero-Sanz et al. 2017, 2019).

The control of the steel surface state (oxidation, decarburizing), which has influence in the ductility and fatigue limit of the wire rods, is very important during the hot rolling and heat treatment of patenting. Apart from the ductility of the wire rod, the capacity of deformation is strongly influenced by the content, distribution, size, and nature of the inclusions (cleaning of the steel, avoid hard, and brittle inclusions of Al_2O_3, method of refinement and casting of the steel). A good lubrication is fundamental in the wire drawing, especially in the case of high carbon steels. A deficient lubrication can give origin to important local heating, and an extreme situation takes place when re-austenization of a fine peripheral layer transforms during the cooling into martensite and, in this way, the wire can experience cracks and break. The coefficient of friction, with a good solid lubricant (soaps), is in the range 0.01 and 0.05. In the wire drawing, reductions in each pass are typically around 20% and the semi-angle α of the draw plates (draw dies) is 3–10°. The minimum diameters of the wire are around 0.1 mm. The cumulated deformation, without intermediate heat treatments, can be very high: ε = 10 (extra-mild steel) and ε = 4 (eutectoid steel), which is equivalent to reductions in diameter within 85% and 99%, respectively (they are limit deformations in the industrial practice). Deformation rate, at room temperature, is very high, between 10^2 and 10^5 s^{-1}: as the wire rate increases in each pass, it can reach thousands of meters per minute. The wear resistance of the draw dies requires to use sintered WC (widia) in their manufacture, and they are fabricated in diamond in the case of diameters smaller than 0.15 mm. These materials (with very great wear resistance) are characterized by their high Young's modulus and low coefficient of thermal expansion, but also by their great hardness and thermal shock resistance.

6.4.3 *Mechanics of the Wire Drawing*

In an energy balance in the zone of deformation for any forming process involves three components. Per unit of volume:

$$W = W_h + W_f + W_\phi \tag{6.129}$$

W_h is the homogeneous plastic deformation minimum to produce an external change of shape, W_f is the work employed to bear the friction, and W_ϕ is the redundant plastic work, used to produce plastic deformation in the external required change, but that is unavoidable due to the geometry of the operation.

Therefore, it is possible to define the efficiency of the forming operation as:

$$\eta = \frac{W_h}{W} \tag{6.130}$$

and, if it is possible, we will try to optimize this efficiency.

Starting from an elemental analysis (mainly considering the reference of Hosford and Backofen 1964), we will try to delimitate these three components of the wire drawing (and, for that reason, of wire drawing stress).

Minimum term for the change of shape.

The external change of shape searched in the wire drawing is the pure axisymmetric elongation. In absence of friction, if the deformation was carried out homogeneously (as happens in the tensile stress-strain test):

$$W_h = \int_0^{\overline{\varepsilon}_n} \overline{\sigma} \cdot d\overline{\varepsilon} = \overline{\overline{\sigma}} \cdot \overline{\varepsilon}_n \tag{6.131}$$

where:

$$\overline{\varepsilon}_n = \ln\left(\frac{A_0}{A_1}\right) \tag{6.132}$$

is the nominal deformation in the pass, and $\overline{\overline{\sigma}}$ is the average stress of plasticity in each pass.

The work per unit of volume performed by the stretching force, F_h, is:

$$\frac{F_h}{A_1} = \left(\sigma_{\text{wire drawing}}\right)_h = W_h = \overline{\overline{\sigma}} \cdot \overline{\varepsilon}_n \tag{6.133}$$

$$F_h = \overline{\overline{\sigma}} \cdot \overline{\varepsilon}_n \cdot A_1 \tag{6.134}$$

Terms due to the friction.

We assume now that there is also friction between the wire and conic die, where μ is the friction coefficient.

If $\overline{p}$ is the average pressure performed by the wire on the lateral surface, S, of contact:

$$S \cdot \sin\alpha = A_0 - A_1 \tag{6.135}$$

that is calculated from Fig. 6.41, as follows:

$$S_0 = \pi \cdot r_0 \cdot h_0 = \frac{\pi \cdot r_0^2}{\sin\alpha} \tag{6.136}$$

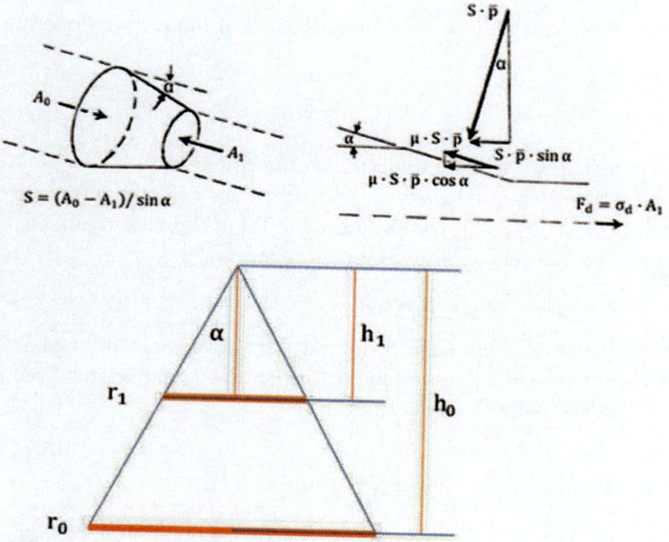

Fig. 6.41 Forces acting on a conical element (upper image). Geometrical considerations (lower image)

where:

$$r_0 = h_0 \cdot \sin \alpha \tag{6.137}$$

$$S_1 = \pi \cdot r_1 \cdot h_1 = \frac{\pi \cdot r_1^2}{\sin \alpha} \tag{6.138}$$

where:

$$r_1 = h_1 \cdot \sin \alpha \tag{6.139}$$

and, finally:

$$S_0 - S_1 = \frac{A_0 - A_1}{\sin \alpha} = S \text{ (surface area of contact with the die)} \tag{6.140}$$

Then, from the equilibrium of forces in the axial direction, Fig. 6.41, we have:

$$\begin{aligned} F_{h+f} &= S \cdot \bar{p} \cdot (\sin \alpha + \mu \cdot \cos \alpha) = S \cdot \bar{p} \cdot \sin \alpha \cdot (1 + \mu \cdot \cot \alpha) \\ &= (A_0 - A_1) \cdot \bar{p} \cdot (1 + \mu \cdot \cot \alpha) \end{aligned} \tag{6.141}$$

But when there is no friction ($\mu = 0$),

$$(F_{h+f})_{\mu=0} = F_h = (A_0 - A_1) \cdot \bar{p} = \bar{\bar{\sigma}} \cdot \bar{\varepsilon}_n \cdot A_1 \quad (6.142)$$

Consequently, it is possible to demonstrate—and deduce considering the criterion of plasticity—that, for the usual values of μ, $\bar{p}(\mu = 0) \simeq \bar{p}(\mu > 0)$ and $\bar{p}(\mu = 0) > \bar{p}(\mu > 0)$, and it is possible to obtain the following equation (where the stretching force is overestimated):

$$F_{h+f} = \bar{\bar{\sigma}} \cdot \bar{\varepsilon}_n \cdot A_1 \cdot (1 + \mu \cdot \cot \alpha) \quad (6.143)$$

$$\left(\sigma_{\text{wire drawing}}\right)_{h+f} = W_{h+f} = \bar{\bar{\sigma}} \cdot \bar{\varepsilon}_n \cdot (1 + \mu \cdot \cot \alpha) = W_h \cdot (1 + \mu \cdot \cot \alpha) \quad (6.144)$$

That is to say,

$$n_f = \frac{1}{1 + \mu \cdot \cot \alpha} \quad (6.145)$$

where n_f is the efficiency of the operation with friction.

Term due to the redundant plastic work. Final equation for the wire drawing stress.

The deformation of the wire takes place in a localized zone, after passing through a converging channel. The trespassing of the plastic border at the entrance of this channel and the exit of the plastic zone involve relative displacements tangential to the material overlapped to those longitudinal, indicate that in the wire drawing there are unnecessary shearing deformation terms for the change of shape by elongation. The redundant deformation is "visible" in the distortion of the orthogonal reticulate net previously marked in the average section of the material. Definitely, the real equivalent deformation exceeds the nominal or apparent one:

$$\bar{\varepsilon}_{real} > \bar{\varepsilon}_n \quad (6.146)$$

and it is possible to define a deformation efficiency:

$$\eta_\phi = \frac{\bar{\varepsilon}_n}{\bar{\varepsilon}_{real}} = \frac{1}{\phi} \quad (6.147)$$

The sense of this efficiency and a method to measure it can be understood with the scheme of Fig. 6.42 (Gil Sevillano 1982).

The calculation of ϕ requires sophisticated methods, but it is possible to empirically demonstrate that:

$$\Delta < 1, \phi = 1 \quad (6.148)$$

$$\Delta \geq 1, \phi = 1 + C \cdot (\Delta - 1) \quad (6.149)$$

Fig. 6.42 Procedure for determining redundant deformation of drawn wire

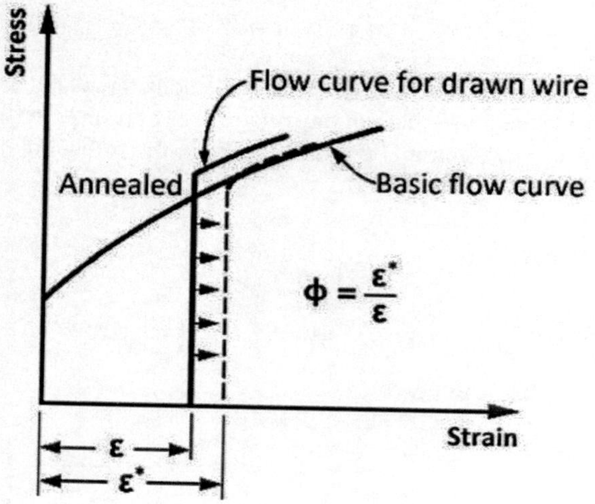

with $C \simeq 0.12$.

The parameter Δ characterizes the zone of deformation, as it is defined as the relation between the average height of the deformation zone and the length of the area of contact tool-material.

$$\Delta = \frac{\bar{h}}{L} \simeq \frac{\alpha}{r} \cdot \left(1 - \sqrt{1 - r}\right)^2 \tag{6.150}$$

where r is the reduction of section:

$$r = \frac{A_0 - A_1}{A_0} \rightarrow 1 - r = \frac{A_1}{A_0} = \frac{d_1^2}{d_0^2} \tag{6.151}$$

Finally, we have:

$$F = \frac{F_h}{\eta} = \frac{F_h}{n_f \cdot \eta_\phi} = \bar{\bar{\sigma}} \cdot \bar{\varepsilon}_n \cdot A_1 \cdot (1 + \mu \cdot \cot \alpha) \cdot [1 + C \cdot (\Delta - 1)] \tag{6.152}$$

Optimization of a wire drawing pass from the efficiency point of view.

$$\frac{1}{\eta} = \frac{\sigma_{\text{wire drawing}}}{\bar{\bar{\sigma}} \cdot \bar{\varepsilon}_n} = \frac{W}{W_h} = (1 + \mu \cdot \cot \alpha) \cdot [1 + C \cdot (\Delta - 1)]$$
$$\simeq 1 + C \cdot (\Delta - 1) + \frac{\mu}{\alpha} \tag{6.153}$$

(a)

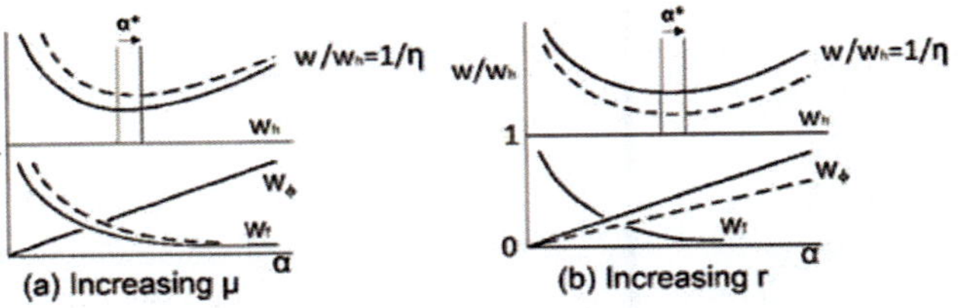

(b)

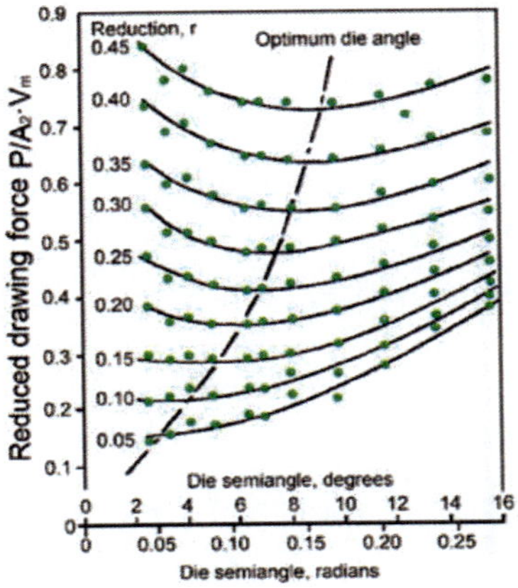

Fig. 6.43 **a** Effects of (**a**) increasing μ, and (**b**) increasing r on the die angle, α*, for maximum deformation efficiency (Note that all terms are relative to a w_h of unity, which accounts for w_ϕ being lowered by increasing r in (**b**)). **b** Variation of pull with die angle

For a given reduction, r, $1 + \mu \cdot \cot\alpha$ decreases with α and $[1 + C \cdot (\Delta - 1)]$ grows, and this explains the existence of an optimum angle in each pass (Fig. 6.43). This way, for the maximum efficiency in conic draw dies:

$$\alpha^* \simeq \frac{\mu \cdot r}{1 - r} \tag{6.154}$$

Exercise 6.10 Demonstrate that the deformation-zone geometry parameter which characterizes the deformation zone and the reductant work is equal to:

$$\Delta = \frac{\bar{h}}{L} = \frac{\alpha}{r} \cdot \left[1 + \sqrt{\frac{A_1}{A_0}}\right]^2 \tag{6.155}$$

Solution starts:

We know that:

$$r = \frac{A_0 - A_1}{A_0} = 1 - \left(\frac{d_1}{d_0}\right)^2 \tag{6.156}$$

and:

$$\bar{h} = \frac{d_0 + d_1}{2} \tag{6.157}$$

The value L is approximately:

$$L \simeq \frac{d_0 - d_1}{\alpha} \tag{6.158}$$

and, thus:

$$\frac{\bar{h}}{L} \simeq \frac{\alpha \cdot (d_0 + d_1)}{d_0 - d_1} \tag{6.159}$$

We multiply and divide by $d_0 + d_1$:

$$\frac{\bar{h}}{L} \simeq \frac{\alpha \cdot (d_0 + d_1)}{d_0 - d_1} \cdot \frac{(d_0 + d_1)}{(d_0 + d_1)} = \frac{\alpha \cdot (d_0 + d_1)^2}{d_0^2 - d_1^2} = \alpha \cdot \frac{d_0^2 + 2 \cdot d_0 \cdot d_1 + d_1^2}{d_0^2 \cdot \left(1 - \frac{d_1^2}{d_0^2}\right)} \tag{6.160}$$

Considering the value of r:

$$\frac{\bar{h}}{L} \simeq \frac{\alpha}{r} \cdot \left[1 + \left(\frac{d_1}{d_0}\right)^2 + 2 \cdot \frac{d_1}{d_0}\right] = \frac{\alpha}{r} \cdot \left(1 + \frac{d_1}{d_0}\right)^2 = \frac{\alpha}{r} \cdot \left(1 + \sqrt{1 - r}\right)^2 \tag{6.161}$$

Δ increases when α increases and reduction decreases. By the contrary, Δ decreases when α decreases and reduction increases. For example:

- If $\alpha = 5°$ and $r = 0.2$:

$$\Delta = \frac{\bar{h}}{L} = \frac{5}{0.5} \cdot \frac{\pi}{180} \cdot \left(1 + \sqrt{1 - 0.2}\right)^2 = 0.63 \tag{6.162}$$

Reduction is (when $C = 1.2$ and $\mu = 0.02$, see Eq. 6.148)

$$\frac{1}{\eta} \simeq (1 + \mu \cdot \cot \alpha) \rightarrow \eta = \frac{1}{(1 + \mu \cdot \cot \alpha)} \cdot 100 = \frac{1}{(1 + 0.02 \cdot \cot 5°)} \cdot 100 = 81.40\% \tag{6.163}$$

– If $\alpha = 10°$ and $r = 0.5$:

$$\Delta = \frac{\bar{h}}{L} = \frac{10}{0.5} \cdot \frac{\pi}{180} \cdot \left(1 + \sqrt{1 - 0.5}\right)^2 = 1.02 \tag{6.164}$$

Reduction is (when $C = 1.2$ and $\mu = 0.02$)

$$\begin{aligned}\frac{1}{\eta} &\simeq (1 + \mu \cdot \cot\alpha) \cdot [1 + C \cdot (\Delta - 1)] \rightarrow \eta \\ &= \frac{1}{(1 + \mu \cdot \cot\alpha) \cdot [1 + C \cdot (\Delta - 1)]} \cdot 100 \\ &= \frac{1}{(1 + 0.02 \cdot \cot 10°) \cdot [1 + 1.2 \cdot (1.02 - 1)]} \cdot 100 = 87.71\%\end{aligned} \tag{6.165}$$

– If $\alpha = 10°$ and $r = 0.2$:

$$\Delta = \frac{\bar{h}}{L} = \frac{10}{0.2} \cdot \frac{\pi}{180} \cdot \left(1 + \sqrt{1 - 0.2}\right)^2 = 3.14 \tag{6.166}$$

Reduction is (when $C = 1.2$ and $\mu = 0.02$)

$$\begin{aligned}\frac{1}{\eta} &\simeq (1 + \mu \cdot \cot\alpha) \cdot [1 + C \cdot (\Delta - 1)] \rightarrow \eta \\ &= \frac{1}{(1 + \mu \cdot \cot\alpha) \cdot [1 + C \cdot (\Delta - 1)]} \cdot 100 \\ &= \frac{1}{(1 + 0.02 \cdot \cot 10°) \cdot [1 + 1.2 \cdot (3.14 - 1)]} \cdot 100 = 25.17\%\end{aligned} \tag{6.167}$$

For a given reduction, r, $(1 + \mu \cdot \cot\alpha)$ decreases with α and $[1 + C \cdot (\Delta - 1)]$ grows. Thus, there is an optimal angle for each pass (see Fig. 6.43, Gil Sevillano 1982). The maximum efficiency in conic draw dies is:

$$\alpha^* = \frac{\mu \cdot r}{1 - r} \tag{6.168}$$

where μ is the coefficient of friction.

We now shall examine the limit of drawability for steady wire drawing. The equation:

$$\sigma_{ax} = \phi \cdot \bar{\sigma} \cdot \ln\left(\frac{A_0}{A_1}\right) \cdot (1 + \mu) \tag{6.169}$$

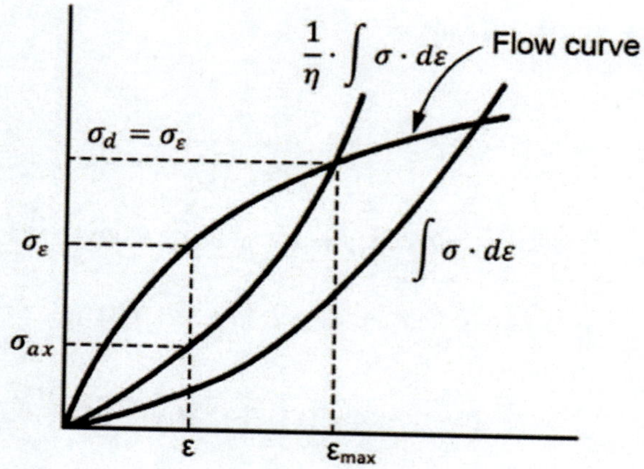

Fig. 6.44 Development of limit on drawability

can be expressed most simply by:

$$\sigma_{ax} = \frac{1}{\eta}\int\overline{\sigma}\cdot d\varepsilon \tag{6.170}$$

$\eta = W_h/W$, the efficiency of the deformation, Fig. 6.44 (Gil Sevillano 1982) plots the flow curve of the material $\int\overline{\sigma}\cdot d\varepsilon$ for an ideal process without friction and reductant work, and $(1/\eta)\int\overline{\sigma}\cdot d\varepsilon$ for the actual drawing process. For a given strain $\ln(A_0/A_1)$ produced by the die, the values of σ_{ax} and flow stress σ_ε are indicated. The weakest link is the wire which has exited from the die. On being deformed through the die the material is strain hardened. For a severely strain hardened material (steel), the formation of a neck is quickly followed by fracture. Therefore, the drawing limit is reached when $\sigma_d = \sigma_\varepsilon$ as shown in Fig. 6.44. If the material follows a power-law hardening relationship:

$$\sigma = k\cdot\varepsilon^n \tag{6.171}$$

(Ludwik-Hollomon model), the equation:

$$\sigma_{ax} = \frac{1}{\eta}\int\overline{\sigma}\cdot d\varepsilon \tag{6.172}$$

becomes:

$$\sigma_d = \frac{k\cdot\varepsilon^{n+1}}{\eta\cdot(n+1)} = \frac{\sigma_\varepsilon\cdot\varepsilon}{\eta\cdot(n+1)} \tag{6.173}$$

Substituting the criterion for the maximum drawing strain in a single pass, that is $\sigma_d = \sigma_\varepsilon$:

$$\varepsilon_{max} = \eta \cdot (n + 1) \tag{6.174}$$

and since:

$$\varepsilon = -\ln\left(\frac{A_0}{A_1}\right) \tag{6.175}$$

thus:

$$\left(\frac{A_0}{A_1}\right)_{max} = \exp[-\eta \cdot (n + 1)] \tag{6.176}$$

and, by definition the reduction is:

$$r = 1 - \left(\frac{A_0}{A_1}\right) \rightarrow r_{max} = 1 - \exp[-\eta \cdot (n + 1)] \tag{6.177}$$

For wire subjected to repeated reductions through a series of dies, n will decrease toward 0 and the allowable reduction will decrease. A comparison of the maximum strain in drawing with that in stretching (tension) shows the important role played by the strain hardening component:

$$\frac{\varepsilon(\text{drawing})}{\varepsilon(\text{tension})} = \frac{\eta \cdot (n + 1)}{n} \tag{6.178}$$

Typically, the limited strain in drawing is at least twice that in pure tension.

End of the exercise.

Exercise 6.11 Demonstrate that a pass of wire drawing, for a strong hardened material, $n \rightarrow 0$, and an operation efficiency $\eta \simeq 0.75$, the maximum reduction available is approximately 0.5.

Solution starts:

From the equation:

$$\bar{\varepsilon}_{max} = \eta \cdot (n + 1) \tag{6.179}$$

if $n \rightarrow 0$, thus:

$$\bar{\varepsilon}_{max} = \eta = 0.75 \rightarrow \ln\left(\frac{A_0}{A_1}\right) = 0.75 \rightarrow \frac{A_0}{A_1} = 2.1 = \frac{1}{1 - r} \rightarrow r = 0.52(\simeq 0.5) \tag{6.180}$$

Note: For that reason, it is convenient to make after several passes of wire drawing, $n \to 0$, a recrystallization annealing to recover the strain hardening coefficient value is necessary.

End of the exercise.

Exercise 6.12 Demonstrate that the limit strain in drawing is at least twice that in pure tension ($\eta = 0.75$; $n = 0.2$; $\alpha = 10°$).

Solution starts:

Taking into account the formula:

$$\frac{\varepsilon_d(\text{drawing})}{\varepsilon_t(\text{tension})} = \frac{\eta \cdot (n+1)}{n} = \frac{0.75 \cdot (0.2+1)}{0.2} = 4.5 > 2 \quad (6.181)$$

Note: The maximum reduction each pass is increased:

(a) Increasing the yield (decreasing the friction, by optimizing the profile of the draw die, Δ).
(b) Increasing the coefficient n of hardening by deformation.

End of the exercise.

6.4.4 Ductility. Internal Fractures. Cavitation (Structural Damage)

A wire drawing pass can fail even when the reduction imposed was smaller than the maximum admissible deduced in the previous section. The reason is that under the stresses and deformations generated in the zone of plastic deformation, it is possible to find internal fractures (cavitation) that weaken the section and will produce the breaking at the exit of the draw die under the wire drawing stress, or the wire will be damaged and the mechanical properties will not be sufficient (Gutiérrez-Stampa et al. 1983; Fernández et al. 1986). These fractures, for instance, those identified in a pearlitic steel wire of 5 mm in diameter, have a pointed shape (as the point of an arrow) and receive different names (cuppy-fracture, arrowhead cracking, chevron fractures, etc.) and are usually placed in the axis of the wire. These fractures nucleate in defects (pores, inclusions, segregations, etc.) located in the vicinity of the axis of the wire: nucleation occurs in presence of triaxial states of tension (hydrostatic tension) associated to values greater than 1 for the deformation zone geometry parameter or $\Delta = \bar{h}/L$ (great α angles and small reductions r in each pass). The fracture grows due to shearing and produces the conic shape that is characteristic of the "fracture surface." The understanding of this phenomenon requires the knowledge, even basic, of the so-called "ductile fracture."

In the ductile fracture, it is possible to distinguish three stages:

(1) Nucleation of the fracture nuclei (cracks, voids).
(2) Growth of the voids.
(3) Coalescence of the voids until the final fracture.

We describe now these three stages:

(1) *Nucleation of the ductile fracture* (nucleation of the voids): the centers of fracture are mainly non-metallic inclusions and second phases of different plasticity than that of the matrix. There are also, obviously, preexisting pores, segregations, micro-cracks, weak interfaces, etc. Very small deformations are sufficient to create the initial holes (Fig. 6.45).
(2) *Growth of the holes*: the volume fraction of inclusions, f_v, and the average distance, l, between them characterize the number of "nucleated" holes, although the change in the shape strongly depends, as it is possible to deduce, on the hydrostatic state of stresses; negative pressures increase the porosity; positive pressures decrease the porosity. The relation between f_v, l, and r is:

$$f_v \simeq f_A \simeq f_p \simeq \left(\frac{2 \cdot r}{1 + 2 \cdot r}\right)^3 \tag{6.182}$$

l being the average tangent distance between randomly distributed inclusions. The fractions of points (f_p), area (f_A), and volume (f_v) occupied by a second phase (inclusion) theoretically coincide (Sellars, Quantitative Metallurgy).

(3) *Coalescence of voids. Deformation until fracture.* Voids grow and elongate preferentially in the principal direction of the plastic flow until a moment when the ligaments-fibers of the matrix that separates them collapse due to "breaking"

Fig. 6.45 Structural damage due to decohesion in the matrix-inclusion interface and inclusion-inclusion of a mixed particle of Al_2O_3-MnS

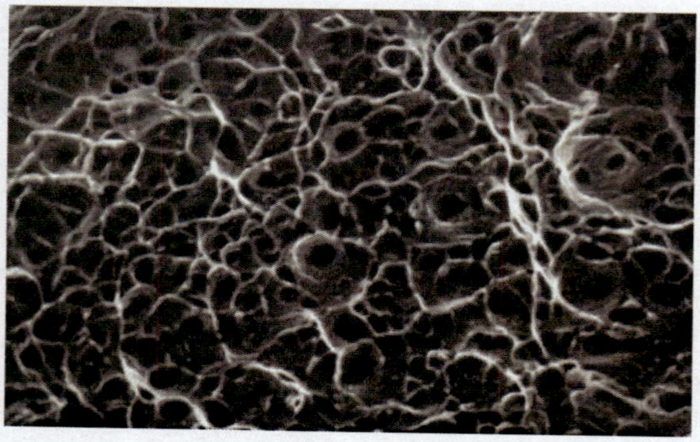

Fig. 6.46 Voids resulted from ductile fracture

(100% of plastic deformation) and motivates the microscopic fracture in whose surface it is possible to appreciate images of the voids (domes or "dimples") that usually contain in their bottom the particles that gave origin to this holes (Fig. 6.46)

Without using a complicated model of ductile fracture, we consider the scheme of the enclosed Fig. 6.47 (Gil Sevillano 1982). After the nucleation there are n holes in the section A_0, separated a distance l:

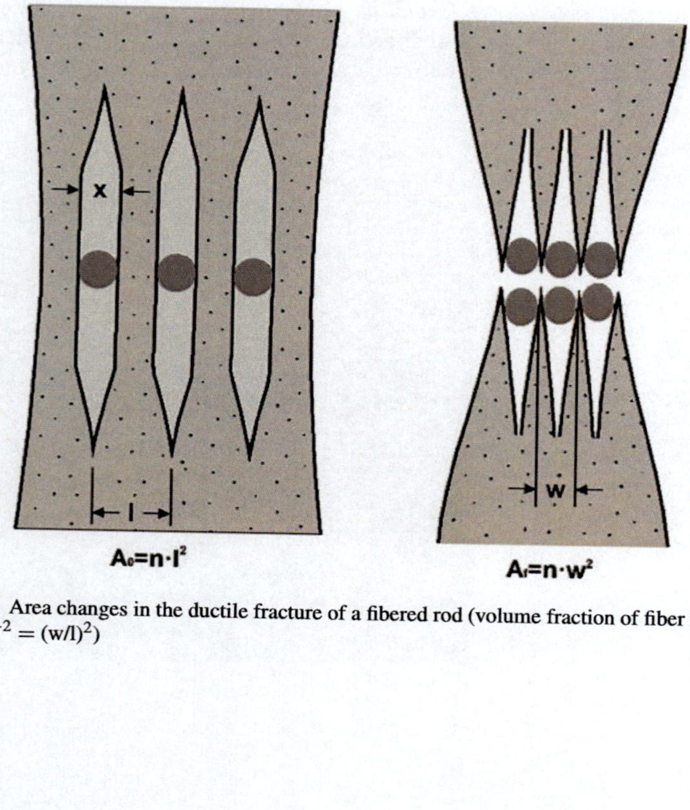

Fig. 6.47 Area changes in the ductile fracture of a fibered rod (volume fraction of fiber is $f_v = n \cdot w^2 \cdot n^{-1} \cdot l^{-2} = (w/l)^2$)

$$A_0 = n \cdot l^2 \tag{6.183}$$

If the holes have an average width w in the cross direction to the maximum applied stress (main component), when the collapse by fracture of the ligaments, the apparent section will be:

$$A_f \simeq n \cdot w^2 \tag{6.184}$$

And the deformation when breaking is:

$$\varepsilon_f \simeq 2 \cdot \ln\left(\frac{l}{w}\right) \tag{6.185}$$

Although, we have:

$$f_v = \frac{n \cdot w^2}{n \cdot l^2} = \left(\frac{w}{l}\right)^2 \tag{6.186}$$

where f_v is the fraction in volume of holes that, in a non-pre-deformed material, will coincide with that of particles. Definitely:

$$\varepsilon_f \simeq \ln\left(\frac{1}{f_v}\right) \tag{6.187}$$

that is the typical relation between f_v and ε_f.

Another equivalent criterion, although more accurate, is:

$$\varepsilon_f \cdot \overline{\sigma}_t = \text{constant} \tag{6.188}$$

$\overline{\sigma}_t$ is the component of stress that is transversal to the direction of maximum extension. The constant, to be determined, will depend on the operation variables (temperature, deformation rate, fraction in volume of inclusions, and orientation of the inclusions). For a certain fixed state of stresses, $\overline{\sigma}_t$ and σ_1 are related to the yield criterion (Von Mises):

$$2 \cdot \overline{\sigma}^2 = 2 \cdot (\sigma_1 - \overline{\sigma}_t)^2 \rightarrow \sigma_1 = \overline{\sigma} + \overline{\sigma}_t \tag{6.189}$$

being $\overline{\sigma}_t \simeq \sigma_t$ (hydrostatic pressure in the axis of the wire $p = -\sigma_t$). In other words, to avoid the ductile fracture of the wire during the wire drawing, it is convenient, as it was previously detailed, to have low values of Δ, avoid low reductions r and having clean steel ($f_v < 0.5‰$) with low fraction in volume of inclusions, specially in the axis of the wire.

6.4.5 Relation Structure-Mechanical Properties

In mild steels—ferritic structure,—for strains $\varepsilon > 0.5$, the mechanical deformation grows linearly with the true strain, with a slope of approximately 150 MPa per unit of deformation:

$$\mathrm{UTS}(\mathrm{Fe}_{\alpha})_{\varepsilon>0.5} = 470 + 150 \cdot \varepsilon(\mathrm{MPa}) \tag{6.190}$$

On the other hand, in eutectoid steels, for deformations $\varepsilon > 0.5$, they have initial resistances three times greater than that of the mild steel (Pero-Sanz et al. 2017, Chapter VIII) and it grows with the deformation with an initial slope twice than that. However, what is more noticeable is that the slope of the curve grows with the deformation with an exponential law:

$$\mathrm{UTS}(\text{lamellar pearlite})_{\varepsilon>0.5} \propto \exp(\varepsilon/4)(\mathrm{MPa}) \tag{6.191}$$

$$\mathrm{UTS}(\text{fine lamellar pearlite})_{\varepsilon>0.5} \simeq 70 + 1330 \cdot \exp(\varepsilon/4)(\mathrm{MPa}) \tag{6.192}$$

This exponential growth of the resistance of the wire drawn eutectoid steels makes them to reach, after strong reductions, that is to say, for small diameters of the wire, resistances of around 3000 MPa, and in steels with low nitrogen and few inclusions, ultimate tensile strengths of 4000 MPa (around 0.10 mm in diameter), which situate this steel in values close to their ideal ultimate tensile strength. On another note, ductility (reduction of section in tension) exhibits a singular behavior: it initially grows until strains of $\varepsilon \simeq 1.5$, and then it decreases continuously (on the contrary to wire drawn ferritic steels). This behavior is attributed to the progressive alignment of the cementite lamellas with respect to the axis of the wire as a consequence of the deformation.

The resistance (ultimate tensile strength) and ductility of hypoeutectoid steels (0.02% < %C < 0.8%) are intermediate between those of the mild steel and those of the eutectoid steel. This way ($\varepsilon > 0.5$):

$$\mathrm{UTS}(\varepsilon) = \mathrm{f}_{\mathrm{v}}^{\alpha} \cdot \mathrm{UTS}^{\alpha}(\varepsilon) + \left(1 - \mathrm{f}_{\mathrm{v}}^{\alpha}\right) \cdot \mathrm{UTS}^{\mathrm{P}}(\varepsilon) \tag{6.193}$$

In a general manner, grains of a polycrystalline aggregate deform, grosso modo, in a uniformly way, that is to say, similarly to the polycrystalline aggregate as a whole. The wire drawing of carbon steels constitutes, however, a flagrant exception to this rule.

Deformation due to wire drawing—axisymmetric elongation—in BCC metals (as ferrite) produces a monocomponent texture <1 1 0> : all the crystals have a crystallographic direction <1 1 0> parallel to the wire drawing axis. A BCC monocrystal [1 1 0] subjected to wiring, deforms by plane elongation, without contraction in the direction $[1\ \bar{1}\ 0]$, perpendicular to the main deformation. Constricted to deform axis-symmetrically ($\varepsilon_1 > 0$; $\varepsilon_2 = \varepsilon_3 < 0$), the work per unit of volume and unit of deformation is 1.5 times greater than the work of plane deformation ($\varepsilon_1 > 0$; $\varepsilon_3 = -\varepsilon_1$; $\varepsilon_2 = 0$), Fig. 6.7. This difference is sufficient to achieve that the grains of a BCC polycrystal (ferrite/pearlite), strongly deformed (wire drawn), tend to deform by plane elongation and they curve to maintain the compatibility of deformations (those of each crystal are equivalents to those of the polycrystalline aggregate). This is the reason for the peculiar appearance of the cross section of the carbon steel wires (Fig. 6.48, Gil Sevillano 1982, Leslie 1982).

In pearlitic steel, the ultimate tensile strength is directly proportional (and inversely proportional to the square root of the interlamellar distance) to $S_0^{-1/2}$ (Pero-Sanz et al. 2017). As:

$$\varepsilon = 2 \cdot \ln\left(\frac{D_0}{D}\right) \simeq 2 \cdot \ln\left(\frac{S_0}{S}\right) = -2 \cdot \ln\left(\frac{S}{S_0}\right) \quad (6.194)$$

The diameter of the wire, D, is proportional to the interlamellar distance, S.

$$\frac{S}{S_0} \propto \exp\left(-\frac{\varepsilon}{2}\right) \quad (6.195)$$

As:

$$\sigma(\varepsilon) \propto S^{-\frac{1}{2}} \propto \left[S_0 \cdot \exp\left(-\frac{\varepsilon}{2}\right)\right]^{-\frac{1}{2}} \propto \frac{1}{S_0^{\frac{1}{2}}} \cdot \exp\left(\frac{\varepsilon}{4}\right) \quad (6.196)$$

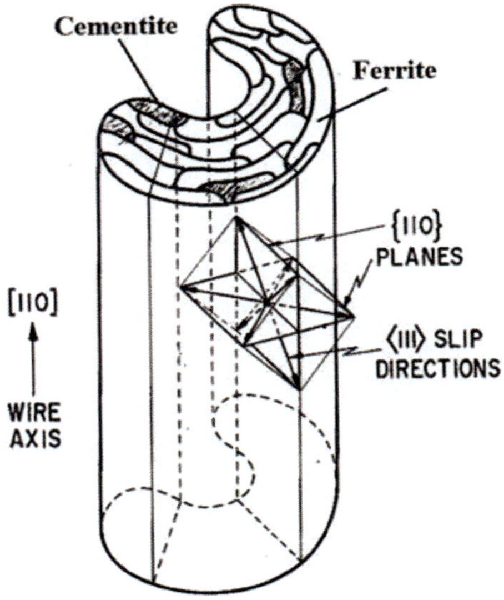

Fig. 6.48 Schematic representation of grain and cell shape in (1 1 0) textured bcc wire

similar to the equation previously indicated and that also predicts the requirement of starting from very fine pearlite (around 0.1 μm) obtained by patenting or controlled cooling from the austenitic state to obtain wires with the same resistance.

Exercise 6.13 Plot a curve of the ratio of draw stress to flow stress versus reduction for $B = \mu \cdot \cot \alpha = 0$, 1 *and* 2, $\alpha = 10°$.

Solution starts:

Applying the uniform deformation energy method:

$$\delta \cdot W = P \cdot \delta \cdot L = \overline{\sigma} \cdot A \cdot \delta \cdot L \quad (6.197)$$

where $\overline{\sigma}$ is the mean stress.

The deformation energy per unit volume is:

$$\frac{\delta \cdot W}{V} = \frac{\overline{\sigma} \cdot A}{A} \cdot \frac{dL}{L} \quad (6.198)$$

The plastic work of deformation per unit volume is:

$$U_P = \overline{\sigma} \int d\varepsilon = \overline{\sigma} \int \frac{dLL}{=} \overline{\sigma} \cdot \ln\left(\frac{L_B}{L_A}\right) \quad (6.199)$$

If we apply this to draw a cylindrical wire from area A_A down to area A_B:

$$W = P_B \cdot L_B = U_P \cdot V = A_B \cdot L_B \cdot \overline{\sigma} \cdot \ln\left(\frac{L_B}{L_A}\right) \quad (6.200)$$

$$\frac{P_B}{A_B} = \sigma_{ax}(\text{axysimetric stress}) = \overline{\sigma} \cdot \ln\left(\frac{A_A}{A_B}\right) = \overline{\sigma} \cdot \ln\left(\frac{1}{1-r}\right) \quad (6.201)$$

– No friction ($\mu = 0$):

$$\frac{\sigma_{ax}}{\overline{\sigma}} = \ln\left(\frac{1}{1-r}\right) \quad (6.202)$$

We represent $\boldsymbol{\sigma_{ax}/\overline{\sigma}}$ as a function of **r** (B = 0) (Fig. 6.49):

– Friction:

$$\frac{\sigma_{ax}}{\overline{\sigma}} = \ln\left(\frac{1}{1-r}\right) \cdot (1 + B) \quad (6.203)$$

We represent $\sigma_{ax}/\overline{\sigma}$ as a function of r when B = 1 (Fig. 6.50).
We represent $\sigma_{ax}/\overline{\sigma}$ as a function of r when B = 2 (Fig. 6.51).

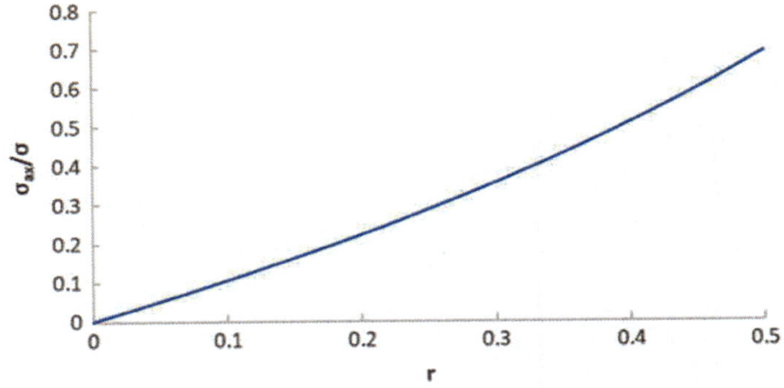

Fig. 6.49 $\sigma_{ax}/\overline{\sigma}$ as a function of r (B = 0)

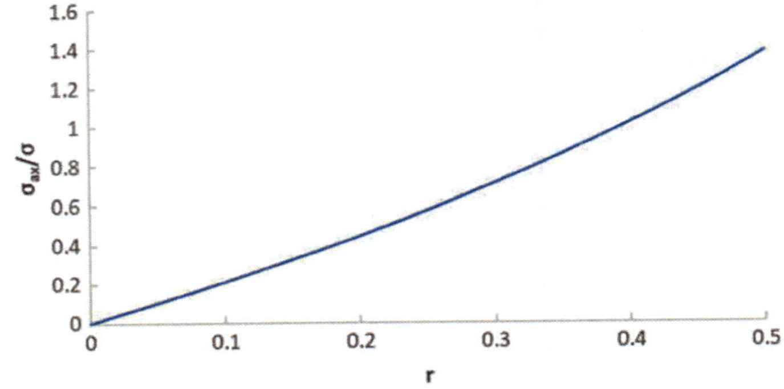

Fig. 6.50 $\sigma_{ax}/\overline{\sigma}$ as a function of r when B = 1

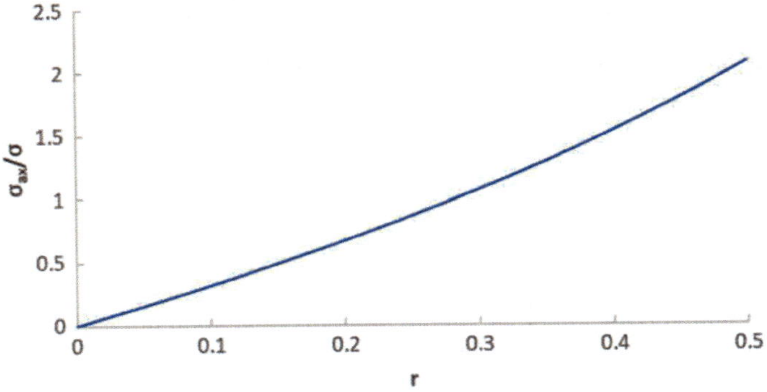

Fig. 6.51 $\sigma_{ax}/\overline{\sigma}$ as a function of r when B = 2

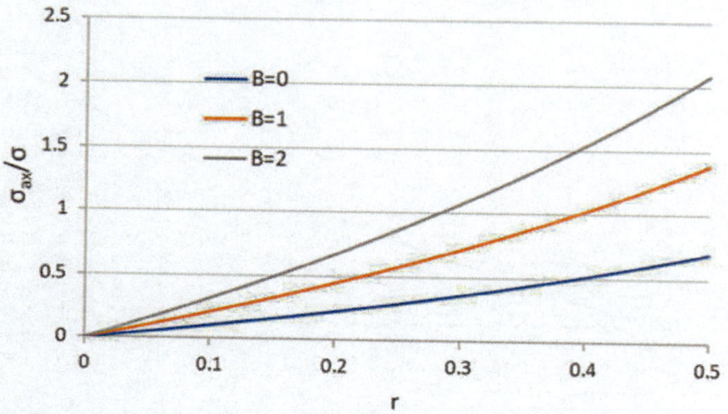

Fig. 6.52 $\sigma_{ax}/\overline{\sigma}$ as a function of r for different values of B

For the same reduction, the value of the stress grows with the friction between the draw die and the wire. We plot in Fig. 6.52 $\sigma_{ax}/\overline{\sigma}$ as a function of r for the different values of μ considered in the exercise.

End of the exercise.

Exercise 6.14 Plot a curve of the ratio of draw stress to flow stress versus semiangle die for the conditions $\mu = 0.05$, $r = 0.4$.

Solution starts:

As the value of B is calculated as follows:

$$B = \mu \cdot \cot \alpha \tag{6.204}$$

and as we know the value of μ:

$$B = 0.05 \cdot \cot \alpha \tag{6.205}$$

Considering the Eq. 6.203:

$$\frac{\sigma_{ax}}{\overline{\sigma}} = \ln\left(\frac{1}{1-r}\right) \cdot (1 + B) \tag{6.206}$$

we have:

$$\frac{\sigma_{ax}}{\overline{\sigma}} = \ln\left(\frac{1}{1-0.4}\right) \cdot (1 + 0.05 \cdot \cot \alpha) \tag{6.207}$$

We represent $\sigma_{ax}/\overline{\sigma}$ as a function of α (Fig. 6.53).
From angles of around 5°, the drawing stress remains more or less constant.

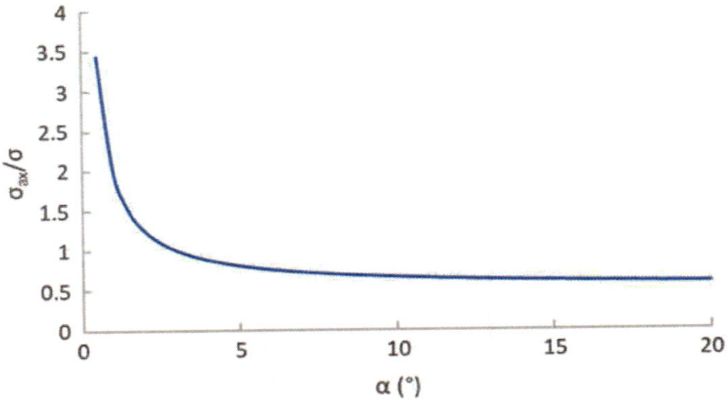

Fig. 6.53 $\sigma_{ax}/\bar{\sigma}$ as a function of α

End of the exercise.

Exercise 6.15 Show that the strain rate of wire been pulled through a die can be expressed by:

$$\dot{\varepsilon} = \frac{4 \cdot D \cdot \tan\alpha}{D_0^2 \cdot (1 - r)} \cdot \frac{dx}{dt} \tag{6.208}$$

What is the maximum strain rate for a 1 mm diameter wire pulled at 600 m/min (10000 mm/s) through a 10° semiangle die? Data: r = 0.2

Solution starts:

We know that:

$$\varepsilon = \ln\left(\frac{1}{1 - r}\right) \rightarrow -\varepsilon = \ln(1 - r) \tag{6.209}$$

If we derivate, we obtain:

$$-\frac{d\varepsilon}{dt} = -\dot{\varepsilon} = \frac{-dr/dt}{1 - r} \tag{6.210}$$

Since:

$$D = D_0 - 2 \cdot \tan\alpha \rightarrow D_0 - D = 2 \cdot \tan\alpha \tag{6.211}$$

then, as we know how the reduction, r, is determined, we calculate dr/dt:

$$r = 1 - \frac{D^2}{D_0^2} \rightarrow \frac{dr}{dt} = \frac{-2 \cdot D \cdot (dD/dt)}{D_0^2} = \frac{-4 \cdot D \cdot \tan\alpha \cdot (dx/dt)}{D_0^2} \tag{6.212}$$

And considering Eq. (6.210), we obtain:

$$\dot{\varepsilon} = \frac{-dr/dt}{1-r} = \frac{-\left[\frac{-4\cdot D\cdot \tan\alpha\cdot(dx/dt)}{D_0^2}\right]}{1-r} = \frac{4\cdot D\cdot \tan\alpha\cdot(dx/dt)}{D_0^2\cdot(1-r)} \tag{6.213}$$

If we replace the variables by their corresponding values, we can calculate the value of the strain rate:

$$\dot{\varepsilon} = \frac{4\cdot D\cdot \tan\alpha\cdot(dx/dt)}{\left(-\frac{D^2}{r-1}\right)\cdot(1-r)} = \frac{4\cdot 1\ \text{mm}\cdot \tan 10^\circ\cdot 10000\ \text{mm/s}}{\left[-\frac{(1\text{mm})^2}{0.2-1}\right]\cdot(1-0.2)} = 7.05\cdot 10^3\ \text{s}^{-1} \tag{6.214}$$

End of the exercise.

6.5 Coatings

Coated steel products represent the technological answer of the steelmaking industry to the problem of corrosion, and to all those situations related to the surface engineering. Sobolev et al. (2004), Puértolas et al. (2010) and Vázquez (2000) provide a detailed knowledge of the protecting layers developed by interaction or reaction of the substrate that must be protected with gaseous or aqueous environments (natural layers), or causing this interaction or reaction by precipitation or growth of materials in an artificial way (artificial layers) via different technologies of surface finishing.

Corrosion is the electrochemical process of materials' degradation. Generally, it is a slow process and causes weight losses in the part, but also leads to the mechanical properties' deterioration and, on some occasions, surface changes. It is possible to distinguish the following mechanisms of corrosion in steels:

- Uniform: all the part is uniformly corroded, and it is hardly possible to notice the corrosion in the surface of the material.
- Localized: as in the case of steel in contact with saltwater, which is performed by means of the classical pit of the material that finally acquires a very rough surface.
- Intergranular: very usual in the chemical industry. This mechanism is based on the propagation of corrosion through the grain boundaries with severe consequences because there is neither alteration of the surface of the metal nor any other visual evidence that could indicate corrosion problems (intergranular cracking, pit corrosion of the stainless steels, etc.).

Even though there is not an infallible and durable technology that could completely eradicate the corrosion, there are several mechanisms of surface protection that can ensure durable protection of steel. It is possible to mention three types of surface protection:

A. Metallic coatings, with zinc, tin, aluminum-zinc alloys, chromium-chromium oxide compounds, aluminum alloys, or chromium alloys.
B. Organic coatings, by means of organic polymers (varnishes and organic paints).
C. Non-metallic coatings, using enamels, inorganic paints (ZnO, TiO_2) or chromium oxide, phosphates, and metallic silicate paints.

One of the most used metallic protections of the steel is that of zinc, which (due to the series of reduction potentials) is more electronegative than the iron and protects the base metal by means of its dissolution or acting as anode in the electrochemical system Zn-steel (cathodic protection).

Industrially, it is possible to electrolytically deposit zinc on the surface of the steel (zinc plating lines) or it is also possible to cover steel by immersion of the steel strip in a bath of liquid zinc (galvanizing lines):

- In the zinc plating process of the steel sheet, the cold-rolled, annealed, and tempered coil is used as starting material (Fig. 6.54). It is important to achieve a good adherence between the zinc layer and the base steel for which it is necessary an excellent surface cleaning to ensure that there are not particles of oxide or oil in contact with the steel.
- Galvanizing lines of steel sheet use cold-rolled coil coming from the tandem mill or the four-high reversing mill.

The electrolytic pickling of the zinc plating line previous to the electrodeposition of the zinc is replaced by thermal pickling whose aim is, the same as the electrolytic, preparing and activating the steel for optimal adhesion of the zinc (Fig. 6.55).

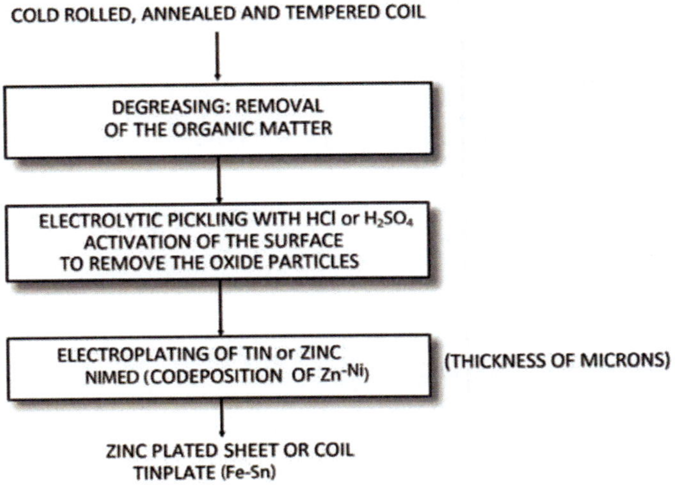

Fig. 6.54 Scheme of the surface protection of a steel strip with electroplated zinc and tin

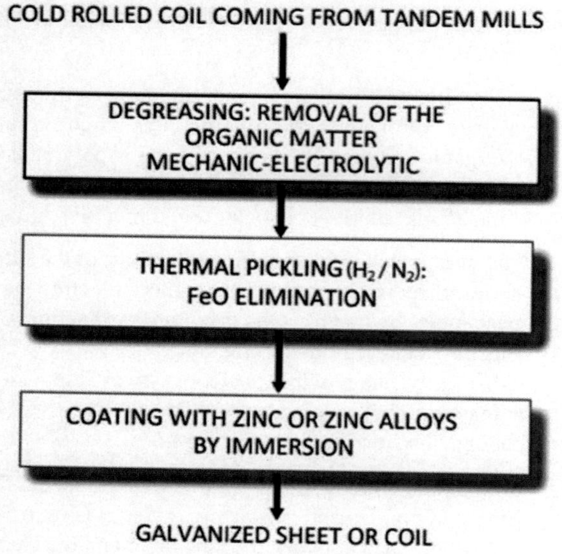

Fig. 6.55 Scheme of the galvanizing process of a steel coil

One of the advantages of the zinc-plated sheet is the possibility of the differential coating of the steel (with different thickness) and allowing the protection of the steel sheet either in one or in both faces of the strip. Equally, the thickness that can be electrolytically applied is smaller and with more strict tolerance than that obtained by immersion. By electrolysis, it is possible to apply zinc layers with 100 g·m^{-2} ($\simeq$15 μm), while by galvanizing the common thickness is around 500 g·m^{-2} ($\simeq$500 μm). However, the surface characteristics of the electrolytic deposits are typical from high purity zinc while in the case of galvanized steel there is a zinc layer and another intermetallic layer in contact with the base steel (Fig. 6.56). Even though the adherence of the electrolytic layers to the steel substrate is smaller than that obtained with the galvanizing: there is in this second case intermediate intermetallic compounds that allow a certain degree of alloy with the base steel (Fig. 6.56). On the contrary, the plasticity and levels of deep drawing obtained by the zinc plated layers are greater than those obtained by galvanizing.

One of the advantages of the galvanizing process is the covering with different zinc alloys, from which we can mention:

- Galfan: 95% Zn-5% Al.
- Lavegal: 70% Zn-30% Al.
- Galvalume and Algafort: 45% Zn-55% Al.
- Galvanneal: 90% Zn-10% Fe.

The Nimed coating was developed (by codeposition of a layer of this metal around 11% Ni) to improve the corrosion resistance of the zinc electrolytic deposits.

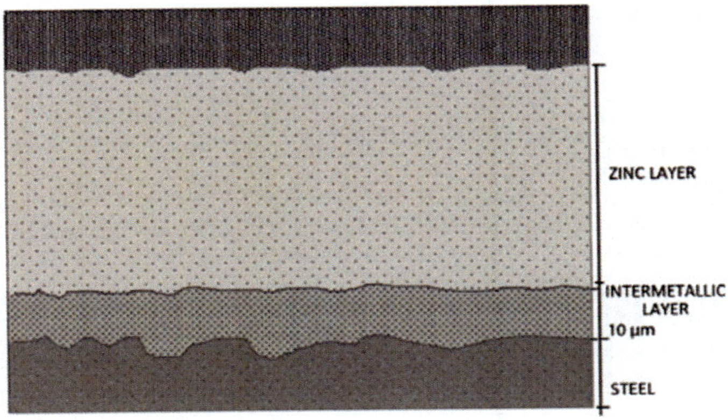

Fig. 6.56 Surface structures of the coatings obtained by hot immersion in zinc bath

The galvanizing of steelmaking products is not only circumscribed to the steel strip, but there are also many other finished products that are susceptible of being protected by immersion in a bath of zinc.

Grey coatings or dark zones can appear in the galvanizing of steel parts that are usually thicker than in the normal situation and that have a lower adherence. The Sandelin curve (Fig. 6.57) shows the two intervals where the hot galvanizing gives good results, that is to say, shiny and with good adherence:

- Interval I: 0.030–0.040% Si.
- Interval II: 0.150–0.250% Si.

Thus, standards stablished for steels that can be galvanized define two classes of steels (Table 6.8).

From the ironmaking point of view, manufacturing steel with silicon contents in the range 0.15–0.25% is much easier than in the case of those with 0.03–0.04% Si. Even though this idea is not sufficiently corroborated, it seems that to hot galvanize, if low silicon (%Si $\leq$ 0.03–0.04) and limited phosphorus content (first interval of the

Fig. 6.57 Sandelin curve

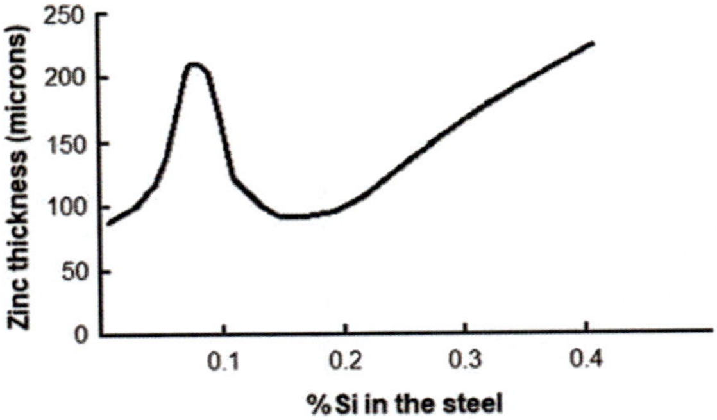

Table 6.8 Classes of steel to galvanize

Class	%Si	%Si + 2.5%P
I	≤0.03	≤0.09
II	≤0.09	≤0.11

Sandelin curve) are used, the quality and the market of the products of galvanized steel are improved with respect to the steels used of the second interval of the curve (0.15–0.25% Si).

It is nowadays more difficult to produce low silicon steel than silicon-killed steel, although two decades ago this was not the situation. Most of the liquid iron is formed by means of continuous casting, which implies deoxidizing the steel with aluminum or silicon. Deoxidizing with aluminum has the following problems:

- Cause the obstruction of the nozzles of the tundish for products of small dimensions (for instance, billets).
- As aluminum is an active metal, it attacks the matrix constituent of the alumina refractories and incorporates silicon into the liquid iron.
- It has operating costs slightly greater than in the case of the ferrosilicon (the least expensive deoxidizing element is the Fe-75% Si).

The use of tin as anodic protecting agent of the steel is produced as a consequence of the slightly basic environment that appears in the canned food of vegetables, fish, and carbonated beverages. Despite the standard potential of reduction of the tin in acid medium is very higher than that of the iron, the half-reaction in a basic medium (Ballester et al. 2000, p. 209; Sienko and Plane 1970, p. 137):

$$Sn(OH)_3^-(aq) + 2e^- \leftrightarrow Sn(s) + 3(OH)^-(aq) \quad (6.215)$$

$$E^0 = -0.91 \text{ V} \quad (6.216)$$

has a reductant potential slightly smaller than that of the half-reaction of the iron (Fig. 6.58):

$$Fe(OH)_2(s) + 2e^- \leftrightarrow Fe(s) + 2(OH)^-(aq) \quad (6.217)$$

$$E^0 = -0.88 \text{ V} \quad (6.218)$$

Even though the coating of the steel can be performed by immersion in molten tin, the optimization of the thicknesses (in the range 1.0–11.2 $g{\cdot}m^{-2}$, 0.14–1.5 μm) and the possibility of applying selective or differential protections make from the electrolytic process the most widely used alternative for the production of tinplate on a large scale (Fig. 6.54). The suitable steels to manufacture cans for carbonated beverages in deep drawn and stretched tinplate have the following characteristics:

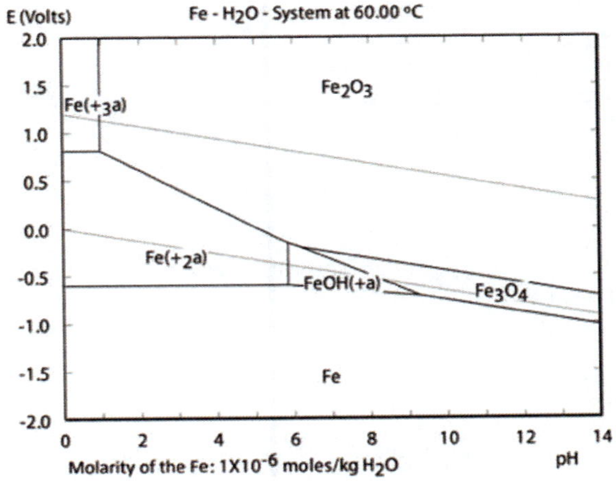

Fig. 6.58 E(V)-pH diagram for the Fe-O-H system at 60 °C (a, aqueous)

- Low or ultralow carbon.
- Low number of inclusions per unit of volume.
- Vacuum treated.
- Hot-rolled, cold-rolled, annealed, and tempered under severe quality controls to guarantee an optimal flatness and constant thickness.

A can for beverages has a thickness on its bottom of around 0.22 mm that is equal to the thickness of the tinplate used as raw material, and the thickness of the walls of the can is smaller than 70 μm.

For some packing applications, an alternative to the tinplate is the chrome-plated steel plate. The main characteristics of the electrolytic coatings of chromium-chromium oxide are similar to those of the tinplate; however, the plate covered with the compound metal-ceramic is characterized by:

- Better abrasion resistance.
- Better behavior for adherence of lacquers and paints.
- Worse plasticity than the tinplate.

Other metallic coatings, apart from those previously mentioned (zinc, zinc alloys, tin, and the chromium-chromium oxide) are:

- Aluminum.
- Aluminum-10% silicon.
- Lead-15% tin (terne coating).
- Special coatings resistant to wear, to oxidation, or to electrochemical corrosion based on metallic carbides (wolfram-chromium) dispersed in a metallic matrix.

From the possible alternatives for the coating of steel with metallic materials, polymers, and paints, it is possible to design multilayer products with a sequence of materials different between the inner and the external zones of the part (Fig. 6.59).

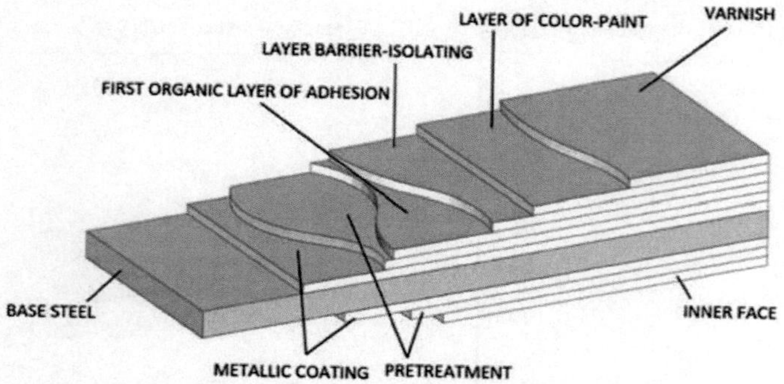

Fig. 6.59 Detail of the different protecting layers that a steel sheet can have

With the aim of developing an automobile whose passenger compartment was only and exclusively manufactured in steel, the following programs were carried out:

- PNGV (*Partnership for New Generation of Vehicles*) in the United States of America.
- ULSAB (*Ultra-Light Steel Auto Body*) as an international consortium of steel producers and automobile manufacturers whose main requirement for the compliance of the proposed objectives was having a steel quality: easily formable, with good surface characteristics, weldable, and conveniently protected against corrosion.

Steel must have the following characteristics to ensure good plasticity, from the chemical point of view:

- A carbon content $\leq 0.02\%$.
- Nitrogen contents of 0.005% (50 ppm N).
- Low contents in residual elements as nickel, cobalt, copper, etc.

Ferritic grain size must be the smallest as possible to simultaneously increase the formability and resistance. Thus, it is essential a suitable control of the hot rolling, cold rolling, and annealing processes.

Increases in the yield strength, σ_Y, of the steel sheet were recently achieved taking advantage of the artificial aging of the metal during the final curing of the paint layer (*bake hardening* steel) when operating at temperatures in the range 100–120 °C. The drying temperature of the paint layer is sufficiently high to allow the diffusion of carbon and nitrogen atoms toward the surroundings of the slipping planes of the edge dislocations (Cottrell atmospheres) that allow pinning the dislocations with the corresponding increase of the yield strength of the material (~30–40 MPa).

6.5.1 Solved Exercises

Exercise 6.16 Calculate the partial pressure and the volume composition of NH_3 in equilibrium with the compounds of a gas formed by N_2, H_2, H_2O, O_2, and NH_3 at the temperatures of 500 and 1000 °C, when the total pressure is one atmosphere and the volume composition of H_2 and H_2O are 20.00 and 0.20%, respectively.

Solution starts:

The reaction of dissociation of the ammonia, NH_3, is:

$$NH_3(g) \leftrightarrow \frac{3}{2}H_2(g) + \frac{1}{2}N_2(g) \tag{6.219}$$

which is thermodynamically favorable, as it is observed as follows:

$$\Delta G^0(T) = 49.3 - 0.109 \cdot T(K) kJ \cdot mol^{-1} \tag{6.220}$$

- at 500 °C (773 K):

$$\Delta G^0(773K) = 49.3 - 0.109 \cdot 773 = -34.957\,kJ \cdot mol^{-1} \tag{6.221}$$

- at 1000 °C (1273 K)

$$\Delta G^0(1273\,K) = 49.3 - 0.109 \cdot 1273 = -89.457\,kJ \cdot mol^{-1} \tag{6.222}$$

The equilibrium constant is:

- at 500 °C:

$$\begin{aligned}\Delta G^\circ(773K) &= -R \cdot T \cdot \ln k_{NH_3/N_2} \rightarrow k_{NH_3/N_2} \\ &= \exp\left(-\frac{\Delta G^\circ(773\,K)}{R \cdot T}\right) \\ &= \exp\left(-\frac{-34957\,J \cdot mol^{-1}}{8.314\,J \cdot mol^{-1} \cdot K^{-1} \cdot 773\,K}\right) = 230.2\end{aligned} \tag{6.223}$$

- at 1000 °C:

$$\begin{aligned}\Delta G^\circ(1273\,K) &= -R \cdot T \cdot \ln k_{NH_3/N_2} \rightarrow k_{NH_3/N_2} \\ &= \exp\left(-\frac{\Delta G^\circ(1273\,K)}{R \cdot T}\right) = \exp\left(-\frac{\Delta G^\circ(1273\,K)}{R \cdot T}\right)\end{aligned}$$

$$= 4684.0 \tag{6.224}$$

Besides, the total pressure of the system, P_T, is:

$$P_T = P_{N_2} + P_{H_2} + P_{H_2O} + P_{O_2} + P_{NH_3} \tag{6.225}$$

From the equilibrium of partial pressures of hydrogen and water vapor:

$$H_2(g) + \frac{1}{2}O_2(g) \leftrightarrow H_2O(g) \tag{6.226}$$

it is possible to obtain the value of the partial pressure of the oxygen:

- at 500 °C:

$$P_{O_2} = 2.01 \cdot 10^{-32}\ \text{atm} \tag{6.227}$$

- at 1000 °C:

$$P_{O_2} = 3.32 \cdot 10^{-19}\ \text{atm} \tag{6.228}$$

The equilibrium of the ammonia dissociation is used to obtain the NH_3 partial pressure in the gas (Eq. 6.219):

$$k_{NH_3/N_2} = \frac{P_{H_2}^{3/2} \cdot P_{N_2}^{1/2}}{P_{NH_3}} \rightarrow k_{NH_3/N_2}^2 = \frac{P_{H_2}^3 \cdot P_{N_2}}{P_{NH_3}^2} \tag{6.229}$$

where the nitrogen partial pressure is:

$$P_{N_2} = P_T - \left(P_{H_2} + P_{H_2O} + P_{O_2}\right) - P_{NH_3} \tag{6.230}$$

When we replace P_{N_2} in the equilibrium of the ammonia dissociation, we obtain the following second-grade equation:

$$a \cdot P_{NH_3}^2 + b \cdot P_{NH_3} + c = 0 \tag{6.231}$$

where:

$$a = \left(k_{NH_3/N_2}\right)^2 \tag{6.232}$$

$$b = P_{H_2}^3 \tag{6.233}$$

Table 6.9 Pressure and volume composition of the compounds

	500 °C		1000 °C	
	P (atm)	%(vol.)	P(atm)	%(vol.)
NH_3	$3.47 \cdot 10^{-4}$	0.03	$1.71 \cdot 10^{-5}$	0.00
H_2	0.2000	20.00	0.2000	20.00
N_2	0.7977	79.77	0.7980	79.80
O_2	0.0000	0.00	0.0000	0.00
H_2O	0.0020	0.20	0.0020	0.20

$$c = -\left(P_T - P_{H_2} - P_{H_2O} - P_{O_2}\right) \cdot P_{H_2}^3 \tag{6.234}$$

Both the partial pressure and the volume composition of the ammonia, as well as those of the rest of the compounds are shown in Table 6.9, at 500 and 1000 °C (total pressure of one atmosphere).

Commentaries: ammonia-dry gas (without H_2O) atmospheres are usually employed in annealing heat treatments in batch and in continuous to achieve a shiny finishing of the sheet free of surface carbon (bright annealing). On the other hand, nitriding atmospheres to obtain surface hard layers usually contain ammonia-gas at temperatures of around 500 °C (Pero-Sanz 2004, pp. 227–243).

End of the exercise.

Exercise 6.17 Calculate the driving force that determines the corrosion of the iron when contacts with an aqueous environment of pH = 7 at 25 °C with a concentration of Fe^{2+}(aq) of 10^{-5} mol per kilogram of H_2O.

Data: the standard potentials of reduction, E^0_{red}, for the discharge of oxygen and Fe^{2+}, at 25 °C, are respectively 1.23 and −0.44 V. The constant of Faraday is 96485 C·g equivalent^{-1}.

Solution starts:

The fundamental equations of electrochemistry are (Ballester et al. 2000, p. 210):

– The first relates the electric potential or electric energy, E(V) and the quantity of electrons exchanged, n, with the chemical energy associated to the reaction, ΔG:

$$\Delta G = -n \cdot F \cdot E \tag{6.235}$$

– The second relates the electric potential of reduction for a reaction out of the standard conditions, E_{red}, with its value in standard conditions of operation, E^0_{red} (Nernst equation):

$$E_{red} = E^0_{red} + \frac{R \cdot T}{n \cdot F} \cdot \ln\left(\frac{\text{reagents}^r}{\text{products}^p}\right) \tag{6.236}$$

Iron corrosion (steel) in aqueous medium takes place by means of an oxidation reaction, which provides electrons in a local anode:

$$Fe(s) \leftrightarrow Fe^{2+}(aq) + 2e^- \tag{6.237}$$

and, on the other hand, the discharge of the oxygen, which takes place in a cathodic zone of the iron (aerated water), where the reduction with consumption of electrons happens, is:

$$\frac{1}{2}O_2(g) + H^+(aq) + 2e^- \leftrightarrow H_2O \tag{6.238}$$

The global process involves the sum of reactions (6.237) and (6.238):

$$\frac{1}{2}O_2(g) + 2H^+(aq) + Fe(s) \leftrightarrow H_2O + Fe^{2+}(aq) \tag{6.239}$$

Precisely, the value of the free energy associated with the reaction 6.239 constitutes the driving force of the aqueous corrosion of the iron. To calculate the electric potential of the reaction 6.239, it is necessary to obtain the values of each one of the oxidation reactions (reaction 6.237) and reduction reactions (reaction 6.238). As the active species involved in the corrosion process are far from the state of reference (molar concentration equal to one and pressure of one atmosphere), it is necessary to use Nerst's equation to correct these deviations. The necessary alteration that is necessary for that the concentration of the Fe^{2+} cation was of 10^{-5} mol $\cdot$ (kg $H_2O)^{-1}$, on the standard potential of reduction of the iron (2+), is:

$$E_{red} = E^0_{red} + \frac{R \cdot T}{n \cdot F} \cdot \ln\left(Fe^{2+}\right) \tag{6.240}$$

$$\frac{R \cdot T}{n \cdot F} = \frac{8.314 \cdot 298}{96485 \cdot n} \cdot \ln 10 = \frac{0.0591}{n} \tag{6.241}$$

$$E_{red} = -0.44 + \frac{0.0591}{n} \cdot \log\left(10^{-5}\right) = -0.59 \text{ V} \tag{6.242}$$

The estimated Nerst's potential is consumed in the reduction of the Fe(2 +), which is contrary to that of the reaction 6.237. Consequently, the value of the potential associated with the reaction 6.237 will be the same, but of opposite sign, and the free energy of this reaction is:

$$\Delta G_{\text{reaction 6.237}} = -2 \cdot 96485 \cdot 0.59 = -113434.4 \text{ J} \cdot (\text{mol Fe})^{-1} \tag{6.243}$$

Nerst's potential associated to the reaction of discharge of oxygen (reaction 6.238), under condition out of the standard state, is:

$$E_{red} = 1.23 + \frac{0.0591}{2} \cdot \log\left\{\left(10^{-7}\right)^2 \cdot (0.21)^{0.5}\right\} = 0.81\ \text{V} \tag{6.244}$$

In this case, the free energy associated with the reaction 6.238 is:

$$\Delta G_{\text{reaction 6.238}} = -2 \cdot 96485 \cdot 0.81 = -155548.4\text{J} \cdot (\text{mol O})^{-1} \tag{6.245}$$

Consequently, the free energy of the iron corrosion process is:

$$\Delta G_{\text{reaction 6.239}} = \Delta G_{\text{reaction 6.237}} + \Delta G_{\text{reaction 6.238}} = -268980.6\ \text{J} \cdot (\text{mol Fe})^{-1} \tag{6.246}$$

Commentaries: if we analyze the variables that can have an influence on the final value of the free energy of the iron corrosion process, it is possible to conclude that the process is conditioned by the temperature, pressure of O_2, concentration of iron in the water, and pH of the medium.

It is necessary for the inexistence of any variable that may impede the electronic transportation from the anodes to the cathodes (local zones) to achieve that the anodic and cathodic reactions could take place.

End of the exercise.

Exercise 6.18 Calculate the standard potential of reduction and the corresponding generalized equation that could express the variation of the redox potential with the pH at 25 °C, for the following equation of reduction (Fig. 6.58):

$$Fe_3O_4(s) + 8H^+(aq) + 8e^- \leftrightarrow 4H_2O + Fe(s) \tag{6.247}$$

Solution starts:

The following reactions are considered to obtain the reduction reaction proposed in the Eq. 6.247:

(a) Formation of Fe_3O_4 from its elements (Appendix 1):

$$2Fe(s) + 2O_2(g) \leftrightarrow Fe_3O_4(s) \tag{6.248}$$

$$\Delta G^0_{\text{reaction 6.248}} = -1105.0 + 0.328 \cdot T\ \text{kJ} \cdot (\text{mol } Fe_3O_4)^{-1} \tag{6.249}$$

(b) Discharge of oxygen, with standard potential of reduction of 1.23 V:

$$\frac{1}{2}O_2(g) + 2H^+(aq) + 2e^- \leftrightarrow H_2O \quad (6.250)$$

The standard free energy associated with the reaction 6.247 at 25 °C is:

$$\Delta G^0_{\text{reaction 6.247}} = -\Delta G^0_{\text{reaction 6.248}} + 4 \cdot \Delta G^0_{\text{reaction 6.250}} \quad (6.251)$$

$$\Delta G^0_{\text{reaction 6.248}} = -1007256.0\ \text{J} \cdot (\text{mol Fe}_3\text{O}_4)^{-1} \quad (6.252)$$

$$\Delta G^0_{\text{reaction 6.250}} = -2 \cdot 96485 \cdot 1.23 = -237353.1\left[\text{J} \cdot (\text{mol H}_2\text{O})^{-1}\right] \quad (6.253)$$

$$\Delta G^0_{\text{reaction 6.247}} = 57843.6\ \text{kJ} \cdot (\text{mol Fe}_3\text{O}_4)^{-1} \quad (6.254)$$

The value of the standard potential of reduction associated to the reaction 6.247 is:

$$E^0_{red} = -\frac{57843.6}{8 \cdot 96485} = -0.075\ \text{V} \quad (6.255)$$

Finally, the variation of the redox potential with the pH for the reaction 6.247 at 25 °C is:

$$E_{red} = -0.075 - 0.0591\ \text{pH} \quad (6.256)$$

which is obtained considering the Eqs. 6.236 and 6.240:

$$E_{red} = -0.075 + \frac{R \cdot T}{8 \cdot F} \cdot \ln\left[H^+\right]^8 \quad (6.257)$$

$$E_{red} = -0.075 + \frac{0.0591}{8} \cdot 8 \cdot \log\left[H^+\right] \quad (6.258)$$

$$E_{red} = -0.075 - 0.0591\ \text{pH} \quad (6.259)$$

where:

$$\text{pH} = -\log\left[H^+\right] \quad (6.260)$$

Commentaries: the equilibriums of the Fe-O-H aqueous system at 60 °C are represented in the Pourbaix diagram (E/pH) of Fig. 6.58. It is possible to analyze what would be the variation of the standard potential of reduction of the enunciation at this temperature. The equations that must be considered (Upadhyaya and Dube 1977, p. 187) are:

$$\left(\Delta c_p\right)_{\text{reaction 6.247}} = n \cdot F \cdot T \cdot \left(\frac{\delta^2 E}{\delta T^2}\right)_P \tag{6.261}$$

$$\left(\frac{\delta E}{\delta T}\right)_P = \frac{\left(\Delta c_p\right)_{\text{reaction 6.247}}}{n \cdot F} \cdot \ln\left(\frac{T_2}{T_1}\right) \tag{6.262}$$

Considering the data of Fine and Geiger (1993) (pp. 431–432) for the iron and for the ferrous-ferric oxide, the variation of the specific heat associated to the reaction 6.247, $\left(\Delta c_p\right)_{\text{reaction 6.247}}$ (difference between the specific heats of the products and reagents), is -79.28 J·mol^{-1} · K^{-1}. When the value of T_1 is 298 K and that of T_2 is 333 K, the result for the gradient of standard voltage is:

$$\left(\frac{\delta E^0}{\delta T}\right)_P = -1.411 \cdot 10^{-5}\ V \cdot K^{-1} \tag{6.263}$$

Consequently, the value of the standard potential for the reaction 6.247 at 60 °C must be corrected decreasing the value of −0.075 V at 25 °C in 0.0004 V, that is to say, it does not affect significantly the calculated value at 25 °C (298 K).

End of the exercise.

Exercise 6.19 The corrosion rate of a steel sheet of area S and length Δl exposed to an aqueous medium can be expressed as the decrease that its thickness Δe suffers after a time Δt. Derive the generalized equation to calculate the wear as a function of the electrochemical potential, ΔE, of the surface assigned to the cathodic, S_C, and anodic, S_A, regions as well as of the resistance R to the flow of electrons.

Solution starts:

The anodic current flow $J_{Q/A}$ can be expressed as follows:

$$J_{Q/A} = \frac{Q}{\Delta t \cdot S_A}\ C \cdot s^{-1} \cdot m^{-2} \tag{6.264}$$

where Q is the charge (in C) that must be transported in a certain time Δt through a certain anodic section S_A. If we know that to dissolve a mol of any metal (atomic weight divided by the number of exchange electrons n) 96485 Coulombs are required and we know that the atomic weight of the iron is 55.85 g·mol^{-1}, the anodic current flow can be expressed as:

$$\frac{Q}{\Delta t \cdot S_A} = \frac{n \cdot F \cdot m_{Fe}}{55.85} \cdot \frac{\Delta l}{\Delta t \cdot S_A} \tag{6.265}$$

where m_{Fe} is the quantity of iron that was dissolved (with density ρ_{Fe}) that at the same time is equal to:

$$m_{Fe} = \Delta e \cdot S \cdot \rho_{Fe} \tag{6.266}$$

On the other hand, Ohm's law indicates:

$$\frac{Q}{\Delta t \cdot S_A} = I = \frac{\Delta E}{R} \tag{6.267}$$

where R is the resistance (in Ω) of the system to the flow of the amperage I and ΔE is the corrosion potential (in V).

At the same time, the resistance of the medium can be related both with the properties of the aqueous solution (electrolytic conductor) and with those of the metallic conductor (the material that experiences the corrosion):

$$R = \rho_e \cdot \frac{\Delta l}{S} \tag{6.268}$$

where ρ_e is the electrical resistivity of the aqueous solution, Δl is the length of the iron plate, and S (width·thickness) is the section of the metal that is in contact with the electrolyte.

Thus, a method of expressing the variation that the thickness of the iron plate experiences with the time will be:

$$\frac{\Delta e}{\Delta t} = \frac{55.85 \cdot 10^{-3}}{n \cdot F \cdot \rho_{Fe}} \cdot \frac{\Delta E}{R} \cdot \frac{r}{S_C} \tag{6.269}$$

where the corrosion rate of the iron sheet is estimated in m·s^{-1}, when the variables ΔE, R, ρ_{Fe} and S_C are expressed in units of measurement of the International System (V, Ω, kg·m^{-3}, and m). However, in this last equation we add a new variable of remarkable relevance in the conservation of large steel structures in contact with the marine environment (oil platforms, wind turbines, or ships), which is the relation between the cathodic S_C and S_A surfaces of the structure to be protected by means of galvanic anodes (aluminum alloys and zinc):

$$r = \frac{S_C}{S_A} \tag{6.270}$$

End of the exercise.

Exercise 6.20 Calculate the corrosion rate that a steel sheet of 1.0 meters in length (l) and 1.0 m^2 in section (S = width · thickness) can experience, under a corrosion potential of 1.20 V when it is in contact with aqueous medium with the characteristics indicated in Table 6.10.

Solution starts:

Using the derived equation of the previous exercise, it is possible to estimate the corrosion rate of the steel sheet in all the proposed cases. However, it is necessary to consider:

Table 6.10 Characteristics of the corrosion process

Case	Resistivity (Ω·m)	Conditions
A	10^5	Deionized water free of contaminated ions
B	10^3	Water with low concentration of contaminated salts
C	7.0	Aqueous solution 0.01 molar of potassium chloride (KCl)
D	0.78	Aqueous solution 0.1 molar of potassium chloride (KCl)
E	$9.00 \cdot 10^{-2}$	Aqueous solution 1.0 molar of potassium chloride (KCl)

(a) That the relation r between the cathodic (S_C) and anodic surfaces (S_A), in all the situations, is equal to 1.
(b) As the length and the section of the plate are numerically the same, also the resistivity of the aqueous medium ρ_e coincides with the resistance R of the system ($\rho_e = R$).

We collect in Table 6.11 the data about the corrosion rate of the iron plate for the five cases that were described in the enunciation (Table 6.10).

Commentaries: while the corrosion process of the iron plate in deionized water free of contaminants is nearly nonexisting, when the concentration of the saline medium is high, the homogeneous corrosion rate in the surface of the plate, which is formed by local anodes/cathodes, is catastrophic.

End of the exercise.

Exercise 6.21 A film generated by oxidation can be protective for the metal or alloy that is covering. The relation between the volume of produced oxide, V_{oxide}, and the volume of metal that was oxidized, V_{metal}, is known as Pilling–Bedworth ratio, P—B *ratio* (Smith 1998, p. 627):

$$\text{P—B ratio} = \frac{V_{oxide}}{V_{metal}} \tag{6.271}$$

When this relation is nearly one, the film oxide is protective for the base metal.

Calculate the P—B ratio in the oxidation of the alpha-iron and in the oxidation of the aluminum at 600 °C.

Table 6.11 Corrosion rate for the six situations presented in the enunciation

Case	Corrosion rate m·s^{-1} (cm·year^{-1})
A	$4.41 \cdot 10^{-16}$ ($1.39 \cdot 10^{-6}$)
B	$4.41 \cdot 10^{-14}$ ($1.39 \cdot 10^{-4}$)
C	$6.22 \cdot 10^{-12}$ ($1.96 \cdot 10^{-2}$)
D	$5.65 \cdot 10^{-11}$ (0.18)
E	$4.90 \cdot 10^{-10}$ (1.55)

Data: according to TAPP, the densities of alpha-iron and FeO (stoichiometric wustite), at 600 °C, are, respectively, 7689 and 5747 kg·m^{-3}. Equally, the densities of aluminum and alumina, at 600 °C, are 2584 and 3930 kg·m^{-3}, respectively.

Solution starts:

Considering the oxidation of 10.00 g of alpha-iron, and if we take into account that the product of iron oxidation is wustite, FeO, the reaction that takes place is:

$$Fe(s) + \frac{1}{2}O_2(g) \leftrightarrow FeO(s) \quad (6.272)$$

We see that the oxidation of 1.0 mol of alpha-iron generates 1.0 mol of FeO. For 10.00 g of alpha-iron:

$$10.00\ g\ Fe - \alpha \cdot \frac{71.85\ g\ FeO}{55.85\ g\ Fe} = 12.8648\ g\ FeO \quad (6.273)$$

The volume associated to 12.8648 g of FeO is:

$$V_{FeO} = 12.8648\ g\ FeO \cdot \frac{cm^3\ FeO}{5.747\ g\ FeO} = 2.2385\ cm^3\ FeO \quad (6.274)$$

On the other hand, the volume occupied by 10.00 g of alpha-iron at 600 °C is:

$$V_{Fe} = 10.00\ g\ Fe \cdot \frac{cm^3\ Fe}{7.689\ g\ Fe} = 1.3006\ cm^3\ Fe \quad (6.275)$$

Consequently, the Pilling-Bedworth ratio of the iron at 600 °C will be:

$$P - B\ ratio\ (FeO/Fe)_{600\ °C} = \frac{2.2385}{1.3006} = 1.72 \quad (6.276)$$

For the aluminum, the calculation procedure is similar, although the oxidation reaction that must be considered is:

$$2Al(s) + \frac{3}{2}O_2(g) \leftrightarrow Al_2O_3(s) \quad (6.277)$$

The production of aluminum oxide from 10.00 g of metal is:

$$10.00\ g\ Al \cdot \frac{101.96\ g\ Al_2O_3}{54.96\ g\ Al} = 18.8955\ g\ Al_2O_3 \quad (6.278)$$

The volume associated to 18.8955 g Al_2O_3 at 600 °C is 4.8080 cm^3, and the corresponding to 10.00 g of aluminum is 3.9246 cm^3. Thus, the P – B ratio for the aluminum/alumina system at 600 °C is 1.23.

Commentaries: even though the protecting action of the oxide layer is effective when values of the P – B ratio close to 1.0 (as in the case of the aluminum) are obtained, we must take into account some other important variables:

- Adherence between the metal and the oxide.
- High melting point and low vapor pressure of the deposited oxide.
- Coefficient of thermal expansion of both the metallic substrate and the oxide.
- Plasticity of the protecting layers.
- High resistance to the diffusional transfer and the electrons transfer in the protecting layer.

Finally, in the case of the iron and for a strongly oxidizing atmosphere (air), the FeO can be unstable and can oxidize to Fe_2O_3 (completely oxidized form of the iron). In this case, the value of P – B ratio is 2.13.

End of the exercise.

Exercise 6.22 High-Velocity Oxygen Fuel Spraying (HVOF).

Solution starts:

One of the methods of surface protection to minimize the negative effects of the electrochemical corrosion, the wear as a result of abrasive fluids or the breaking due to fatigue, is the creation of protecting monolayers or multilayers. Using as base material structural steel, it is possible to think of covering the surface of the steel with composite materials of either ceramic or metallic matrix (Puértolas et al. 2010; Sobolev et al. 2004; Verdeja et al. 2014; Béranger et al. 1994). Technologies used to carry out the surface protection are:

- HVOF: High-Velocity Oxygen Fuel Spraying.
- PLD: Pulsed Laser Deposition.
- CVD: Chemical Vapor Deposition.

Starting from a powder of the following chemical characteristics: 9.30% C, 20.50% Ni, and 70.20% Cr, we will try to determine the chemical and structural characteristics of the projected layer. In this case, the working hypotheses are the following ones:

1. X-ray diffraction demonstrates that there are three different phases in the powder used in the experiments: two corresponding to chromium carbides, Cr_7C_3 and Cr_3C_2, and another metallic of high nickel content.
2. All the carbon is associated to chromium carbides, $(\%C)^0$.
3. The chromium percentage in the powders, $(\%Cr)^0$, can distribute between the metallic phase and carbides, $(\%Cr)^C$:

$$(\%\mathrm{Cr})^0 = (\%\mathrm{Cr})^C + (\%\mathrm{Cr})^m \tag{6.279}$$

4. All the nickel available in the powders will be part of the metallic matrix of the coating, $(\%\mathrm{Ni})^0$.
5. The atomic and molecular weights of the phases involved in the process are Ni, 58.71 g/mol; C, 12.01 g/mol; Cr, 51.996 g/mol; Cr_7C_3, 400.002 g/mol; Cr_3C_2, 180.008 g/mol. Three equations can be considered, two associated with the balance to the carbon and elemental chromium, and the other associated with the totality of the phases-constituents of the coating: two ceramics (Cr_7C_3 and Cr_3C_2) and one metallic

This way, the chromium associated with the carbides,$(\%\mathrm{Cr})^C$, is:

$$\begin{aligned}(\%\mathrm{Cr})^C = &\left(\frac{1\ \mathrm{mol\ Cr_7}C_3}{400.002\ \mathrm{g\ Cr_7}C_3}\cdot\frac{7\ \mathrm{mol\ Cr}}{1\mathrm{mol\ Cr_7}C_3}\cdot\frac{51.996\ \mathrm{g\ Cr}}{1\ \mathrm{mol\ Cr}}\right)\cdot(\%\mathrm{Cr_7}C_3)\\ &+\left(\frac{1\ \mathrm{mol\ Cr_3}C_2}{180.008\ \mathrm{g\ Cr_3}C_2}\cdot\frac{3\ \mathrm{mol\ Cr}}{1\ \mathrm{mol\ Cr_3}C_2}\cdot\frac{51.996\ \mathrm{g\ Cr}}{1\ \mathrm{mol\ Cr}}\right)\cdot(\%\mathrm{Cr_3}C_2)\end{aligned} \tag{6.280}$$

and the carbon associated with the carbides, $(\%C)^0$, is:

$$\begin{aligned}(\%C)^0 = &\left(\frac{1\ \mathrm{mol\ Cr_7}C_3}{400.002\ \mathrm{g\ Cr_7}C_3}\cdot\frac{3\ \mathrm{mol}\ C}{1\ \mathrm{mol\ Cr_7}C_3}\cdot\frac{12.01\ \mathrm{g}\ C}{1\ \mathrm{mol}\ C}\right)\cdot(\%\mathrm{Cr_7}C_3)\\ &+\left(\frac{1\ \mathrm{mol\ Cr_3}C_2}{180.008\ \mathrm{g\ Cr_3}C_2}\cdot\frac{2\ \mathrm{mol}\ C}{1\ \mathrm{mol\ Cr_3}C_2}\cdot\frac{12.01\ \mathrm{g}\ C}{1\ \mathrm{mol}\ C}\right)\cdot(\%\mathrm{Cr_3}C_2)\end{aligned} \tag{6.281}$$

We know that:

$$100 = (\%\mathrm{Cr_7}C_3) + (\%\mathrm{Cr_3}C_2) + (\%\mathrm{Ni})^0 + (\%\mathrm{Cr})^m \tag{6.282}$$

The calculation of the $\%\mathrm{Cr_7}C_3$ is:

$$A\cdot(\%\mathrm{Cr_7}C_3) = B \tag{6.283}$$

where A and B are:

$$\begin{aligned}A = &\frac{1\,\mathrm{mol\,Cr_7}C_3}{400.002\,\mathrm{g\,Cr_7}C_3}\cdot\frac{7\,\mathrm{mol\,Cr}}{1\,\mathrm{mol\,Cr_7}C_3}\cdot\frac{51.996\,\mathrm{g\,Cr}}{1\,\mathrm{mol\,Cr}} - \frac{1\,\mathrm{mol\,Cr_7}C_3}{400.002\,\mathrm{g\,Cr_7}C_3}\\ &\cdot\frac{3\,\mathrm{mol}\,C}{1\,\mathrm{mol\,Cr_7}C_3}\cdot\frac{12.01\,\mathrm{g}\,C}{1\,\mathrm{mol}\,C}\cdot\frac{1\,\mathrm{mol}\,C}{12.01\,\mathrm{g}\,C}\cdot\frac{1\,\mathrm{mol\,Cr_3}C_2}{2\,\mathrm{mol}\,C}\\ &\cdot\frac{180.008\,\mathrm{g\,Cr_3}C_2}{1\,\mathrm{mol\,Cr_3}C_2}\cdot\frac{1\,\mathrm{mol\,Cr_3}C_2}{180.008\,\mathrm{g\,Cr_3}C_2}\cdot\frac{3\,\mathrm{mol\,Cr}}{1\,\mathrm{mol\,Cr_3}C_2}\end{aligned}$$

$$\cdot \frac{51.996\,\text{g Cr}}{1\,\text{mol Cr}} = 0.3250 \tag{6.284}$$

$$B = (\%\text{Cr})^c - (\%C)^0 \cdot \frac{1\,\text{mol}C}{12.01\,g\,C} \cdot \frac{1\,\text{mol Cr}_3C_2}{2\,\text{mol}C} \cdot \frac{180.008\,\text{g Cr}_3C_2}{1\,\text{molCr}_3C_2} \cdot \frac{1\,\text{mol Cr}_3C_2}{180.008\,\text{g Cr}_3C_2} \cdot \frac{3\,\text{mol Cr}}{1\,\text{mol Cr}_3C_2} \cdot \frac{51.996\,\text{g Cr}}{1\,\text{mol Cr}} \tag{6.285}$$

The calculation of the $\%\text{Cr}_3C_2$ is:

$$C \cdot (\%\text{Cr}_3C_2) = D \tag{6.286}$$

where C and D are:

$$C = \frac{1\,mol\;Cr_3C_2}{180.008g\;Cr_3C_2} \cdot \frac{2\,mol\;C}{1\,mol\;Cr_3C_2} \cdot \frac{12.01g\;C}{1\,mol\;C} = 0.1334 \tag{6.287}$$

$$D = (\%C)^0 - \frac{1\,\text{mol Cr}_7C_3}{400.002\,\text{g Cr}_7C_3} \cdot \frac{3\,\text{mol}\;C}{1\,\text{mol Cr}_7C_3} \cdot \frac{12.01\text{g}C}{1\,\text{mol}\;C} \cdot (\%\text{Cr}_7C_3) \tag{6.288}$$

In the case where the metallic phase is formed only by nickel, the structure of the coating will be formed by a metallic matrix with 25.50% Ni and a disperse constituent formed by 30.17% Cr_7C_3 and 49.33% Cr_3C_2.

If we detected that 100% of the total chromium is dissolved in the metallic phase of nickel (solid solution):

$$(\%\text{Cr})^0 = 70.20\% = (\%\text{Cr})^c + 1.00\% \rightarrow (\%C)^c = 69.20\% \tag{6.289}$$

In this case, the structure of the coating would be formed by a metallic matrix (20.50% Ni and 1.00% Cr, in solid solution, Pero-Sanz et al. 2017) and disperse formed by 27.09% Cr_7C_3 and 51.41% Cr_3C_2.

End of the exercise.

References

Ballester A, Verdeja LF, Sancho JP (2000) Metalurgia extractiva: fundamentos. Ed Síntesis, Madrid, Spain

Béranger G, Henry G, Sanz G (1994) Le Livre de l'Acier, Ed Lavoisier Tec & Doc, London, UK

Blickwede DJ (1984) The new look of sheet steels. ISIJ Int 24(9):687–701

Bocos-González JL (2002) Evolución Microestructural y Desarrollo de la Textura durante el Recocido Continuo en Aceros de Bajo Carbon, PhD thesis, University of Navarra (Supervisor: Isabel Gutiérrez Sanz), San Sebastián, Spain, Feb 2002

Cullity BD (1959) Elements of X-ray diffraction, 2nd printing. Addison-Wesley Publishing Company Inc., Reading, MA, USA

Dieter GE (1981) Mechanical metallurgy, 2nd edn. McGraw-Hill International Book Company, Tokyo, Japan
Dinda S, James KF, Keeler SP, Stine PA (1981) How to use circle grid analysis for Die Tryout. American Society for Metals, Metals Park, OH, USA
Fernández J, Riba J, Verdeja JI (1986) Estudio del daño estructural en una chapa de acero microaleado con niobio. Revista de Metalurgia de Madrid 22(3):154–162
Fine HA, Geiger GH (1993) Handbook on material and energy balance calculations in Metallurgical, Processes edn. TSM, Warrendale, USA
Gil-Sevillano J (1982) Trefilado de Aceros al Carbono. Ensidesa-Formación, Avilés, Asturias
Goodman SR, Hu H (1970) Effect of hot-rolling texture on the plastic strain ratio of low-carbon steels. Metall Trans 1:1629–1640
Gourgues-Lorenzon AF (2007) Application of electro backscatter diffraction to the study of phase transformations. Int Mater Rev 52(2):65–128
Gutierrez-Stampa M, Verdeja-González JI, Pero-Sanz Elorz JA (1983) El "daño estructural" en las chapas de aceros no aleados para embutición profunda. Revista de Metalurgia del CENIM 19(5):253–260
Hosford WF, Backofen WA (1964) Strength and plasticity of textured metals in Fundamentals of Deformation processing. Syracuse University Press, pp 259–291
Lacombe P (1963) Polishing and etching methods and their applications to optical metallography. Metallography 80:50–120
Leslie WC (1982) The physical metallurgy of steels, 1st edn. McGraw-Hill International Book Company, London, UK
Martínez-Perez VJ (2003) Aceros libres de intersticiales y de esmaltería destinados a operaciones de embutición. Tesis Doctoral Universidad de Oviedo (Supervisor: José Ignacio Verdeja González), Dpto de Ciencia de los Materiales e Ingeniería Metalúrgica, Universidad de Oviedo, Oviedo, Asturias
Pero-Sanz JA (1996) Ciencia e Ingeniería de Materiales. Estructura, Transformaciones, Propiedades y Selección, 3rd edition, Ed. Dossat 2000, Madrid, Spain
Pero-Sanz JA (2004) Aceros. Metalurgia Física. Selección y Diseño. Ed Dossat-2000, Madrid, Spain
Pero-Sanz JA, Quintana-Hernández MJ, Verdeja LF (2017) Solidification and solid-state transformations of metals and alloys, 1st edn. Elsevier, Boston, USA
Pero-Sanz JA, Fernández-González D, Verdeja LF (2018) Physical metallurgy of cast irons, 1st edn. Springer International Publishing, Cham, Switzerland
Pero-Sanz JA, Fernández-González D, Verdeja LF (2019) Structural materials: properties and selection, 1st edn. Springer International Publishing, Cham, Switzerland
Puértolas JA, Ríos R, Castro M, Casals JM (2010) Tecnología de Superficies en Materiales. Ed Síntesis, Madrid, Spain
Romano-Sobrino G (2000) Aceros para Hojalata y Embutición Profunda. Estructura, Textura y Propiedades Mecánicas, PhD thesis, University of Oviedo (Supervisor: José Ignacio Verdeja González), Oviedo, Asturias, Nov 2000
Sienko MJ, Plane RA (1970) Química – Física, Inorgánica edn. Reverté, Barcelona, Spain
Smith WF (1998) Fundamentos de la Ciencia e Ingenieria de Materiales, 3rd ed. Ed McGraw Hill International, Madrid, Spain
Sobolev VV, Guilemany JM, Nutting J (2004) High velocity oxy-fuel spraying. Maney Publishing and Institute of Metals (IOM), London, UK
Stuwe HP, Faustmann J (1969) Introducción a las Texturas de los Materiales Metálicos, Ed Montecorvo, Madrid, Spain
Turkdogan ET (1996) Fundamentals of steelmaking. Ed The Institute of Materials (IOM), University Press, Cambridge, UK
Upadhyaya GS, Dube RK (1977) Problems in metallurgical thermodynamics and kinetics. Ed Pergamon Press, Oxford, UK

Vázquez AJ, Damborenea JJ (2000) Ciencia e Ingeniería de la superficie de los materiales metá-licos. Consejo Superior de Investigaciones Científicas (CSIC), Madrid, Spain
Verdeja JI, Fernández G (1992) Texturas de materiales metálicos. Revista de Minas 7:77–85
Verdeja LF, Ballester A, Sancho JP, González R (2014) Refractory and ceramic materials. Ed Síntesis, Madrid, Spain
Verdeja JI, Fernández-González D, Verdeja LF (2020) Operations and basic processes in ironmaking, 1st edn. Springer International Publishing, Cham, Switzerland
Whiteley RL (1960) The importance of directionality in drawing quality sheet steel. Trans ASM 52:154–169

Chapter 7
Strengths and Uncertainties of the Steel as Structural Material

Abstract This last chapter is aimed at studying, as the own title indicates, the strengths and uncertainties of the steel as structural material. We indicate the different big changes or revolutions in the field of the steel production, and the direction toward which the research will direct in the future. Additionally, we analyze the conflictive aspects that difficulty the utilization of the steel and how to solve them. The sustainable and environmentally friendly production of steel will arise as an important factor to increase the competitivity of the production of this structural material. On the other side, there are several environmental questions that might negatively affect the production of steel. CO_2 emissions are going to be a problem for the iron and steelmaking in the next years, the same as the other pollutants and residues generated in the production of iron and steel. The same way, we also analyze the efficient utilization of the energy in the iron and steelmaking process.

7.1 Introduction

It is possible to consider the steel as one of the outstanding protagonists of the industrial revolutions that have historically taken place from the beginning of the nineteenth century until the present, see Sect. 1.1:

- The first industrial revolution (Comellas 2000), identified as 1.0 Industry according to the current terminology, had started during the first half of the nineteenth century. It is possible to indicate, as relevant fact for its beginning, the utilization of the steam machine in boats and locomotives (engine traction), which allowed to facilitate the displacement of people and goods.
- The second industrial revolution, identified with the appearance of the Bessemer process (1856), which meant the beginning of the development of the 2.0 Industry.
- The third industrial revolution, 3.0 Industry, which started after the end of the Second World War, whose relevant fact might be associated with the development of the Linz-Donawitz process of conversion of pig iron into a low carbon metallic melt (1956).

L. F. Verdeja González et al., *Operations and Basic Processes in Steelmaking*, Topics in Mining, Metallurgy and Materials Engineering,
https://doi.org/10.1007/978-3-030-68000-8_7

- Finally, the fourth industrial revolution, identified with the first achievements reached by the Artificial Intelligence (AI), at around the year 2010, where the aim is to use the great quantity of data-variables that were generated and have been being accumulated as a consequence of the third industrial revolution. The final-main objective pursued by the transformation of the 3.0 Industry into the 4.0 Industry is to achieve that the automatism was autonomous (Paramore and Kirmse 2020).

Along the six chapters into which we have divided the operations and basic processes associated with the production of steel, we have tried to develop, as didactically as possible to facilitate the learning, a manuscript with a double direction:

- From the bottom to the top, from the particular and singular of each one of the presented exercises toward the global of each considered process.
- From up to bottom that, considering as starting point complex processes that habitually can provide a great quantity of phenomenological data, big data, might carry on and facilitate the construction of easy models and solutions. Definitely, the objective is to use the minimum number of variables to reach a consistent knowledge and control of the operations and processes of the steel metallurgy.

Without losing sight that the true progress of the scientific-technical research is to find "the truth," we point out three reasons that might endanger the scientific-technical research development:

1. When we try to study-research the problems in such close and detailed way, on many occasions, we do not have the capacity of analyzing "the whole." The "whole" can be equally more true than the detail considered in an isolated way. Possibly, the gains in precision are lost in the knowledge of "the truth."
2. There are many people or research groups that do high-quality scientific-technical research. They have means and skills that can be developed with undoubtedly lucidity and efficiency. However, when these skills and means tend to be chaotic in such a way that can be non-useful for the rest, the consequences might be serious. They focus on their own field and they simply ignore the importance of other different projects or points of view. Consequently, they think that their activity is the single in the world and the only one that should be developed, and they feel a kind of disdain for other projects that handle the topic in a different manner.
3. It is not frequent that a single person or group could have the possibility of doing the full research in a certain topic. The knowledge of the scientific-technical "truth" is a cumulative process that includes different stages and points of view. The research field started by certain people is usually finished by others. The error, in some cases, is the only path to reach the "truth" because the "truth" is never against the "truth." It is possible that erroneous concepts could arise for some time, but the "truth" finally imposes. Definitely, the presence of errors in the scientific community would be aimed at finding the "truth."

In this final chapter of the book about steel metallurgy, and with losing the sight that the final objective is the production of steel, we are going to review as clear and concretely possible the main strengths, weaknesses, and uncertainties associated with the manufacture of this material. The objective of this everything is to try to approach a bit more to the scientific-technical "truth" and to provide objective data that could support the R&D and management decisions that might be convenient.

Finally, we could ask the following question about the 4.0 Industry (Paramore and Kirmse 2020): what is the real future that is expected for the steel? To answer this question, it is possible to consider the data of the World Health Association that expects, for the next decades, that the world population in certain zones-regions of the world (Asia and Africa) to grow in 4000 million people (we should remember that the world population reached the 1000 million people in 1835). For a world steel consumption per capita of 225 kilograms (see Chap. 1 in Verdeja et al. 2020), we might see an increase of steel consumption of 900 million tons, which is around 55% of the current consumption of steel.

7.2 Favorable Circumstances Associated to the Production of Iron and Steel

The *Book of Deuteronomy* (Chapter 8, verses 9–11) associates the material prosperity of the people to have "a land where bread will not be scarce and you will lack nothing; a land where the rocks are iron and you can dig copper out of the hills. When you have eaten and are satisfied, praise the Lord your God for the good land he has given you" (Biblia de Navarra 2008).

Nowadays, without going to such extremes, it is not reasonable to disregard the idea of living in a land or region that could provide abundant harvest and with mountains plenty of high-quality hematite. However, other reasons can be provided to explain that iron should be still considered as an essential element for the development and prosperity of the countries. The facts that could be indicated are the following ones:

- Iron is the fourth element most abundant in the Earth's crust. The availability and abundance of iron mineral resources in the planet is big. On the other hand, the presence of military or social conflicts in different regions and countries seem to have influence neither in the supply nor in the price of the raw materials. Finally, and as opposed to other metallic elements, the mined iron mineral resources (Habashi 1996) are characterized because of being of high purity in iron and low presence of those non-desired elements: sulphur, phosphorus, and alkaline elements.
- Historical facts related with both the invention of the steam machine and its utilization in boats and maritime transport (first industrial revolution, 1807) and the Bessemer process (second industrial revolution, 1856) are closely related with the steel as structural product par excellence.

- The recycling of any product manufactured in steel makes possible the development of the concept of "circular economy." We should think that around 35% of the world steel production has as raw material the recycled steel (popularly known as "scrap"). It is possible that the habitual terminology used to define the recycled steel, scrap, favors neither the attractive image of the steel in the market in particular nor in the "circular economy" of the steel in the new materials in general.
- Steel, together with concrete and timber, are the most used materials. On the other hand, steel is an essential product both in mature and developed economies and in those countries or regions in developing countries.
- The "sustainable development" of the iron and steel metallurgy is closely related with both a continuous improvement of the processes and operations efficiency and the quality of the finished products. The only manner of surviving in this "relentlessly competitive" environment is to develop a continuous improvement of processes and products offering to the market materials with better characteristics at lower prices. That is to say, reaching the concept of "virtuous circle of competitivity" for the steel that consists of offering products with better properties at lower prices.

7.3 Conflictive Aspects that Difficulty the Utilization of Steel

Even when we have indicated that steel is one material with the most important structural applications, we want to revise some of its less favorable aspects and properties that limit its utilization in certain markets:

1. Its great density, 7950 $kg \cdot m^{-3}$, which impedes its classification as a light metal-material and, consequently, limits its utilization in a massive manner in the transport industry, where light materials with great resistance and resilience are usually employed.
2. Problems derived from the aqueous corrosion are being adequately solved with either the utilization of protecting coatings, mechanisms of anodic and cathodic protection, or the utilization of stainless steel.
3. The lack of functional characteristics in steels can be in part compensated with magnetic properties that allow distinguishing or separating with easiness and efficiency of carbon steels from austenitic stainless steels.
4. The image of a traditional and old industry (the first blast furnaces started to work in Central Europe at the middle of the fifteenth century) is counteracted with the greater and greater application of operations and productive processes connected with technologies at the vanguard as: Artificial Intelligence, Big Data, Blockchain or Additive Manufacturing, which are supported in the development of the 4.0 Industry (Bolton 2019; McDonald 2019). In January 2019, the World Economic Forum (WEF) announced the inclusion of the steel plant of Tata

Steel Europe of Ijmuiden (close to Amsterdam, the Netherlands) in the list of "lighthouses," distinction that is given to those installations that are leaders in technology of the 4.0 Industry (Diaz 2012).

5. Other negative aspect associated to the science and technology of traditional materials is that both the processes and properties of materials and products associated to the ironmaking and steelmaking process are well-known. For example, the mass or heat transfer properties as the viscosity or the thermal conductivity, the surface or interface energies as well as the enthalpies associated with solid-state reactions are, on many occasions, unknown. Equally, the exactness of some thermodynamic data, such as the melting point or the fusion enthalpies are subjected to significant levels of uncertainty: ±20 °C in the case of the melting point and ±10 °C in the case of the enthalpy of fusion, see Appendix 6 (properties of the most important elements of the ferroalloys in molten state).
6. The recovery and decontamination of soils and gases that carry organic molecules with benzene rings (associated to sintering plants and coke batteries, Habashi 2002) is carried out with the utilization of modern techniques of biotechnology and photocatalysis.

Finally, we want to indicate that the above-mentioned concepts: Artificial Intelligence, Blockchain or Additive Manufacturing are leading to unjustifiable emotional reactions and exaggerated expectations in the mass media. In general, in the scientific-technical field, we consider:

A. that the Artificial Intelligence is a tool in which researchers and technicians have been working for decades and that, in this moment, we are collecting some of the results due to the improvement in the capacity of storing data and the greater rate of processing variables. Definitely, the Artificial Intelligence is only a manner of trying to adequately correlate the variables that take part in a process, according to their physical and chemical fundamentals.
B. that the Blockchain is an instrument to try to solve problems that, with the knowledge of one variable, it was possible to determine other variable, or also that by means of the knowledge of a block of variables, it would be possible to access to the knowledge of other variables related with the fundamentals and control of processes.
C. that the Additive Manufacturing in the ironmaking and steelmaking field might be identified with the manufacture-solidification step by step of a continuous casting ingot.

Definitely, the weight and importance of these concepts in the future of the ironmaking and steelmaking industry will help to strengthen the steel inside of the "virtuous circle of the competitiveness" with respect to the potential competitors (nonferrous metals, polymers (natural or artificial) and ceramics) more than weakening the expectations of development. For example, we want to briefly analyze two situations where the Artificial Intelligence, in the short and medium term, could provide solutions in the efficiency of processes and in the quality of the finished products:

- Inspection and reprocessing of hot-rolled coils.
- Do not use mechanical tests in the certification of the quality of coils, slabs, profiles, and rods.

In the first situation, the Artificial Intelligence can help to improve, in the short term, the inspection and reprocessing of defective coils before their shipping. In the second case, it is still necessary to obtain more consistent correlations between chemical composition, microstructure, variables of the process, and mechanical properties. In any case, there would be solutions given by a "weak Artificial Intelligence" that are being developed in operations and processes with limits-boundaries very defined and with reduced importance.

Definitely, it is convenient to remember and consider with the aim of achieving the automatization of the automatisms of the 3.0 Industry what are the stages of the learning process: the obtaining of the data, the understanding of the information, the remembering and the relationship. From all of them, the most important stage to consider that a system has reached an autonomous learning in the Artificial Intelligence is the fourth stage: the relationship.

7.4 Recycling of Non-metallic Materials

Putting the recycling of steel (scrap, Pero-Sanz et al. 2019) aside, which does not represent any problem, there are other residues of non-metallic nature that necessarily require attention due to the environmental consequences if they are not treated. We might consider three types of residues:

(a) Solids: slags, powders, and sludges which have different names depending on either the process where they were generated or the granulometry that they have (the name sludge is given to the particles with a size smaller than 100 μm).
(b) Liquids: aqueous liquors coming from operations of pickling of the sheet with sulphuric or hydrochloric acids. Aqueous solutions used for the obtaining of protecting passivating layers in coatings with zinc or tin are also a liquid residue. There are also aqueous emissions associated to the cleaning of fumes-gases, particularly in the case of the gases generated in the refinement and fusion processes in converters and electric furnaces or in the cooling of slags with water.
(c) Gases, released in those operations and ironmaking and steelmaking processes that, apart from suspended solid particles, can contain inorganic compounds of sulphur or nitrogen and aromatic halogenated organic compounds that are dangerous for the health. The concentration of these suspended solids in the gases released through the chimney is one of the environmental parameters most controlled by the environmental legislation. Until a few years ago, it was quite common to find in the emissions of chimneys, concentrations of these solids greater than 100 mg·Nm^{-3}. Nowadays, it is common to detect concentrations of around 2.0 mg·Nm^{-3} in developed countries.

Habitually, the protocols that are applied to the industrial solid residues are:

1. Finding, if possible, the possible methods of exploitation and valorization of valuable elements.
2. Considering the full or partial recycling of the waste in the ironmaking and steelmaking process.
3. Studying their possible utilization in the production of other materials (for instance, composites).
4. Subjecting the residue to inertization treatments.
5. Having available space in dumps.

On the other hand, it is necessary to remember that all the human activity, in general, and all those of the ironmaking and steelmaking, in particular, comprise a group of operations and processes that never reach the equilibrium state. Consequently, unstable residues are generated, and they tend to react-interact with the atmosphere and the aqueous environment that is around them. In this case, it is possible to consider two alternatives:

1. To inertize the residue using hydrometallurgical methods by solving the most unstable thermodynamically elements, making the residue a material inert-stable and inactive with the environment. However, this method generates a new problem that is the management of the aqueous liquors produced in the lixiviation of the residue, where it is not usual to find valuable elements.
2. To inertize the residue using thermal methods. A new residue is not produced by means of this option, and it is possible to achieve-reach the thermodynamic stabilization of the solid. The main inconvenience of this method is the cost of the energy even when it is possible to generate a product with certain commercial value. In this case, it is possible, and thinking in future developments, to use concentrated solar thermal energy to inertize or give value to unstable ironmaking and steelmaking residues (Fernández-Gonzalez et al. 2018a, b, c, 2019a, b, c).

The fact by which it is not possible to reach the thermodynamic equilibrium in the processes might lead to think about the effects of the variable *time*:

- With short times of operation, it is not possible to reach materials free of defects: point, linear, surface, or volume.
- It is necessary to remember again that, although it was already indicated in the last century, except in limited occasions (for instance, in some whiskers), the theoretical yield strength cannot be reached because the plastic deformation or failure takes place earlier.
- As it is indicated by Cottrell 1962 (p. 7), the energy formulation, ΔE, of the uncertainty principle of Heisenberg:

$$\Delta E \cdot \Delta t = \frac{h}{2 \cdot \pi} \tag{7.1}$$

where h is the Planck's constant, $6.626 \cdot 10^{-34}$ J·s, determines that small energy magnitudes, such as those that might be associated to the nucleation of crystalline clusters with perfect ordering, would be carried out in times, Δt, so long that the process would be unacceptable for the energy and economic efficiencies of the industrial processes. On the contrary, in the case of ultrafast cooling processes that would favor disordered structures (vitreous metals) or very fine-grained structures (that could exhibit either high yield strength or superplastic behavior) can be favored by the consequences of the energy formulation of the uncertainty principle.

7.5 Energy Efficiency

The energy efficiency has been one of the aspects that, in the last decades, was optimized in the group of industrial operations and basic processes. In the particular case of the iron and steel production, it is estimated an energy consumption of $25.00 \cdot 10^{18}$ J·year^{-1} (25.00 EJ·year^{-1}), see Cavaliere 2019 (pp. 6-12). The main contributors that allow reaching the above-mentioned value are:

1. Energy consumed by the blast furnace (BF) and basic converter (BOF, Basic Oxygen Furnace) to obtain a molten product, E_1, was studied in detail in Chap. 6 of Verdeja et al. 2020 and in Chap. 1 of the present book.
2. Energy required to obtain a molten product in the electric furnace from recycled steel, E_2.
3. Energy that is necessary to supply to the reheating furnaces located before the hot plastic deformation (hot rolling) of the steel (1000–1350 °C), E_3.
4. Energy associated to the annealing furnaces used to regenerate the crystalline structure consequence of the processes of finishing-cold rolling (550–650 °C), E_4.
5. Work-mechanical energy associated to the reduction of thickness in both the hot rolling process and the cold rolling process of steels, E_5.

The working hypothesis that was considered to justify the energy consumption associated to the production of steel are the following ones:

- 65% of the world production of steel, estimated as 1650 million of tons, is obtained via blast furnace + basic oxygen furnace, that is to say: 1072.5 million tons.
- 35% of the world production of steel is obtained by means of the electric furnace and processes of direct reduction of iron oxides with gas, DRI: 577.5 million tons.
- The thermal energy required for the production of a metallic melt by means of the blast furnace and basic oxygen furnace is 168.46 kJ·(mol Fe)$^{-1}$ (see Verdeja et al. 2020).
- The consumption of energy associated with the production of a metallic melt in an electric furnace (electric steelworks) is estimated as 400 kWh·t^{-1} of steel (even when there are steelworks with energy consumptions of 300 kWh·t^{-1}, not all of them have the same energy efficiency).

- The pressure of rolling required to reduce in a factor of reduction of 1.2 (1.2 times the unit of length), for instance, to pass from a thickness of 3.0 mm to 2.5 mm involves a stress of 36.72 MPa. It is possible to assume as working hypothesis that the factor of reduction of the mean thickness associated with the steel rolling processes is of 1250, which assumes a reduction from a dimension of the thickness of 250 mm to other of 0.20 mm.
- The density of the steel is 7850 kg·m^{-3}.
- It is assumed that 33% of the 25.00 EJ·year^{-1} is required to heat the flat or long formats of the continuous casting (it is assumed that all the solidification of the steel is made by continuous casting without the participation of the casting in cast iron ingot molds, Pero-Sanz et al. 2018). That is to say, we are considering a value for the E_3 of 8.25 EJ·year^{-1}. From this working hypothesis, we can argue that the value that can be given to E_4 would be precisely 50% of the value that is assumed for E_3: the temperature of the furnaces used to regenerate the crystalline structure is half the temperature that is necessary to reach in the furnaces of reheating of full austenization of the formats of continuous casting. Consequently, the value given to E_4 would be 4.13 EJ·year^{-1}.

The consumption of energy associated with the term E_1 considering that the iron mass is the same as the steel mass, we have:

$$\frac{168.46\,\mathrm{kJ}}{\mathrm{mol\ Fe}} \cdot \frac{1\,\mathrm{mol\ Fe}}{55.85\,\mathrm{g\ Fe}} \cdot \frac{1\,\mathrm{g\ Fe}}{1\,\mathrm{g\ steel}} \cdot \frac{10^6\mathrm{g\ Fe}}{1\,\mathrm{t}} \cdot \frac{1.0725\cdot 10^9\mathrm{t\ steel}}{\mathrm{year}} \tag{7.2}$$

$$E_1 = 3.23\,\mathrm{EJ}\cdot\mathrm{year}^{-1} = 3.23\cdot 10^{18}\,\mathrm{J}\cdot\mathrm{year}^{-1} \tag{7.3}$$

To calculate the energy associated to the production of a metallic melt in an electric furnace, E_2 term, would be necessary to operate as follows:

$$\frac{400\,\mathrm{kWh}}{\mathrm{t\ steel}} \cdot \frac{3.6\cdot 10^6\,\mathrm{J}}{\mathrm{kWh}} \cdot \frac{577.5\cdot 10^6\,\mathrm{t\ steel}}{\mathrm{year}} \tag{7.4}$$

$$E_2 = 0.83\,\mathrm{EJ}\cdot\mathrm{year}^{-1} = 0.83\cdot 10^{18}\,\mathrm{J}\cdot\mathrm{year}^{-1} \tag{7.5}$$

Finally, to calculate the energy stored in the steel (see Boustead 1980, p. 43), consequence of the processes of plastic deformation (in this data, we also consider the energy that is dissipated as heat), we proceed as follows:

$$\frac{37\,\mathrm{MPa}}{1.2\mathrm{times}} \cdot 1250\,\mathrm{times} = 38.54\cdot 10^9\,\mathrm{J}\cdot\mathrm{m}^{-3} \tag{7.6}$$

$$\frac{38.54\cdot 10^9\,\mathrm{J}}{\mathrm{m}^3} \cdot \frac{\mathrm{m}^3}{7850\,\mathrm{kg\ steel}} \cdot \frac{1650\cdot 10^9\,\mathrm{kg\ steel}}{\mathrm{year}} \tag{7.7}$$

$$E_5 = 8.10\,\mathrm{EJ}\cdot\mathrm{year}^{-1} = 8.10\cdot 10^{18}\,\mathrm{J}\cdot\mathrm{year}^{-1} \tag{7.8}$$

Consequently, the addition of the five contributions in which we have divided the total consumption of energy associated to the steel metallurgy is 24.54 EJ, which represents a negative deviation of 1.83% with respect to the value of 25.00 EJ given by Cavaliere (2019).

In view of the results, it is possible to conclude that only around 13% of the energy used in the ironmaking and steelmaking process is consumed in the manufacture of primary iron coming from the oxidized raw materials. Certainly, it is possible to reduce the consumptions in any of the five blocks previously considered. However, those blocks that can experience a further reduction are related with the processes of reheating (homogenization heat treatment), rolling, and annealing. One of the works and research, nowadays in progress, is to reduce the energy consumption associated with the reheating of the solidified products obtained by continuous casting through one of the two following lines of work:

(a) Increasing the heat transfer toward the steel and achieving an increase of the energy efficiency of the process without losing quality.
(b) Trying to connect the processes of continuous casting of the steel to avoid that the solidified products could be cooled until reaching the room temperature.

7.6 CO_2 Emissions

CO_2 emissions are one of the topics about which there are nowadays significant discussions and research, especially in those industrial sectors whose processes require a great quantity of energy to manufacture their products.

Considering the data provided by Cavaliere (2019, pp. 13–22), the production of iron and steel contributes to 2.00 Gt of CO_2 every year. If we assume that the production of steel by the route blast furnace and basic oxygen furnace requires 400 kilograms of carbon per ton of pig iron, CO_2 emissions in this line would be.

$$\frac{1075.5\,\text{Mt steel}}{\text{year}} \cdot \frac{0.400\,\text{t C}}{\text{t pig iron}} \cdot \frac{1.11\,\text{t pig iron}}{\text{t steel}} \cdot \frac{44.01\,\text{g CO}_2}{12\,\text{g C}} = 1.75\,\text{Gt of CO}_2 \cdot \text{year}^{-1} \tag{7.9}$$

Consequently, the rest of the emissions can be attributed to the electric steelworks and electric furnace: 0.25 Gt of CO_2 every year. That is to say, 87.5% of the CO_2 emissions are identified with the production of primary steel (starting from iron ore) via blast furnace and basic oxygen converter furnace route.

There is not a single solution to try to reduce CO_2 emissions in the production of iron and steel. The alternatives that are nowadays considered to reduce CO_2 emissions can be summarized as follows:

1. Capture and storage of CO_2.
2. Replacement of the carbon as single reductant reagent by hydrogen or electric power (aqueous electrolysis).

3. Introduction of renewable energy sources such as the wind power, the biomass or the solar thermal energy in the operations and processes aimed at the manufacture of iron and steel.

Finally, we want to make some considerations about the implications of carbon dioxide in the environment and production of materials:

(a) It is necessary to consider and reiterate that the presence of CO_2 in the atmosphere is necessary for the growth of plants and generation of biomass.
(b) CO_2 concentration in the atmosphere is around 380 ppm. It is estimated that in the Jurassic period, characterized by an abundant vegetation in the planet, CO_2 concentration might have been as great as 2000 ppm.
(c) During the period comprised within the years 1940 and 1975, there was a slight cooling of the planet, even when CO_2 concentration in the atmosphere has increased during that period.
(d) It is possible that the increase of CO_2 concentration in the atmosphere might be also related with the presence of other compounds—macromolecules, organic radicals, or non-metallic oxides that by itself might potentiate the influence of CO_2 in the atmosphere above its nominal concentration. Equally, the presence of CO_2 might potentiate-activate the chemical capacity of other chemical species dissolved or emulsified in the atmosphere and that act, definitely, as true responsible of the "climate change."
(e) The alternation of wet periods with other drought periods are collected even in the Bible. For instance, in Chapters 17 and 18 of the First Book of Kings it is indicated that during the period of the prophet Elias (874 BC) there was a long drought in all Palestine (Biblia de Navarra 2008). It does not seem that in that period the values of CO_2 concentration in the atmosphere were greater than the current ones.
(f) The massive manufacture of different qualities of carbon fibers is closely related with the coke production: reductant material, up to now, par excellence in the metallurgical industry. A drastic reduction in the coke production (Schwartz 2018) could be harmful and seriously alter the manufacture of carbon fiber composites, every day more and more used in the transportation industry (ground, air, and sea) and goods.

References

Biblia de Navarra (2008) Ediciones Universitarias de Navarra S. A. (ENUNSA), Pamplona, Navarra, Spain
Bolton J (2019) Green steel and future for the sector in UK. Mater World Mag 27(8):21
Boustead I (1980) Working solids. Unit 12. Ed. The Open University Press, Milton Keynes, UK
Cavaliere P (2019) Clean Ironmaking and Steelmaking, Process edn. Springer Nature Switzerland, Cham, Switzerland
Comellas JL (2000) Historia Breve del Mundo, Contemporáneo edn. Rialp, Madrid, Spain
Cottrell AH (1962) Metalurgia, Física edn. Reverté, Barcelona, Spain

Díaz M (2012) Ingeniería de Bioprocesos. Ed. Paraninfo, Madrid, Spain
Fernández-González D, Prazuch J, Ruiz-Bustinza I, González-Gasca C, Piñuela-Noval J, Verdeja LF (2018a) Solar synthesis of calcium aluminates. Sol Energy 171:658–666
Fernández-González D, Prazuch J, Ruiz-Bustinza I, González-Gasca C, Piñuela-Noval J, Verdeja LF (2018b) Iron metallurgy via concentrated solar energy. Metals 8(11), art. 873
Fernández-González D, Ruiz-Bustinza I, González-Gasca C, Piñuela-Noval J, Mochón-Castaños J, Sancho-Gorostiaga J, Verdeja LF (2018c) Concentrated solar energy applications in materials science and metallurgy. Sol Energy 170(8):520–540
Fernández-González D, Prazuch J, Ruiz-Bustinza I, González-Gasca C, Piñuela-Noval J, Verdeja LF (2018d) Transformations in the Mn-O-Si sytstem using concentrated solar energy. Sol Energy 184:148–152
Fernández-González D, Prazuch J, Ruiz-Bustinza I, González-Gasca C, Piñuela-Noval J, Verdeja LF (2018e) Transformations in the Si-O-Ca system: Silicon-calcium via solar energy. Sol Energy 181:414–423
Fernández-González D, Prazuch J, Ruiz-Bustinza I, González-Gasca C, Piñuela-Noval J, Verdeja LF (2018f) The treatment of Basic Oxygen Furnace (BOF) slag with concentrated solar energy. Sol Energy 180:372–382
Habashi F (1996) Pollution problems in the mineral and metallurgical industries. Ed. Metallurgie Extractive Quebec, Quebec, Canada
Habashi F (2002) Textbook of pyrometallurgy. Ed. Metallurgie Extractive Quebec, Quebec, Canada
McDonald Ch (2019) Steel innovations. Mater World Mag 27(5):52–55
Paramore A, Kirmse R (2020) Industry 4.0: myths and methods. Mater World Mag 28(6): 31–33
Pero-Sanz JA, Fernández-González D, Verdeja LF (2018) Physical metallurgy of cast irons. Ed. Springer Nature Switzerland, Cham, Switzerland
Pero-Sanz JA, Fernández-González D, Verdeja LF (2019) Structural materials: properties and selection. Ed. Springer Nature Switzerland, Cham, Switzerland
Schwartz M (2018) Coking coal needed. Mater World Mag 26(10):52–54
Verdeja JI, Fernández-González D, Verdeja LF (2020) Operations and basic processes in Ironmaking. Ed. Springer Nature Switzerland, Cham, Switzerland

Appendix 1

Standard free energies for some iron and steelmaking reactions, $\Delta_r G^0$.

$\Delta_r G^0 (kJ \cdot mol^{-1}) = A + B \cdot T(K)$	Linear correlation coefficient				Ref. (*)
	−A	B	(+/−)kJ	°C	
Aluminum					
$2Al(s) + 3/2O_2(g)=Al_2O_3(s)$	1672.0	0.313	6.9	25–659	1,2,3
$2Al(l) + 3/2O_2(g)=Al_2O_3(s)$	1680.0	0.324	4.5	659–1700	1,2,3,4
$Al(s) + 1/2N_2(g)=AlN(s)$	319.0	0.101	6.6	25–659	1,2,3
$Al(l) + 1/2N_2(g)=AlN(s)$	328.0	0.112	19.8	659–1700	1,2,3,4
$4Al(s) + 3C(s)=Al_4C_3(s)$	213.0	0.042	4.1	25–659	1,2,3
$4Al(l) + 3C(s)=Al_4C_3(s)$	283.0	0.107	4.5	659–1700	1,2,3
$2Al(l) + 1/2O_2(g)=Al_2O(g)$	193.0	−0.046	0.2	1227–1727	2,3,4
$Al(l) + 1/2O_2(g)=AlO(g)$	−42.5	−0.059	0.1	1227–1727	2,3
Barium					
$Ba(s) + 1/2O_2(g)=BaO(s)$	554.0	0.094	11.1	25–704	1,2,3
$Ba(l) + 1/2O_2(g)=BaO(s)$	561.0	0.102	8.1	704–1638	1,2,3,4
$3Ba(s) + N_2(g)=Ba_3N_2(s)$	352.0	0.234	8.9	25–704	1,3
$BaO(s) + SiO_2(s)=BaSiO_3(s)$	135.0	0.004	22.4	25–1300	1,3
Beryllium					
$Be(s) + 1/2O_2(g)=BeO(s)$	604.0	0.098	8.3	25–1283	1,2,3
$Be(l) + 1/2O_2(g)=BeO(s)$	605.0	0.100	8.4	1283–1700	1,2,3,4
$3Be(s) + N_2(g)=Be_3N_2(s)$	578.0	0.175	14.9	25–700	1,2,3
Boron					
$2B(s) + 3/2O_2(g)=B_2O_3(s)$	1272.0	0.264	0.8	25–450	1,2,3
$2B(s) + 3/2O_2(g)=B_2O_3(l)$	1225.0	0.208	3.5	450–1700	1,2,3
$B(s) + 1/2N_2(g)=BN(s)$	252.0	0.088	1.0	25–900	1,2,3

(continued)

L. F. Verdeja González et al., *Operations and Basic Processes in Steelmaking*, Topics in Mining, Metallurgy and Materials Engineering,
https://doi.org/10.1007/978-3-030-68000-8

(continued)

$\Delta_r G^0(kJ \cdot mol^{-1}) = A + B \cdot T(K)$	Linear correlation coefficient				Ref. (*)
	−A	B	(+/−)kJ	°C	
$4B(s) + C(s) = B_4C(s)$	60.6	0.003	6.2	25–900	1,2,3
$B(s) + 1/2N_2(g) = BN(g)$	−648.0	−0.112		1227–1727	4
$BaO(s) + SiO_2(s) = BaSiO_3(s)$	135.0	0.004	22.4	25–1300	1,3
Calcium					
$Ca(s) + 1/2O_2(g) = CaO(s)$	633.0	0.103	1.5	25–850	1,2,3
$Ca(l) + 1/2O_2(g) = CaO(s)$	640.0	0.109	2.9	850–1487	1,2,3,4
$Ca(g) + 1/2O_2(g) = CaO(s)$	788.0	0.193	1.6	1487–1700	1,2,3,4
$Ca(s) + 1/2S_2(g) = CaS(s)$	517.0	0.072	27.7	25–850	1,2,3
$Ca(l) + 1/2S_2(g) = CaS(s)$	549.0	0.106	3.5	850–1487	1,2,3,4
$Ca(g) + 1/2S_2(g) = CaS(s)$	696.0	0.189	3.1	1487–1700	2,4
$3Ca(s) + N_2(g) = Ca_3N_2(s)$	437.0	0.155	31.6	25–850	1,3
$Ca(s)a + 2C(s) = CaC_2(s)$	57.0	−0.024	0.3	25–400	1,3
$Ca(s)b + 2C(s) = CaC_2(s)$	49.4	−0.035	0.5	400–850	1,3
$Ca(l) + 2C(s) = CaC_2(s)$	60.6	−0.025	1.2	850–1487	1,3,4
$Ca(g) + 2C(s) = CaC_2(s)$	212.0	0.058	13.3	1487–1900	1,3,4
$Ca(s) + Si(s) = CaSi(s)$	150.0	0.002	–	25–850	1
$Ca(l) + Si(s) = CaSi(s)$	107.0	−0.028	–	850–1444	1
$2Ca(l) + Si(s) = Ca_2Si(s)$	178.0	−0.019	–	850–1444	1
$3CaO(s) + Al_2O_3(s) = Ca_3Al_2O_6(s)$	18.9	−0.028	11.1	25–1550	1,3,5
$12CaO(s) + 7Al_2O_3(s) = Ca_{12}Al_{14}O_{33}(s)$	72.8	−0.210	–	25–1500	1
$CaO(s) + Al_2O_3(s) = CaAl_2O_4(s)$	20.4	0.017	4.2	25–1600	1,3,5
$CaO(s) + CO_2(g) = CaCO_3(s)$	168.0	0.143	–	25–880	1
$2CaO(s) + Fe_2O_3(s) = Ca_2Fe_2O_5(s)$	38.3	−0.010	–	600–1435	1
$2CaO(s) + Fe_2O_3(s) = Ca_2Fe_2O_5(l)$	−31.5	−0.051	–	1435–1600	1
$4CaO(s) + P_2(g) + 5/2O_2(g) = Ca_4P_2O_9(s)$	2348.0	0.600	–	1300–1600	1
$3CaO(s) + P_2(g) + 5/2O_2(g) = Ca_3P_2O_8(s)$	2306.0	0.600	–	1300–1600	1
$2CaO(s) + SiO_2(s) = Ca_2SiO_4(s)$	103.0	−0.020	18.7	25–400	1,5
$CaO(s) + SiO_2(s) = CaSiO_3(s)\alpha$	88.7	0.001	–	25–1210	1
$CaO(s) + SiO_2(s) = CaSiO_3(s)\beta$	82.9	−0.003	–	1210–1543	1

(continued)

(continued)

$\Delta_r G^0(kJ \cdot mol^{-1}) = A + B \cdot T(K)$	Linear correlation coefficient				Ref. (*)
	−A	B	(+/−)kJ	°C	
$CaO(s) + SiO_2(s)=CaSiO_3(s)$	76.7	−0.003	–	25–727	5
$3CaO(s) + 2SiO_2(s)=Ca_3Si_2O_7(s)$	202.0	−0.013	–	25–727	5
$3CaO(s) + SiO_2(s)=Ca_3SiO_5(s)$	115.0	−0.023	–	25–727	5
$CaO(s) + 2Al_2O_3(s)=CaAl_4O_7(s)$	16.8	−0.034	3.8	25–1765	3,5
$CaO(s) + 6Al_2O_3(s)=CaAl_{12}O_{19}(s)$	22.5	−0.032	–	25–727	5
$2CaO(s) + SiO_2(s) + Al_2O_3(s)=Ca_2SiAl_2O_7(s)$	96.0	−0.050	31.5	25–1590	3,5
$CaO(s) + 2SiO_2(s) + Al_2O_3(s)=CaSi_2Al_2O_8(s)$	55.1	−0.035	35.4	25–1553	3,5
$CaO(s) + MgO(s) + 2SiO_2(s)=CaMgSi_2O_6(s)$	37.6	−0.015	29.6	25–1392	3
$CaO(s) + MgO(s) + SiO_2(s)=CaMgSiO_4(s)$	99.2	−0.009	14.3	25–1227	3,5
$2CaO(s) + MgO(s) + 2SiO_2(s)=Ca_2MgSi_2O_7(s)$	157.0	−0.030	24.8	25–1454	3,5
$3CaO(s) + MgO(s) + 2SiO_2(s)=Ca_3MgSi_2O_8(s)$	204.0	−0.032	–		5
Carbon					
$C(s) + 2H_2(g)=CH_4(g)$	95.6	0.113	1.6	25–2000	1,3,4
$C(s) + 1/2O_2(g)=CO(g)$	118.0	−0.084	3.7	25–2000	1,2,3,4
$C(s) + O_2(g)=CO_2(g)$	395.0	0.001	5.1	25–2000	1,2,3,4
$C(s) + 1/2S_2(g)=CS(g)$	−176.0	−0.067	10.7	1600–1800	1,2,3,4
$CO(g) + 1/2O_2(g)=CO_2(g)$	277.0	0.085			
$C(s) + S_2(g)=CS_2(g)$	−26.5	−0.050	43.4	25–1300	1,2,3
$CO(g) + 1/2S_2(g)=COS(g)$	132.0	0.065	98.2	25–1200	1,2,3
$2C(s) + H_2(g)=C_2H_2(g)$	−217.0	−0.049	–	1227–1727	4
Cerium					
$2Ce(s) + 3/2O_2(g)=Ce_2O_3(s)$	1805.0	0.309	2.0	25–804	1,3
$2Ce(l) + 3/2O_2(g)=Ce_2O_3(s)$	1810.0	0.325	68.4	804–1700	1,3,4
$Ce(s) + O_2(g)=CeO_2(s)$	1053.0	0.207	35.5	25–804	1,3
$Ce(l) + O_2(g)=CeO_2(s)$	1046.0	0.211	52.8	804–1700	1,3,4
Chromium					
$2Cr(s) + 3/2O_2(g)=Cr_2O_3(s)$	1129.0	0.264	11.6	25–1898	1,2,3
$2Cr(s) + 1/2N_2(g)=Cr_2N(s)$	117.0	0.067	11.4	25–1898	1,2,3
$Cr(s) + 1/2N_2(g)=CrN(s)$	113.0	0.075	4.8	25–1898	1,2,3

(continued)

(continued)

$\Delta_r G^0(kJ \cdot mol^{-1}) = A + B \cdot T(K)$	Linear correlation coefficient				Ref. (*)
	−A	B	(+/−)kJ	°C	
$23Cr(s) + 6C(s) = Cr_{23}C_6(s)$	352.0	−0.045	55.1	25–1400	1,2,3
$7Cr(s) + 3C(s) = Cr_7C_3(s)$	164.0	−0.026	10.5	25–1200	1,2,3
$3Cr(s) + 2C(s) = Cr_3C_2(s)$	84.7	−0.011	5.4	25–1700	1,2,3
Cobalt					
$Co(s) + 1/2O_2(g) = CoO(s)$	235.0	0.074	1.4	25–1495	1,2,3
$3CoO(s) + 1/2O_2(g) = Co_3O_4(s)$	198.0	0.158	10.3	25–1000	1,2,3
$9Co(s) + 4S_2(g) = Co_9S_8(s)$	1316.0	0.634	18.7	25–778	1,3
$2Co(s) + C(s) = Co_2C(s)$	−16.5	−0.009		25–900	1
Copper					
$2Cu(s) + 1/2O_2(g) = Cu_2O(s)$	168.0	0.074	2.9	25–1084	1,2,3
$2Cu(l) + 1/2O_2(g) = Cu_2O(s)$	182.0	0.082	1.9	1084–1230	1,2,3
$1/2Cu_2O(s) + 1/4O_2(g) = CuO(s)$	72.1	0.054	3.5	25–1000	1,2,3
$2Cu(l) + 1/2O_2(g) = Cu_2O(l)$	129.0	0.049	6.0	1229–1727	2,3,4
$Cu(l) + 1/2O_2(g) = CuO(l)$	154.0	0.088	1.6	1447–1727	2,4
Hydrogen					
$H_2(g) + 1/2O_2(g) = H_2O(g)$	249.0	0.057	1.0	25–1700	1,2,3,4
$H_2(g) + 1/2S_2(g) = H_2S(g)$	87.3	0.045	1.4	25–1500	1,2,3
$3/2H_2(g) + 1/2N_2(g) = NH_3(g)$	49.3	0.109	0.5	25–700	1,3
Iron					
$Fe(s) + 1/2O_2(g) = FeO(s)-\alpha$	265.0	0.065	6.0	25–1371	1,2,3
$0.95Fe(s) + 1/2O_2(g) = Fe_{0.95}O(s)$	261.3	0.065	–	25–1371	–
$Fe(s) + 1/2O_2(g) = Wustita(s)-\beta$	31.2	0.019	–	25–1371	5
$Fe(l) + 1/2O_2(g) = FeO(l)$	244.0	0.050	4.4	1537–1700	1,2
$3Fe(s) + 2O_2(g) = Fe_3O_4(s)$	1105.0	0.328	14.0	25–560	1,2,3
$3FeO(s) + 1/2O_2(g) = Fe_3O_4(s)$	296.0	0.112	14.6	560–1371	1,2,3
$2/3Fe_3O_4(s) + 1/6O_2(g) = Fe_2O_3(s)$	80.2	0.045	2.0	25–1400	1,2,3
$Fe(s) + 1/2S_2(g) = FeS(s)\alpha$	164.0	0.080	6.0	25–140	1,2,3
$Fe(s) + 1/2S_2(g) = FeS(s)\beta$	138.0	0.038	12.3	140–906	1,2,3
$FeS(s)b + 1/2S_2(g) = FeS_2(s)$	135.0	0.134	23.9	300–800	1,2,3
$4Fe(s) + 1/2N_2(g) = Fe_4N(s)$	7.7	0.044	1.9	25–600	1,3
$3Fe(s) + 1/2P_2(g) = Fe_3P(s)$	224.0	0.066	5.6	25–1170	1,3

(continued)

(continued)

$\Delta_r G^0(kJ \cdot mol^{-1}) = A + B \cdot T(K)$	Linear correlation coefficient				Ref. (*)
	−A	B	(+/−)kJ	°C	
$3Fe(s) + C(s) = Fe_3C(s)$	−26.5	−0.023	0.7	25–190	1,3
$3Fe(s) + C(s) = Fe_3C(s)$	−28.1	−0.027	0.5	190–840	1,3
$3Fe(s) + C(s) = Fe_3C(s)$	−9.8	−0.010	–	840–1537	1,3
$FeO(s) + Al_2O_3(s) = FeAl_2O_4(s)$	47.1	0.017	15.3	25–1371	1,3
$FeO(s) + Cr_2O_3(s) = FeCr_2O_4(s)$	33.2	−0.006	18.9	35–1371	1,3
$2FeO(s) + SiO_2(s) = Fe_2SiO_4(s)$	55.0	0.028	0.1	25–1217	1,3,5
$2FeO(s) + SiO_2(s) = Fe_2SiO_4(l)$	−62.0	−0.048	–	1217–1371	1
$2FeO(l) + SiO_2(s) = Fe_2SiO_4(l)$	−14.4	−0.019	–	1371–1700	1
$FeO(s) + TiO_2(s) = FeTiO_3(s)$	11.2	−0.003	7.4	900–1371	1,3
Lanthanum					
$2La(s) + 3/2O_2(g) = La_2O_3(s)$	1823.0	0.283	33.1	25–700	1,3
$La(s) + 1/2N_2(g) = LaN(s)$	299.0	0.097	3.3	25–700	1,3
Magnesium					
$Mg(s) + 1/2O_2(g) = MgO(s)$	600.0	0.105	1.6	25–650	1,2,3
$Mg(l) + 1/2O_2(g) = MgO(s)$	609.0	0.115	2.5	650–1120	1,2,3
$Mg(g) + 1/2O_2(g) = MgO(s)$	732.0	0.205	5.6	1120–1700	1,2,3,4
$Mg(s) + 1/2S_2(g) = MgS(s)$	391.0	0.070	28.5	25–650	1,2,3
$Mg(l) + 1/2S_2(g) = MgS(s)$	419.0	0.103	1.6	650–1120	1,2,3
$Mg(g) + 1/2S_2(g) = MgS(s)$	548.0	0.196	3.2	1120–1700	1,2,3,4
$3Mg(s) + N_2(g) = Mg_3N_2(s)$	460.0	0.199	3.0	25–650	1,2,3
$3Mg(l) + N_2(g) = Mg_3N_2(s)$	487.0	0.228	3.9	650–1120	1,2,3
$MgO(s) + CO_2(g) = MgCO_3(s)$	106.0	0.158	14.1	25–700	1,2,3
$3MgO(s) + P_2(g) + 5/2O_2(g) = Mg_3P_2O_8(s)$	2047.0	0.576	12.2	1000–1250	1,2,3
$2MgO(s) + SiO_2(s) = Mg_2SiO_4(s)$	60.1	−0.002	7.1	650–1120	1,2,3,5
$MgO(s) + SiO_2(s) = MgSiO_3(s)$	37.8	0.003	2.8	25–1300	1,2,3,5
$Mg(g) + 2C(s) = MgC_2(s)$	30.9	0.072	37.9	122–1727	2,4
$3Mg(g) + N_2(g) = Mg_3N_2(l)$	865.0	0.507	2.1	1227–1727	2,4
$MgO(s) + Al_2O_3(s) = MgAl_2O_4(s)$	17.0	−0.015	7.7	25–1727	3,5

(continued)

(continued)

$\Delta_r G^0(kJ \cdot mol^{-1}) = A + B \cdot T(K)$	Linear correlation coefficient				Ref. (*)
	−A	B	(+/−)kJ	°C	
Manganese					
$Mn(s) + 1/2O_2(g)=MnO(s)$	383.0	0.073	1.0	25–1244	1,3
$Mn(l) + 1/2O_2(g)=MnO(s)$	402.0	0.086	3.2	1244–1700	1,3,4
$Mn(s) + 1/2S_2(g)=MnS(s)$	274.0	0.066	4.3	25–1244	1,3
$Mn(l) + 1/2S_2(g)=MnS(s)$	291.0	0.078	9.0	1244–1530	1,3,4
$Mn(l) + 1/2S_2(g)=MnS(l)$	265.0	0.064	9.3	1530–1700	1,3,4
$3Mn(s) + C(s)=Mn_3C(s)$	14.0	−0.001	0.2	25–740	1,3
$MnO(s) + SiO_2(s)=MnSiO_3(s)$	25.4	0.004	5.7	25–1300	1,3,5
$3Mn(l) + 2O_2(g)=Mn_3O_4(\beta)$	1411.0	0.366	7.0	1243–1560	3,4
$7Mn(l) + 3C(s)=Mn_7C_3(s)$	65.8	−0.058	–	1243–1727	4
$2MnO(s) + SiO_2(s)=Mn_2SiO_4(s)$	55.6	0.016	9.0	25–1345	3,5
$MnO(s) + Al_2O_3(s)=MnAl_2O_4(s)$	45.2	0.010	2.0	25–1727	3,5
Molybdenum					
$Mo(s) + O_2(g)=MoO_2(s)$	584.0	0.177	1.4	25–1000	1,3
$MoO_2(s) + 1/2O_2(g)=MoO_3(s)$	158.0	0.074	1.6	25–1000	1,2,3
$2Mo(s) + 3/2S_2(g)=Mo_2S_3(s)$	560.0	0.242	6.7	850–1200	1,2,3
$2Mo(s) + 1/2N_2(g)=Mo_2N(s)$	82.2	0.090	0.1	25–1000	1,3
$2Mo(s) + C(s)=Mo_2C(s)$	36.8	0.004	17.3	25–1000	1,3
$Mo(s) + C(s)=MoC(s)$	−40.4	−0.058	–	1227–1727	4
$Mo(s) + 3/2O_2(g)=MoO_3(g)$	392.0	0.057	87.9	1280–1727	2,3,4
$Mo(s) + S_2(g)=MoS_2(s)$	379.0	0.181	41.4	1227–1727	2,3,4
Nickel					
$Ni(s) + 1/2O_2(g)=NiO(s)$	236.0	0.088	1.7	25–1452	1,3
$Ni(l) + 1/2O_2(g)=NiO(s)$	249.0	0.093	0.5	1452–1900	1,3,4
$Ni(s) + 1/2S_2(g)=NiS(s)$	135.0	0.057	24.2	300–580	1,2,3
$3Ni(s) + C(s)=Ni_3C(s)$	−63.5	−0.055	–	25700	1
Niobium					
$2Nb(s) + 2O_2(g)=Nb_2O_4(s)$	1560.0	0.336	–	1227–1727	4
$2Nb(s) + 5/2O_2(g)=Nb_2O_5(s)$	1860.0	0.404	5.8	1227–1453	2,3,4
$2Nb(s) + 5/2O_2(g)=Nb_2O_5(l)$	1741.0	0.337	6.4	1453–1727	2,3,4

(continued)

(continued)

$\Delta_r G^0(kJ \cdot mol^{-1}) = A + B \cdot T(K)$	Linear correlation coefficient				Ref. (*)
	−A	B	(+/−)kJ	°C	
Phosphorus					
$1/2P_2(g) + 1/2O_2(g)=PO(g)$	104.0	− 0.001	11.3	1227–1727	2,4
$2P_2(g) + 5O_2(g)=P_4O_{10}(g)$	3125.0	0.980	38.9	1227–1727	2,3,4
Platinum					
$Pt(s) + 1/2S_2(g)=PtS(s)$	16.0	−0.005	0.3	1227–1727	3
Silicon					
$Si(s) + 1/2O_2(g)=SiO(g)$	102.0	−0.072	6.4	25–1410	1,2,3
$Si(l) + 1/2O_2(g)=SiO(g)$	157.0	−0.047	5.5	1410–1700	1,2,3,4
$Si(s) + O_2(g)=SiO_2(s)$cristob.	903.0	0.175	3.6	400–1410	1,2,3
$Si(l) + O_2(g)=SiO_2(s)$cristob.	945.0	0.198	5.7	1410–1700	1,2,3,4
$3Si(s) + 2N_2(g)=Si_3N_4(s)$	738.0	0.326	14.2	25–1410	1,2,3
$3Si(l) + 2N_2(g)=Si_3N_4(s)$	874.0	0.406	3.0	1410–1700	1,2,3,4
$Si(s) + C(s)=SiC(s)\beta$	66.3	0.006	11.2	25–1410	1,2,3
$Si(l) + C(s)=SiC(s)\beta$	113.0	0.036	11.1	1410–1700	1,2,3,4
$2SiO_2(s) + 3Al_2O_3(s)=Si_2Al_6O_{13}(s)$	−22.0	−0.032	5.3	25–1850	3,5
Sulphur					
$1/2S_2(g)=S(g)$	−213.0	−0.057	1.3	25–700	1,2,3
$1/2S_2(g) + 1/2O_2(g)=SO(g)$	61.7	−0.006	3.7	25–1700	1,2,3,4
$1/2S_2(g) + O_2(g)=SO_2(g)$	361.0	0.072	1.3	25–1700	1,2,3,4
$1/2S_2(g) + 3/2O_2(g)=SO_3(g)$	455.0	0.151	10.7	25–1500	1,2,3
Tantalum					
$2Ta(s) + 5/2O_2(g)=Ta_2O_5(s)$	1991.0	0.395	5.0	1227–1727	2,3
$2Ta(s) + C(s)=Ta_2C(s)$	179.0	0.007	35.9	1227–1727	3,4
$Ta(s) + C(s)=TaC(s)$	157.0	0.006	6.5	1227–1727	3,4
$Ta(s) + 1/2N_2(g)=TaN(s)$	235.0	0.078	2.5	1227–1727	3,4
Thorium					
$Th(s) + O_2(g)=ThO_2(s)$	1226.0	0.185	3.5	25–1500	1,3
$3Th(s) + 2N_2(g)=Th_3N_4(s)$	1298.0	0.364	11.0	25–1700	1,3
$Th(s) + 2C(s)=ThC_2(s)$	156.0	0.001	29.0	25–2000	1,3
Tin					
$Sn(l) + O_2(g)=SnO_2(s)$	584.0	0.213	1.7	500–700	1,3
$Sn(l) + 1/2O_2(g)=SnO(s)$	268.0	0.089	13.1	1227–1727	3,4
$Sn(l) + 1/2O_2(g)=SnO(g)$	14.0	−0.045	11.5	1227–1727	3,4
$Sn(l) + O_2(g)=SnO_2(l)$	511.0	0.168	–	1227–1727	4
Titanium					

(continued)

(continued)

$\Delta_r G^0(kJ \cdot mol^{-1}) = A + B \cdot T(K)$	Linear correlation coefficient				Ref. (*)
	−A	B	(+/−)kJ	°C	
$Ti(s) + 1/2O_2(g) = TiO(s)$	530.0	0.092	18.3	300–1700	1,2,3
$2TiO(s) + 1/2O_2(g) = Ti_2O_3(s)$	448.0	0.087	32.0	25–1700	1,2,3
$3/2Ti_2O_3(s) + 1/4O_2(g) = Ti_3O_5(s)$	176.0	0.032	7.3	400–1700	1,2,3
$1/3Ti_3O_5(s) + 1/6O_2(g) = TiO_2(s)$	116.0	0.031	15.5	25–1850	1,2,3
$Ti(s)\alpha + 1/2N_2(g) = TiN(s)$	337.0	0.095	1.6	25–882	1,2,3
$Ti(s)\beta + 1/2N_2(g) = TiN(s)$	336.0	0.094	1.3	882–1200	1,2,3
$Ti(s)\alpha + C(s) = TiC(s)$	183.0	0.011	0.9	25–882	1,2,3
$Ti(s)\beta + C(s) = TiC(s)$	187.0	0.014	1.3	882–1700	1,2,3
$Ti(s) + 1/2O_2(g) = TiO(g)$	−32.4	−0.079	1.3	1227–1667	3,4
Vanadium					
$V(s) + 1/2O_2(g) = VO(s)$	425.0	0.081	11.0	600–1500	1,2,3
$2VO(s) + 1/2O_2(g) = V_2O_3(s)$	393.0	0.065	58.2	550–1112	1,2,3
$V(s) + C(s) = VC(s)$	102.0	0.009	1.7	900–1100	1,3
$2V(s) + 2O_2(g) = V_2O_4(s)$	1380.0	0.294	2.7	1227–1545	2,4
$2V(s) + 2O_2(g) = V_2O_4(l)$	1256.0	0.225	1.9	1545–1727	2,4
$2V(s) + 5/2O_2(g) = V_2O_5(l)$	1442.0	0.320	9.1	1227–1727	2,3,4
$V(s) + 1/2N_2(g) = VN(s)$	195.0	0.079	28.8	1227–1727	2,3,4
Wolfram					
$W(s) + C(s) = WC(s)$	38.8	0.002	1.3	25–1700	1,3
$W(s) + O_2(g) = WO_2(s)$	567.0	0.164	2.3	1227–1727	2,3,4
$W(s) + 3/2O_2(g) = WO_3(s)$	815.0	0.232	3.8	1227–1470	2,3,4
$W(s) + 3/2O_2(g) = WO_3(l)$	736.0	0.186	4.0	1470–1727	2,3,4
Zirconium					
$Zr(s)\alpha + O_2(g) = ZrO_2(s)$	1092.0	0.191	11.3	25–870	1,2,3
$Zr(s)\beta + O_2(g) = ZrO_2(s)$	1076.0	0.177	11.9	1205–1865	1,2,3,4
$Zr(s)\alpha + 1/2N_2(g) = ZrN(s)$	365.0	0.094	2.6	25–862	1,2,3
$Zr(s)\beta + 1/2N_2(g) = ZrN(s)$	365.0	0.093	3.9	862–1200	1,2,3
$Zr(s) + C(s) = ZrC(s)$	196.0	0.010	11.6	25–1900	1,2,3

(*) In the elaboration of this table we took as reference the work of the Association of Iron and Steel Engineers (AISE) previously included in the database of Janaf, Knacke, Pehlke and Gaye (according to Verdeja, L.F., Alfonso, A. and Suárez, M.: Energías libres de formación de compuestos siderúrgicos. Revista de Minas, No. 15–16, 109–112,1997). The numbers of the last column are contributions to the values of A and B:

1. Lankford WT (1985) The making, shaping and treating of steel. Association of Iron and Steel Engineers, 10ª edición. Pittsburgh, EEUU.
2. Chase MW Jr, Davies CA, Downey JR Jr., Frurip DJ, McDonald RA, Syverud AN (1986) JANAF Thermochemical Tables. American Chemical Society y American Institute of Physics for the National Bureau Standards, 3ª edition, New York, USA.

3. Knacke O, Kubaschewski O, Hesselmann K (1991) Thermochemical properties of inorganic substances. Springer. 2ª edition, Berlin, Germany.
4. Pehlke RD (1975) Unit processes of extractive metallurgy. Elsevier, 2ª edición. New York, USA.
5. Gaye HY, Welfringer J (1984) Modelling of the thermodynamic properties of complex metallurgical slags. Proceedings Second International Symposium on Metallurgical Slags and Fluxes. Editors Fine, U.A. and Gaskell, D.R., Metallurgical Society of AIME, Warrendale, USA.

Appendix 2

Standard enthalpies of several iron and steelmaking reactions, $\Delta_r H^0$.

$\Delta_r H^0 (kJ \cdot mol^{-1}) = A + B \cdot T(K)$	Linear correlation coefficient				Ref. (*)
	−A	B	(+/−)kJ	°C	
Aluminum					
$2Al(s) + 3/2O_2(g)=Al_2O_3(s)$	1676.0	0.0016	0.6	25–659	1,2
$2Al(l) + 3/2O_2(g)=Al_2O_3(s)$	1710.0	0.0149	0.2	659–1700	1,2
$Al(s) + 1/2N_2(g)=AlN(s)$	318.0	−0.0026	0.4	25–659	1,2
$Al(l) + 1/2N_2(g)=AlN(s)$	331.0	0.0014	03	659–1700	1,2
$4Al(s) + 3C(s)=Al_4C_3(s)$	212.0	−0.0005	3.5	25–659	1,2
$4Al(l) + 3C(s)=Al_4C_3(s)$	257.0	−0.0058	7.7	659–1700	1,2
$2Al(l) + 1/2O_2(g)=Al_2O(g)$	157.0	−0.0203	0.1	1227–1727	1,2
$Al(l) + 1/2O_2(g)=AlO(g)$	−55.0	−0.0074	0.1	1227–1727	1,2
Barium					
$Ba(s) + 1/2O_2(g)=BaO(s)$	547.0	−0.0016	0.5	25–704	1,2
$Ba(l) + 1/2O_2(g)=BaO(s)$	564.0	0.0040	0.6	704–1638	1,2
Beryllium					
$Be(s) + 1/2O_2(g)=BeO(s)$	608.0	−0.0003	0.2	25–1283	1,2
$Be(l) + 1/2O_2(g)=BeO(s)$	627.0	0.0069	3.5	1283–1700	1,2
$3Be(s) + N_2(g)=Be_3N_2(s)$	599.0	0.0311	10.5	25–700	1,2

(continued)

L. F. Verdeja González et al., *Operations and Basic Processes in Steelmaking*, Topics in Mining, Metallurgy and Materials Engineering,
https://doi.org/10.1007/978-3-030-68000-8

(continued)

$\Delta_r H^0 (kJ \cdot mol^{-1}) =$ $A + B \cdot T(K)$	Linear correlation coefficient				Ref. (*)
	−A	B	(+/−)kJ	°C	
Boron					
$2B(s) + 3/2O_2(g)=B_2O_3(s)$	1273.0	0.0020	0.4	25–450	1,2
$2B(s) + 3/2O_2(g)=B_2O_3(l)$	1251.0	0.0172	0.2	450–1700	1,2
$B(s) + 1/2N_2(g)=BN(s)$	251.0	−0.0012	0.8	25–900	1,2
$4B(s) + C(s)=B_4C(s)$	62.7	0.0003	0.2	25–900	1,2
$B(s) + 1/2O_2(g)=BO(g)$	−8.0	−0.0108	0.1	1227–1727	1,2
Calcium					
$Ca(s) + 1/2O_2(g)=CaO(s)$	636.0	0.0043	0.1	25–850	1,2
$Ca(l) + 1/2O_2(g)=CaO(s)$	647.0	0.0030	0.1	850–1487	1,2
$Ca(g) + 1/2O_2(g)=CaO(s)$	823.0	0.0184	0.1	1487–1700	1,2
$Ca(s) + 1/2S_2(g)=CaS(s)$	503.0	−0.0068	32.2	25–850	1,2
$Ca(l) + 1/2S_2(g)=CaS(s)$	548.0	0.0016	0.1	850–1487	1,2
$Ca(g) + 1/2S_2(g)=CaS(s)$	724.0	0.0168	0.1	1487–1700	1,2
Carbon					
$C(s) + 1/2O_2(g)=CO(g)$	104.0	−0.0075	0.2	25–2000	1,2
$C(s) + O_2(g)=CO_2(g)$	393.0	−0.0017	1.0	25–2000	1,2
$C(s) + 1/2S_2(g)=CS(g)$	−224.0	−0.0074	0.1	1600–1800	1,2
$CO(g) + 1/2S_2(g)=COS(g)$	142.0	−0.0196	9.2	25–1200	1,2
Chromium					
$2Cr(s) + 3/2O_2(g)=Cr_2O_3(s)\beta$	1143.0	0.0180	3	25–1898	1,2
$2Cr(s) + 1/2N_2(g)=Cr_2N(s)$	128.0	0.0071	0.6	25–1898	1,2
$Cr(s) + 1/2N_2(g)=CrN(s)$	120.0	0.0088	0.2	25–1898	1,2
$23Cr(s) + 6C(s)=Cr_{23}C_6(s)$	342.0	0.0458	2	25–1400	1,2
$7Cr(s) + 3C(s)=Cr_7C_3(s)$	167.0	0.0209	0.4	25–1200	1,2
$3Cr(s) + 2C(s)=Cr_3C_2(s)$	89.9	0.0151	0.3	25–1700	1,2
Cobalt					
$Co(s) + 1/2O_2(g)=CoO(s)$	241.0	0.0110	0.3	25–1495	1,2
$3CoO(s) + 1/2O_2(g)=Co_3O_4(s)$	192.0	−0.0298	5.0	25–1000	1,2
Copper					
$2Cu(l) + 1/2O_2(g)=Cu_2O(s)$	10.5	−0.1230	9	1084–1230	1,2

(continued)

(continued)

$\Delta_r H^0(kJ \cdot mol^{-1}) = A + B \cdot T(K)$	Linear correlation coefficient				Ref. (*)
	−A	B	(+/−)kJ	°C	
$1/2Cu_2O(s) + 1/4O_2(g){=}CuO(s)$	72.4	0.0057	0.1	25–1000	1,2
$2Cu(l) + 1/2O_2(g){=}Cu_2O(l)$	149.0	0.0158	0.2	1229–1727	1,2
Hydrogen					
$H_2(g) + 1/2O_2(g){=}H_2O(g)$	245.0	−0.0034	0.2	25–1700	1,2
$H_2(g) + 1/2S_2(g){=}H_2S(g)$	82.0	−0.0097	0.2	25–1500	1,2
$3/2H_2(g) + 1/2N_2(g){=}NH_3(g)$	40.8	−0.0176	0.3	25–700	1,2
Iron					
$Fe(s) + 1/2O_2(g){=}FeO(s){-}\alpha$	271.0	0.0080	3	25–1371	1,2
$3Fe(s) + 2O_2(g){=}Fe_3O_4(s)$	1129.0	0.0338	3.6	25–560	1,2
$3FeO(s) + 1/2O_2(g){=}Fe_3O_4(s)$	292.0	0.0035	4.4	560–1371	1,2
$2/3Fe_3O_4(s) + 1/6O_2(g){=}Fe_2O_3(s)$	78.6	−0.0011	1	25–1400	1,2
$Fe(s) + 1/2S_2(g){=}FeS(s)\alpha$	175.0	0.0188	4	25–140	1,2
$Fe(s) + 1/2S_2(g){=}FeS(s)\beta$	170.0	0.0149	4.6	140–906	1,2
$FeS(s)b + 1/2S_2(g){=}FeS_2(s)$	135.0	−0.0066	3	300–800	1,2
$4Fe(s) + 1/2N_2(g){=}Fe_4N(s)$	10.7	−0.0006	0.4	25–600	1,2
Magnesium					
$Mg(s) + 1/2O_2(g){=}MgO(s)$	600.0	0.0024	2	25–650	1,2
$Mg(l) + 1/2O_2(g){=}MgO(s)$	607.0	0.0007	2.4	650–1120	1,2
$Mg(g) + 1/2O_2(g){=}MgO(s)$	755.0	0.0157	3	1120–1700	1,2
$Mg(s) + 1/2S_2(g){=}MgS(s)$	376.0	−0.0074	32.2	25–650	1,2
$Mg(l) + 1/2S_2(g){=}MgS(s)$	607.0	0.0007	2.4	650–1120	1,2
$Mg(g) + 1/2S_2(g){=}MgS(s)$	563.0	0.0140	05	1120–1700	1,2
$3Mg(s) + N_2(g){=}Mg_3N_2(s)$	461.0	−0.0012	0.5	25–650	1,2

(continued)

(continued)

$\Delta_r H^0(kJ \cdot mol^{-1}) = A + B \cdot T(K)$	Linear correlation coefficient				Ref. (*)
	−A	B	(+/−)kJ	°C	
$3Mg(l) + N_2(g)=Mg_3N_2(s)$	477.0	−0.0108	1.3	650–1120	1,2
$MgO(s) + CO_2(g)=MgCO_3(s)$	112.0	0.0057	6.4	25–700	1,2
$3MgO(s) + P_2(g) + 5/2O_2(g)=Mg_3P_2O_8(s)$	2155.0	0.0902	0.3	1000–1250	1,2
$2MgO(s) + SiO_2(s)=Mg_2SiO_4(s)$	66.0	−0.0027	4	650–1120	1,2
$MgO(s) + SiO_2(s)=MgSiO_3(s)$	37.2	−0.0032	1.9	25–1300	1,2
$MgO(s) + Al_2O_3(s)=MgAl_2O_4(s)$	31.5	0.0068	20.7	25–1727	1,2
Molybdenum					
$MoO_2(s) + 1/2O_2(g)=MoO_3(s)$	159.0	0.0043	0.1	25–1000	1,2
$2Mo(s) + 3/2S_2(g)=Mo_2S_3(s)$	597.0	0.1430	195	850–1200	1,2
$Mo(s) + 3/2O_2(g)=MoO_3(g)$	337.0	−0.0080	0.3	1280–1727	1,2
$Mo(s) + S_2(g)=MoS_2(s)$	407.0	0.0112	0.1	1227–1727	1,2
Nickel					
$Ni(s) + 1/2S_2(g)=NiS(s)$	98.7	−0.0356	38.4	300–580	1,2
Niobium					
$2Nb(s) + 5/2O_2(g)=Nb_2O_5(s)$	1909.0	0.0300	1.2	1227–1453	1,2
$2Nb(s) + 5/2O_2(g)=Nb_2O_5(l)$	1898.0	0.0821	1.3	1453–1727	1,2
Phosphorus					
$2P_2(g) + 5O_2(g)=P_4O_{10}(g)$	3228.0	0.0618	0.8	1227–1727	1,2
Silicon					
$Si(s) + 1/2O_2(g)=SiO(g)$	97.6	−0.0060	1.1	25–1410	1,2
$Si(l) + 1/2O_2(g)=SiO(g)$	147.0	−0.0084	1.1	1410–1700	1,2
$Si(s) + O_2(g)=SiO_2(s)$cristob.	912.0	0.0082	0.7	400−1410	1,2
$Si(l) + O_2(g)=SiO_2(s)$cristob.	962.0	0.0093	1	1410−1700	1,2
$3Si(s) + 2N_2(g)=Si_3N_4(s)$	741.0	−0.0121	0.4	25–1410	1,2
$3Si(l) + 2N_2(g)=Si_3N_4(s)$	934.0	0.0296	3	1410–1700	1,2
$Si(s) + C(s)=SiC(s)\beta$	72.6	0.0006	0.9	25–1410	1,2

(continued)

(continued)

$\Delta_r H^0(kJ \cdot mol^{-1}) =$ $A + B \cdot T(K)$	Linear correlation coefficient				Ref. (*)
	−A	B	(+/−)kJ	°C	
$Si(l) + C(s)=SiC(s)\beta$	125.0	0.0017	0.8	1410–1700	1,2
Sulphur					
$1/2S_2(g)=S(g)$	−211.0	0.0058	0.1	25–700	1,2
$1/2S_2(g) + 1/2O_2(g)=SO(g)$	57.8	−0.0008	0.4	25–1700	1,2
$1/2S_2(g) + O_2(g)=SO_2(g)$	363.0	0.0009	0.1	25–1700	1,2
$1/2S_2(g) + 3/2O_2(g)=SO_3(g)$	459.0	−0.0025	0.5	25–1500	1,2
Tantalum					
$2Ta(s) + 5/2O_2(g)=Ta_2O_5(s)$	2068.0	0.0424	0.2	1227–1727	1,2
$Ta(s) + C(s)=TaC(s)$	146.0	0.0027	0.2	1227–1727	1,2
Titanium					
$Ti(s) + 1/2O_2(g)=TiO(s)$	546.0	0.0082	0.2	300–1700	1,2
$2TiO(s) + 1/2O_2(g)=Ti_2O_3(s)$	441.0	0.0180	1	25–1700	1,2
$3/2Ti_2O_3(s) + 1/4O_2(g)=Ti_3O_5(s)$	165.0	−0.0176	5.9	400–1700	1,2
$1/3Ti_3O_5(s) + 1/6O_2(g)=TiO_2(s)$	120.0	−0.0094	5.4	25–1850	1,2
$Ti(s)\alpha + 1/2N_2(g)=TiN(s)$	339.0	0.0027	0.4	25–882	1,2
$Ti(s)\beta + 1/2N_2(g)=TiN(s)$	345.0	0.0068	1.4	882–1200	1,2
$Ti(s)\alpha + C(s)=TiC(s)$	185.0	0.0022	0.2	25–882	1,2
$Ti(s)\beta + C(s)=TiC(s)$	186.0	0.0004	1.4	882–1700	1,2
Vanadium					
$V(s) + 1/2O_2(g)=VO(s)$	438.0	0.0136	0.4	600–1500	1,2
$2VO(s) + 1/2O_2(g)=V_2O_3(s)$	350.0	−0.0048	3	550–1112	1,2
$2V(s) + 5/2O_2(g)=V_2O_5(l)$	1482.0	0.0237	1.1	1227–1727	1,2
$V(s) + 1/2N_2(g)=VN(s)$	219.0	0.0062	0.2	1227–1727	1,2
Wolfram					
$W(s) + O_2(g)=WO_2(s)$	606.0	0.0227	0.7	1227–1727	1,2
$W(s) + 3/2O_2(g)=WO_3(s)$	852.0	0.0221	0.4	1227–1470	1,2
$W(s) + 3/2O_2(g)=WO_3(l)$	817.0	0.0443	0.4	1470–1727	1,2
Zirconium					
$Zr(s)\alpha + O_2(g)=ZrO_2(s)$	1101.0	0.0080	1.7	25–870	1,2
$Zr(s)\beta + O_2(g)=ZrO_2(s)$	1095.0	0.0068	1.8	1205–1865	1,2

(continued)

(continued)

$\Delta_r H^0 (kJ \cdot mol^{-1}) = A + B \cdot T(K)$	Linear correlation coefficient				Ref. (*)
	−A	B	(+/−)kJ	°C	
$Zr(s)\alpha + 1/2N_2(g)=ZrN(s)$	368.0	0.0042	1.6	25–862	1,2
$Zr(s)\beta + 1/2N_2(g)=ZrN(s)$	375.0	0.0078	2.4	862–1200	1,2
$Zr(s) + C(s)=ZrC(s)$	203.0	0.0035	5.9	25–1900	1,2

(*) The numbers indicated in the last column of the table are the contributions that were considered to calculate the constants A and B of the linear regression line to obtain the standard enthalpies of the above mentioned reactions:

1. Chase MW Jr, Davies CA, Downey JR Jr, Frurip DJ, McDonald RA, Syverud AN (1986) JANAF Thermochemical Tables. American Chemical Society y American Institute of Physics for the National Bureau Standards, 3ª edition, New York, USA.
2. Knacke O, Kubaschewski O, Hesselmann K (1991) Thermochemical properties of inorganic substances. Springer. 2ª edition, Berlin, Germany.

Appendix 3

Free energies associated to the change of standard state: from Raoult's activity one to Henry's activity one (*).

$\Delta_r G^0 (J \cdot mol^{-1}) = A + B \cdot T(K)$	Linear correlation coefficient			Ref
	A	B	(+/−) J	
Ag(l)	82400	−43.8	36	1, 2
Al(l)	−67100	−23.7	4857	1, 2
B(s)	−65300	−21.5	17	1, 2
C(s)	19900	−41.1	565	1, 2
Ca(g)	−39455	49.37		1
Co(l)	1004	−38.74		1
Cr(l)	0	−37.7	2	1, 2
Cr(s)	19200	−46.9	23	1, 2
Cu(l)	33500	−39.4	16	1, 2
$1/2H_2(g)$	36500	30.5	12	1, 2
Mn(l)	4084	−38.16		1
Mo(l)	0	−42.8		1
Mo(s)	27614	−52.38		1
$1/2N_2(g)$	6757	22	698	1, 2
Nb(l)		−42.68		1
Nb(s)	23000	−52.3	6	1, 2
Ni(l)	−20900	−31.1	47	1, 2
$1/2O_2(g)$	−117000	−3.14	500	1, 2
$1/2P_2(g)$	−140000	−6.92	7782	1, 2

(continued)

L. F. Verdeja González et al., *Operations and Basic Processes in Steelmaking*,
Topics in Mining, Metallurgy and Materials Engineering,
https://doi.org/10.1007/978-3-030-68000-8

(continued)

$\Delta_r G^0 (J \cdot mol^{-1}) = A + B \cdot T(K)$	Linear correlation coefficient			Ref
	A	B	(+/−) J	
Pb(l)	212547	−106.27		1
$1/2S_2(g)$	−130000	21	609	1, 2
Si(l)	−132000	−17.2	1	1, 2
Sn(l)	15983	−44.43		1
Ti(l)	−46024	−37.03		1
Ti(s)	−13690	−44.98		1
U(l)	−56100	−50.3	113	1, 2
V(l)	−42300	−36.0	39	1, 2
V(s)	−20700	−45.6	11	1, 2
W(l)		48.1	11	1, 2
W(s)	31380	−63.6		1
Zr(l)	−51000	−42.4	26	1, 2
Zr(s)	−34700	−50	13	1, 2

*According to: (1) Sigworth, G. K., and Elliot, J. F., 1994, The thermodynamic of liquid dilute iron alloy, *Metal Science*, 8, pp. 298–310; (2) The Japanese Society for the Promotion of Science. The 19th Committee on Steelmaking: Steelmaking data sourcebook. Gordon and Breach Science, New York, USA, 1988.

Appendix 4

Interaction parameters of the Henry's law at 1600 °C in the liquid according to (a) Sigworth and Elliot, 1974, and (b) The Japanese Society for the Promotion of Science (1988).

Solutes *i*	Solutes *j*							
	C	Si	Mn	P	S	N	O	H
C	0.14 (a) 0.243 (b)	0.08 (a) 0.08 (b)	−0.012 (a) −0.008 (b)	0.051 (a) 0.051 (b)	0.046 (a) 0.044 (b)	0.11 (a) 0.11 (b)	−0.34 (a) −0.32 (b)	0.67 (a) 0.67 (b)
Si	0.18 (a) 0.18 (b)	0.11 (a) 0.10 (b)	0 (a) −0.015 (b)	0.11 (a) 0.09 (b)	0.056 (a) 0.066 (b)	0.092 (b)	−0.23 (a) −0.12 (b)	0 (a) 0.64 (b)
Mn	−0.07 (a) −0.054 (b)	0 (a) −0.033 (b)	0 (a) 0 (b)	−0.035 (a) −0.054 (b)	−0.048 (a) −0.047 (b)	−0.091 (a) −0.091 (b)	−0.083 (a) −0.083 (b)	−0.31 (a) −0.34 (b)
P	0.13 (a) 0.126 (b)	0.12 (a) 0.099 (b)	0 (a) −0.032 (b)	0.062 (a) 0.054 (b)	0.028 (a) 0.034 (b)	0.094 (a) 0.13 (b)	0.13 (a) 0.13 (b)	0.021 (a) 0.33 (b)
S	0.11 (a) 0.12 (b)	0.063 (a) 0.075 (b)	−0.026 (a) −0.027 (b)	0.24 (a) 0.035 (b)	−0.028 (a) −0.046 (b)	0.01 (a) 0.01 (b)	−0.27 (a) −0.27 (b)	0.12 (a) 0.41 (b)
O	−0.45 (a) −0.421 (b)	−0.131 (a) −0.066 (b)	−0.021 (a) −0.021 (b)	0.07 (a) 0.07 (b)	−0.133 (a) −0.133 (b)	0.057 (a) −0.14 (b)	−0.20 (a) −0.17 (b)	−3.1 (a) 0.73 (b)

(continued)

L. F. Verdeja González et al., *Operations and Basic Processes in Steelmaking*, Topics in Mining, Metallurgy and Materials Engineering, https://doi.org/10.1007/978-3-030-68000-8

(continued)

Solutes *i*	Solutes *j*							
	C	Si	Mn	P	S	N	O	H
H	0.06 (a) 0.06 (b)	0.027 (a) 0.027 (b)	−0.0014 (a) −0.002 (b)	0.011 (a) 0.015 (b)	0.008 (a) 0.017 (b)		−0.19 (a) 0.05 (b)	0 (a) 0 (b)

(a) Sigworth, G. H.; Elliot, J. M. (1974): Thermodynamic of dilute iron alloys, Metals Science, Vol. 8, pp. 298–310. (b) The Japan Society for the Promotion of Science. The 19th Committee on Steelmaking (1988): Steelmaking Data Sourcebook, Ed. Gordon and Breach Science Publishers, New York, USA.

Data of the optical basicity, $\Lambda_{M,i}$, of the metallic oxides of the slags according to (a) Springorum 1995 and (b) McCloy et al., 2011.

Oxide	$\Lambda_{M,i}$(a)	$\Lambda_{M,i}$(b)	Oxide	$\Lambda_{M,i}$(a)	$\Lambda_{M,i}$(b)
Li_2O	1.00	0.84	CoO	0.95	0.98
Na_2O	1.15	1.11	NiO	0.95	0.92
K_2O	1.40	1.32	CuO	0.90	1.10
Cs_2O	1.70	1.52	ZnO	0.90	0.80
MgO	0.78	0.95	PbO	1.15	1.18
CaO	1.00	1.00	Cr_2O_3	0.80	0.80
SrO	1.10	1.08	Fe_2O_3	0.80	0.80
BaO	1.15	1.33	TiO_2		0.91
B_2O_3	0.42	0.40	V_2O_3		1.04
Al_2O_3	0.605	0.61	ZrO_2		0.85
SiO_2	0.48	0.48	SnO_2		0.85
P_2O_5	0.40	0.40	SO_3		0.33
MnO	1.00	0.95	SrO		1.08
FeO	1.00		MoO_3		1.07

(a) Springorum, D. (1995): Slag Atlas, Ed. Verein Deutscher Eisenhüttenleute (VDeh), Düsseldorf, Germany. (b) McCloy, J. S.; Schweiger, M. J.; Rodríguez, C. P.; Vienna, J. D. (2011): Nepheline crystallization in nuclear waste glasses. Progress toward acceptance of high-alumina formulations, International Journal of Applied Glass Science, Vol. 2, No. 3, pp. 201–214.

Appendix 5

Electronegativity of the atom of the Periodic Table relatives to the hydrogen (Sienko and Plane 1970, p. 72).

L. F. Verdeja González et al., *Operations and Basic Processes in Steelmaking*, Topics in Mining, Metallurgy and Materials Engineering,
https://doi.org/10.1007/978-3-030-68000-8

Legend: 1,007 ← Atomic weight; H ← Symbol; Electronegativity → 2,1

1,007 **H** 2,1																
6,939 **Li** 1,0	9,012 **Be** 1,5											10,811 **B** 2,0	12,01 **C** 2,5	14,007 **N** 3,0	15,999 **O** 3,5	18,998 **F** 4,0
22,989 **Na** 0,9	24,312 **Mg** 1,2											26,982 **Al** 1,5	28,086 **Si** 1,8	30,974 **P** 2,1	32,064 **S** 2,5	35,453 **Cl** 3,0
39,102 **K** 0,8	40,08 **Ca** 1,0	44,956 **Sc** 1,3	47,90 **Ti** 1,5	50,94 **V** 1,6	51,996 **Cr** 1,6	54,938 **Mn** 1,5	55,847 **Fe** 1,8	58,933 **Co** 1,8	58,71 **Ni** 1,8	63,54 **Cu** 1,9	65,37 **Zn** 1,6	69,72 **Ga** 1,6	72,59 **Ge** 1,8	74,922 **As** 2,0	78,96 **Se** 2,4	79,909 **Br** 2,8
85,47 **Rb** 0,8	87,62 **Sr** 1,0	88,905 **Y** 1,2	91,22 **Zr** 1,4	92,906 **Nb** 1,6	95,94 **Mo** 1,8	98 **Tc** 1,9	101,07 **Ru** 2,2	102,905 **Rh** 2,2	106,4 **Pd** 2,2	107,870 **Ag** 1,9	112,40 **Cd** 1,7	114,82 **In** 1,7	118,69 **Sn** 1,8	121,75 **Sb** 1,9	127,60 **Te** 2,1	126,904 **I** 2,5
132,905 **Cs** 0,7	137,34 **Ba** 0,9	**La-Lu** 1,1-1,2	178,49 **Hf** 1,3	180,948 **Ta** 1,5	183,85 **W** 1,7	186,2 **Re** 1,9	190,2 **Os** 2,2	192,2 **Ir** 2,2	195,09 **Pt** 2,2	196,967 **Au** 2,4	200,59 **Hg** 1,9	204,37 **Tl** 1,8	207,19 **Pb** 1,8	208,980 **Bi** 1,9	210 **Po** 2,0	210 **At** 2,2
223 **Fr** 0,7	226 **Ra** 0,9	227 **Ac** 1,1	232,038 **Th** 1,3	231 **Pa** 1,5	238,03 **U** 1,7	**Np-Lw** 1,3										

Bibliography

Sienko MJ, Plane RA (1970) Química-física inorgánica. Ed. Reverté, Barcelona, Spain.

Appendix 6

Physical and thermal properties of the most typical molten metals of the ferroalloys measured at around their melting point.

Metal	Atomic weight (g)	Surface energy ($nM \cdot m^{-1}$)	Enthalpy of fusion ($kJ \cdot kg^{-1}$)	Thermal conductivity ($W \cdot m^{-1} \cdot K^{-1}$)
Iron (Fe)	55.847	1972	277.0	64
Manganese (Mn)	54.938	1090	244.0	874
Silicon (Si)	28.086	865	1793.0	83.5
Chromium (Cr)	51.996	1700	362.4	936
Titanium (Ti)	47.900	1650	393.0	14
Molybdenum (Mo)	95.940	2250	374.5	444
Nickel (Ni)	58.710	1778	300.0	35
Vanadium (V)	50.940	1951	443.5	931
Niobium (Nb)	92.906	1900	284.3	450
Tungsten (W)	183.850	2498	192.4	294

Metal	Specific heat ($J \cdot kg^{-1} \cdot K^{-1}$)	Density ($kg \cdot m^{-3}$)	Melting point (K)	Viscosity (Pa·s)
Iron (Fe)	824.0	6746	1811	$6.930 \cdot 10^{-3}$
Manganese (Mn)	837.2	5560	1518	$1.000 \cdot 10^{-3}$
Silicon (Si)	938.2	2368	1685	$0.683 \cdot 10^{-3}$

(continued)

L. F. Verdeja González et al., *Operations and Basic Processes in Steelmaking*, Topics in Mining, Metallurgy and Materials Engineering,
https://doi.org/10.1007/978-3-030-68000-8

(continued)

Metal	Specific heat ($J \cdot kg^{-1} \cdot K^{-1}$)	Density ($kg \cdot m^{-3}$)	Melting point (K)	Viscosity (Pa·s)
Chromium (Cr)	846.8	6278	2158	$1.000 \cdot 10^{-3}$
Titanium (Ti)	967.0	4109	1941	$2.200 \cdot 10^{-3}$
Molybdenum (Mo)	392.5	9430	2891	$1.000 \cdot 10^{-3}$
Nickel (Ni)	734.0	7902	1728	$1.000 \cdot 10^{-3}$
Vanadium (V)	862.4	5545	2160	$2.400 \cdot 10^{-3}$
Niobium (Nb)	449.5	7765	2745	$1.000 \cdot 10^{-3}$
Tungsten (W)	193.5	17600	3675	$1.000 \cdot 10^{-3}$

Note The above-indicated values were obtained-estimated from the references indicated below. However, when the viscosity around the melting point is $1.000 \cdot 10^{-3}$ Pa·s, we want to indicate that this value was not found in any of the considered references and it was indicated as a reference of its order of magnitude of the viscosity for the considered metal in the molten state.

Bibliography

Weast RC (1974): Handbook of chemistry and physics. 57th CRC Press, Cleveland, Ohio, USA. ISBN: 087819-454-1.

TAPP Database. Thermochemical and Physical Properties. (1994). E. S. Microware Inc, Cincinnati, Ohio, USA.

HSC Database Chemistry. (2002). Outokumpu Research Oy, Version 5.1. Finland.

Knacke O, Kubaschewski OY, Hesselmann K (1991) Thermochemical properties of inorganic substances. Vol. I and II. Second Edition, Springer, Düsseldorf, Germany. ISBN: 3-514-00363-7.

Chase MW, Davies CA, Downey JR, Frurip DJ, McDonald RA, Syverud AN (1986) JANAF thermochemical tables. Part I (Al-Co). Part II (Cr-Zr). Third Edition. American Chemical Society and American Institute of Physics. New York, USA. ISBN: 0-88318-473-7.

The Japan Society for the Promotion of Science. The 19th Committee on Steelmaking (1988): *Steelmaking Data Sourcebook*. Gordon and Breach Science Publishers, New York, USA. ISBN: 2-88124-153-0.

Kawai Y, Shiraishi Y (1988): Handbook of physico-chemical properties at high temperatures. The Iron and Steel Institute of Japan, Committee of Japan Society for Promotion of Science, Tokyo, Japan.

Appendix 7

Values of the erf function.

z	erf(z)	z	erf(z)	z	erf(z)
0	0.00000	0.55	0.56332	1.20	0.91031
0.05	0.05637	0.60	0.60386	1.30	0.93401
0.10	0.11246	0.65	0.64203	1.40	0.95229
0.15	0.16800	0.70	0.67780	1.50	0.96611
0.20	0.22270	0.75	0.71116	1.60	0.97635
0.25	0.27633	0.80	0.74210	1.70	0.98379
0.30	0.32863	0.85	0.77067	1.80	0.98909
0.35	0.37938	0.90	0.79691	1.90	0.99279
0.40	0.42839	0.95	0.82089	2.00	0.99532
0.45	0.47548	1.00	0.84270	2.50	0.99959
0.50	0.52050	1.10	0.88021	3.00	0.99997

The following analytical equation can be also used in the calculation of the error function:

$$\begin{aligned} \mathrm{erf(z)} &= \mathrm{A + B \cdot z + C \cdot z^2 + D \cdot z^4 + E \cdot z^5 + F \cdot z^6 + G \cdot z^7} \\ &= -0.0191 + 1.17 \cdot \mathrm{z} - 0.206 \cdot \mathrm{z}^2 - 0.338 \cdot \mathrm{z}^4 \\ &\quad + 0.303 \cdot \mathrm{z}^5 - 0.0967 \cdot \mathrm{z}^6 + 0.0108 \cdot \mathrm{z}^7 \end{aligned}$$

values are collected in the following table.

L. F. Verdeja González et al., *Operations and Basic Processes in Steelmaking*, Topics in Mining, Metallurgy and Materials Engineering,
https://doi.org/10.1007/978-3-030-68000-8

z	erf(z)	z	erf(z)	z	erf(z)
0	−0.00191	0.55	0.56332	1.20	0.90849
0.05	0.05607	0.60	0.60386	1.30	0.93162
0.10	0.11300	0.65	0.64203	1.40	0.94922
0.15	0.16878	0.70	0.67780	1.50	0.96243
0.20	0.22331	0.75	0.71116	1.60	0.97235
0.25	0.27640	0.80	0.74210	1.70	0.97997
0.30	0.32781	0.85	0.77067	1.80	0.98607
0.35	0.37729	0.90	0.79691	1.90	0.99120
0.40	0.42449	0.95	0.82089	2.00	0.99569
0.45	0.46905	1.00	0.84270	2.50	1.00571
0.50	0.51054	1.10	0.88021	3.00	1.03039